AF700552

Bioorganic Synthesis

BIOORGANIC SYNTHESIS AN INTRODUCTION

Gary W. Morrow

OXFORD
UNIVERSITY PRESS

OXFORD
UNIVERSITY PRESS

Oxford University Press is a department of the University of Oxford. It furthers the University's objective of excellence in research, scholarship, and education by publishing worldwide. Oxford is a registered trade mark of Oxford University Press in the UK and certain other countries.

Published in the United States of America by Oxford University Press
198 Madison Avenue, New York, NY 10016, United States of America.

© Oxford University Press 2016

All rights reserved. No part of this publication may be reproduced, stored in a retrieval system, or transmitted, in any form or by any means, without the prior permission in writing of Oxford University Press, or as expressly permitted by law, by license, or under terms agreed with the appropriate reproduction rights organization. Inquiries concerning reproduction outside the scope of the above should be sent to the Rights Department, Oxford University Press, at the address above.

You must not circulate this work in any other form
and you must impose this same condition on any acquirer.

Library of Congress Cataloging-in-Publication Data
Names: Morrow, Gary W., 1951–
Title: Bioorganic synthesis : an introduction / Gary W. Morrow.
Description: New York, NY : Oxford University Press, [2016] | Includes index.
Identifiers: LCCN 2016003452 | ISBN 9780199860531
Subjects: LCSH: Organic compounds—Synthesis. | Biosynthesis. | Chemistry, Organic.
Classification: LCC QD262 .M744 2016 | DDC 572/.45—dc23 LC record available at http://lccn.loc.gov/2016003452

9 8 7 6 5 4 3 2 1

Printed by Sheridan Books, Inc., United States of America

Contents

Acknowledgments

A book like this is the result of the input, support, and effort of so many individuals that properly thanking them all would require yet another chapter. Nevertheless, I would like to acknowledge a global host of scientists in various fields whose research efforts over many decades provided the foundation for all that is found in this book. Among the many essential reference materials listed at the end of the text, I especially wanted to acknowledge the important works of Paul Dewick, John Mann, and Richard Silverman; their volumes proved to be inspirational and indispensible guides for me throughout the preparation of my own. I would also like to thank all those at Oxford University Press who assisted in the development of this book at various stages, especially Jeremy Lewis who encouraged the initial ideas and then waited patiently over a number of years for the work to finally be completed. I am also deeply indebted to the University of Dayton for providing a sabbatical leave during which several portions of the book were completed. Most importantly, I wanted to thank my wife Barbara Smith for her abundant patience, encouragement, and support throughout the writing process.

GWM

May 2016

Introduction

Nearly twenty years ago while on a sabbatical at Ohio State University, I decided to sit in on an advanced graduate level course that dealt with a topic I knew almost nothing about, namely the biosynthesis of organic compounds. Though I was (and still am) a traditional organic chemist, I nevertheless found many of the topics in that course to be fascinating, especially the remarkable similarities between the mechanisms of long-established organic reactions and those of cellular bioorganic processes. In the years that followed, I gradually accumulated additional research materials with the thought of eventually developing an undergraduate-level course in bioorganic synthesis for students who I believed would have an interest in this topical area. I subsequently taught such a course for a number of years to many interested students, majors in chemistry, biology, premedicine, and bioengineering, and their positive response to the course and the great fun I had in teaching it were my prime motivation for converting most of the lecture materials into this textbook. To help set the stage for further understanding of what the book is all about, a brief look at several important introductory points would seem to be in order, so here goes.

THE UNIQUE ROLE OF CARBON

It all begins with carbon, the essential element of all organic and biological chemistry. As illustrated in Fig. I.1, the **carbon cycle** is basically a description of the processes involved in the storage and transport of carbon throughout the planet. Though carbon is of relatively low abundance in the earth's crust (~0.03% by weight) compared to elements such as silicon (~45%) and oxygen (~29%), its importance cannot be overstated since it is the fundamental element essential for the existence of life. Understanding how carbon is distributed and stored throughout the environment and how it is converted from one form to another is a good place to start in gaining a deeper understanding of the unique relationship between carbon and life.

CARBON STORAGE AND THE ROLE OF METABOLISM

The group of processes that converts carbon from one form into another in living systems is called **metabolism**. **Anabolic** metabolism uses energy derived from

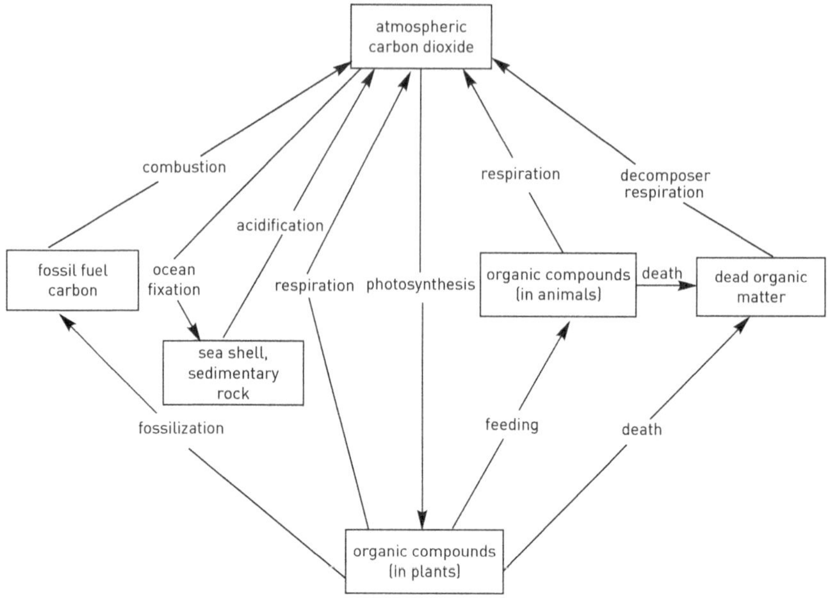

FIGURE I.1
The carbon cycle, illustrating the various fates of carbon in the biosphere.

sunlight or other sources to synthesize complex organic molecules from simpler ones. **Autotrophs** (mainly plants) are organisms that carry out anabolic processes such as **photosynthesis**, for which the fundamental source of energy is sunlight and the source of carbon is atmospheric carbon dioxide (carbon fixation). In terms of the carbon cycle, the most important autotrophs are trees and phytoplankton, which fix carbon by carrying out photosynthesis, converting carbon dioxide and water into **glucose** ($C_6H_{12}O_6$), one of the basic organic materials from which most other organic compounds essential for living organisms are derived. Overall, this is a reductive process, converting carbon from an oxidized form (CO_2) to a reduced form (glucose) via the familiar photosynthesis equation: $6CO_2 + 6H_2O + \text{light energy} \rightarrow C_6H_{12}O_6 + 6O_2$.

When atmospheric carbon dioxide dissolves in the oceans, it is used in photosynthesis, but may also be converted to bicarbonate (HCO_3^-) and carbonate (CO_3^{2-}); marine organisms utilize calcium ion for the formation of shell material ($CaCO_3$). The accumulation of seabed shell deposits over millions of years ultimately leads to significant carbon storage in the form of sedimentary rock deposits such as chalk, limestone, and dolomite. Considerable amounts of carbon are also stored within fossil fuels such as petroleum and natural gas. These are believed to have formed over time by decomposition of large quantities of marine organisms in oxygen-depleted waters in combination with high heat and pressure. Coal, another significant carbon source, dates from Earth's carboniferous period and was presumably formed by similar processes over the ages through decomposition of terrestrial plant material.

For organisms which cannot fix carbon to make their own organic compounds (**heterotrophs**), the consumption of other organisms or their parts (such as seeds or fruits) is necessary to obtain both carbon and the energy required for the anabolic processes used to transform it. The fermentation or decay of dead organic matter by fungi or bacteria also transfers carbon: these are examples of **catabolism**, degradative metabolic processes which involve the breaking down of more complex organic materials (derived from anabolic processes) into simpler ones and the principal path by which most carbon leaves the biosphere. **Aerobic** (with oxygen) catabolism (**respiration**) releases carbon back to the environment mainly by oxidative degradation of organic compounds to give carbon dioxide and water in what is essentially the reverse of the photosynthesis equation (with energy released in forms other than light). **Anaerobic** respiration (without oxygen) may also occur, releasing significant amounts of carbon back into the environment in the form of methane (as in marsh gas or cattle flatulence). Additional release of carbon dioxide into the atmosphere also occurs through the burning of fossil fuels and other combustion processes and by acidification of carbonate sources. With the subsequent conversion of atmospheric CO_2 back to glucose by photosynthesis, the carbon cycle, in which living organisms play a key role, continues.

DISTINGUISHING PRIMARY VERSUS SECONDARY METABOLISM

From a chemistry standpoint, both autotrophs and heterotrophs vary a great deal in their ability and capacity to synthesize and transform complex organic compounds in a set of anabolic processes collectively referred to as **biosynthesis**. As noted above, plants are efficient at constructing glucose via photosynthesis using simple inorganic materials (CO_2 and H_2O) derived from the environment. By contrast, most other organisms, from animals to simple microorganisms, must rely primarily on degradation of plant or animal material in the diet in order to obtain both the energy and the raw materials necessary to carry out biosynthesis. But in all cases, the chemical reaction mechanisms and pathways employed by living cells for synthesizing or modifying the basic organic compounds essential for life, such as carbohydrates, fats, proteins, or nucleic acids, are essentially identical in all known organisms, owing to their common evolutionary ancestry. Collectively, this set of processes constitutes the core of biochemistry and is known as **primary metabolism**. The compounds derived from such processes, all of which are essential for life, are generally referred to as **primary metabolites** (Fig. I.2).

SECONDARY METABOLITES AND NATURAL PRODUCTS

Alternatively, **secondary metabolism** in organisms employs metabolic pathways for the production of certain complex organic compounds which are unrelated to

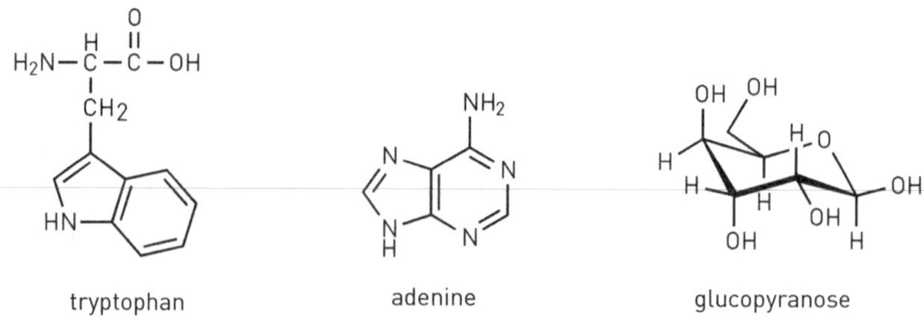

FIGURE I.2
Three examples of primary metabolites: an amino acid, a DNA base, and a carbohydrate.

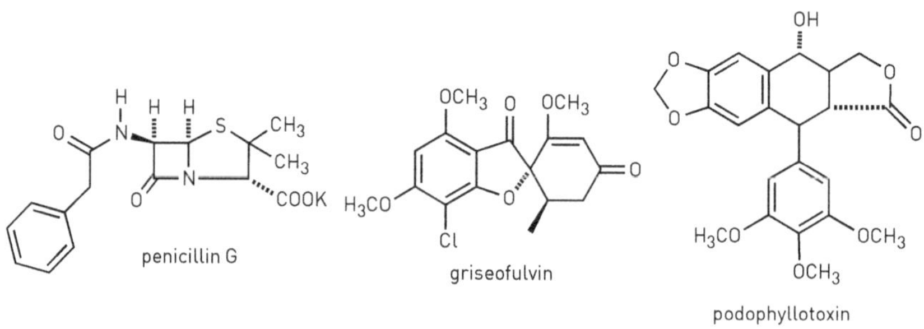

FIGURE I.3
Three examples of secondary metabolites: an antibiotic, an antifungal, and an anticancer agent.

and therefore unnecessary for an organism's growth, development, or reproduction. Such **secondary metabolites** are usually unique to a specific organism or group of organisms and may be thought of as fundamental expressions of the organism's identity, even though these compounds may be produced only under specific conditions. Such **organic natural products** have functions that are frequently obscure; that is, the benefits to the producing organism may be entirely unknown, though in some cases they have been found to be associated with some specific ecological activity, such as acting as chemical agents for the direction of sexual attraction or social behavior in insects or for the suppression of pathogens or predators in plants.

Most importantly, since many secondary metabolites and their derivatives have been found to possess powerful antibacterial (e.g., penicillin G), antifungal (e.g., griseofulvin), or anticancer (e.g., podophyllotoxin) activity, such natural products are of extraordinary interest and significance in medicine and human health (Fig. I.3).

NATURAL PRODUCTS IN ORGANIC CHEMISTRY AND MEDICINE

Humans have been using organic compounds derived from plant sources for the treatment of various health conditions for well over 4000 years, but it was not until the 1800s that many of the active principals from various plants were actually isolated and studied. This important work helped to establish a scientific basis for the often-powerful action of organic compounds as medicines, toxins, or hallucinogens, moving them for the first time out of the realms of superstition, magic, or folklore. In fact, up until the early 1900s, the study of organic chemistry was essentially the study of natural products, and many of the now familiar techniques for isolation and characterization of organic compounds were originally developed as part of such studies.

In modern times, natural products still play an essential role in medicine and drug discovery. Of the nearly 120 plant-derived organic compounds in use as medicinal agents around the world today, almost three-fourths were derived from studies of the active principals isolated from plants used in traditional folk medicine. Although the pharmaceutical industry had begun in the 1990s to move away from drug discovery modeled on bioactive natural product structures in favor of rapid biological screening of enormous libraries of organic compounds that could be produced in small quantities via automated or combinatorial synthetic methods, it had become apparent by the middle of the first decade of the 21st century that this newer approach to drug discovery had met with only limited success, leading to a renewed interest in natural products chemistry as a complementary source of lead compounds. Today, screening assays continue to uncover new organic compounds from nature that in some cases reveal previously unknown modes of biological action. Others may someday hold the key to treating conditions we are currently unable to adequately address with known medicinal agents. Such "bioprospecting" continues to expand into areas involving marine organisms and microbes—relatively untapped sources of new organics as compared to plant sources—making it increasingly important for scientists from a variety of different fields to have a better understanding of the fundamental organic chemistry used by organisms to produce such important bioactive compounds.

THE ORGANIC CHEMISTRY OF SECONDARY METABOLITE BIOSYNTHESIS

Most students are probably aware of the role that cholesterol-reducing drugs such as *lovastatin* (Mevacor) have played in the control of heart disease in recent years, but even chemistry majors might be surprised to learn that many statin drugs are organic compounds isolated from nature and that cells use an enzymatic Diels–Alder reaction to form the two 6-membered rings at the structural core of this life-saving compound (Fig. I.4).

Familiar organic reactions such as this one constitute the fundamental core of methods used by all organisms to construct the complex organic compounds associated with life on our planet. For this reason, development of a book focusing exclusively on such organic reactions and their mechanisms, with special attention given to compounds of medicinal and pharmacological interest, seemed a reasonable way

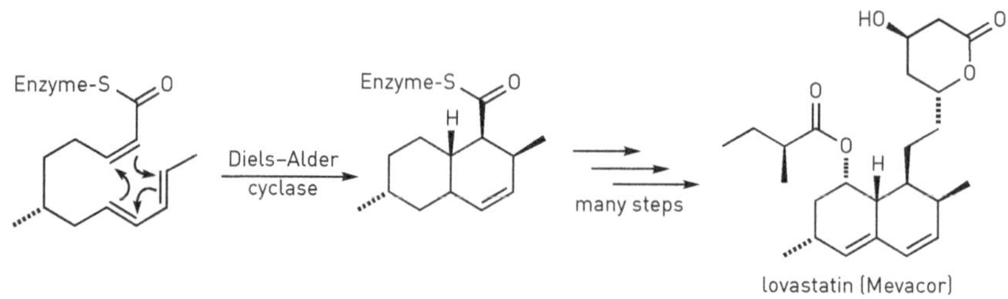

FIGURE I.4
Diels–Alder cyclization in the biosynthesis of lovastatin.

to introduce undergraduate students to the elegantly simple organic chemistry used in various biosynthetic pathways.

GOALS AND STRUCTURE OF THIS BOOK

My hope is that *Bioorganic Synthesis: An Introduction* will prove useful in supporting various advanced undergraduate courses for chemistry, biochemistry, biology, premedicine, bioengineering, and related majors whose students could benefit from a deeper understanding of the chemical logic of reactions carried out in organisms and the origins and uses of the biologically active compounds they often produce. Importantly, the book assumes no prior background in biochemistry, though students who have had an introductory biochemistry course should find the material enlightening, and all students with the necessary organic chemistry prerequisites should find the book supportive of their previous studies as it guides them through the development of deductive skills to identify the relationship between specific organic compound structural features and the biosynthetic mechanisms and pathways involved in their production.

The topics and sequence differ somewhat from the many excellent available monographs on biosynthesis and natural products chemistry by focusing primarily on the reaction mechanisms involved in the production of representative and pharmacologically significant compounds from each of the major biosynthetic pathways known to us. Most importantly, the book is designed specifically for undergraduate students, rather than as an exhaustive reference work for graduate students or professional researchers. Most organic chemistry instructors at the undergraduate level will find their own background in organic chemistry more than sufficient for the development of a meaningful course based on the text.

The book consists of eight chapters, beginning with a brief review of relevant topics from introductory organic chemistry. This is followed by a detailed presentation in Chapter 2 of an essential set of organic and biochemical reactions used throughout the book as well as a brief introduction to the function and chemistry of the most important coenzymes.

Chapter 3 covers basic carbohydrate chemistry and the biosynthesis of amino acids for those students who may have had only a brief introduction to these topics from their previous organic studies. This material is essential for a proper understanding of the role these important compounds play as building blocks in natural product biosynthesis via glycoside and polypeptide formation and other processes. The next four chapters are each devoted to a particular biosynthetic pathway.

The terpenoid pathway in Chapter 4 is the starting point for biosynthesis studies, since carbocations are the principal intermediates in the mechanisms of this pathway and so should be most familiar to students from an introductory organic chemistry sequence. The chapter covers biosynthesis of all the important classes of terpenoids, with a detailed description of the biosynthesis and biological relevance of the steroids.

This is followed by the acetate pathway presented in Chapter 5. This chapter begins with examination of the biosynthesis of saturated and unsaturated fatty acids and prostaglandins and their biological relevance, a discussion which provides the basis for subsequent understanding of the biosynthetic origins of acetate-derived polyketide natural products through cyclizations and transformations of intermediates via familiar aldol, Claisen, and related carbonyl condensation reactions.

Chapter 6 introduces the shikimate pathway. The biosynthesis of aromatic amino acids and other shikimate-derived metabolites such as coumarins, lignans, and products derived from the intersection of shikimate metabolites and the acetate pathway are also examined. This material is followed by an introduction to alkaloids and related nitrogenous compounds in Chapter 7.

Finally, Chapter 8 considers some of the obstacles faced by synthetic and medicinal chemists who tackle the challenge of recreating intricate molecules like *lovatstin* in the laboratory *from scratch*. We will be left to marvel at just how difficult and intellectually challenging it can be to replicate many of the molecular construction projects that our cellular friends seem to achieve with such ease. But we will also see how knowledge of biosynthetic pathways can also lead to new ways of thinking about chemical problems and the development of new methods for organic synthesis. And we'll find that occasionally we can even employ the direct assistance of simple organisms and their enzymes in the laboratory to help us accomplish goals in synthetic and medicinal organic chemistry.

In the end, it is my sincere hope that both students and instructors of organic chemistry will find the book to be a useful introductory guide to further exploration of the exciting world of biological organic chemistry.

1 Brief Organic Review

As soon as we touch the complex processes that go on in a living thing, be it plant or animal, we are at once forced to use the methods of this science [chemistry].
—John Jacob Abel (*co-founder of* The Journal of Biological Chemistry, *1905*)

REVIEW OF FUNCTIONAL GROUPS, STEREOCHEMISTRY, AND CONFORMATIONAL ANALYSIS

In addition to simple hydrocarbon structures (alkanes, alkenes, alkynes, and aromatic systems) and alkyl groups (methyl, ethyl, propyl, isopropyl, etc.), this text assumes a familiarity with the most common functional groups associated with organic chemical structures and their basic reactivity patterns. Table 1.1 summarizes the names and structures of some of the more important functional groups we will be dealing with throughout the remainder of the book. It is important to remember that functional groups containing O or N with nonbonding electrons have an affinity for both protic and Lewis acids and are important participators in H-bonding. Groups containing a carbonyl (C=O) function are especially important, as these bonds are strongly polarized ($^{\delta+}C{=}O^{\delta-}$), the C atom being electron deficient and the O atom electron excessive; this strong polarization is mainly responsible for the familiar reactivity patterns associated with carbonyl compounds.

Table 1.1 Common Organic Functional Groups

Name	Structure	Name	Structure
alcohol	ROH	monophosphate	$RO-P(=O)(O^{\ominus})O^{\ominus}$ (ROP)
phenol	ArOH	diphosphate	$RO-P(=O)(O^{\ominus})-O-P(=O)(O^{\ominus})O^{\ominus}$ (ROPP)
ether	R-O-R, 	acid	$R-C(=O)-OH$
amine	$R-NH_2$, R_2NH, R_3N, 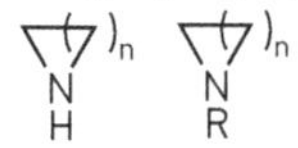	ester	$R-C(=O)-OR'$
thiol	RSH	lactone	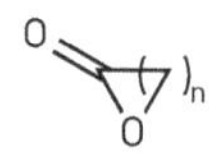
sulfide	R-S-R	thioester	$R-C(=O)-SR'$
disulfide	R-S-S-R	amide	$R-C(=O)-NH_2$, $-NHR$, $-NR_2$
aldehyde	$R-C(=O)-H$	lactam	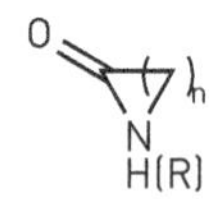
ketone	$R-C(=O)-R'$ 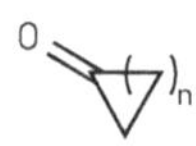	imine (Schiff's base)	$R-C(=NH(R))-H(R)$

STEREOCHEMISTRY REVIEW: THE SHAPES OF THINGS TO COME

Figure 1.1 depicts the standard classification of isomers in organic chemical structures. Recall that **constitutional isomers** are compounds with the same molecular formula but different atom connectivity, such as 1-butanol versus 2-butanol (Fig. 1.2). **Stereoisomers**, on the other hand, are compounds with the same formula and the *same* atom connectivity, differing from one another only in the three-dimensional orientation of their atoms in space. These are divided into two groups: **enantiomers** and **diastereomers**.

Enantiomers are nonsuperimposable mirror image molecules whose **asymmetry** is usually the result of a tetrahedral carbon atom with four *different* atoms or groups attached to it, as in the 2-butanol enantiomers. Such **chiral** molecules rotate the plane of polarized light either (+) or (–) and so are said to be **optically active**. Achiral molecules, such as 1-butanol, do *not* rotate the plane of polarized light and so are **optically inactive**.

A standard formalism for representation of a chiral center is to use bond line drawings with two of the four atoms or groups lying in the plane of the paper, a third projecting outward (wedge bond), and the fourth projecting inward (dashed bond).

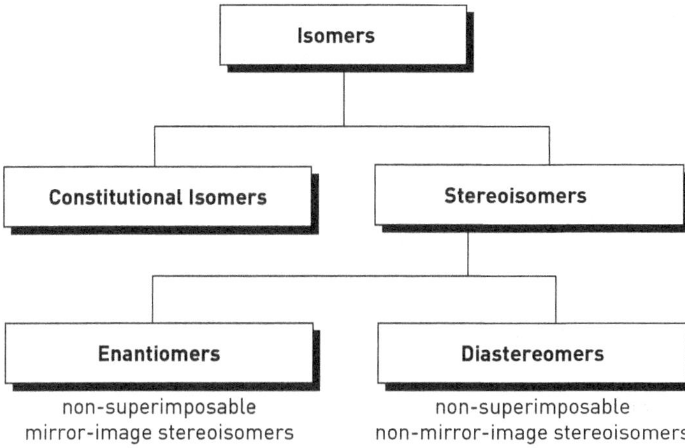

FIGURE 1.1
Classification of isomers in organic chemical structures.

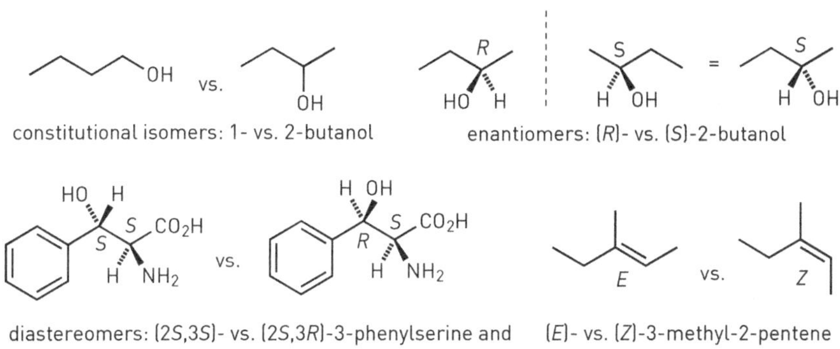

FIGURE 1.2
Comparison of constitutional isomers, enantiomers, and diastereomers.

Note that an enantiomeric pair may be drawn formally as reflected mirror images, though it is useful to remember that an enantiomer may be generated simply by interchanging any two of the four atoms or groups attached to the chiral carbon (compare (*R*)- versus (*S*)-2-butanol in Fig. 1.2). The stereochemical prefixes *S* and *R* are added to the chemical names of enantiomers to distinguish one from the other and are assigned on the basis of the *priorities* established for the four atoms or groups attached to the chiral carbon atom, using the Cahn–Ingold–Prelog (CIP) rules. When the molecule is oriented so that it may be viewed looking down the bond between the chiral carbon and the *lowest-ranked* atom or group, the three remaining atoms or groups are arranged much as in a three-spoked wheel. If the order of decreasing priority of these three traces a counter-clockwise path, the **absolute configuration** of the molecule is assigned as *S*. If the path is clockwise, the configuration is assigned as *R*. Recall that *R* versus *S* and (+) versus (−) are not necessarily related: some *R* compounds are (+) while others are (−). Optical rotation, both magnitude and sign, must be measured and cannot be predicted simply on the basis of absolute configuration assignment. Compounds with more than one stereocenter will have more than two stereoisomers. In general, n stereocenters give 2^n stereoisomers. Thus, 3-phenylserine (Fig. 1.2) has two stereocenters and a total of four stereoisomers, but while (2*R*,3*R*) and (2*S*,3*S*) 3-phenylserine are enantiomers (nonsuperimposable mirror image stereoisomers), the (2*S*,3*S*) and (2*S*,3*R*) stereoisomers are related as **diastereomers** (nonsuperimposable, nonmirror image stereoisomers). Unlike enantiomers, diastereomers have different chemical and physical properties.

Recall that the relationship between *cis* and *trans* stereoisomers is also diastereomeric and most commonly occurs among both alkenes and disubstituted cycloalkane systems. For alkenes, the *cis* or *trans* stereochemical designation is unambiguous only for disubstituted alkenes, while the *E*,*Z* designation is more general and may be applied to all stereochemically different alkenes. The *E*,*Z* designation, as in (*E*)- versus (*Z*)-3-methyl-2-pentene (Fig. 1.2) assigns *priorities* to the two atoms or groups attached to each end of the double bond (one higher priority, one lower priority for each atom or group attached to C). These priorities are assigned one end at a time, using the same CIP rules as for chiral carbons. The *Z* designation results when the higher priority atoms or groups are on the same side of the double bond; *E* when they are on opposite sides.

Certain compounds with multiple stereocenters may nevertheless be achiral (optically inactive) if the molecule possesses an internal plane of symmetry. Such achiral molecules are known as **meso compounds** and can often be recognized as such by finding a mirror plane that reflects one half of the molecule in the other half, as shown for 2,3-dichlorobutane isomers and 1,2-dimethylcyclopentane isomers (Fig. 1.3). A meso compound and its mirror image are superimposable and therefore identical, even though two (or more) stereocenters are involved.

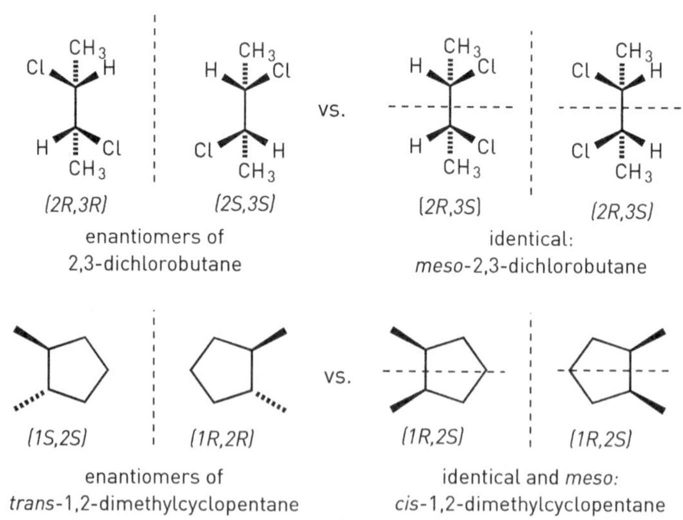

FIGURE 1.3
Comparison of enantiomers, diastereomers, and meso compounds.

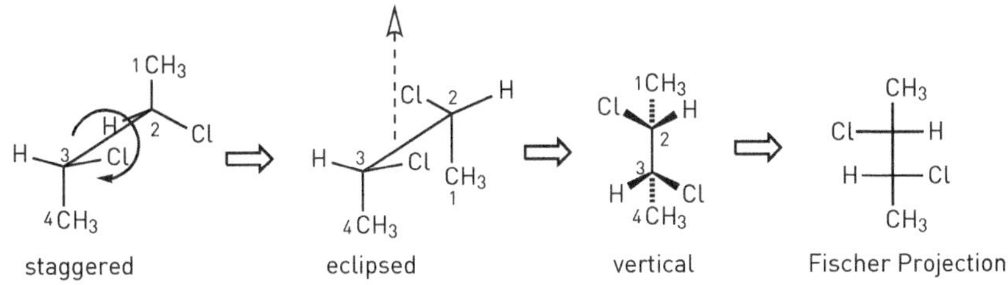

FIGURE 1.4
Conversion of a staggered conformation into a Fischer projection.

A convenient way to represent the configuration of certain acyclic compounds with one or more chirality centers without drawing wedge and dashed bond structures is through the use of **Fischer projections**. The process for generating such structures from more conventional representations is illustrated for (2*R*,3*R*)-2,3-dichlorobutane (Fig. 1.4). Beginning with a staggered conformer, the molecule is then rotated about the C2–C3 bond to generate an eclipsed conformer which, when viewed from above, is redrawn as a *vertical chain* with correct use of wedge and dashed bonds, with the lowest numbered atom of the compound (from the IUPAC name, *not* from CIP priorities) at the top of the chain. Finally, the wedge and dashed bonds are converted to simple lines. When drawn this way, it is *understood* that in a Fischer projection horizontal bonds project outward while vertical bonds project inward, with chirality centers at the points of intersection. Such projections are especially useful in the representation of carbohydrate structures, as will be seen in Chapter 3.

PROCHIRAL RELATIONSHIPS: ONE STEP FROM CHIRALITY

By **prochiral** carbon we mean an achiral carbon atom in a molecule which, through a chemical transformation, may be converted into a chiral carbon. One such case involves prochiral carbon with so-called **enantiotopic** atoms (or groups). For example, if replacement of one of two identical atoms (or groups) at an achiral carbon by an arbitrary atom X produces a chiral enantiomer as product, such an achiral carbon is said to be *prochiral* and the atom replaced is said to have been *enantiotopic* (Fig. 1.5).

Furthermore, assuming that X has a CIP priority higher than that of the atom replaced but lower than that of the other two groups, if the enantiomer produced has the *R* configuration, then the enantiotopic atom replaced by X is called the pro-*R* atom; replacement of the other enantiotopic atom would produce the *S* enantiomer and so would be the pro-*S* atom. In the example shown, H_R and H_S are identical in all respects (chemical and physical properties), but will differ in their reactions with *chiral* reagents. Thus, the chiral enzyme that catalyzes the oxidation (dehydrogenation) of ethanol to acetaldehyde in the liver is selective for removal of the pro-*R* hydrogen atom of ethanol (Fig. 1.6).

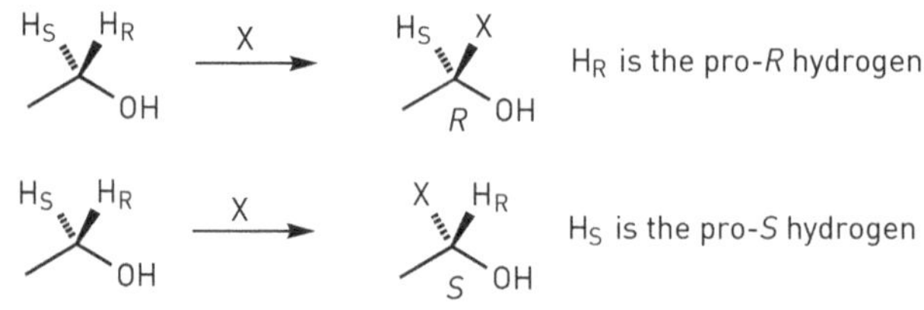

FIGURE 1.5
Enantiotopic hydrogen atoms on the prochiral carbon of an alcohol.

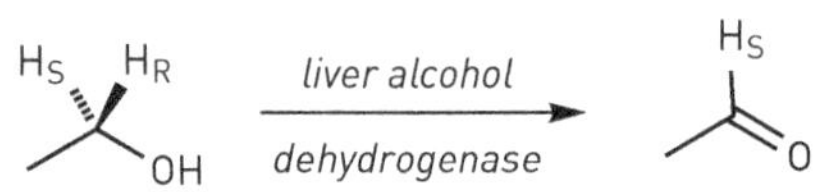

FIGURE 1.6
Enantiotopic proton selectivity in an enzymatic alcohol dehydrogenation reaction.

PROCHIRAL Π-SYSTEMS: "TWO-FACED" REACTION CENTERS

In appropriately substituted systems, an sp^2 carbon may also be prochiral. In such cases, we find that in reactions in which an achiral sp^2 carbon is converted to a chiral sp^3 carbon by addition of a new atom or group, different enantiomers will be formed depending on which "face" of the planar π system is approached when the new bond is formed. These two prochiral faces are distinguished from one another by the stereochemical terms *Si* and *Re* which are readily assigned to a prochiral sp^2 center by application of the CIP rules to the three different atoms or groups around the sp^2 carbon, as shown in Fig. 1.7. Thus, we see that reaction of achiral 2-butanone with phenylmagnesium bromide produces chiral 2-phenyl-2-butanol by nucleophilic attack from the front or the back of the planar ketone π-system. While achiral reactants (such as PhMgBr) cannot distinguish between *Si* and *Re* faces (and so produce racemic mixtures), chiral reagents or catalysts such as enzymes *can* distinguish between them and may result in formation of enantiomerically pure products from such prochiral π-systems.

Note that attack at a *Si* face does not necessarily indicate subsequent generation of the *S* enantiomer of the resulting chiral product and vice versa. The *Si* and *Re* stereochemical assignments are used only to distinguish between the two prochiral π-system faces.

One example of addition to a prochiral alkene carbon to produce an enantiomerically pure product is the enzyme-catalyzed hydration of the π-bond of fumaric acid to give only (*S*)-malic acid (Fig. 1.8). Note that in addition of the elements H–O–H across the π-bond, the O atom approaches only the *Si* face of the prochiral carbon. Since the *fumarase* enzyme binding the substrate is chiral, it can distinguish between the two faces, blocking one and facilitating the addition of water to the other, producing a single enantiomer product.

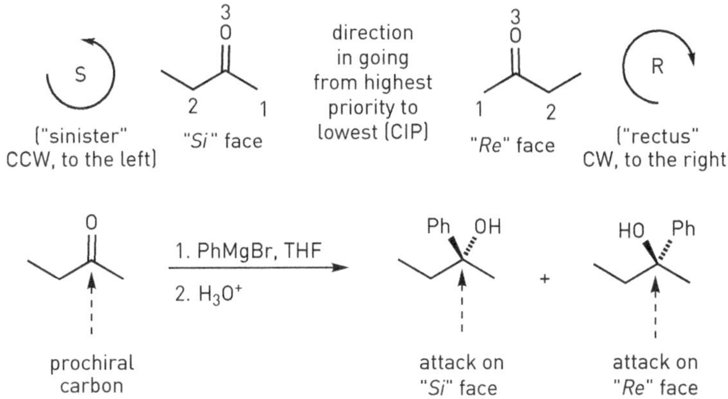

FIGURE 1.7
Assignment of *Si* and *Re* faces at a prochiral carbonyl center.

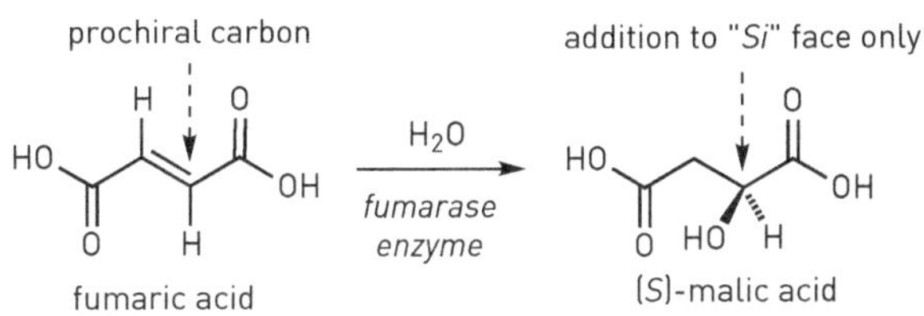

FIGURE 1.8
Face selectivity of fumarase enzyme in the hydration of fumaric acid to (*S*)-malic acid.

DIASTEREOTOPIC ATOMS AND GROUPS: ONE STEP FROM A DIASTEROEOMER

If replacement of one of two identical atoms (or groups) at an achiral carbon by an arbitrary atom X produces a *diastereomer* as product, such atoms on the prochiral carbon are said to be *diastereotopic* (Fig. 1.9).

Unlike enantiotopic atoms, diastereotopic atoms are *not* in identical chemical environments and so are distinguishable by both chiral *and* achiral reagents. For example, the enzyme *phenylalanine ammonia lyase* is selective for the diastereotopic H_S proton in its catalysis of the elimination of ammonia to yield *trans*-cinnamic acid (Fig. 1.10).

Finally, we note that when substitution occurs at a chiral carbon, several possible stereochemical outcomes are possible, depending on the mechanistic details and reactive intermediates involved. These outcomes are summarized in Fig. 1.11, using examples in which a new atom X replaces atom D at a chirality center. Thus, we may refer to a reaction at chiral carbon as proceeding with retention of configuration, inversion of configuration, or racemization (complete or partial) at the chirality center (Fig. 1.11).

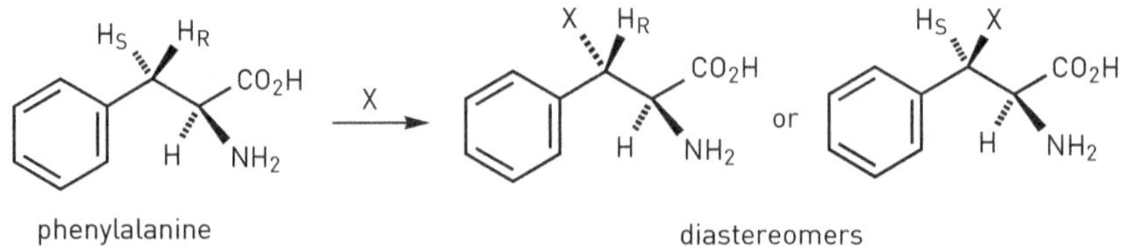

FIGURE 1.9
Diastereotopic protons on the prochiral carbon of phenylalanine.

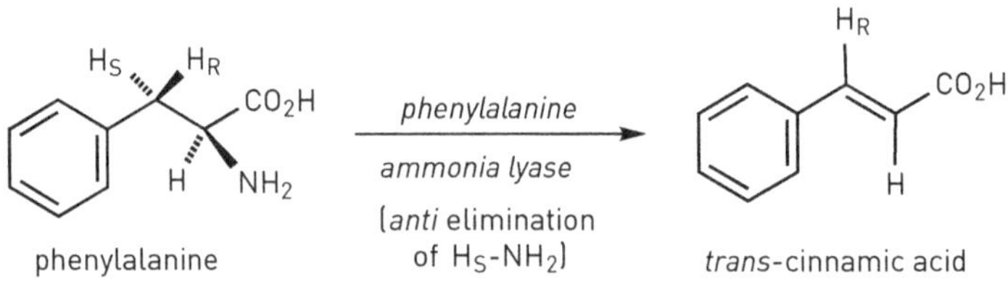

FIGURE 1.10
Diastereotopic proton selectivity in an enzyme-catalyzed antielimination reaction.

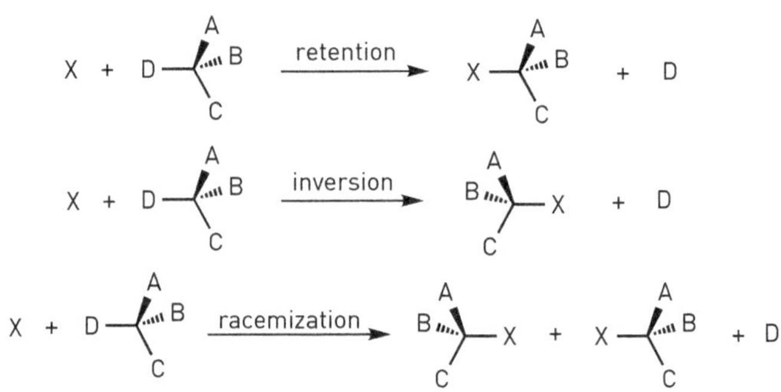

FIGURE 1.11
Possible stereochemical outcomes at a chiral reaction center.

RING SYSTEM CONFORMATIONS: FROM LYING FLAT TO SITTING IN A CHAIR

Cyclopropanes are the only truly flat cycloalkane rings. These rings possess significant angle strain due to their 60° bond angles versus 109.5° for an ideal sp^3 tetrahedral angle, so the C–C bonds have poor sp^3–sp^3 orbital overlap. Along any C–C bond, adjacent C–H bonds are completely eclipsed leading to a total ring strain (angle strain plus torsional strain) of about 27 kcal/mol. (Fig. 1.12).

Cyclobutanes are slightly puckered rather than flat with a total ring strain of about 26 kcal/mol, though the strain is distributed over four carbons rather than three as in cyclopropane (Fig. 1.13). The vicinal C–H bonds are no longer completely eclipsed, leading to less torsional strain.

The cyclopentane ring adopts a folded "envelope" conformation that rapidly equilibrates with only ~6 kcal/mol of ring strain. While the base of the envelope

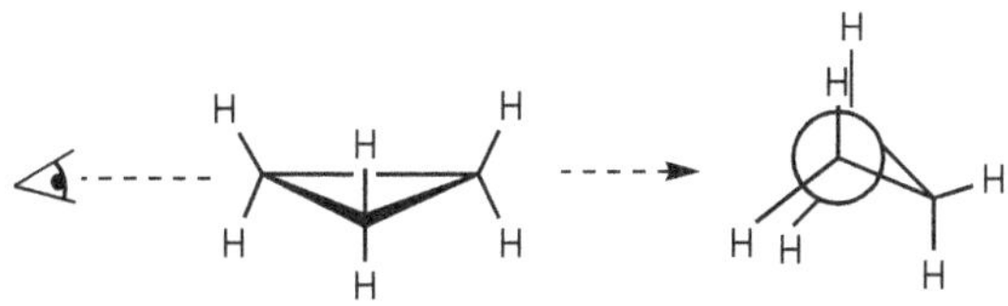

FIGURE 1.12
Strained planar geometry of cyclopropane ring systems.

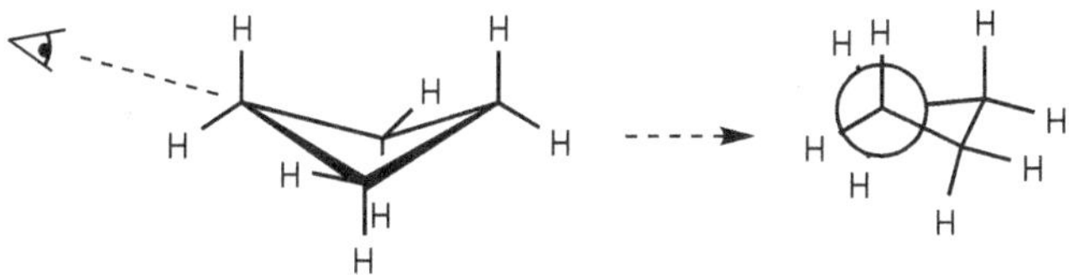

FIGURE 1.13
Puckered cyclobutane ring conformation.

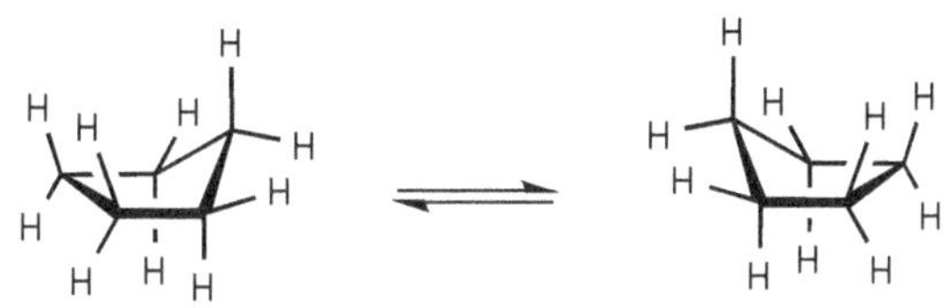

FIGURE 1.14
Envelope conformation of the cyclopentane ring system.

has some torsional strain, the vicinal C–H bonds in the "flap" have minimal strain (Fig. 1.14).

The cyclohexane ring has essentially no ring strain due to its rapidly interconverting "chair" conformations. Interconversion between chair conformers is sometimes called a "ring flip" and is important because this conformational equilibrium interchanges all axial and equatorial positions on the ring. Ring flipping involves tipping one end of the ring up into a higher energy "half-chair" conformation (~7 kcal/mol higher than chair) which can relax to a slightly lower energy "twist-boat" conformation (about 1.5 kcal/mol lower in energy than the boat) and so on (Fig. 1.15).

MONOSUBSTITUTED CYCLOHEXANES: FAVORING EQUATORIAL POSITIONS

The more stable of two chair conformers for a monosubstituted cyclohexane is one in which the substituent occupies an equatorial rather than an axial position. Axial substituents introduce steric strain via "1,3-diaxial" interactions. This strain is relieved via a ring flip to the lower energy equatorial conformer (Fig. 1.16).

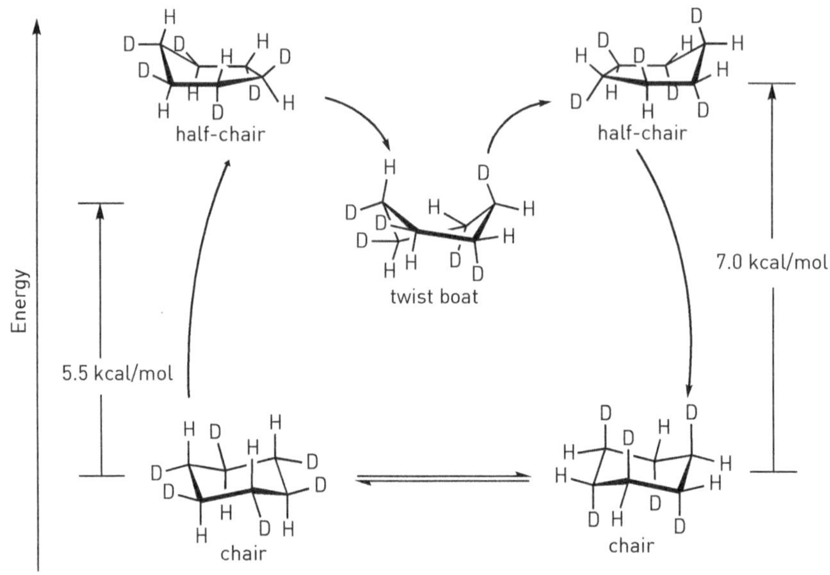

FIGURE 1.15
Chair-chair interconversion ("ring flipping") of cyclohexane conformers.

FIGURE 1.16
Equatorial preference of monosubstituted cyclohexanes and 1,3-diaxial interactions.

DISUBSTITUTED CYCLOHEXANES: EQUIVALENT AND NONEQUIVALENT COMBINATIONS

For 1,4-dimethylcyclohexane, the *trans* stereoisomer is more stable than *cis*, even though the methyl groups are clearly too far apart to crowd one another. Both methyls become equatorial after a ring flip. The *cis* stereoisomer has one axial methyl regardless of conformation. Similar stabilities are observed with the 1,2-dimethylcyclohexanes, with the *trans* stereoisomer more stable in the diequatorial conformer than either conformer of the *cis* (Fig. 1.17).

For 1,3-dimethylcyclohexane, the *cis* stereoisomer is more stable than *trans*; one conformer of *cis* has both methyl groups equatorial (Fig. 1.18).

In some cases, factors other than steric crowding may influence conformational stability. For example, the diaxial conformation of *cis*-1,3-cyclohexanediol is more stable in nonpolar environments due to intramolecular H-bonding (Fig. 1.19).

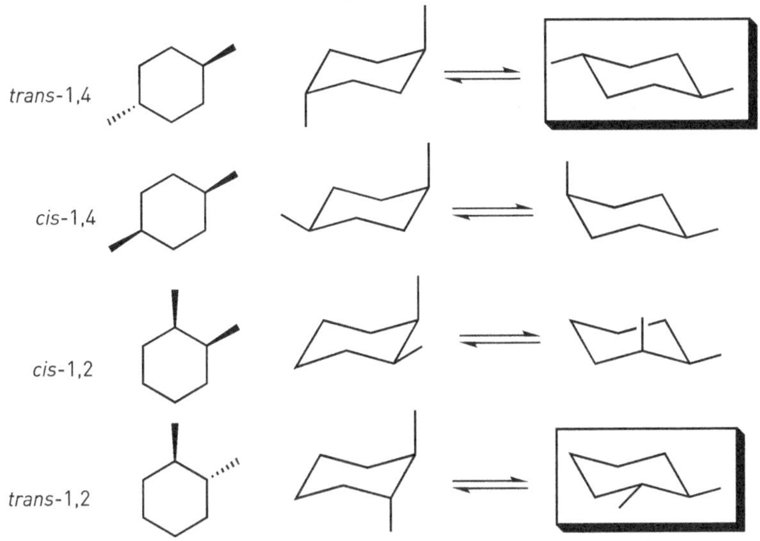

FIGURE 1.17
Conformational preferences in 1,4- and 1,2-disubstituted cyclohexanes.

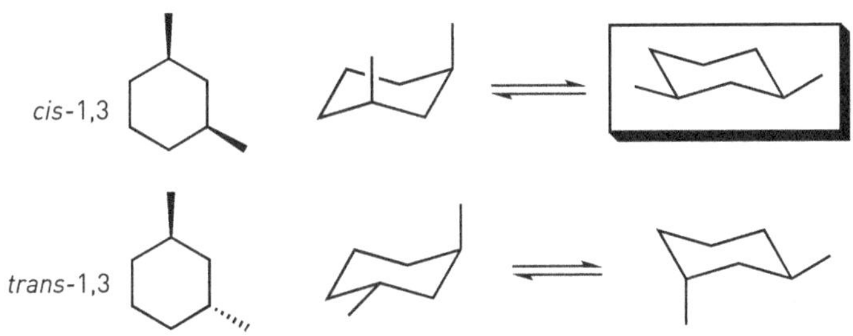

FIGURE 1.18
Conformational preferences in 1,3-disubstituted cyclohexanes.

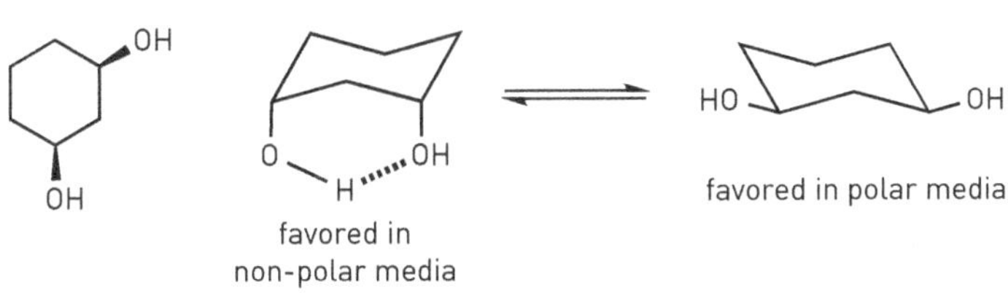

FIGURE 1.19
Intramolecular H-bonding in *cis*-1,3-cyclohexanediol.

BICYCLIC SYSTEMS: JOINING OF RINGS

Of the fused bicyclic ring systems, the 6-5 and 6-6 ring junctions are the most frequently observed. The 6-5 or **hydrindane** ring fusion (bicyclo[4.3.0]nonane) has both *cis* and *trans* ring junctures, with the methyl-substituted *trans* junction a common feature of steroid structures (Fig. 1.20).

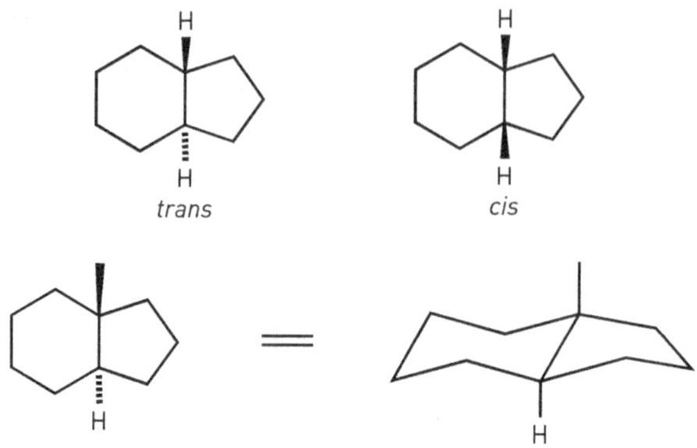

FIGURE 1.20
Cis and *trans* hydrindane (6-5) ring systems.

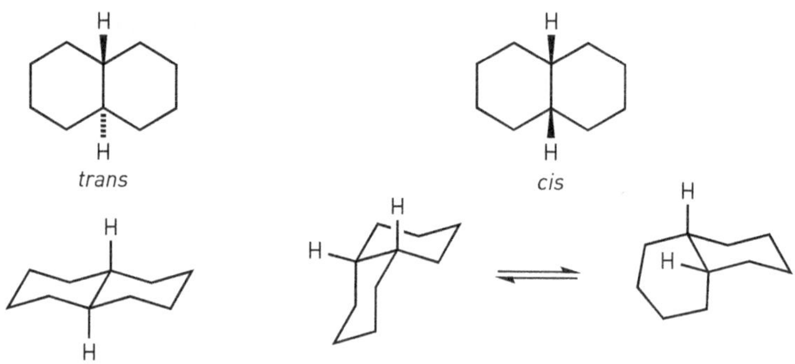

FIGURE 1.21
Cis and *trans* decalin (6-6) ring systems.

The *cis* and *trans* decalin ring systems (bicyclo[4.4.0]decanes) are especially noteworthy. The *trans* junction is conformationally immobile, while the *cis*-fused ring system exists as equilibrating chair forms (Fig. 1.21). Both are also important structural components of the tetracyclic steroids.

HETEROCYCLIC RING SYSTEMS: ONE ATOM MAKES ALL THE DIFFERENCE

Many of the medicinal agents important to natural products chemistry contain one or more heterocyclic rings. Examples include the powerful analgesic N-heterocycle *morphine* (from poppy seeds), O-heterocycles such as *catechin*, a flavonoid compound with antioxidant and antiHIV activities, and *lipoic acid*, an important S-heterocyclic cofactor in many enzymatic systems as well as an over-the-counter dietary supplement (Fig. 1.22).

A brief review of the most important heterocyclic ring systems and their common names is found in Table 1.2. While all the names used here are not rigorously correct in terms of IUPAC nomenclature, we will use them when convenient to refer to

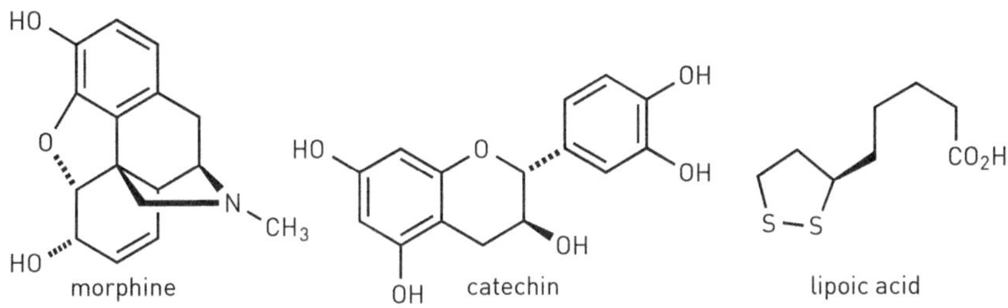

FIGURE 1.22
Some natural products containing N-, O-, and S-heterocycles.

Table 1.2 Some Common Heterocyclic Ring Systems

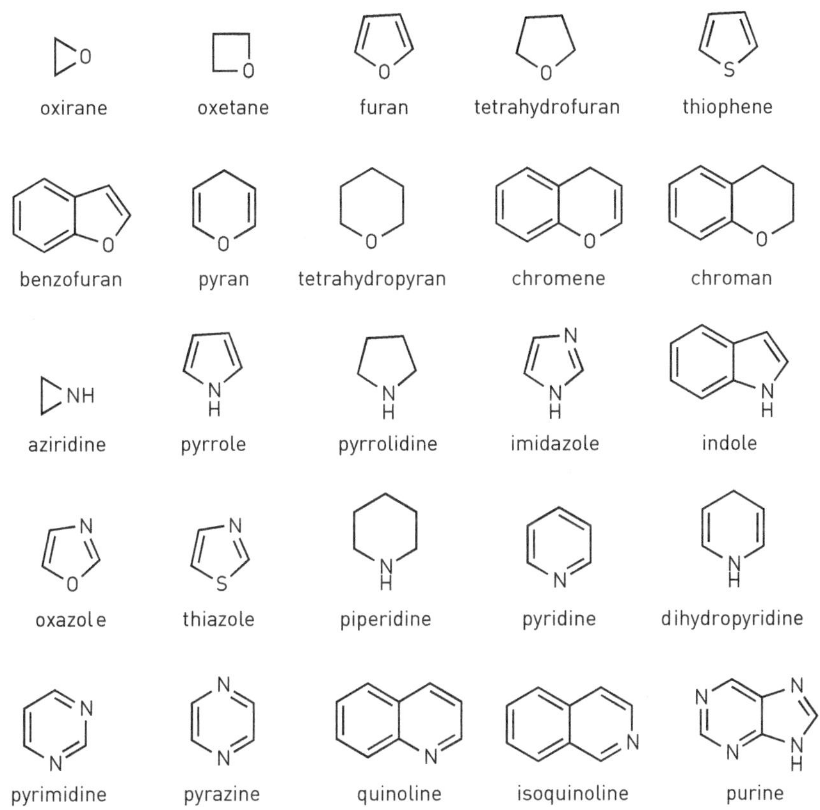

portions of more complex organic structures which may contain one or more such heterocyclic components.

NUCLEOPHILES AND ELECTROPHILES: GIVE AND TAKE IN BOND MAKING

Since one of our main topical focuses will be organic reaction mechanisms in a biological context, a brief review of the basics of how organic mechanisms are depicted

$HO:^-$ $Br:^-$ $HS:^-$ — anionic nucleophiles

$H_2O:$ $:NH_3$ $H_2S:$ — neutral nucleophiles

FIGURE 1.23
Examples of charged and uncharged nucleophilic species.

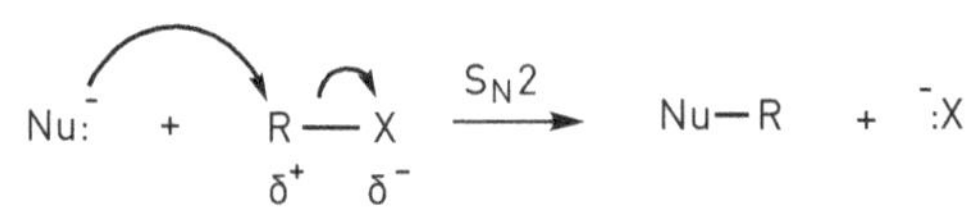

FIGURE 1.24
Nucleophilic attack on an electrophilic atom in a polarized covalent bond.

is in order. Recall that an electron pair that initiates the "attack" in an organic reaction usually comes from a nonbonding pair on an atom. The atom can be negatively charged (like a hydroxide ion) or neutral (like NH_3). For simplicity's sake, only one nonbonding pair is shown on the nucleophilic atoms in Fig. 1.23. Electrons from a two-electron σ-bond or a π-bond can also participate in some cases (we'll see specific examples of this later on). When an attacking base forms a new covalent bond in the resulting organic reaction product, the base is called a **nucleophile** ("nucleus-loving").

Bases and nucleophiles are always **electron-rich**: they have at least one pair of electrons to share (or a readily available pair from a two-electron bond). The terms "nucleophile" and "base" are often used interchangeably. Species that are attacked by nucleophiles are called **electrophiles** ("electron-loving"). Electrophiles are always **electron-poor** or **electron-deficient.** This characteristic of electrophiles can be due to any one of several factors: i) presence of a full positive charge on an atom with an empty orbital (i.e., a cation or formal positive charge); ii) electrically neutral but electron-deficient due to an empty orbital (lack of an octet, such as in boron or aluminum compounds); 3) *partially* positive due to bonding to a more electronegative atom (part of a "polar" covalent bond). Polarized bonds are common electrophilic components in many organic reactions, as in the familiar S_N2 reaction (Fig. 1.24). The reaction of the nucleophile ($Nu:^-$) with an alkyl halide (R–X) is an example of nucleophilic attack by an electron-rich atom (nonbonding electron pair, negatively charged) on an electron-deficient atom (R is partially positive due to its covalent bond to the more electronegative halogen atom X).

Understanding and use of curved arrow notation (or arrow pushing formalism) is fundamental to the depiction of the mechanistic details of organic and biological reactions. The better you understand this notation, the greater your understanding of the mechanism will be, which will help you develop the intuition needed to predict reaction products. The simplest use of the arrow-pushing formalism is to show the "movement" of electron pairs in the resonance forms of molecules or ions (Fig. 1.25). Note the use of the double-headed arrow rather than equilibrium arrows between resonance forms. This is especially useful in evaluating where electron pairs and negative or positive charges may reside in structures.

FIGURE 1.25
Curved arrows showing electron pair movement in resonance forms.

BOND MAKING AND BREAKING: HAVE PAIR, WILL SHARE; NEED TWO FROM YOU

In almost all organic chemistry reactions, two-electron covalent bonds are made and broken. To represent these bond making/breaking processes, we use the curved arrows to show how the electrons are moving or "flowing" in the mechanism. We represent two-electron covalent bond formation resulting from an electron pair "attacking" an electron-deficient atom by using a curved arrow pointing *from* the electron pair *toward* the electron-deficient atom (never the other way around). Such curved arrows are *always* meant to represent the movement of *one electron pair*. They are *never* used to represent anything else. A simple example would be the reaction of ammonia with hydrogen ion to give ammonium ion (Fig. 1.26).

Bond breaking often occurs in conjunction with bond making. In addition to the use of a curved arrow to show where a bond is forming, arrow pushing will also be used to show where an electron pair is going when a bond breaks. In this case, the arrow will be pointing *away* from the two-electron bond that is breaking and *toward* the atom on which the electron pair will ultimately reside, as in the reaction of a base $B:^-$ with an acid H–A shown in Fig. 1.27.

In some instances, a curved arrow will be used to show the formation of a two-electron bond that becomes one of the bonds in a multiple bond (double or triple bond). In such cases, the arrow will be pointing away from a two-electron bond that is breaking and toward the area between two other atoms where the *new* two-electron bond will form. In the E2 elimination reaction depicted in Fig. 1.28, a base $B:^-$ attacks and breaks a C–H bond.

In concert with C–H bond breaking, a B–H bond forms and the electron pair from the broken C–H bond moves between the two carbon atoms, pushing the electron pair of the C–Br bond onto Br, thereby breaking the C–Br bond and forming bromide ion. Note that *three* electron pairs are moving when going from left to right in the equation, so *three* curved arrows are needed to show how and where the electrons have moved. In any correctly written arrow-pushing mechanism (with charges correctly assigned), the mass and net charge on both sides of the equation will balance.

BRØNSTED ACID–BASE REACTIONS: PROTON DONORS GLADLY ACCEPTED

Acids that act as proton (H^+) donors and bases that act as proton acceptors are called **Brønsted** acids or bases. These are among the most important processes in

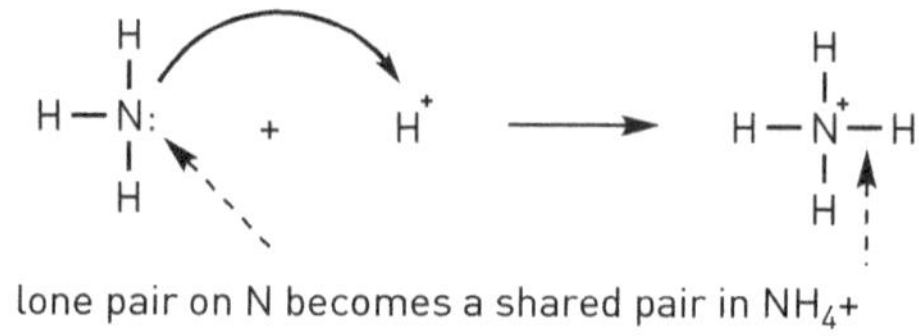

FIGURE 1.26
Curved arrow depiction of a two-electron bond-making process.

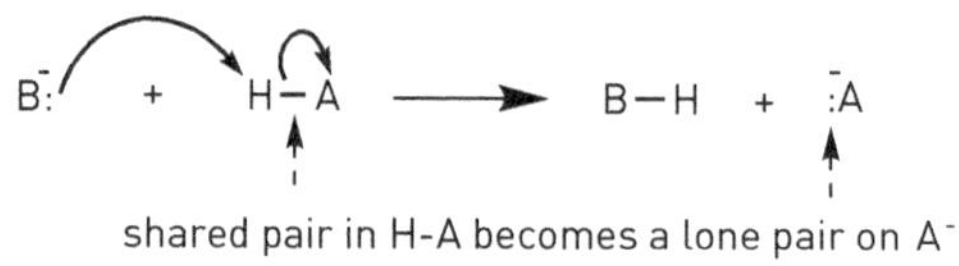

FIGURE 1.27
Curved arrow depiction of simultaneous bond making and bond breaking.

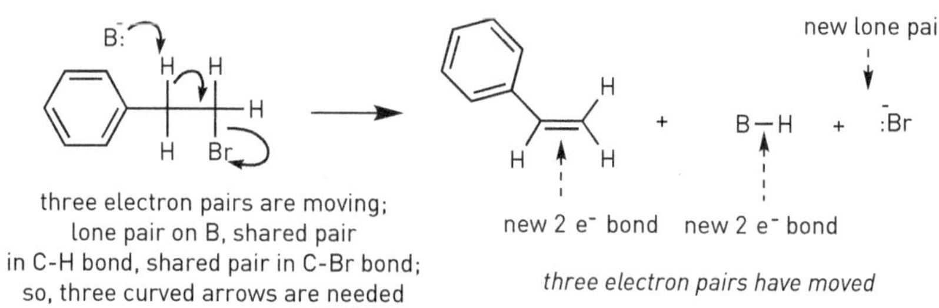

FIGURE 1.28
Curved arrow depiction of electron pair movement in an E2 elimination reaction.

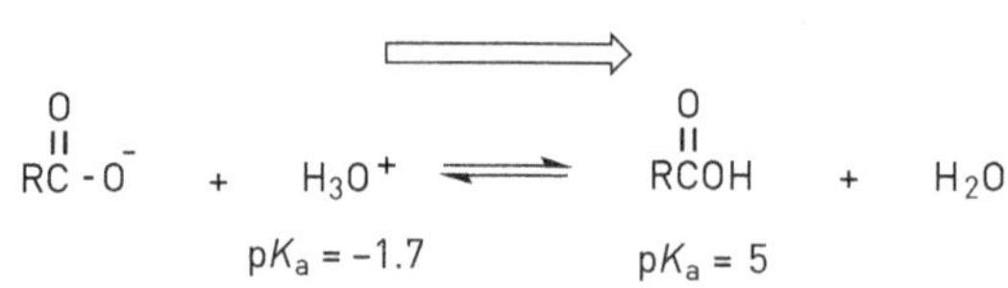

FIGURE 1.29
Acid–base equilibrium position favored in the direction of the weaker acid.

our studies of organic reaction mechanism, and the following trend is an important one to remember for all such equilibria: *acid-base reactions always proceed from the stronger acid/base to the weaker acid/base.* Thus, for the reaction shown in Fig. 1.29, the equilibrium clearly lies to the right (on the side of the carboxylic acid and water), *not* to the left (carboxylate ion and hydronium ion).

To predict the position of an acid–base equilibrium, it is necessary to know the relative strengths of a variety of acids and bases, keeping in mind that the larger the pK_a, the weaker the acid. Table 1.3 provides a brief list of some useful pK_a values for our purposes. While exact numbers can be difficult to keep in mind, it is often useful to have a few simple approximations memorized. The following "skip five" sequence

Table 1.3 Some Common pK_a Values (acidic proton underlined)

Acid	pK_a	Acid	pK_a
$\underline{H}_3O^+$	-1.7	$RNHC(=\overset{+}{N}\underline{H}_2)NH_2$	12.5
$RCO_2\underline{H}$	4–6	$\underline{H}_2O$	15.7
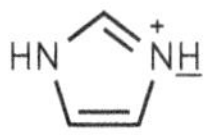	6.9	$RO\underline{H}$	16–18
$RC(=O)C\underline{H}_2C(=O)R$	9	$RC(=O)C\underline{H}_3$	19–20
$ArO\underline{H}$, $RN\underline{H}_3^+$, $RS\underline{H}$	10	$RSC(=O)C\underline{H}_3$	21
$RC(=O)C\underline{H}_2C(=O)OR$	11	$ROC(=O)C\underline{H}_3$	25

of pK_a values, while only approximate, is easily memorized and is frequently useful in evaluating or comparing the acidity of many simple carbon compounds: carboxylic acids ~5; phenols, thiols, and ammonium ions, all ~10; water and alcohols, ~16; ketones and thioesters, ~20; esters and terminal alkynes, ~25. The sequence can be extended to include much weaker acids: ammonia and amines ~35; allylic C–H (RCH=CH–CH_3), benzylic C–H (Ph–CH_3), aryl C–H (Ph–H), and vinyl C–H (R–CH=CH_2) all ~45; and alkyl C–H (R–CH_3) ~50.

ACIDITY TRENDS: WHY THAT PROTON IS OR ISN'T ACIDIC

By referring to the pK_a values from Table 1.3 or other sources, simple trends in acidity can be seen, which can be useful when evaluating or comparing the relative acidity of various species.

i. Within a series of X–H bonds where X does not change, the greater the positive charge on X, the greater the acidity of the X–H bond. Compare the O–H bond pK_a in H_3O^+ (−1.7) versus H_2O (15.7) or the N–H bond pK_a in NH_4^+ (~10) versus and NH_3 (35). The positively charged X "wants" the electrons in the X–H bond more than neutral X does.
ii. Within a series of X–H bonds where X does not change, the greater the resonance delocalization of negative charge on X after loss of H^+, the greater the acidity of the X–H bond. A developing negative charge on X has the ability to spread out from X via resonance, thereby weakening the X–H bond. Compare

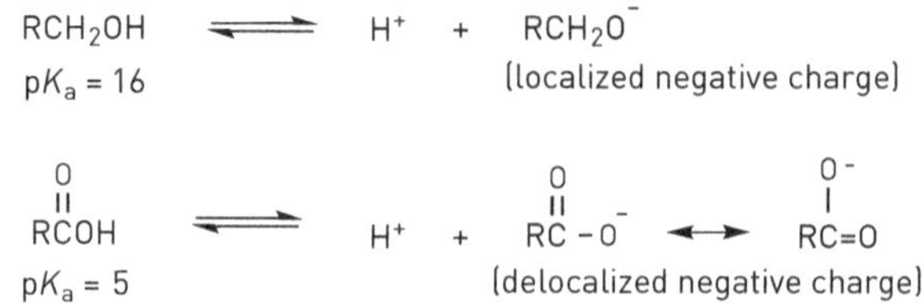

FIGURE 1.30
Charge delocalization effect on O–H bond acidity.

the O–H bond pK_a in an alcohol R–OH (16) versus a carboxylic acid RCO_2H (5) (Fig. 1.30). Likewise, the lower acidity of alkyl C–H bonds in alkanes, R–CH_3 (~50) versus allylic C–H bonds in alkenes, RCH=CH–CH_3 (~45) is due to resonance stabilization of the resulting allylic versus alkyl carbanions.

iii. Within a series of X–H bonds where X has the same charge and is in the same *row* of the periodic table, the more electronegative X is, the greater the acidity of the X–H bond (compare H_3O^+ (−1.7) versus NH_4^+ (~10); H_2O (16) versus NH_3 (35) versus CH_4 (50). The more electronegative X is, the more it "wants" the electrons in the X–H bond.

iv. Within a series of X–H bonds where X has the same charge and is in the same *column* of the periodic table, the greater the size of X, the greater the acidity of the X–H bond. Compare the pK_a values for alcohols, RO–H (16) versus thiols, RS–H (10) or H–F (5) versus H–Cl (−6). The size of X increases going down a column, so the distance of H from the nucleus of X also increases, thus making the X–H bond longer and weaker.

v. Within a series of X–H bonds where X remains constant but changes hybridization, the X–H bond acidity increases as the % "*s*-character" of the hybridization of X increases (*sp* (50% *s*-character) > sp^2 (33% *s*-character) > sp^3 (25% *s*-character)). In this way we can understand the relative pK_a values of terminal alkyne C–H bonds (*sp*), RC≡C–*H* (25) vs alkene vinylic C–H bonds (sp^2), RCH=CH_2 (45) versus alkane C–H bonds (sp^3), RCH_3 (50). Since the energy of hybrid orbitals follows the order $sp < sp^2 < sp^3$, anionic X will follow the same stability trend; the greater the % "*s*-character" of the anion's hybrid orbitals, the lower its energy.

These pK_a values are also useful in evaluating the relative strength of various bases. Rather than using a separate pK_b value for basicity, we need only compare the pK_a values for their respective conjugate acids, keeping in mind that *the stronger the conjugate acid, the weaker the base* and vice versa. For instance, in comparing the relative basicity of imidazole versus methylamine, we need only consider the relative acidity of their conjugate acids (Fig. 1.31). That imidazole is a weaker base than methylamine immediately becomes clear simply by comparing the pK_a values of their respective conjugate acids.

Similarly, we can quickly answer the question regarding the relative basicity of a carboxyate ion versus water (Fig. 1.29). Again, such a process must proceed from

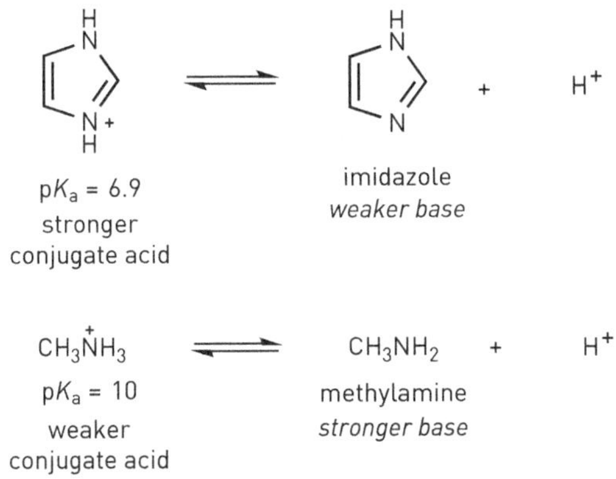

FIGURE 1.31

Comparison of conjugate acid pKa values for evaluation of relative basicities.

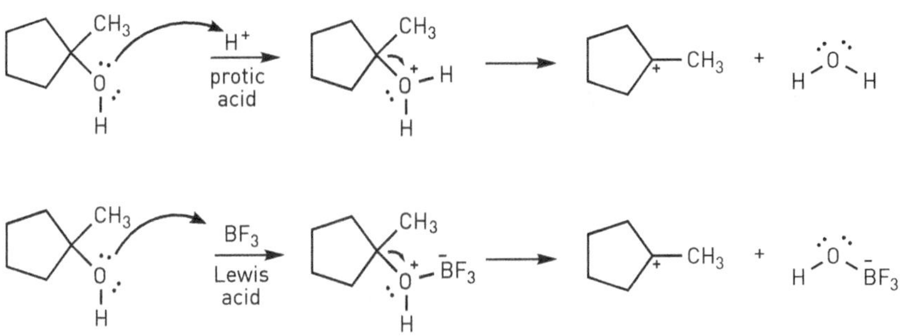

FIGURE 1.32

Comparison of catalytic action of a protic acid vs. a Lewis acid.

the stronger base to the weaker base, and the acid pK_a values clearly reveal that water must indeed be a weaker base than a carboxylate ion.

While many acid–base reactions involve proton transfer from an acid to a base (Brønsted acid–base reactions), many others do not, and a broader definition of acid–base chemistry is often more useful. In **Lewis acid–base** theory, any species which can donate a nonbonding electron pair to an acid can be considered a **Lewis base,** while any species that can accept an electron pair from a base can be considered a **Lewis acid**. This much broader definition of acids and bases means that Lewis acids include many species other than H^+ which may accept an electron pair from a base, the only requirement being that a Lewis acid must have a low-lying, empty orbital into which an electron pair may come to form a bond. Thus, while H^+ is formally a Lewis acid since its 1*s* orbital is empty, so is the boron atom in BF_3, which has an empty *p* orbital perpendicular to its trigonal planar, sp^2 hybridized framework. For many proton-catalyzed reactions, Lewis acid catalysis works just as well (Fig. 1.32). Metal ions such as Fe^{3+}, Zn^{2+}, or Mg^{2+} can be especially important Lewis acid catalysts in certain biological systems, where they can facilitate acid catalysis at physiological pH.

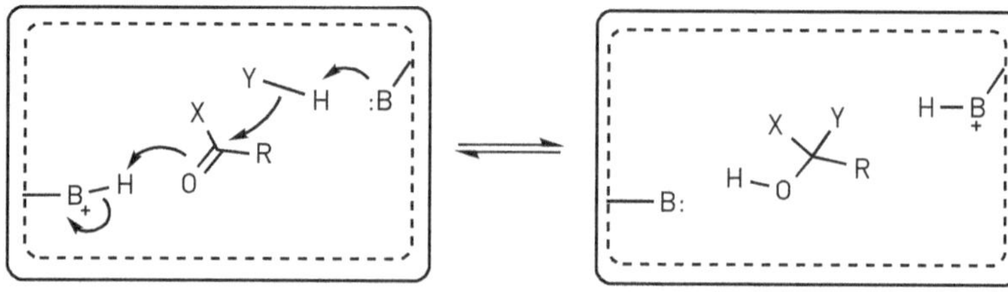

FIGURE 1.33
Simultaneous enzymatic acid–base catalysis.

It is important to note that in biological systems, some enzymes can catalyze acid–base reactions that would be impossible to carry out via conventional solution chemistry. In particular, simultaneous acid and base enzyme catalysis can occur as illustrated in Fig. 1.33. In solution chemistry, having both acid and base together in the same system would lead to simple neutralization.

CARBOCATIONS: THREE BONDS TO CARBON CAN BE A PLUS

As reactive intermediates, carbocations are among the most important species we will encounter in biosynthetic processes. Generally speaking, their ease of formation and stability will follow the usual simple order of 3° (tertiary) > 2° (secondary) > 1° (primary), but it will also be useful to expand this series somewhat to take into account some additional stabilizing features such as the presence of adjacent heteroatoms with lone pairs or conjugation with π-bonds (Fig. 1.34). Such categorizations, while only approximate, serve as useful guides for anticipating the relative ease with which carbocations are likely to form at various carbon centers.

We can quickly summarize some of the most important means for the formation of carbocations in both laboratory and biosynthetic processes by considering four cases (Fig. 1.35). The first is simple loss of a leaving group (such as a phosphate or diphosphate ester) which may include such processes as acid-catalyzed alcohol dehydration. Next is electrophilic additions to π-bonds which may include processes such as simple alkene protonation, reaction of alkenes with carbocations, or other processes such as electrophilic aromatic substitution. The third case derives from the importance of epoxides in biosynthesis, which frequently provide a reliable source of carbocations via acid-catalyzed ring opening reactions. Finally, we consider alkyl radicals which are frequently formed by hydrogen atom abstraction or radical addition to π-bonds and may undergo subsequent one-electron oxidation to afford the corresponding carbocations.

RADICALS: ODD AND REACTIVE

In terms of organic reaction mechanisms, radicals are especially important reactive intermediates in biosynthetic processes, so a brief review of their depiction in

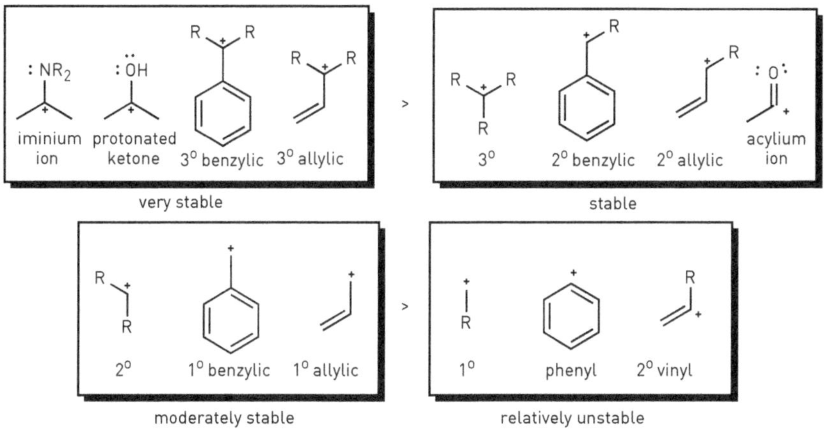

FIGURE 1.34
Classification and comparison of stability of various types of carbocations.

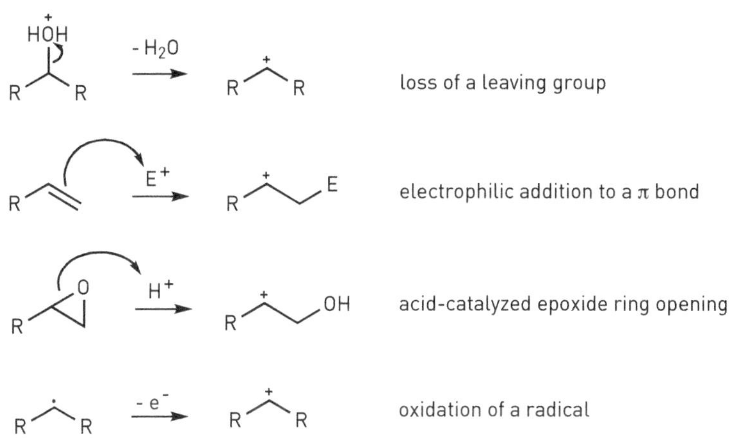

FIGURE 1.35
Typical processes involved in carbocation formation.

mechanistic terms is necessary. Unlike more common two-electron bond-making and breaking processes involving cations or anions where electrons are moving in pairs, radicals are electrically neutral, one-electron species and so cannot be appropriately represented mechanistically via the use of the standard curved arrow formalism. Instead, we use "barbed" arrows to show the movement of only *one* electron. Compare the mechanistic depictions of a two-electron versus a one-electron process in Fig. 1.36. In the two-electron process, an electron pair from the styrene π-bond attacks a carbocation leading to a new two-electron σ-bond and a benzylic carbocation. *One* curved arrow shows how *one* electron pair has moved.

By contrast, in the one-electron radical process, a new two-electron σ-bond is formed as well as a benzylic radical, but only one of the two electrons in the new σ-bond is provided by the π-bond; the other is provided by the initial radical species. Depiction of both σ-bond and radical formation requires a total of *three* barbed arrows, one leading away from the radical toward the π-bond and another leading

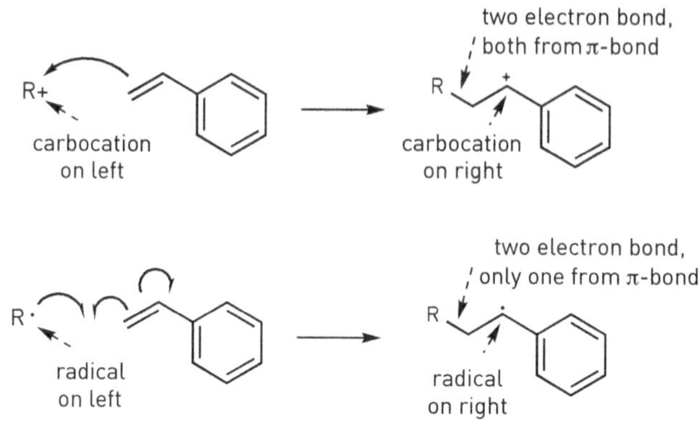

FIGURE 1.36
Comparison of curved arrow formalisms for a carbocation vs. a radical process.

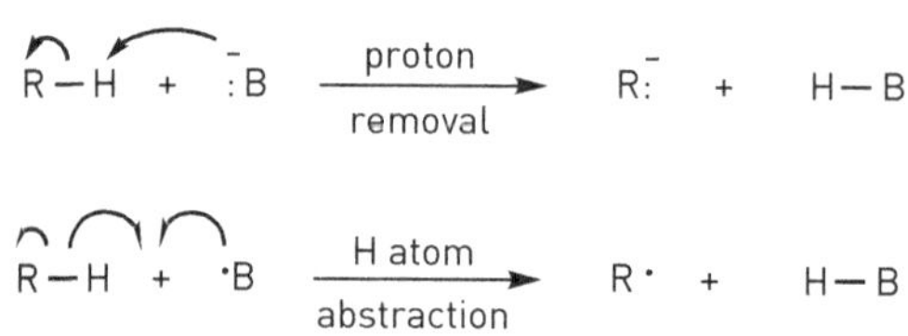

FIGURE 1.37
Mechanistic depiction of proton removal vs. H atom abstraction.

away from the π-bond toward the radical to show how electrons have paired to form the new σ-bond, with a *third* arrow showing how the remaining electron moves to reside on the benzylic carbon. Note that just as charge must balance on both sides in the cationic process, total electron count must balance in the radical process. Finally, note also that radical reactions frequently involve *hydrogen atom abstraction* as opposed to *proton removal* in more familiar acid–base reactions (Fig. 1.37). The facility with which such radical C–H bond-breaking processes will occur depends mainly on homolytic *bond dissociation energies* or BDEs (C–H → C• + H•) as opposed to proton acidity (and pK_a) which at least partially reflects the *heterolytic* bond dissociation energy associated with simple ionization processes (C–H → C:$^-$ + H$^+$). Resonance stabilization of the resulting radical species is also an important consideration regarding the ease with which H atoms are abstracted in such processes.

ELIMINATION REACTIONS: INTRODUCING THE CARBON-CARBON Π-BOND

While we will see a number of new and different ways to introduce carbon–carbon double bonds into molecules, many familiar elimination processes will still be of significance in biosynthetic schemes. The familiar **E1** and **E2** reactions will play an important role in this context as well as the somewhat less familiar **E1cB** process (Fig. 1.38).

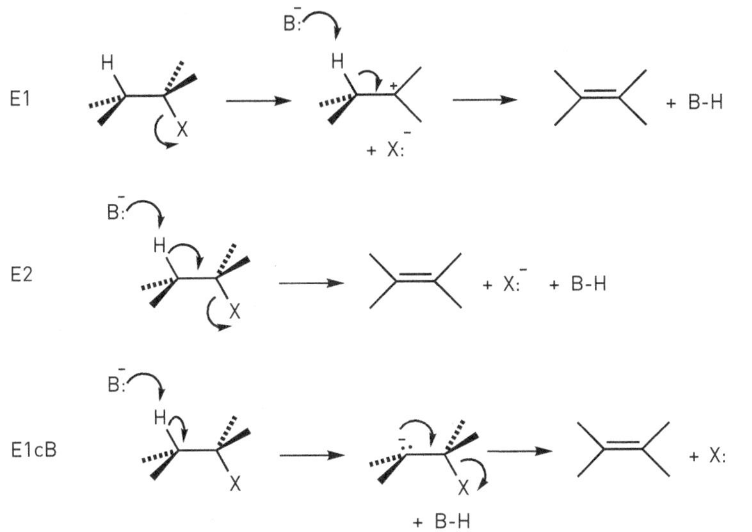

FIGURE 1.38
The three main types of 1,2-elimination processes: E1, E2, and E1cB.

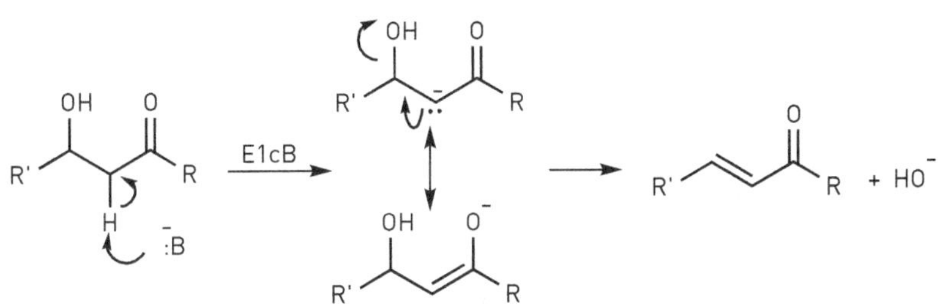

FIGURE 1.39
E1cB mechanism in the dehydration step of an aldol condensation.

Recall that the E1 reaction is a first-order process with carbocation formation as the rate-limiting first step. Subsequent loss of a proton from the carbocation leads to π-bond formation in the second faster step. The E1 reaction is usually observed when loss of a leaving group can readily lead to a stable carbocation (typically tertiary, allylic, or benzylic). This stands in contrast to the single-step E2 reaction, a second order, concerted *anti* elimination process with a specific anticoplanar conformational requirement. This process usually involves strong bases and will be useful when carbocation formation is not favorable. Finally, the E1cB process involves initial proton removal by a suitable base to form an anionic intermediate as the conjugate base (cB) of the substrate. Subsequent expulsion of the leaving group by the anionic lone pair leads to π-bond formation (Fig. 1.39) This process is most often seen when electron withdrawing groups present on the proton-bearing carbon make the proton more acidic. Such groups also help to stabilize the resulting carbanion negative charge, often via resonance. One example of such an E1cB process will be seen later in the base-induced dehydration of an aldol addition product to yield the corresponding α,β-unsaturated carbonyl derivative. Interestingly, in certain biological examples this reaction will be seen to occur as a stereospecific *syn* elimination rather than the usual *anti* orientation seen in most organic elimination reactions.

CARBOCATIONS: REARRANGEMENTS AND FATES

Certain carbocations may undergo internal structural rearrangement in order to increase their stability. These rearrangements are usually achieved by so-called **1,2-shifts** or 1,2-migrations of hydrogen, alkyl, or aryl groups and are known as **Wagner–Meerwein shifts**. 1,3-hydride shifts, though not formally Wagner-Meerwein shifts, are also sometimes observed in biosynthetic processes (Fig. 1.40). The ultimate fate of carbocations, regardless of how they are initially formed, may be loosely summarized as follows: i) loss of a β-proton to a base, resulting in formation of a π-bond; ii) trapping by a nucleophile to form a new σ–bond; and iii) loss of γ–proton to a base, resulting in cyclization to a cyclopropane ring. This latter process, though not common in laboratory organic chemistry, is occasionally observed in certain biological systems. Generally speaking, Wagner–Meerwein shifts occur only when the 1,2-shift process will lower the energy of the carbocation by increasing its degree of substitution, that is, from 1° → 2°/3° or 2° → 3°.

A related and unusual carbocation rearrangement is observed when cyclopropylmethyl carbocation is generated under solvolysis conditions by loss of a leaving group. A mixture of three products is obtained as shown, but interestingly, the same mixture of alcohol products is obtained regardless of whether cyclopropylmethyl, cyclobutyl, or homoallyl substrates are used (Fig. 1.41). This implies a common non-classical carbocation intermediate or rapidly equilibrating carbocations that arise from any one of the three starting structures. This cyclopropylmethyl-cyclobutyl-homoallyl carbocation system is one that has been much studied and is closely

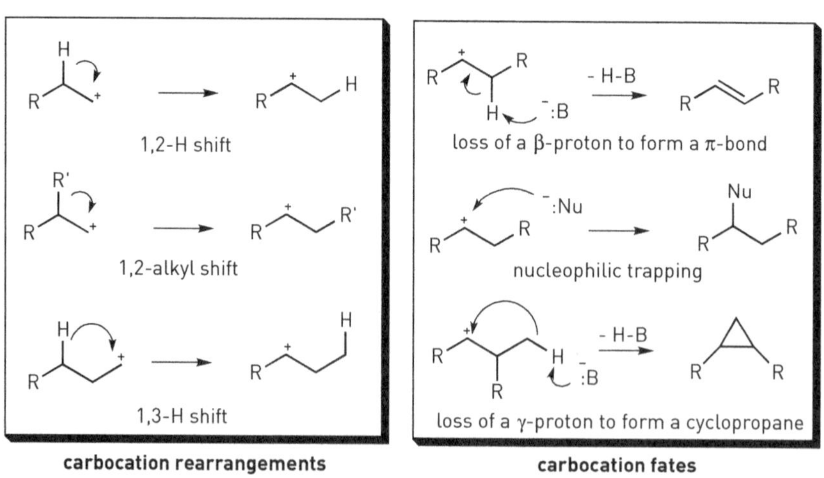

FIGURE 1.40

Examples of the principal rearrangements and fates of carbocation intermediates.

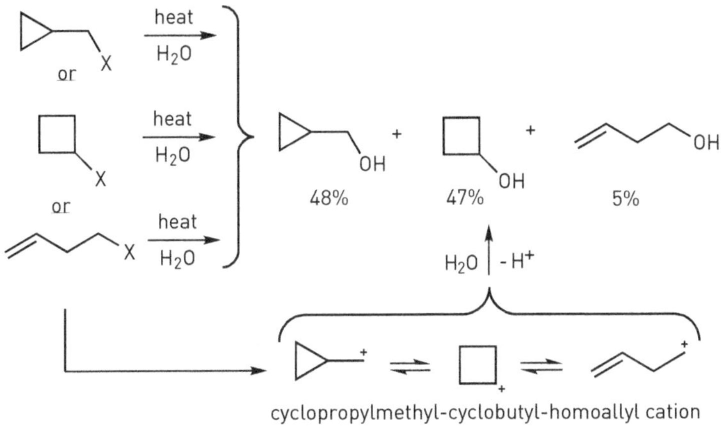

FIGURE 1.41
The cyclopropylmethyl-cyclobutyl-homoallyl carbocation system.

related to a key rearrangement that is just one of a large number carbocation rearrangements and 1,2-shift processes that are of particular mechanistic importance in the biosynthesis of steroids and related structures (Chapter 4).

ELECTROPHILIC ADDITIONS: Π-BONDS AS NUCLEOPHILIC AGENTS

In carbon–carbon double bonds, the electrons in the π-bond are far more accessible than those in the σ-bond and reaction of π-bonds with electrophilic agents (E^+) is an essential bond-making step in many organic reactions. Acid-catalyzed hydration of an alkene to give an alcohol is an illustrative example (Fig. 1.42), with H^+ acting as the electrophile. The familiar Markovnikov orientation of addition is observed here, since initial addition of H^+ to the π-bond occurs in such a way as to give the more stable tertiary carbocation. Nucleophilic attack by water and proton loss complete the process.

Other familiar electrophilic agents that react with π-bonds include the halogens (Cl_2, Br_2, I_2), the active oxygen from peroxyacids (RCO_3H) and many strong Lewis acids (such as BH_3 in alkene hydroboration reactions). In biosynthetic processes, carbocations themselves are among the most important electrophiles that react with alkenes to make new C–C bonds, as shown in the acid-catalyzed reaction of *linalool*, an alcohol found in many flowers and spice plants, to give *α-terpineol*, a fragrant and structurally related alcohol isolated from sources such as pine oil (Fig. 1.43). Interestingly, *α-terpineol* has been shown to suppress the production of some proinflammatory mediators and may impair the growth of certain human melanoma cells.

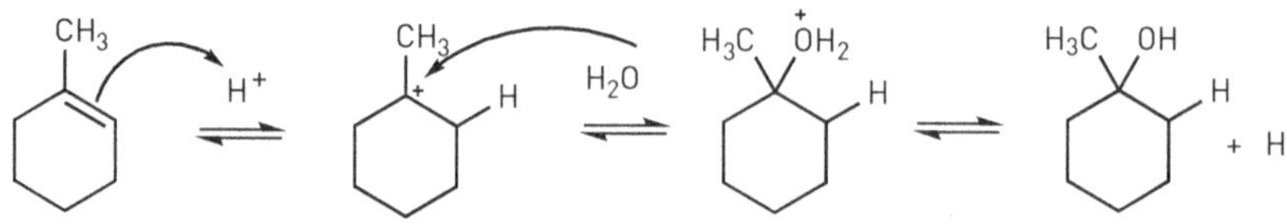

FIGURE 1.42
Typical reaction of an alkene with an electrophilic agent such as H^+.

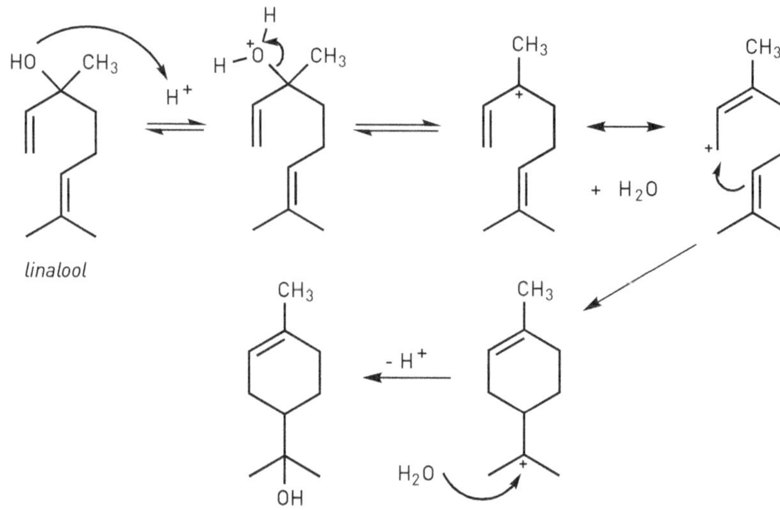

FIGURE 1.43
C–C bond formation via intramolecular trapping of an electrophilic carbocation.

NUCLEOPHILIC SUBSTITUTIONS AND ALKYLATIONS: MAKE OR BREAK FOR C–X BONDS

The two principal types of nucleophilic substitutions are the familiar **S_N1** and **S_N2** reactions. Recall that the S_N1 reaction (substitution, nucleophilic, 1st order) is a two-step process with bond breaking via an initial carbocation formation as the rate-limiting step, followed by nucleophilic bond making in the fast second step (Fig. 1.44). Because of the trigonal planar geometry of the carbocation intermediate, any S_N1 process occurring at a chirality center leads to *racemization.*

Conversely, the S_N2 reaction (substitution, nucleophilic, 2nd order) occurs in a single step with concerted bondmaking and bond-breaking events (Fig. 1.45) and with no reactive intermediates involved (no carbocations or radicals). Nucleophilic attack occurs at the backside of the carbon bearing the leaving group, and inversion of configuration occurs when such substitution takes place at a chirality center.

In the reaction shown in Fig. 1.45, hydrosulfide ion (HS^-) has formed a bond to an alkyl group (the 1-phenylethyl group). Reactions that deliver an alkyl group to a nucleophilic atom are often referred to as *alkylation* reactions. In biological systems,

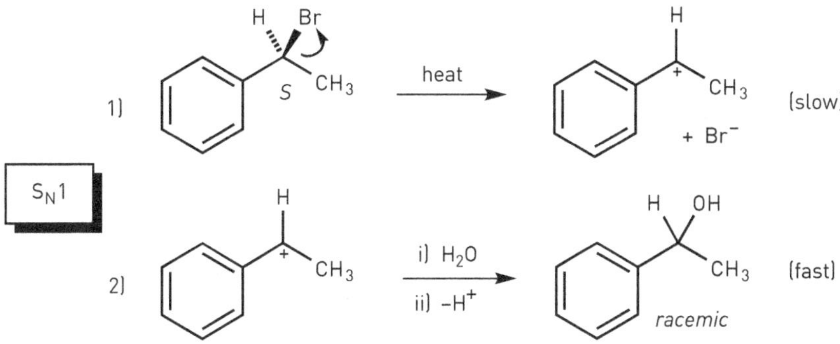

FIGURE 1.44
Stereochemical outcome of S_N1 at a chirality center.

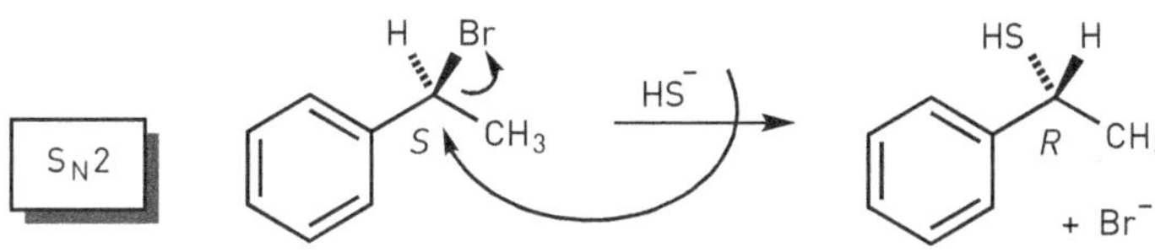

FIGURE 1.45
Stereochemical outcome of S_N2 at a chirality center.

we see alkylation reactions of both the S_N1 and the S_N2 type, depending on the alkyl group involved.

NUCLEOPHILIC CARBONYL ADDITION REACTIONS: C=O π-BOND UNDER ATTACK

The electrophilic character of the carbon atom in the highly polarized carbonyl (C=O) group of aldehydes and ketones makes these compounds very susceptible to attack by many different nucleophiles, forming a variety of potentially useful derivatives (Fig. 1.46). Additions of anionic nucleophiles are usually followed by protonation of the resulting alkoxide intermediate to give a neutral product. For example, if the nucleophile is hydride ion ($H:^-$) or a carbanion ($R:^-$), protonation leads to a stable alcohol. For less reactive neutral nucleophiles such as alcohols (ROH), activation of the carbonyl group by protonation (acid catalysis) may be required to facilitate addition of the nucleophile; in the case or addition of ROH via the mechanism shown in Fig. 1.46, the product would be the familiar (though unstable) hemiacetal product. For purposes of review, we focus on some key carbonyl addition reactions that will be of particular importance in biological transformations.

IMINE FORMATION: MAKING THE ESSENTIAL C=N LINKAGE

In the next chapter we will see some specific examples of **imine** formation (also called **Schiff's bases**) that recur as fundamental components of many of the

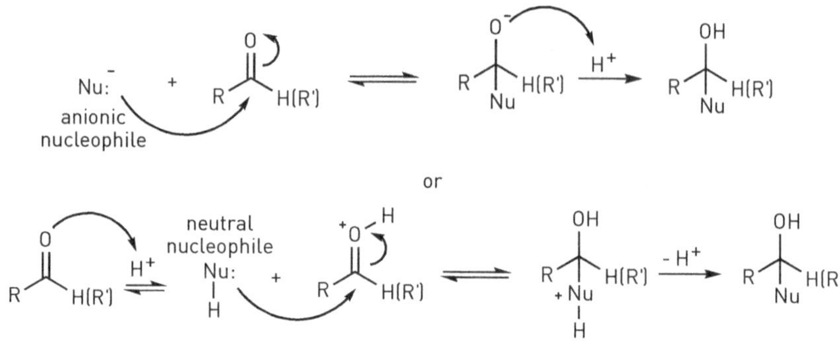

FIGURE 1.46
Anionic vs. neutral nucleophiles in addition reactions with carbonyl groups.

biosynthetic processes treated throughout the remainder of the book. The favorable equilibrium for this reaction of aldehydes or ketones with primary amines is driven in part by the nucleophilicity of the amine nitrogen as well as the thermodynamic stability of the resulting C=N double bond after loss of water. The general mechanism for such condensations is depicted in Fig. 1.47, which shows initial **carbinolamine** formation resulting from addition of the amine N–H bond across the carbonyl C=O bond. This is followed by acid-catalyzed dehydration to give the corresponding imine linkage.

Note also that imines may undergo a tautomerization process similar to keto-enol tautomerism (Fig. 1.47), leading to the structurally related **enamines**. This tautomerization will be an important one in several mechanistic analyses. Note that these processes are reversible. Thus both imines and enamines are subject to acid-catalyzed hydrolysis, reverting to their constituent amines and carbonyls by addition of excess water.

ACETAL FORMATION: PROTECTING ALDEHYDES OR KETONES

Both aldehydes and ketones react with alcohols (usually primary alcohols, always two equivalents) via an acid-catalyzed nucleophilic addition sequence that yields the corresponding **acetal** derivatives. The reaction is more favorable for aldehydes than for ketones and proceeds mechanistically through an intermediate **hemiacetal** for which the equilibrium is generally unfavorable (Fig. 1.48). The equilibrium is driven forward by dehydration of the hemiacetal to a resonance-stabilized carbocation which is subsequently trapped by a second equivalent of alcohol and proton loss to give the acetal product. In acetal form, the reactive carbonyl function has been converted to a much less reactive ether-type linkage. Like imines and enamines, acetals are subject to acid-catalyzed hydrolysis, reverting to their constituent aldehyde/ketone and alcohol components by addition of excess water.

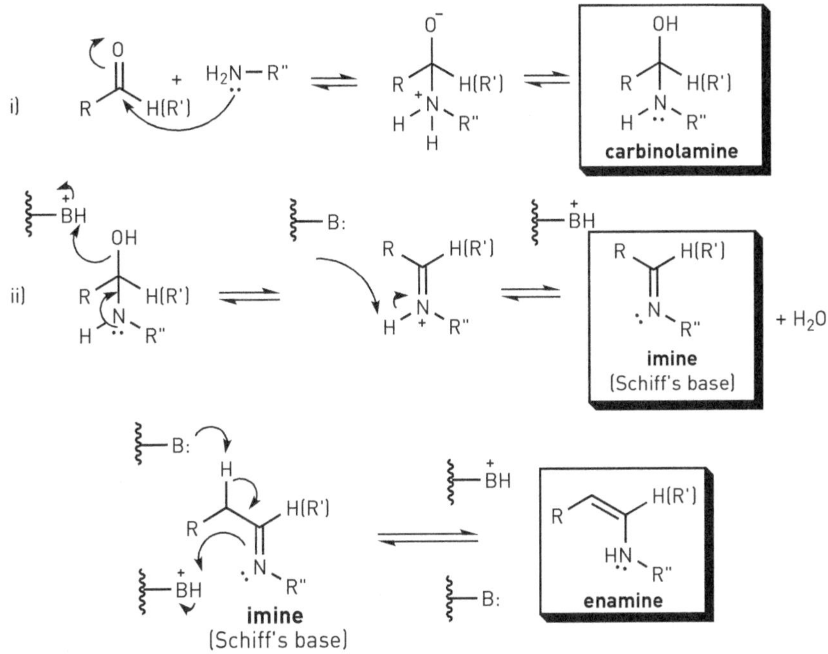

FIGURE 1.47
Conversion of aldehydes or ketones to imines (Schiff's bases) and enamines.

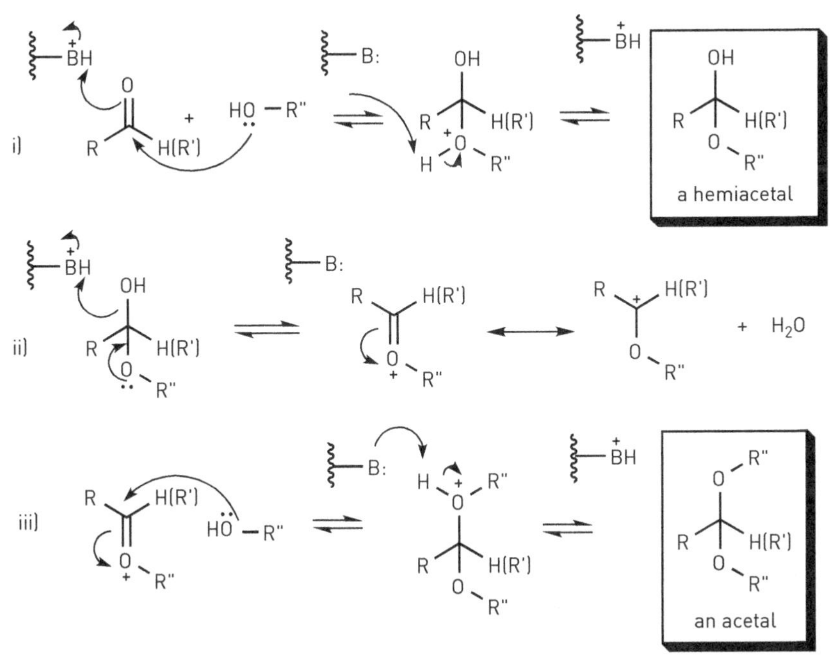

FIGURE 1.48
Conversion of aldehydes or ketones to acetals via a hemiacetal intermediate.

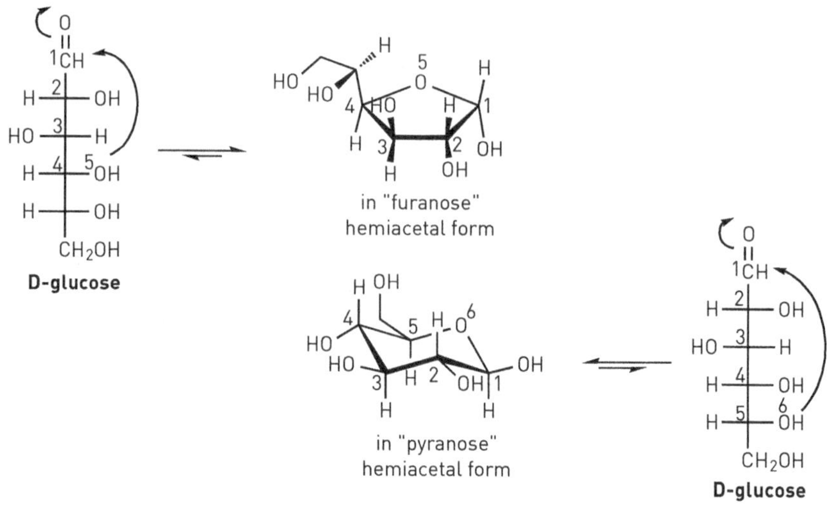

FIGURE 1.49
Formation of furanose and pyranose cyclic hemiacetal forms of glucose.

While the equilibrium for hemiacetal formation from bimolecular reactions of alcohols and aldehydes or ketone carbonyls is generally unfavorable, the equilibrium for the corresponding *intramolecular* reaction is favorable; this means that *cyclic* hemiacetal derivatives (5- or 6-membered rings are best) may often be isolated from hydroxy aldehydes or hydroxy ketones with their alcohol oxygen as the number 5 or number 6 atom relative to the carbonyl carbon (Fig. 1.49). Note the oxygen heterocycle nomenclature employed to describe the two different cyclic hemiacetal forms possible for D-glucose (the pyranose form is favored over the furanose form). Such cyclic hemiacetal structures play an essential role in the chemistry of carbohydrates, a topic we will address in some detail in Chapter 3.

NUCLEOPHILIC 1,4-(CONJUGATE) ADDITION REACTIONS: REMOTE ATTACK ON CONJUGATED CARBONYLS

An especially important component of our review of carbonyl addition reactions focuses on the unique nucleophilic addition chemistry associated with α,β-unsaturated carbonyl compounds. Up to this point, we have been concerned with the direct attack at the carbonyl carbon (C=O) by nucleophiles. For simple aldehydes and ketones, this is the only mode of attack possible, since the carbonyl carbon is the sole electrophilic center in the polarized C=O bond. However, when we consider the dipolar resonance contributors possible for carbonyl compounds in which the C=O bond is in conjugation with a C=C bond, we find two sites available for nucleophilic attack: i) the carbonyl carbon (as usual), and ii) the β-carbon of the C=C π-bond (Fig 1.50). Both carbons are electrophilic centers in such molecules; when direct nucleophilic addition across the C=O π-bond occurs, we refer to this as the **1,2-addition**

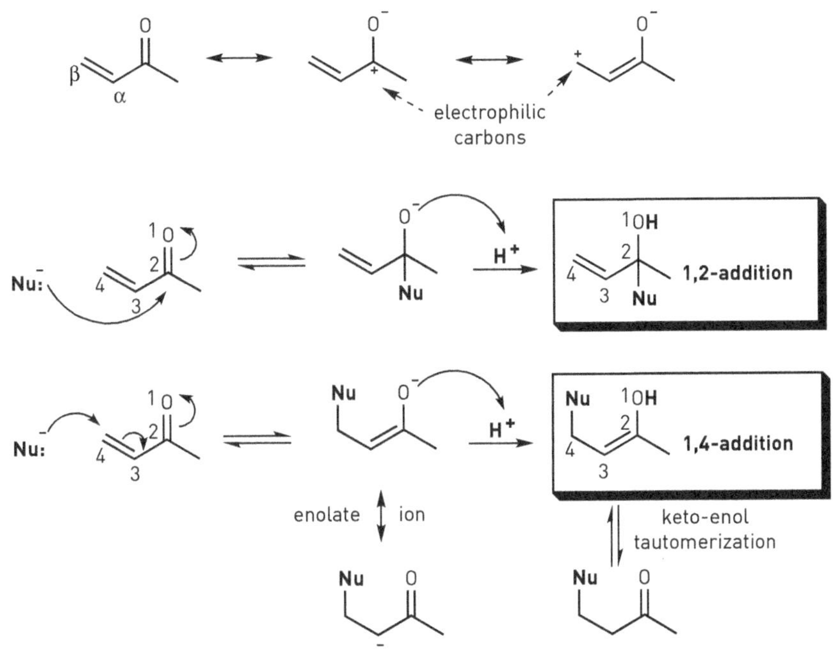

FIGURE 1.50
1,2- vs. 1,4-addition to α,β-unsaturated carbonyl compounds.

mode; when nucleophilic addition occurs at the β-carbon of the C=C π-bond, we refer to this as the **1,4-addition** mode, also called **conjugate addition**. These modes of addition are not limited to aldehydes and ketones but rather may include other α,β-unsaturated carbonyl compounds such as esters, thioesters, or amides.

Note that 1,2-addition of an anionic nucleophile initially leads to an alkoxide ion which is then protonated to give the corresponding alcohol product, whereas in the case of 1,4-addition, an enolate ion is formed after initial nucleophilic addition. If subsequent protonation occurs on oxygen, this leads to an enol product which rapidly tautomerizes to the keto form as shown. Alternatively, protonation of the enolate may occur on carbon, leading directly to the keto form of the product. Generally speaking, strongly basic nucleophiles tend to add to α,β-unsaturated carbonyl systems in a 1,2-fashion, while weakly basic nucleophiles tend to favor the 1,4-mode of addition (there are exceptions). For our purposes, we mainly need to be able to recognize when an opportunity for 1,4-(conjugate) addition exists or to recognize when such an addition has taken place within the context of a biosynthetic scheme.

NUCLEOPHILIC CARBONYL CONDENSATION REACTIONS: MAKING C–C BONDS VIA ENOLATE IONS

Recall that aldehydes, ketones, and thioesters all have pK_a values of ~20, making them sufficiently acidic to be deprotonated by suitable bases. The resulting enolate ions are nucleophilic and can attack the carbonyl groups of aldehydes or ketones to

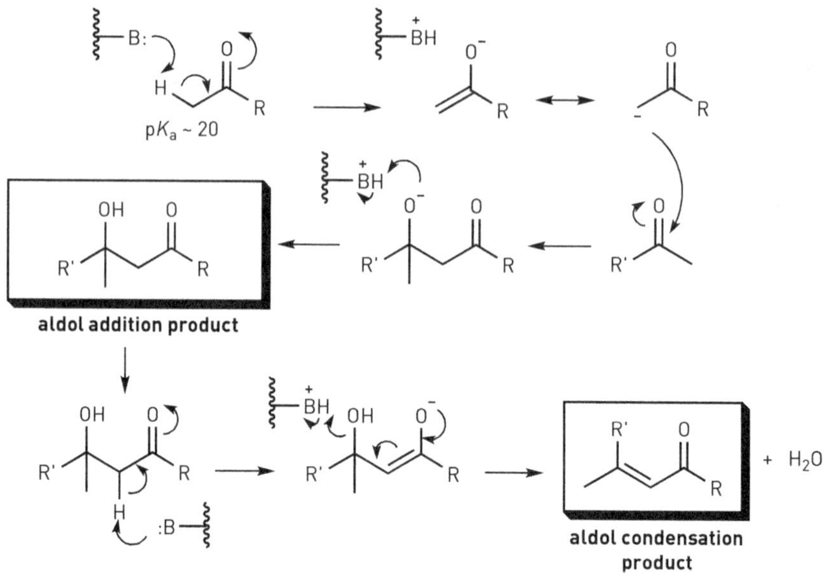

FIGURE 1.51
Aldol addition and aldol condensation reactions.

form alkoxide intermediates which after protonation, afford β-hydroxyaldehydes, ketones, or thioesters, depending on the source of the enolate ion itself (Fig. 1.51). This process is the familiar **aldol addition** reaction. Such aldol addition products can readily lead to the corresponding α,β-unsaturated carbonyl derivatives via a simple dehydration. In the presence of a suitable base, this reaction is again initiated by deprotonation to form an enolate ion, which then expels hydroxide ion to introduce the double bond. When aldol addition is followed by subsequent dehydration, the overall process is known as **aldol condensation.**

We will see many examples of such processes occurring in both a bimolecular and an intramolecular fashion, the latter serving as a particularly useful method for formation of 5- and 6-membered rings. While a synthetic rather than a biosynthetic process, the **Robinson annulation** provides an excellent example of the use of both the conjugate addition of an enolate ion to an α,β-unsaturated carbonyl system and a subsequent intramolecular aldol condensation to form a fused 6-6 bicyclic ring system (Fig 1.52).

An important reaction closely related to the aldol condensation is the **Knoevenagel condensation**, which specifically involves enolate ion formation from relatively acidic 1,3-dicarbonyl compounds such as 1,3-diketones, β-ketoesters, β-cyanoesters, or 1,3-diesters, and their subsequent condensation with aldehydes (or in some cases, ketones) to give α,β-unsaturated carbonyl derivatives. The greater acidity of β-dicarbonyl compounds (pK_a ~9-13) relative to simple aldehydes or ketones (~17–20) means that mild bases can be used to form the β-dicarbonyl enolate ion without initiating a competing self-condensation of the less acidic aldehydes or

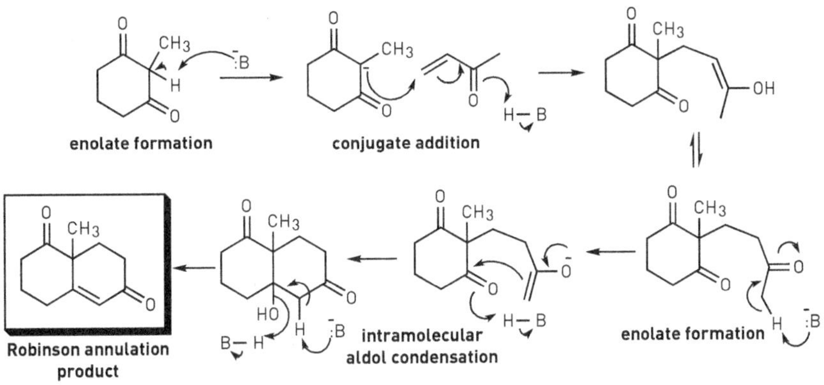

FIGURE 1.52
Conjugate addition and intramolecular aldol condensations in the Robinson annulation reaction.

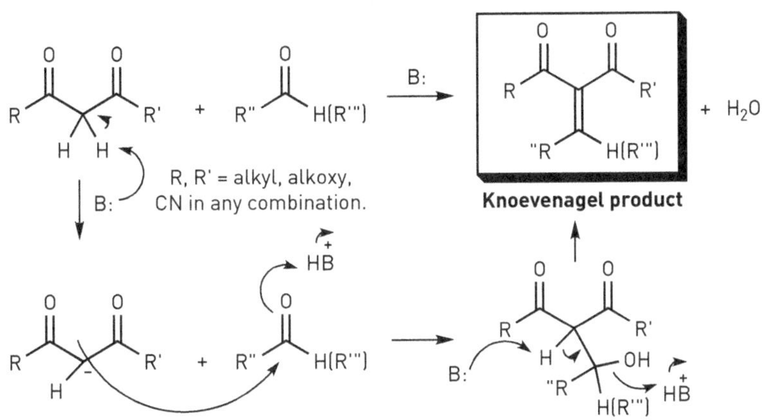

FIGURE 1.53
Reaction of 1,3-dicarbonyls with aldehydes or ketones in the Knoevenagel condensation.

ketones involved (Fig. 1.53). While often more complex mechanistically than the simple condensation shown in Fig. 1.53, we will nevertheless see a number of situations in which it will be useful to invoke a Knoevenagel-type process for depiction of C=C bond formation in certain biosynthetic schemes, and as a particularly powerful component of phenolic ring formation in the polyketide pathway that is the subject of Chapter 5.

Since enolate ions are nucleophilic, we should also expect them to participate in alkylations as well as carbonyl addition reactions, and we will see some specific examples of such alkylations throughout our study of biosynthetic pathways. One related process of particular note in this context is the **Mannich reaction** (Fig. 1.54) which will be of particular significance in alkaloid biosynthesis (Chapter 7). Often referred to as an enolate aminoalkylation process, the reaction consists of condensation of ammonia or primary or secondary amines with a nonenolizable aldehyde to produce the corresponding iminium ion (in the case of

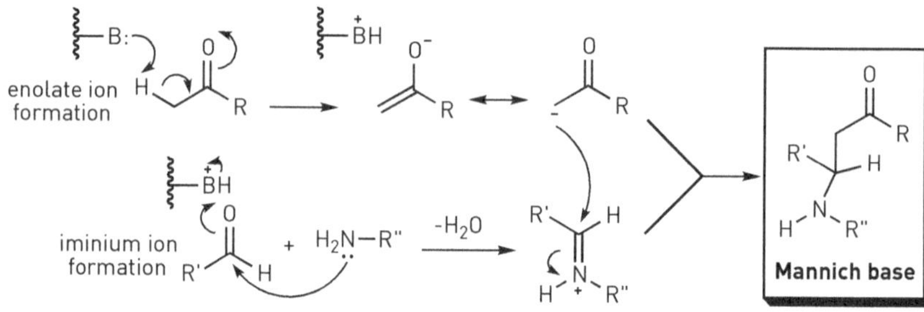

FIGURE 1.54
General mechanistic scheme for the Mannich reaction.

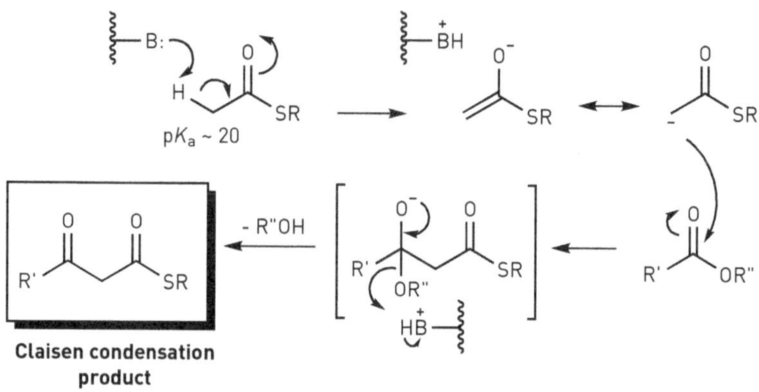

FIGURE 1.55
Reaction of a thioester enolate ion with an ester in a Claisen condensation.

ammonia or primary amines, this product is simply a protonated imine). Attack on the electrophilic carbon of the iminium ion by an enolate carbanion leads to C–C bond formation producing a β-amino-carbonyl compound known as the **Mannich base.**

When the enolate ion of an ester or thioester reacts with the carbonyl group of another ester or thioester rather than one from an aldehyde or ketone, the alkoxide initially formed from such nucleophilic attack is not stable and will collapse with expulsion of RO^- or RS^- from the tetrahedral intermediate, leading to formation of a β-ketoester or β-ketothioester in a process known as the **Claisen condensation.** In the example shown in Fig. 1.55, a thioester enolate ion reacts with a typical ester carbonyl with resultant loss of alkoxide from the tetrahedral intermediate and formation of a β-ketothioester as product. Such chemistry is the basis of the biosynthesis of fatty acids and related products, which is also explored in Chapter5.

Intramolecular versions of the Claisen condensation are also important ring-forming reactions in biosynthetic processes and are sometimes referred to as **Dieckmann condensations.** Using the intramolecular case, we can also make some additional useful observations about enolate ion reaction pathways. For example, we can see that deprotonation of certain α-substituted esters may lead to formation of

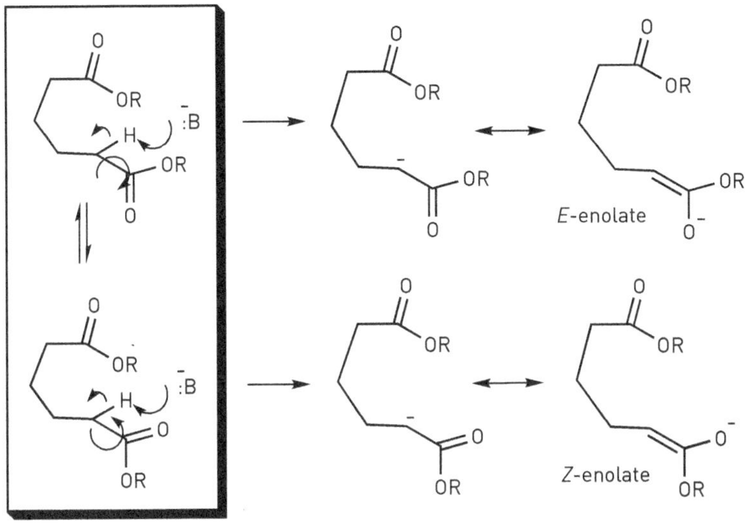

FIGURE 1.56
Isomeric E and Z enolates from deprotonation of a 1,6-diester.

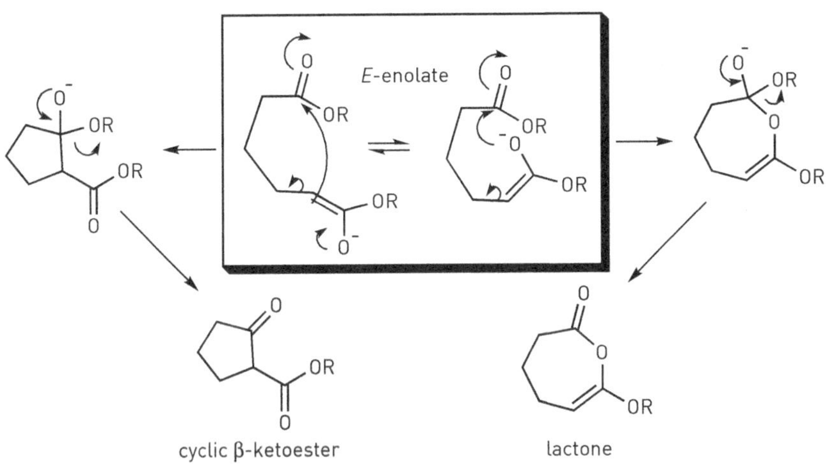

FIGURE 1.57
C vs. O nucleophilic attack and ring closure from an E-enolate.

enolate ions possessing either *E* or *Z* stereochemistry, as shown in the deprotonation of a diester of glutaric acid in Fig. 1.56.

Keeping in mind that enolate ions are ambident nucleophiles, we see that in the intramolecular case, the *E*-enolate may adopt a conformation that favors attack on the proximate ester carbonyl via either its nucleophilic carbanion or its nucleophilic oxyanion form: thus, C-atom nucleophilic attack leads to the expected cyclic β-ketoester (Dieckmann product) while O-atom nucleophilic attack leads to the corresponding lactone structure (Fig. 1.57). Note that lactone formation from the corresponding *Z*-enolate would be geometrically constrained and strongly disfavored.

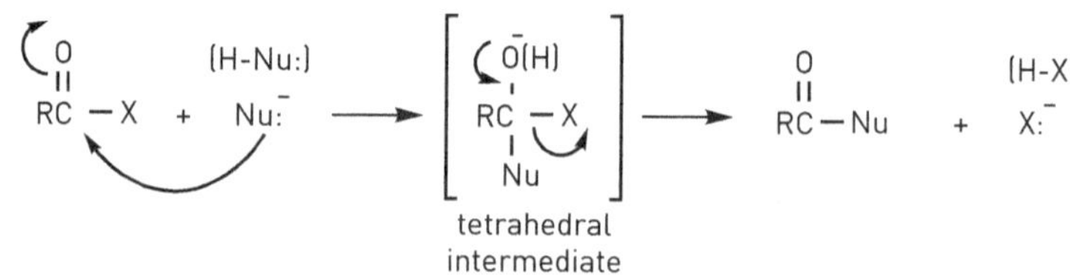

FIGURE 1.58
General mechanistic scheme for nucleophilic acyl substitution reactions.

As we look more closely at the various pathways utilized in biosynthetic processes, we will find numerous examples of the use of all of these important reactions as key C–C bond forming processes for the production of an incredible array of widely different organic structures.

NUCLEOPHILIC ACYL SUBSTITUTION REACTIONS: TURNING ONE ACYL COMPOUND INTO ANOTHER

The Claisen condensation may be thought of as a special case of the more general process we know as *nucleophilic acyl substitution*. In such processes, a nucleophile (anionic or neutral) attacks the carbonyl carbon of a carboxylic acid derivative, leading to a new carboxylic acid derivative via substitution after collapse of the usual tetrahedral intermediate (Fig. 1.58). While we may for simplicity's sake occasionally skip the depiction of this intermediate in abbreviated mechanistic analyses, it is important to remember that direct nucleophilic displacement (as in the S_N2 reaction) does *not* take place in acyl systems and that a tetrahedral intermediate is *always* involved.

In general organic chemistry terms, the reactivity order of carboxylic acid derivatives follows the familiar order of acid chloride > acid anhydride > thioester > ester > amide. In biological systems, the most reactive acyl derivative available is the acyl phosphate (RCOOP), derived from phosphorylation of a carboxylic acid by adenosine triphosphate (ATP), a topic taken up in Chapter 2.

The acyl phosphate is actually a mixed anhydride of a carboxylic acid and phosphoric acid and so is quite reactive toward nucleophilic acyl substitution. Thus, acyl phosphates may be converted to any of the less reactive derivatives by nucleophilic acyl substitution with thiols, alcohols, or amines (Fig. 1.59). The next most reactive derivative, the thioester, may be converted to both esters and amides, while esters may be converted only to amides. All derivatives will yield the corresponding carboxylic acid by hydrolysis, with the acyl phosphate being hydrolyzed most readily and amides being the most resistant to hydrolysis. While the thioester is an uncommon derivative in laboratory organic chemistry compared to alcohol-derived esters, it is a crucial acyl derivative in biosynthesis owing to its greater reactivity toward nucleophilic substitution relative to simple esters. Equally

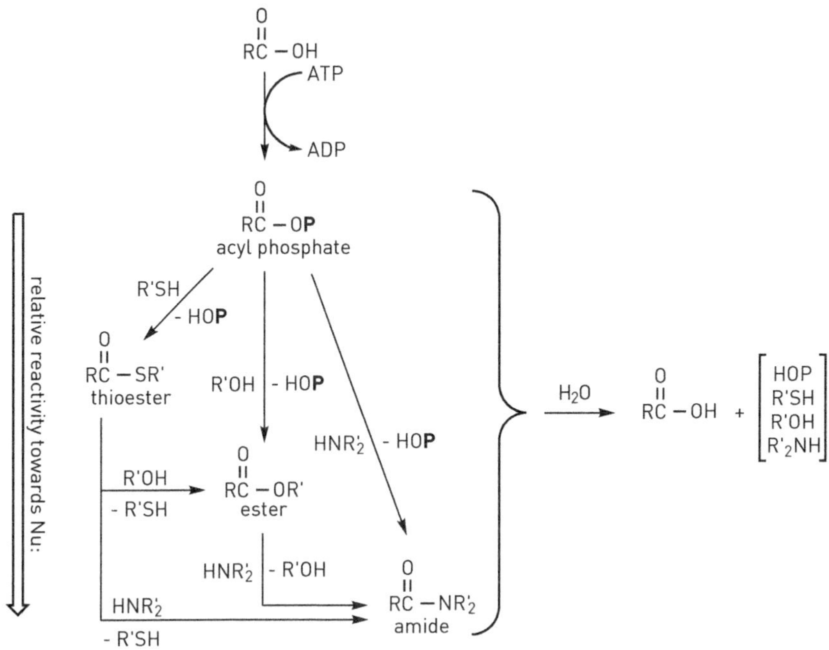

FIGURE 1.59
Relative reactivities and reaction pathways for interconversion of acyl compounds.

significant is the greater acidity of thioester α-hydrogens (pK_a ~20) versus those of simple esters (pK_a ~25), making their enolate ions more accessible and therefore of particular importance as sources of nucleophilic carbon in biosynthetic processes.

LOOKING AHEAD

Now that we have reviewed some essential organic chemistry concepts (and introduced a few new ones), we are in a position to move on to Chapter 2 where we will become familiar with some specific kinds of organic reactions and reagents that are used by biological systems. Let's see how cells achieve the same kinds of transformations or synthesis goals that we ordinarily tackle in our own laboratory-based organic chemistry experiments. The similarities are often surprising.

STUDY PROBLEMS

1. 6-aminopenicillanic acid (6-APA) and 2,6-dimethoxybenzoyl chloride are used to commercially prepare the antibiotic *methicillin*. Give the structure of methicillin and indicate what type of functional group is formed by the reaction shown.

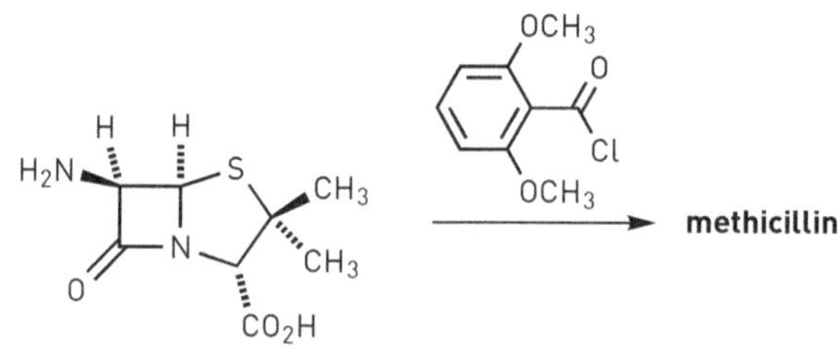

2. Use of pK_b values for comparing relative base strengths has largely been supplanted by comparison of pK_a values for the corresponding conjugate acids of the bases under consideration. Taking note of the relationship: pK_b + pK_a(conj. acid) = 14, rank the following compounds in order of increasing basicity (weakest base first):

   I. guanidine, pK_a(conj. acid) = 13.6 II. ethyl amine, $K_b = 6.5 \times 10^{-4}$

   III. aniline, pK_a(conj. acid) = 4.6 IV. ammonia, pK_b = 4.76

3. Arrange the indicated carbon–hydrogen bonds in order of increasing homolytic bond dissociation energy (weakest bond first). Justify your order using resonance structures.

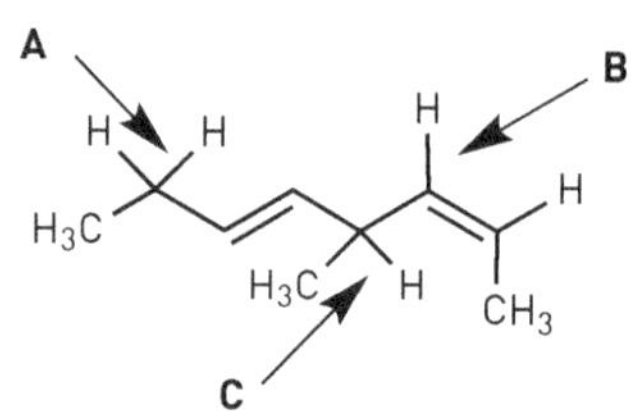

4. In each of the following, label the designated hydrogen atoms as being either pro-*R* or pro-*S*. Are these hydrogens *enantiotopic* or *diasterotopic*? Explain.

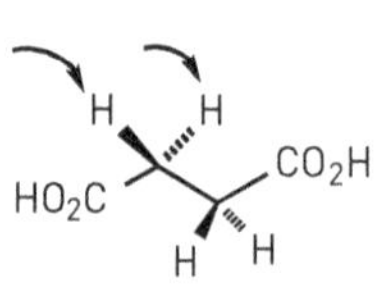

succinic acid

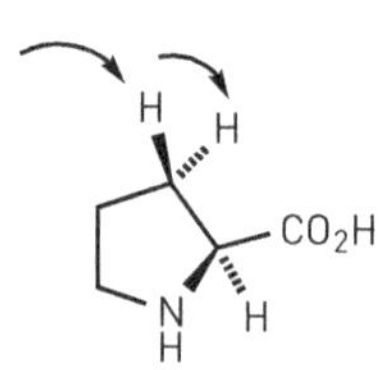

proline

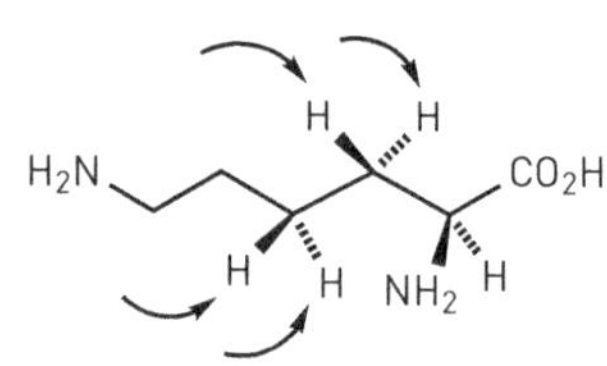

lysine

5. Choose the most acidic hydrogen for the following compound and then explain your choice using appropriate resonance forms.

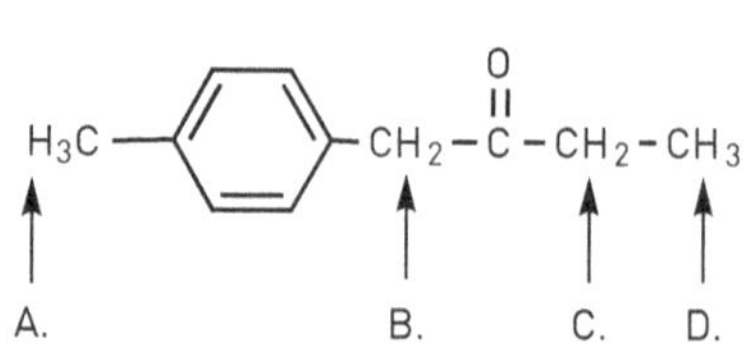

6. Assign *Si* or *Re* to the faces of the prochiral center(s) in each of the following:

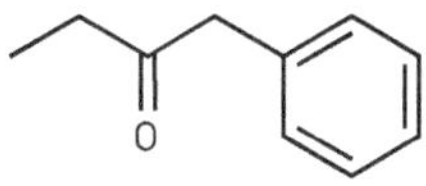

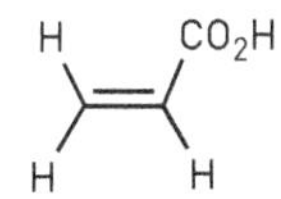

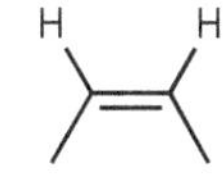

7. A simple bond-line structure for the compound (+)-*juvabione* is shown.

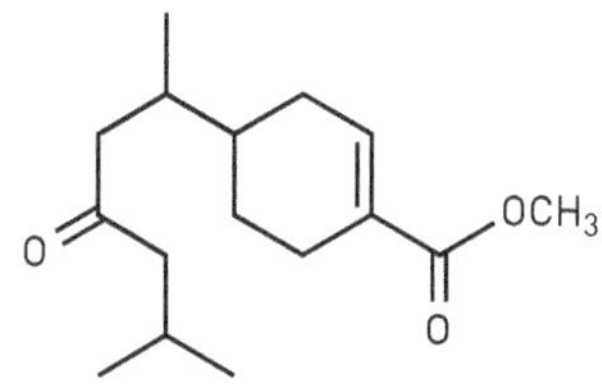

a) What does the (+) in front of the name of the compound indicate?

b) Using wedge and hatched bonds at its two stereogenic centers, draw a stereochemically correct structure of (+)-*juvabione*, given that the natural product has the *R*,*R* configuration.

c) Draw the structure of the alcohol product resulting from hydride addition to the *Re* face of *juvabione* by NADH/H^+.

d) Draw correct structures for the enantiomer and the two diastereomers of (+)-*juvabione*.

8. *Japonilure* is the chemical sex attractant (pheromone) secreted by the female Japanese beetle (*Popillia japonica*) a destructive garden pest of over 200 species of plants in America, where natural predators are lacking. Draw a stereochemically correct structure for Japonilure, given that the natural product has the *R*,*Z* configuration as indicated.

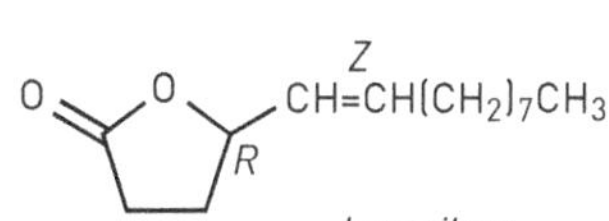

Japonilure

9. Provide reagents and mechanisms for conversion of each of the following substrates into 2-ethylcyclohexanone.

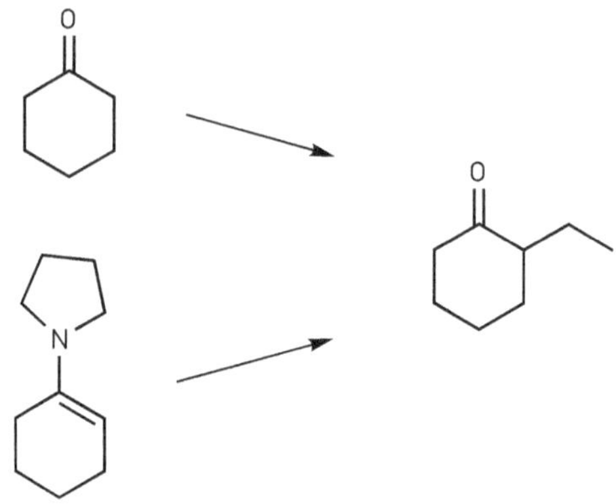

10. What is the correct order of acidity for the numbered hydrogens in the structure shown?

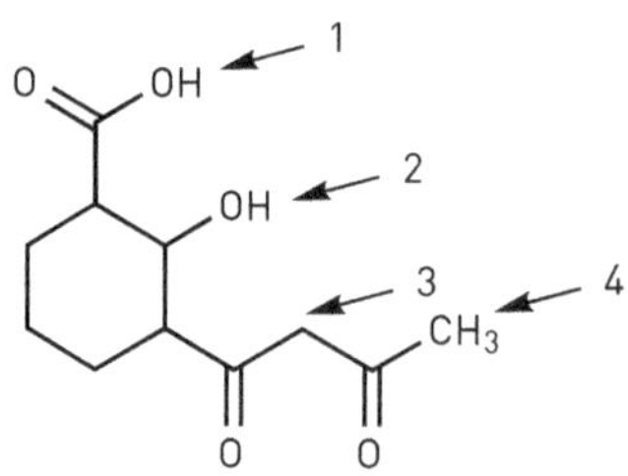

a) 1 > 2 > 3 > 4 b) 4 > 3 > 2 > 1

c) 1 > 3 > 2 > 4 d) 1 > 4 > 3 > 2

11. Which of the following statements describes this first step of the reaction?

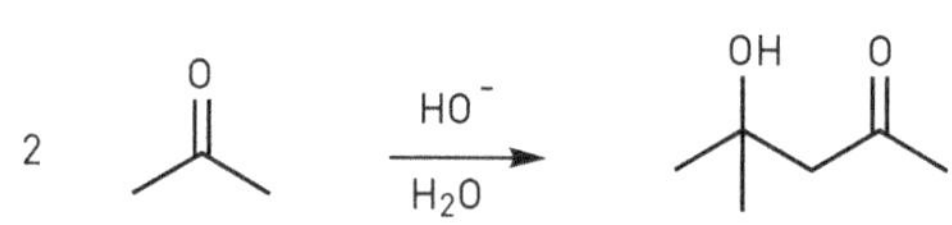

a) The enol form of acetone is formed by proton removal by base.
b) The carbonyl oxygen of acetone is protonated by hydroxide ion.
c) Nucleophilic attack by hydroxide ion on the carbonyl carbon gives an alkoxide.
d) A methyl proton is removed by hydroxide ion to form acetone enolate ion.

12. When 2-methylpropanal is reacted with methyl vinyl ketone in the presence of aqueous base, some 4,4-dimethyl-2-cyclohexen-1-one is formed. Outline in *clear, mechanistic detail* the sequence of reactions involved in this transformation.

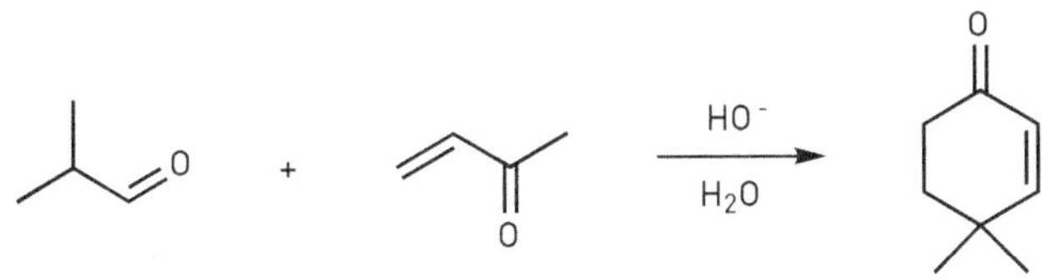

13. Which of the following is also known as a Schiff's base?
    a) an imine b) a cyanohydrin c) a hydrate d) sodium hydroxide
14. Which reagent can be used to convert butanol into butanal?
    a) $LiAlH_4$ b) $Na_2Cr_2O_7$ c) O_3 d) PCC
15. Which of the following conditions can drive the equilibrium of acid-catalyzed acetal formation from reaction of an aldehyde and two moles of an alcohol?
    a) addition of excess H_2O b) removal of H_2O as it is formed c) addition of excess alcohol d) b and d
16. When (*S*)-3-bromo-1-pentene is heated in water, which of the following is produced?
    a) (*S*)-1-penten-3-ol b) (*R*)-1-penten-3-ol c) (*E*)-2-penten-1-ol d) (*Z*)-2-penten-1-ol e) all of these

2 Bioorganic Reactions

Few scientists acquainted with the chemistry of biological systems at the molecular level can avoid being inspired. Evolution has produced chemical compounds exquisitely organized to accomplish the most complicated and delicate of tasks.
—Donald J. Cram (*Nobel Prize in Chemistry, 1987*)

ENZYMES: THE CATALYSTS OF BIOLOGICAL ORGANIC CHEMISTRY

It is not essential to have a background in enzymology or biochemistry to gain at least an introductory-level understanding of many biosynthetic processes, so this book does not deal with enzymology or enzyme structure or function in any significant way, even though much of the chemistry we will be examining depends almost entirely on enzyme catalysis. Nevertheless, we will refer to enzyme catalysis and the names of specific enzymes throughout the text as we examine biosynthetic processes and reactions in significant detail.

So what exactly are enzymes? Simply put, enzymes are naturally occurring proteins that catalyze various biochemical reactions in living systems. As we will see, many of the reactions they catalyze are familiar organic reactions, but have specific purposes and target structures. Generally speaking, enzymes catalyze organic reactions by lowering transition state energies or raising ground state energies of reactants in much the same way as nonenzymatic catalysts in laboratory chemical reactions, though in the case of enzyme catalysis, rate enhancements of as much as 10^{23} have been reported, far exceeding rate enhancements currently achievable by conventional chemical means. Understanding the interaction of enzymes and substrates (reactants) to form an **enzyme–substrate complex** (E–S complex) is fundamental to having some appreciation for how enzymes carry out their work. While overly simplistic, the "lock-and-key" model of enzyme–substrate interaction provides an intuitive context for understanding the remarkable substrate specificity of enzyme-mediated reactions. Thus, so-called enzyme **active sites** or binding sites (the "lock") will only accept certain specific substrate structures (the "key"), with shape, conformation, intermolecular forces, and other factors determining the lock-and-key fit. Enzymes not only catalyze specific kinds of reactions, they can act specifically on certain compounds or classes of compounds or functional groups, often showing remarkable selectivity and stereospecificity, especially in the recognition and/or introduction of chirality centers in organic molecules.

In terms of nomenclature, enzyme names always end with an *ase* suffix and are typically named in accordance with the substrate they act upon and/or the kind of reaction process they catalyze. For example, an enzyme that catalyzes an oxidation process in which only one of the two atoms of O_2 is delivered to a substrate may be named as a "monooxygenase." A more specific example involves "toluene dioxygenase," an enzyme that mediates delivery of *both* oxygen atoms from O_2 to the substrate toluene (Fig. 2.1), with H^+ and the enzyme cofactor NADH providing the diol H atoms. Note that an enantiomerically pure chiral product is produced in this process from an achiral substrate (toluene) and an achiral reactant (O_2). In conventional organic chemistry, such a result would be impossible, since achiral reactants cannot produce optically active products. But the chirality of the enzyme catalyst in this case leads to a chiral binding site for the substrate, with delivery of the two oxygen atoms to only one of the two faces of the toluene π-bond.

Enzymes may be broadly grouped into six main classes, each with a number of subclasses:

- **Oxidoreductases**: include the subclasses *oxidases* (for oxidation), *reductases* (for reductions), and *dehydrogenases* (for π-bond formation via loss of the elements of H_2).

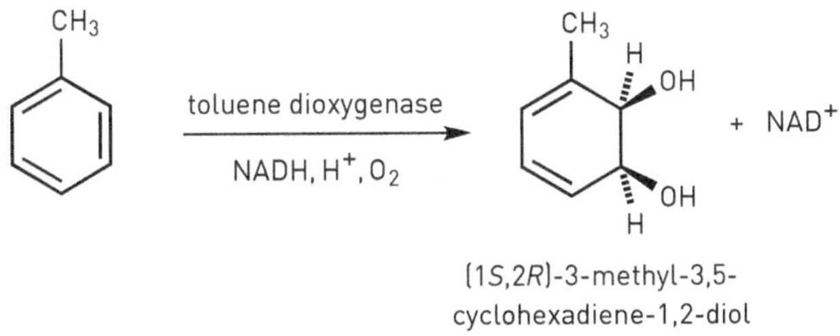

FIGURE 2.1
Enzyme-mediated dioxygenation of toluene.

- **Transferases:** include the subclasses *kinases* (for phosphate transfer), *transaminases* (for amino group transfer), and *transketolases* and *transaldolases* (for ketone or aldehyde group transfer).
- **Hydrolases:** include the subclasses *lipases* (for ester hydrolysis), *proteases* (for amide hydrolysis), and *nucleases* (for phosphate hydrolysis).
- **Lyases:** include *decarboxylases* (for decarboxylation reactions) and *dehydrases* (for dehydration reactions).
- **Isomerases:** include *epimerases* (for chiral center isomerizations), *racemases* (for chiral center racemizations), and *cis–trans isomerases* (for interconversion of *E*/*Z* isomers).
- **Ligases:** include *synthetases* (for new bond formations), *carboxylases* (for additions of CO_2), and *cyclases* (for formation of cyclic compounds).

COFACTORS: ENZYME ASSISTANTS IN BIOORGANIC REACTIONS

Most biosynthetic steps are catalyzed by specific, individual enzymes that perform familiar transformations such as alkylation, hydrolysis, oxidation, reduction, acylation, hydroxylation, elimination, decarboxylation, and other such processes. While very different enzymes may carry out similar reactions in various organisms, most employ a set of common organic or inorganic "assistants" known as **cofactors**. These are molecules or ions which are necessary for enzyme activity, broadly divided into two groups: the organic cofactors (nonprotein organic compounds) and metals ions (such as Fe^{2+}, Mg^{2+}, Zn^{2+}, etc.). Certain enzymes may require multiple cofactors as a combination of organic cofactors and/or metal ions. When loosely bound to an enzyme host, cofactors are sometimes called **coenzymes**, while tightly bound cofactors are often referred to as **prosthetic groups**. Regardless of their role, we will refer to them mainly as cofactors for simplicity's sake. In the sections to follow, we will treat the cofactors of most significance to biosynthetic processes in some detail along with some specific types of bioorganic transformations of importance to our further studies.

Often called the "molecular currency unit" for energy transfer in living cells, **adenosine-5'-triphosphate (ATP)** is used by enzymes and structural proteins for a wide variety of cellular processes. Each ATP molecule has three linked phosphate groups and is produced from either adenosine diphosphate (ADP) or adenosine monophosphate (AMP) and inorganic phosphate (P_i) by the action of the enzyme ATP synthase. The energy derived from the degradation of nutrients is almost always coupled to the stoichiometric production of ATP which, when degraded to lower energy phosphate derivatives (ADP or AMP) by chemical reaction, releases the energy stored in its phosphate bonds to drive other energy-demanding reaction processes. Our interest in ATP is related to one of its main roles in biosynthesis,

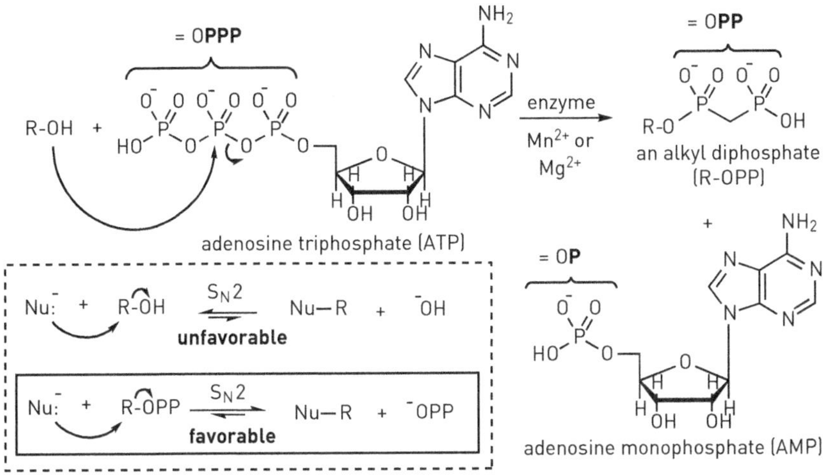

FIGURE 2.2
Conversion of alcohols to diphosphates via reaction with adenosine triphosphate (ATP).

namely the conversion of the -OH group of alcohols (R–OH) or carboxylic acids (R(CO)–OH) into their corresponding mono- or diphosphate derivatives (Fig. 2.2).

Recall that the -OH group of alcohols is a very poor leaving group for nucleophilic substitution (such as S_N1 or S_N2), so to facilitate such reactions, the -OH group must often be converted into a better leaving group. ATP, which is essentially a substituted organic anhydride of phosphoric acid, reacts exothermically ($\Delta G \sim -31$ kJ/mol)) with alcohols to produce phosphate or diphosphate esters in much the same way that alcohols are readily converted to acetate esters by reaction with acetic anhydride. Once an alcohol is converted to its mono- or diphosphate ester derivative (R–OP or R–OPP), the substrate is activated toward subsequent nucleophilic displacement or carbocation formation with loss of monophosphate or diphosphate, depending on which derivative was produced initially. This overall process is analogous to the familiar conversion of alcohols into tosylate or mesylate esters for subsequent nucleophilic substitution reactions (Fig. 2.2).

Similarly, nucleophilic acyl substitution of carboxylic acids (R(C=O)–OH) in biosynthesis may be greatly facilitated by conversion of the acid to an acyl phosphate derivative (R(C=O)–OP) by reaction with ATP, just as the conversion of carboxylic acids to their acid chloride derivatives (R(C=O)–Cl) facilitates such reactions in conventional organic synthesis.

NADH/NADPH: NATURE'S VERSION OF SODIUM BOROHYDRIDE FOR CARBONYL REDUCTION

Redox processes are probably the most common reactions in all of biosynthesis. Living systems carry out hydride-type reductions of carbonyls using either the reduced form of **nicotinamide adenine dinucleotide** (NADH) or its phosphate derivative (NADPH) as hydride donors. In these enzyme cofactors, we focus on the dihydropyridine portion of the molecule that serves as the essential cellular source of hydride ion for the reduction of carbonyl groups to alcohols. Note that unlike laboratory $NaBH_4$ reductions of prochiral ketones which lead to chiral but racemic alcohols, the NADH reduction shown (enzyme-mediated) leads to generation of a single enantiomer product by stereospecific delivery of the pro-*R* hydrogen from NADH to the *Si* face of the β-ketothioester (Fig. 2.3). Such 1,2-hydride additions may also be carried out on thioester carbonyls (RCO–SR'), leading to the corresponding aldehydes with liberation of a thiol, R'–SH. Further reduction of the aldehyde to the primary alcohol may also follow in some instances. Coupled with subsequent protonation, the delivery of hydride by NADH/NADPH to carbonyls is a reductive two-electron transfer reaction ($H:^- + H^+ = 2e^- + 2H^+$). This is the main function of NADH/NADPH, though they may also be used in other cellular processes such as posttranslational modification of proteins. The enzymes involved in NADH/NADPH metabolism are considered important targets for drug discovery because of their significance in such cellular processes.

In addition to reductive 1,2-hydride additions to carbonyls, conjugate (1,4-) additions of hydride may also take place in certain α,β-unsaturated carbonyl systems. For example, one step in the biosynthesis of fatty acids involves an enzyme-mediated reduction of the double bond of an α,β-unsaturated thioester via sequential conjugate addition of hydride from NADPH followed by protonation to give the

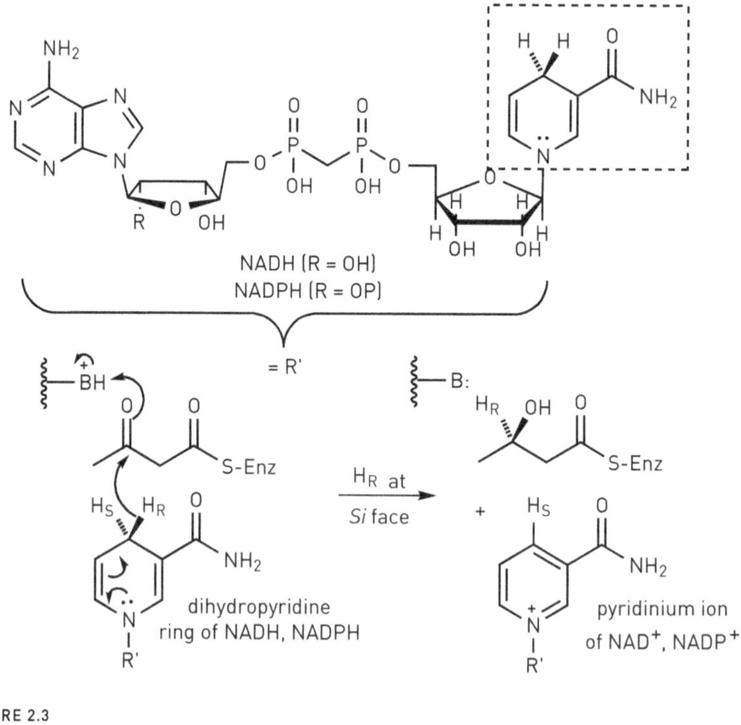

FIGURE 2.3
Carbonyl reduction with NADH or NADPH as hydride source.

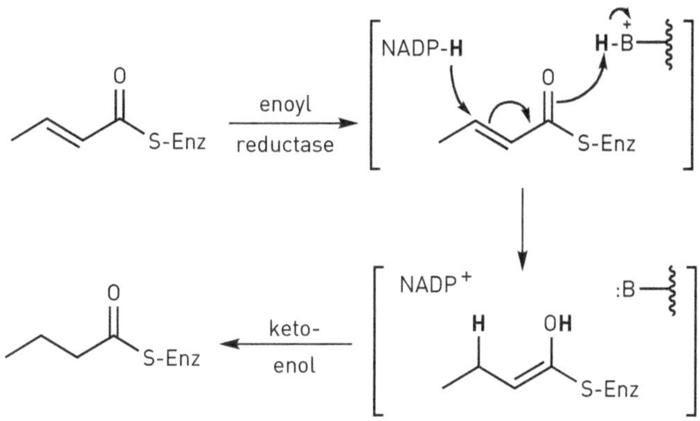

FIGURE 2.4

Conjugate reduction of an α,β-unsaturated carbonyl compound using NADPH.

corresponding saturated thioester (Fig. 2.4). We will examine this process in greater detail in Chapter 5.

NAD$^+$/NADP$^+$: NATURE'S VERSION OF PCC FOR ALCOHOL OXIDATION

Unlike laboratory oxidation and reduction processes which usually lead to spent reagents that often end up as waste products, cellular systems often employ cofactors for reagents that are readily recycled from their oxidized or reduced forms for reuse. Representing models of efficiency for such chemical processes, we often see the oxidized form of a cofactor used by other enzymes to carry out an oxidation, the reverse of the kind of reduction reaction that originally produced the cofactor in *oxidized* form. This leads to an oxidized organic product with simultaneous regeneration of the cofactor in its *reduced* form. Such is the case with NAD$^+$/NADP$^+$, which are the oxidized forms of NADH/NADPH. In oxidized form, these cofactors may act as hydride *acceptors*, and in their reactions with primary and secondary alcohols they produce the corresponding aldehyde or ketone products—the same result obtained from oxidation of such alcohols with the laboratory oxidizer, pyridinium chlorochromate (PCC). While often referred to as alcohol oxidations, conversion of alcohols to aldehydes or ketones by NAD$^+$/NADH$^+$ are formally dehydrogenation reactions (Fig. 2.5). Note the stereospecific delivery of the prochiral hydride (H_R) from the alcohol to the *Re* face of NAD$^+$/NADP$^+$, yielding the corresponding aldehydes and NADH/NADPH. Aldehydes (RCH=O) can also be dehydrogenated by NADP$^+$ via their hydrated form to give carboxylic acids (RCO_2H). Primary or secondary amines (RCH_2NH_2 or RCHNHR') can also be dehydrogenated to the corresponding imines (RCH=NH or RC=NR') in this fashion.

Nature takes advantage of a carefully balanced system in these processes. In oxidized form, the pyridinium ring of NAD$^+$/NADP$^+$ is aromatic but not electrically neutral, while in reduced form, the dihydropyridine ring of NADH/NADPH is

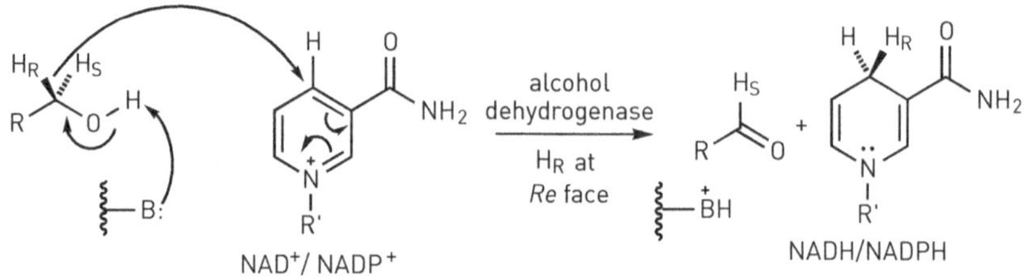

FIGURE 2.5
Alcohol dehydrogenation using $NAD^+/NADP^+$ as hydride acceptor.

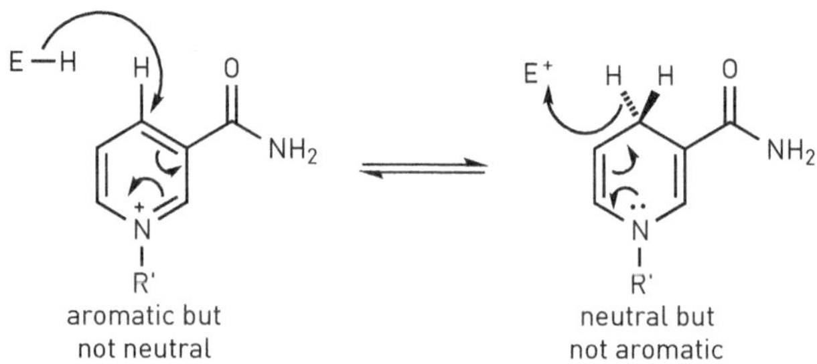

FIGURE 2.6
Redox gain or loss of hydride in NAD^+ NADH interconversion.

electrically neutral but not aromatic (Fig. 2.6). Trapped between the stability associated with electroneutrality and the stability associated with aromaticity, the system is an ideal one for a reversible organic redox process.

FAD: ANOTHER HYDRIDE ACCEPTOR FOR DEHYDROGENATIONS

For reactions that involve dehydrogenation of a carbon chain to give an alkene function, enzymes may use the cofactor **flavin adenine dinucleotide (FAD)**. The various components of this polyfunctional coenzyme include a diphosphate bridge between an AMP unit and a ribose phosphate unit bound via a C–N linkage to the flavin unit that constitutes the heart of FAD's redox functionality (Fig. 2.7). Removal of the AMP unit leads to a structurally simpler and closely related cofactor, **flavin mononucleotide (FMN)**, also called **riboflavin-5'-phosphate**, which functions like FAD in conjunction with certain enzymes. FMN in turn is derived from phosphorylation of **riboflavin** or **vitamin B_2** at the primary –OH group of the ribose unit. Any oxidoreductase enzyme that uses FAD as an electron carrier is called a **flavoprotein**.

In conjunction with such enzymes, FAD can act as a hydride acceptor in a process that is formally a 1,4-addition of hydride across the conjugated –N=C–C=N– π-system of the flavin core. As in the NADH/NAD^+ system, this reversible redox couple delivers two electrons to a conjugated π-system, in this instance by hydride attack at one end and protonation at the other, giving the reduced form of the cofactor, **$FADH_2$** (Fig. 2.8). While in some instances $FADH_2$ is used as a hydride donor for specific

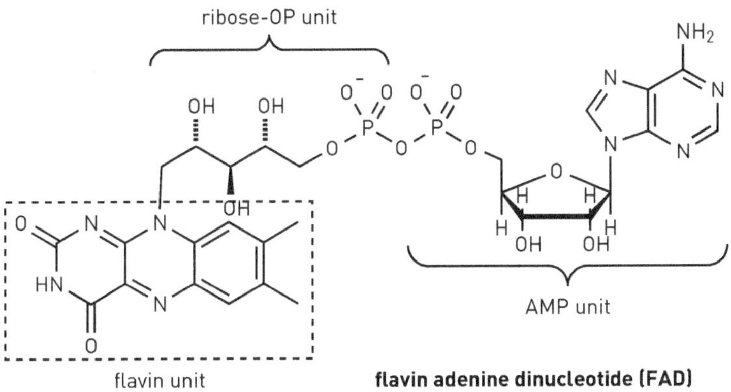

FIGURE 2.7
Flavin, ribose, and AMP components of flavin adenine dinucleotide (FAD).

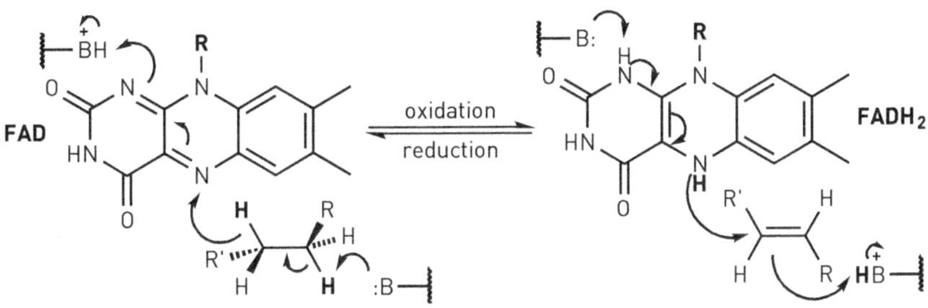

FIGURE 2.8
FAD/FADH2 as hydride acceptor/donor in redox processes.

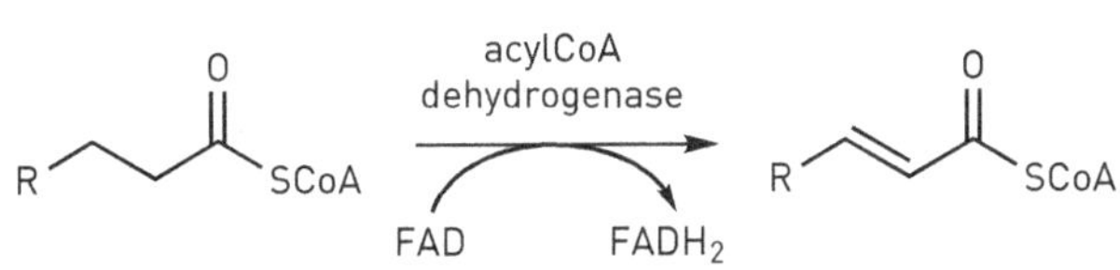

FIGURE 2.9
FAD dehydrogenation of a fatty acid thioester.

reductions in which the reverse process takes place, the principal path for regeneration of FAD involves oxidation via $FADH_2 + O_2 \rightarrow FAD + H_2O_2$, with hydrogen peroxide subsequently reduced to water by the action of other agents.

One example of the action of FAD in dehydrogenation is found in the β-oxidation of fatty acids which begins with dehydrogenation of a saturated fatty acid thioseter to the corresponding α,β–unsaturated derivative (Fig. 2.9).

MONOOXYGENASES: SPECIAL DELIVERY OF ONE O ATOM FROM O_2

An important process to consider in the biosynthetic transformation of organic compounds is the simple oxidative transformation of a C–H bond into a C–O–H

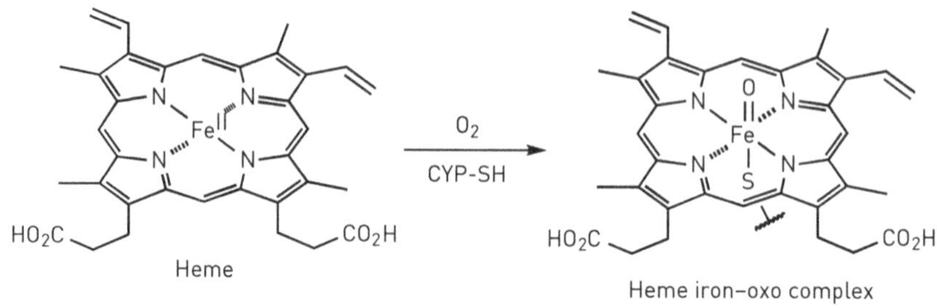

FIGURE 2.10
Formation of the heme iron–oxo complex involved in cytochrome P-450 (CYP) oxidations.

bond. Such processes convert lipid soluble compounds into more water soluble ones. The delivery of oxygen in this manner may occur for alkane C–H bonds (leading to alcohols), alkene C–H bonds (leading from enols to carbonyls) or aromatic C–H bonds (leading to phenols). While simple in conception, such processes are often complex, vary widely from a mechanistic point of view, and may only be partially understood. Nevertheless, since such transformations are fundamental to biosynthetic processes, some simplified representations will be useful. Many of these processes involve an important family of enzymes known as **cytochrome P-450 monooxygenases** (P is for Pigment, 450 for the wavelength of absorption, in nm) or simply **CYP enzymes** of which more than 500 are currently known. At their core is a porphyrin-bound Fe^{3+} that is the key to their function activating molecular oxygen via a heme iron–oxo complex (Fig. 2.10). Initial binding to the heme iron is via a cysteine thiol of the enzyme.

While of great importance, the key intermediates in many of these processes have proven to be among the most elusive in all of mechanistic bioorganic chemistry. For simplicity's sake, a simple diagram representing what are believed to be some of the key steps will suffice for our purposes (Fig. 2.11). The overall oxidation is: **R–H** + NADPH + H^+ + O_2 → **R–OH** + $NADP^+$ + H_2O. Thus, NADPH supplies the electrons required, with the second oxygen atom from O_2 being passed to water. A simple example of a CYP-mediated hydroxylation is the stereospecific conversion of camphor to the corresponding *exo* alcohol by cytochrome $P450_{cam}$ (Fig. 2.11).

Another monoxygenase system of importance is found in the group of amino acid hydroxylases that use the cofactor **tetrahydrobiopterin** (**BH_4**) and a nonheme iron for catalysis. Like some CYP enzymes, this system converts aromatic C–H bonds to phenols and is used in the conversion of phenylalanine to tyrosine as well as in the biosynthesis of neurotransmitters such as serotonin, melatonin, dopamine, norepinephrine and epinephrine (adrenaline), and many other compounds of note. In the reaction, molecular oxygen is incorporated into BH_4 to give BH_4OOH (a hydroperoxide) which delivers its active oxygen to a substrate and is thus reduced to BH_4OH (hydroxy BH_4) in the process, in much the same way that a peroxyacid (RCO_3H) delivers an active oxygen to an alkene substrate (giving an epoxide) while being reduced to the corresponding carboxylic acid (RCO_2H) (Fig. 2.12).

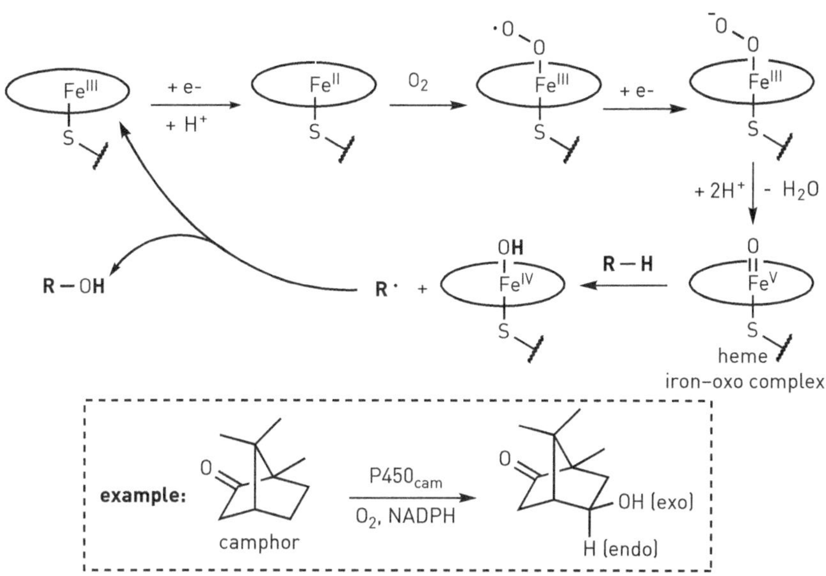

FIGURE 2.11
Simplified P-450 mechanism and example of R–H to R–OH oxidation.

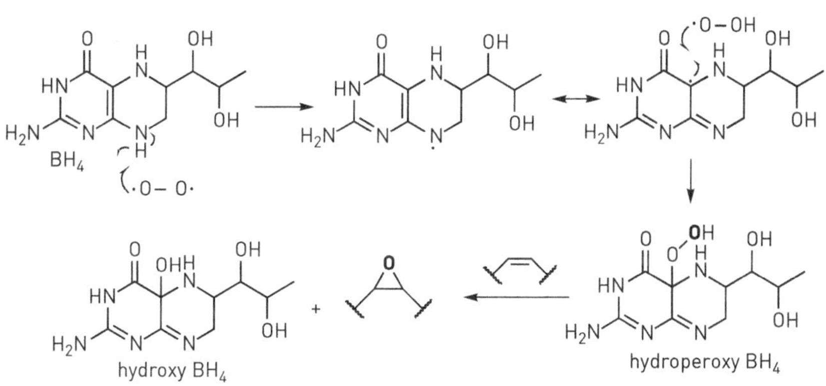

FIGURE 2.12
Simplified mechanistic scheme for tetrahydrobiopterin (BH_4) oxidation of alkenes to epoxides.

For example, in the conversion of phenylalanine to tyrosine (Fig. 2.13), an epoxide known as an **arene oxide** is produced. This epoxide then undergoes an acid-catalyzed ring-opening reaction to yield a carbocation which then undergoes a special kind of 1,2-hydride shift known as the **NIH shift**, named after the National Institutes of Health laboratories where the process was discovered. Tautomerization of the resulting keto form to the aromatic phenol follows.

Given the energy cost associated with loss of aromaticity in arene oxides, such intermediates might seem unlikely, though isolation of an arene oxide from CYP-catalyzed oxidation of naphthalene lends support. NIH-type shifts in aromatic hydroxylations are also observed in certain CYP-catalyzed oxidations, as is epoxide formation from alkenes.

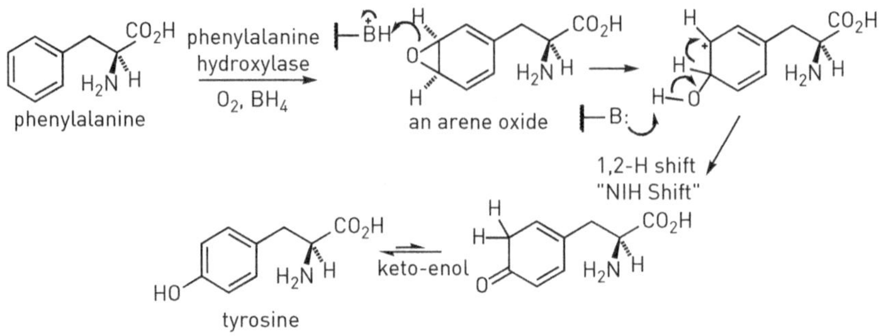

FIGURE 2.13
Arene oxide intermediate and NIH shift in conversion of phenylalanine to tyrosine.

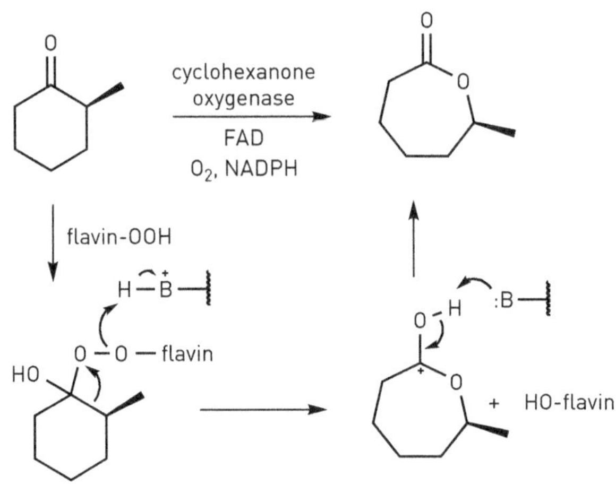

FIGURE 2.14
Flavin-dependent biological Baeyer–Villiger oxidation of 2-methylcyclohexanone.

A final example of some importance in biosynthesis is found in the so-called **Baeyer–Villiger monooxygenases**, a group of flavin-dependent enzymes which catalyze the conversion of ketones to the corresponding esters in much the same way that ketones are converted to esters via peroxyacids in the laboratory version of the Baeyer–Villiger reaction. Studies have shown that the enzymatic version follows the same migratory aptitude for groups attached to the carbonyl (3° > 2° > 1° > methyl) and shows the same stereospecificity, leading to retention of configuration in chiral systems. For example, oxidation of (*S*)-2-methylcyclohexanone with O_2/NADPH in the presence of flavin-dependent cyclohexanone oxygenase yields only the (*S*)-lactone product arising from exclusive migration of the chiral 3° carbon (Fig. 2.14). The oxidized form of the flavin coenzyme is represented here as a hydroperoxide similar in structure to the tetrahydrobiopterin hydroperoxide we previously encountered. Nucleophilic addition of flavin-OOH to the ketone carbonyl (in much the same way that a peroxyacid would add) gives a hemiacetal-like intermediate that cleaves via a stereospecific 1,2-carbon-to-oxygen migration, expelling the reduced flavin-OH and releasing the chiral lactone product.

DIOXYGENASES: DELIVERING BOTH O ATOMS FROM O_2

Enzymes that catalyze the incorporation of both oxygen atoms from O_2 into a single substrate are called **intramolecular dioxygenases**, while those which lead to oxygen atom incorporation into two separate substrates are known as **intermolecular dioxygenases**. Both types require a nonheme form of Fe. Of the intramolecular types, the **catechol dioxygenases** are among the most important because of their role in environmental degradation of aromatic compounds, both natural and man-made, by oxidative cleavage of the catechol ring. While these processes are far more complex mechanistically than is depicted here, the overall results may be usefully summarized by considering the three different cleavage patterns that distinguish one enzyme system from another, namely i) proximal extradiol; ii) intradiol; and iii) distal extradiol cleavages (Fig. 2.15). Dioxygenase-mediated cleavage of nonhydroxylated aromatic rings follows a similar pattern, leading to dialdehydes, diketones, or ketoaldehydes, depending on ring substitution, in a manner analogous to the cleavage of cycloalkene π-bonds via ozonolysis.

Intermolecular dioxygenases usually require **α-ketoglutarate** (glutaric acid in ionized form), which assists the enzyme by being oxidized along with the principal substrate. Thus, such enzymes are frequently referred to as α-ketoglutarate-dependent dioxygenases. In this process, a ligand (L) bound nonheme Fe^{2+} forms a complex with O_2 and α-ketoglutarate, leading to transfer of one O atom to Fe and one O atom to the ketoacid with simultaneous decarboxylation, converting

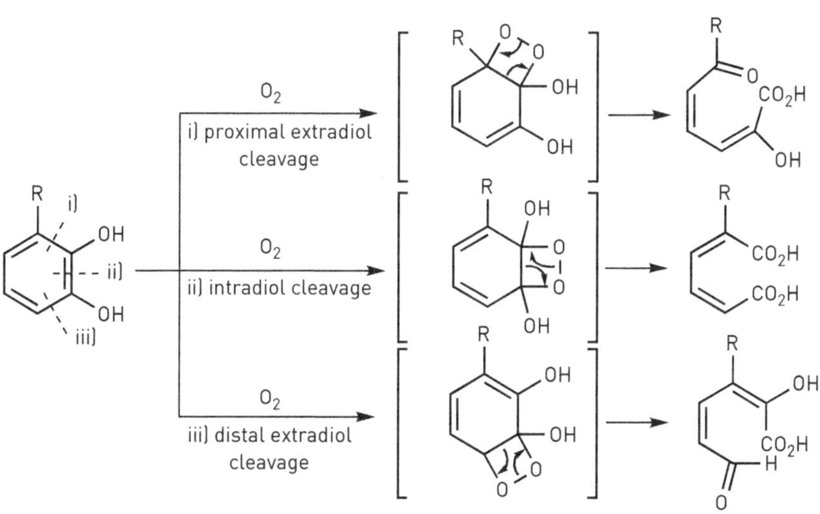

FIGURE 2.15

Different modes of oxidative ring cleavage catalyzed by catechol dioxygenases.

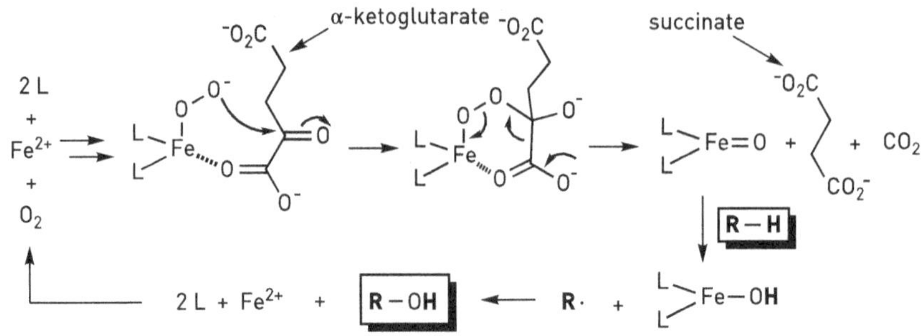

FIGURE 2.16
Simplified mechanism for α-ketoglutarate-dependent oxidations.

α-ketoglutarate to succinate (succinic acid in ionized form). The remaining iron-oxo complex subsequently carries out an oxidation of the R–H to R–OH type in a fashion similar to the P-450-type oxidations we previously encountered. A simplified representation of a possible mechanism for the overall process is shown here and should be sufficient to gain a reasonable sense of how such processes may work (Fig. 2.16).

OTHER OXIDATIONS: HYDROQUINONE AND CATECHOL OXIDATIONS

Certain O_2 oxidations involve hydrogen removal from compounds rather than oxygen incorporation and are catalyzed by **oxidase** enzymes. Passing the hydrogen atoms on to water or hydrogen peroxide, these enzyme systems are especially important in the oxidation of various structures containing hydroquinone or catechol moieties. Mechanisms are complex and not always well understood, but may be generally formulated as $2e^-/2H^+$ redox processes (Fig. 2.17). Thus, oxidation of hydroquinone (1,4-benzenediol) yields *p*-benzoquinone and vice versa, providing an important and useful electron shuttle mechanism in biological systems. Catechol (1,2-benzenediol) is similarly oxidized to the highly reactive *o*-benzoquinone. Catechol is often released when plant tissues are injured, and in the presence of the abundant plant enzyme **catechol oxidase**, it undergoes rapid oxidation to *o*-benzoquinone which then quickly converts to brown-colored polymeric products responsible for most of the dark color of "bruised" fruits like apples or bananas (sometimes called "fruit browning"). The reactivity of *o*-benzoquinone is also related to its toxicity toward bacteria, which can help to protect plants from infections.

Closely related to these oxidations is an important process in biosynthesis known as **oxidative phenolic coupling**. A number of different enzyme systems have been found to participate in these processes, including oxidases, peroxidases, and P-450-type enzymes employing NADPH and O_2 cofactors. The key reactive intermediates in these processes are **phenoxy radicals** (Fig. 2.18), which can participate in intermolecular radical–radical coupling reactions, forming C–O or C–C bonds in the process.

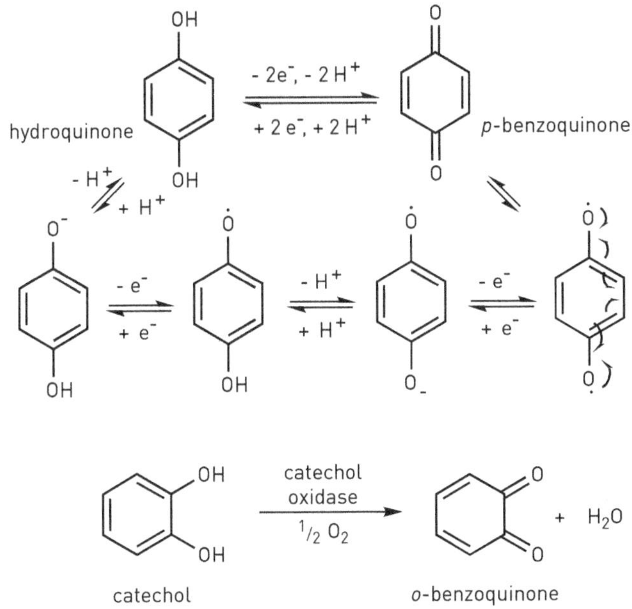

FIGURE 2.17
Oxidation of hydroquinone and catechol to p- and o-benzoquinones.

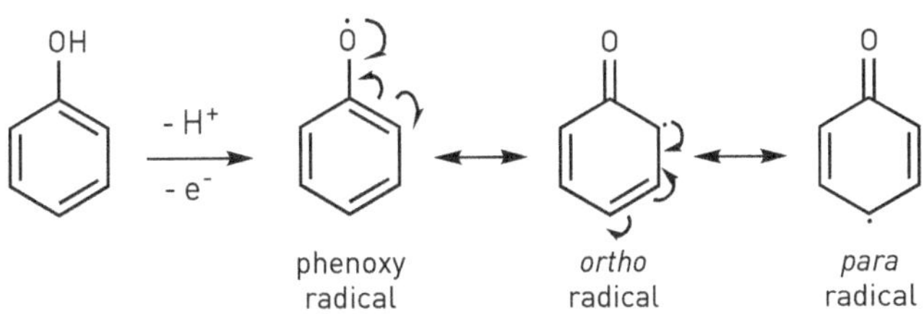

FIGURE 2.18
Resonance forms of the phenoxy radical.

Different resonance forms of the radicals involved may connect in different combinations, leading to *ortho*- or *para*-phenoxyphenol ethers or *ortho–ortho, ortho–para*, and *para–para* C–C coupled diphenolic products (Fig. 2.19). Both C–O and C–C bond-forming processes are important here, but the C–C phenolic coupling reactions are of special significance, as they constitute one of the principal means by which aromatic rings are joined one to another in various biosynthetic processes, occurring in both inter- and intramolecular fashions in a variety of contexts in different pathways.

AMINE OXIDATIONS: FROM IMINES TO CARBONYL COMPOUNDS AND BEYOND

Just as 1° and 2° alcohols may be oxidized (actually dehydrogenated) to give aldehydes and ketones, so 1° and 2° amines can be oxidized to imines which then may be hydrolyzed to give aldehydes or ketones. The oxidation of compounds containing

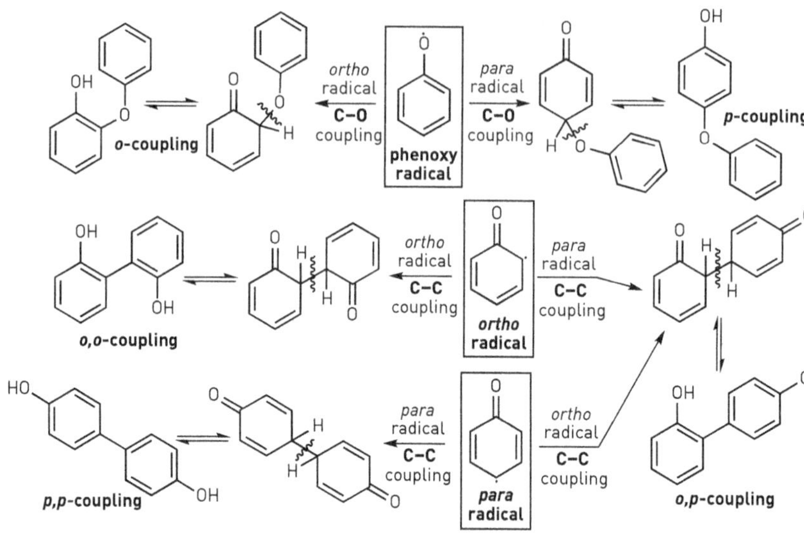

FIGURE 2.19
Various possible C–O and C–C coupling modes for phenoxy radical resonance forms.

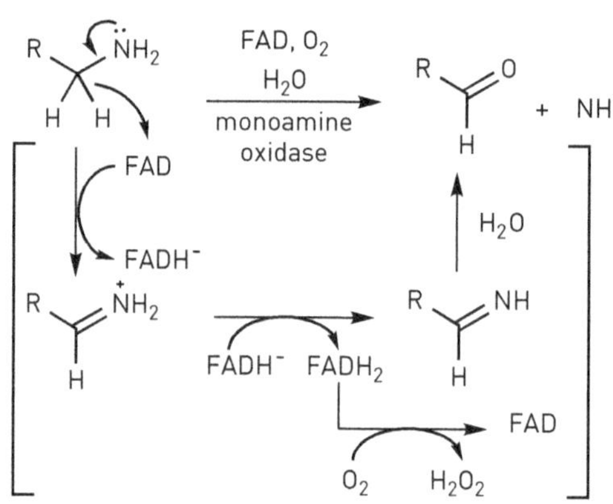

FIGURE 2.20
Monoamine oxidase-mediated amine to imine to aldehyde transformations.

a single amino group is catalyzed by the action of enzymes known as **monoamine oxidases (MAOs)**. A number of different types are known, and a variety of mechanisms have been proposed to explain their action, with a simplified mechanism shown (Fig. 2.20) in which FAD acts as a hydride acceptor, converting the amine to an iminium ion. Loss of a proton to $FADH^-$ yields the imine and $FADH_2$ (which is reoxidized to FAD by O_2 with loss of H_2O_2). Hydrolysis of the imine yields the corresponding carbonyl compound and ammonia.

A similar oxidation of compounds containing two amino groups is catalyzed by the so-called **diamine oxidase** enzymes, although in this process only one of the two amino groups is oxidized. For example, oxidation of the foul-smelling 1,4-diaminobutane (common name: *putrescine*) leads to the corresponding

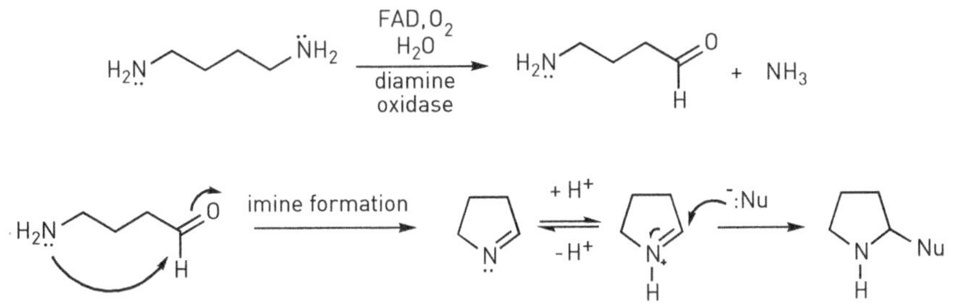

FIGURE 2.21
Diamine oxidase-mediated oxidation followed by cyclization and nucleophilic trapping.

4-aminobutanal. This transformation is an important one in the biosynthesis of many alkaloids containing a pyrrolidine ring, as intramolecular Schiff's base formation from the aminoaldehyde leads to a cyclic imine that may be further elaborated by addition of various nucleophiles (Fig. 2.21).

PLP: TRANSAMINATION AND DECARBOXYLATION OF AMINO ACIDS

Another mechanism for conversion of amino groups into carbonyl groups (and vice versa) but which is especially important in amino acid chemistry is **transamination.** This process uses another essential enzyme cofactor, **pyridoxal-5'-phosphate (PLP).** For simplicity's sake, we show PLP in aldehyde form condensing with an amino acid to form the corresponding imine (Schiff's base) as a first step (Fig. 2.22), though this step actually involves PLP as an enzyme-bound imine which undergoes an imine exchange with the amino acid to form a new imine, with release of the enzyme's amino group. Protonation of the ring nitrogen of PLP enhances the acidity of the amino acid imine α-proton by providing an extended set of conjugated π-bonds over which the electron pair from the α-C–H bond may be delocalized, leading all the way to the protonated PLP ring nitrogen giving, after deprotonation, the "quinonoid" form of the pyridine ring, so-called due to the resemblance of the upper portion of its ring to the conjugated π-system of *p*-benzoquinone. Rearomatization of the pyridine ring via concurrent protonation at the PLP exocyclic π-bond gives an α-ketoacid imine which upon hydrolysis releases the corresponding α-ketoacid and pyridoxamine phosphate. As one might expect, most of the important amino acids have a familiar corresponding α-ketoacid (alanine → pyruvic acid; glutamic acid → α-ketoglutaric acid; phenylalanine → phenylpyruvic acid, etc.) and many of these play an important role as readily available intermediates for various biosynthetic processes.

An alternative pathway for the PLP imine intermediate can be followed by carboxyl group deprotonation of the amino acid rather than removal of an α-proton. The ensuing loss of CO_2 as well as a proton provides the means for conversion of α-amino acids to their corresponding primary amines via the resulting PLP

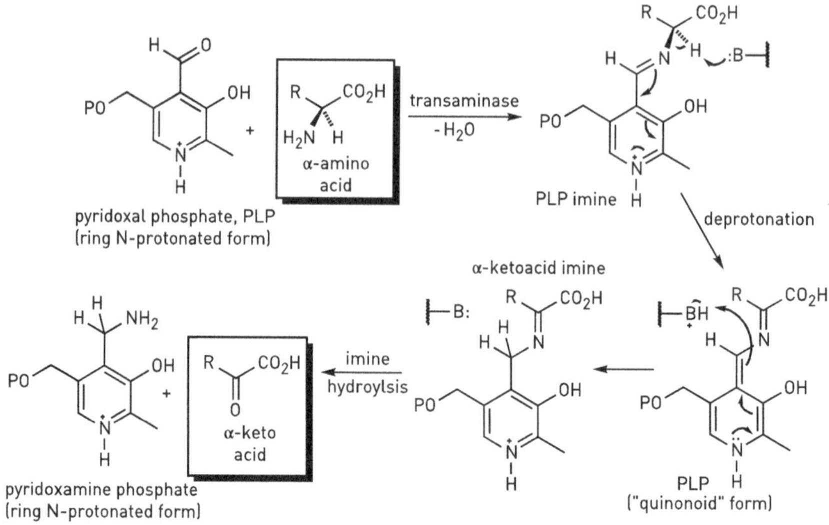

FIGURE 2.22
Mechanism for transamination of α-aminoacids by the action of pyridoxal phosphate (PLP).

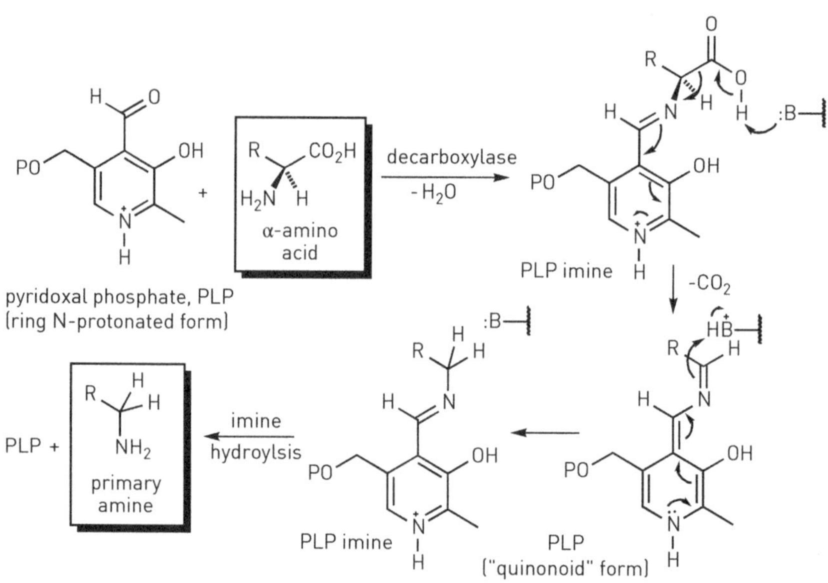

FIGURE 2.23
Mechanism for decarboxylation of α-amino acids by the action of pyridoxal phosphate (PLP).

quinonoid form. Again, rearomatization of the pyridine ring via concurrent protonation at the PLP exocyclic π-bond gives an imine which upon hydrolysis releases the primary amine with concurrent regeneration of PLP (Fig. 2.23). This remarkably versatile cooenzyme also facilitates racemization reactions of amino acids (in conjunction with racemase enzymes) via the PLP imine as well as some specific retroaldol and beta-elimination reactions which we will encounter later on as we move forward.

OTHER IMPORTANT DECARBOXYLATIONS: β-KETO ACIDS, *O*- AND *P*-HYDROXYBENZOIC ACIDS

Recall that the β-ketoester products derived from Claisen condensation reactions are versatile intermediates in laboratory organic synthesis, in part due to the ease with which these β-dicarbonyl compounds can be deprotonated by simple bases ($pK_a \sim 11$) to yield nucleophilic enolate ions which may then be alkylated by simple S_N2 chemistry. Just as important is the chemistry that results from hydrolysis of their ester function. Unlike β-ketoesters, which are quite stable, the ester hydrolysis products (**β-ketoacids**) are thermally unstable, undergoing a spontaneous decarboxylation via a cyclic 6-membered ring transition state to yield the corresponding ketones. This reaction is also an important one in biosynthetic processes with many examples of β-ketoacid decarboxylation providing a simple means of shedding a single carbon atom from a structure. Such β-ketoacids need not be derived from Claisen condensations and the cyclic transition state shown may not necessarily be involved in biosynthetic examples (Fig. 2.24).

Closely related to the decarboxylation of β-keto acids is the decarboxylation of *ortho*- or *para*-hydroxybenzoic acids. While these processes are only rarely encountered in laboratory organic synthesis, they are quite common in biosynthetic processes. Phenols may be considered stable aromatic enol forms of one of two different unstable and nonaromatic keto forms: the conjugated 2,4-cyclohexadienone form or the nonconjugated 2,5-cyclohexadienone form. It is worth noting that cyclohexadienones of these types, regardless of their origin, will always be unstable and will quickly aromatize to the corresponding phenol, provided that there is at least one H atom at the 6-position of a 2,4-cyclohexadienone and/or one H atom at the 4-position of a 2,5-cyclohexadienone (Fig. 2.25).

With an *ortho*-carboxyl group present, the 2,4-cyclohexadiene form is essentially a β-ketoacid and may decarboxylate via the familiar cyclic transition state to yield the phenol. Similarly, phenolic systems with a carboxyl group in the *para* position can decarboxylate to regain aromaticity via the nonconjugated 2,5-cyclohexadienone form (Fig. 2.26). In biosynthesis, the enzymes involved presumably provide a means

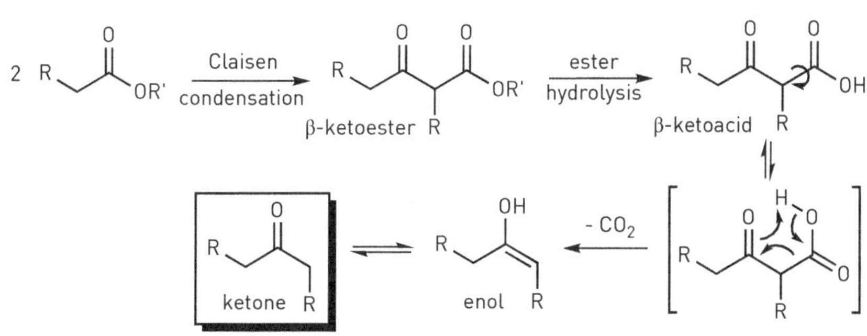

FIGURE 2.24
Spontaneous decarboxylation of β-ketoacids to give ketones.

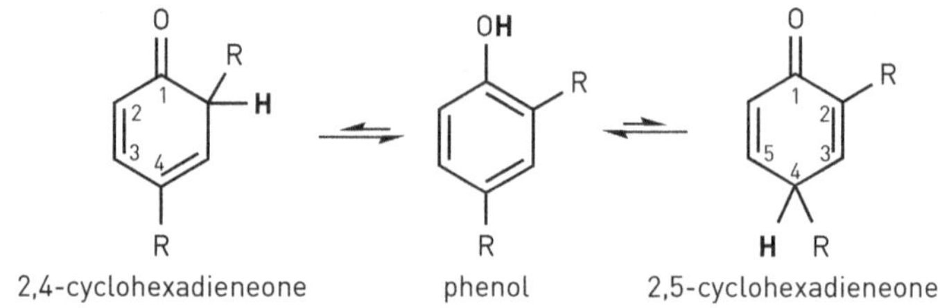

FIGURE 2.25
Keto-enol tautomerization of 2,4- and 2,5-cyclohexadienones to phenols.

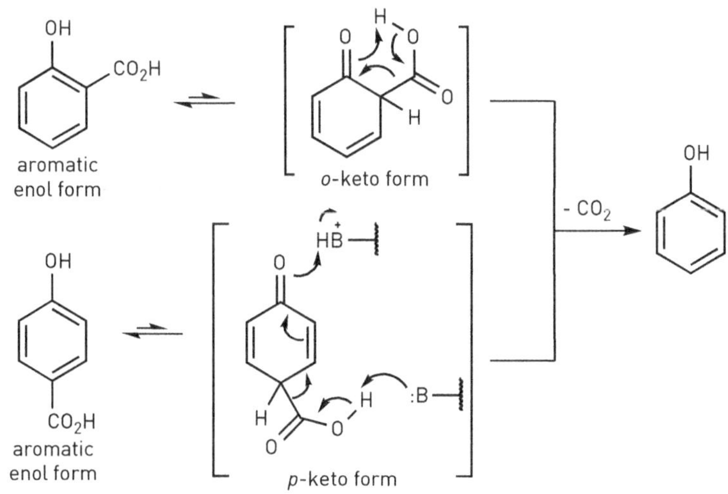

FIGURE 2.26
Decarboxylation of *o*- and *p*-hydroxybenzoic acids.

for stabilizing the unfavorable keto forms, thereby facilitating such decarboxylations. Note that in each case, aromaticity can be restored either via loss of a proton *ortho* or *para* to the ketone function (essentially via 1,2- or 1,4-enolization) or via loss of a proton *and* CO_2 from one or the other of these positions (essentially decarboxylative 1,2- or 1,4-enolization processes).

THIAMINE DIPHOSPHATE (TPP) AND LIPOIC ACID: DECARBOXYLATION AND ACYL TRANSFER

We have seen how PLP facilitates the decarboxylation of amino acids to their corresponding primary amines and also how it participates in the transamination process that converts amino acids to their corresponding α-ketoacids. These keto acids are also available from other pathway sources and their decarboxylation is an important component of various metabolic processes, such as in the decarboxylation of pyruvic acid to ethanal (which then undergoes NADH reduction to ethanol) during yeast fermentation. However, unlike β-ketoacids, decarboxylation of α-ketoacids is not spontaneous and requires the action of a cofactor known as **thiamine diphosphate**

(**TPP**). How TPP works is related in part to the acidity of the proton on the positively charged thiazole ring of its structure (Fig. 2.27).

When its thiazole proton is removed (pK_a ~18) by a base, the resulting anion is called **TPP ylide**. Recall that ylides are dipolar reactive intermediates with a negative charge on carbon and a positive charge on an adjacent heteroatom (N in this case). Thus, TPP ylide (shown in abbreviated form using R and R' groups attached to the core thiazole ring) is similar to the phosphorous ylide of the Wittig reaction in which an ylide acts as a nucleophile toward the carbonyl group of ketones or aldehydes. Similarly, the nucleophilic TPP ylide attacks the ketone carbonyl of an α-ketoacid to initiate the subsequent decarboxylation of the resulting tetrahedral intermediate (Fig. 2.28). After loss of CO_2, a resonance-stabilized carbanion is produced which is first protonated on carbon then deprotonated at its –OH group to release an aldehyde with concurrent regeneration of TPP ylide.

The most important function of TPP in biosynthesis is found in its ability to work in conjunction with two other coenzymes, **lipoic acid (LA)** and the thiol-containing cofactor, **coenzyme A (HSCoA** or **CoASH).** This triad of cofactors couples the

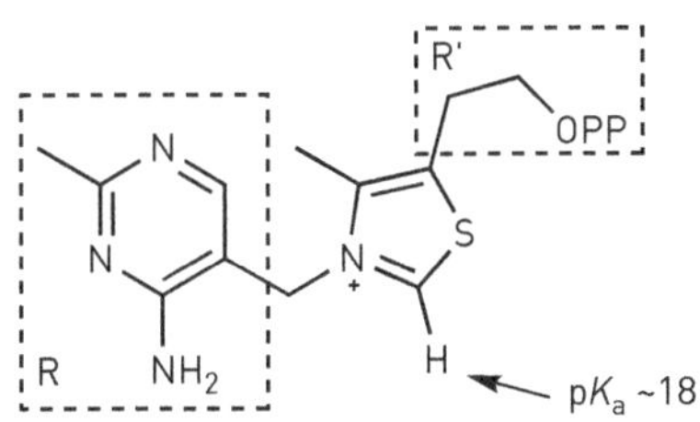

thiamine diphosphate (TPP)

FIGURE 2.27
Acidic proton on the thiazole ring component of thiamine diphosphate (TPP).

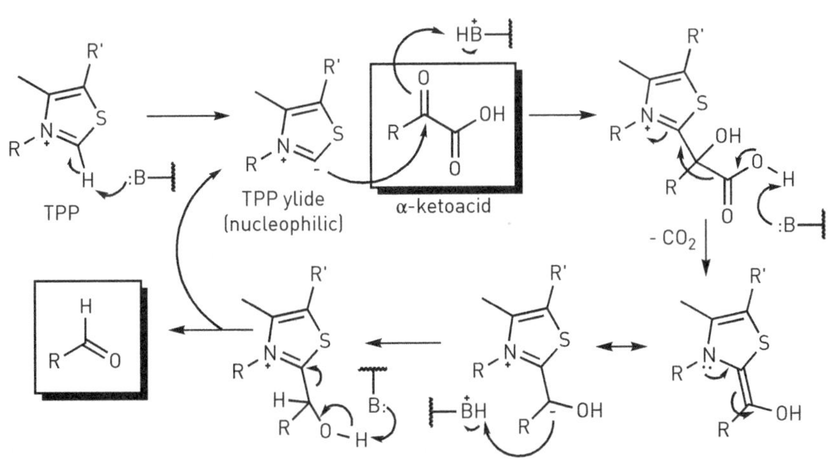

FIGURE 2.28
Mechanism of action of TPP ylide in the decarboxylation of α-ketoacids.

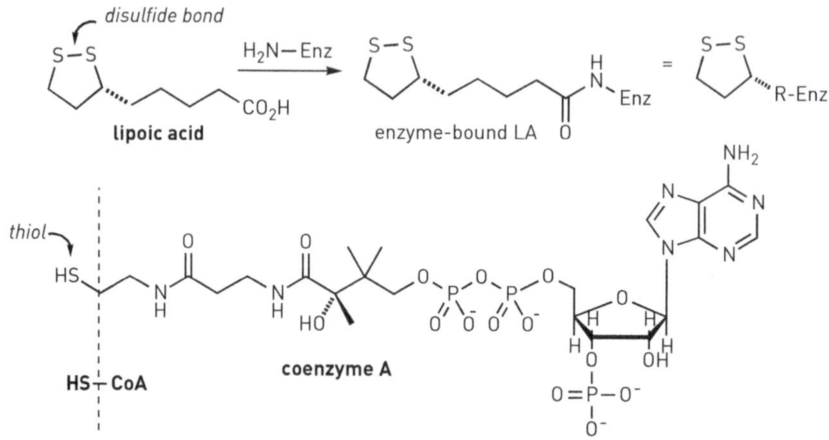

FIGURE 2.29
Structures of the enzyme cofactors lipoic acid (LA) and coenzyme A (CoASH).

decarboxylation of pyruvic acid to its conversion to the important biological acyl group carrier, **acetyl-CoA**. To sort all this out, we begin with the structures of both LA and CoASH, focusing on the key components involved in their reactions with TPP (Fig. 2.29). Of particular note are the thiol component of CoASH and the disulfide linkage of lipoic acid (LA). For the purposes of mechanistic analysis, both structures are usually abbreviated to include only those portions actually transformed during a biosynthetic sequence. In the reactions considered here, LA functions in enzyme-bound form, but usually only the disulfide-containing ring is involved in chemical transformations. The generic abbreviations HSCoA or CoASH are also frequently used as simple structural abbreviations for coenzyme A, since they place all the emphasis on the thiol (HS) that is the most important functional group in its chemical transformations.

In its reaction with pyruvic acid (shown here in nonionized form for simplicity's sake, in Fig. 2.30), TPP ylide leads to the same decarboxylated carbanion intermediate we saw in Fig. 2.28, but its fate will be different in this case.

Rather than attacking a proton source as before, the carbanion formed after decarboxylation attacks a sulfur atom in the disulfide linkage of enzyme-bound LA, leading to formation of a hemithioacetal intermediate, deprotonation of which produces an acetylthioester derivative with concurrent regeneration of TPP ylide (Fig. 2.31). Finally, the acetylthioester undergoes nucleophilic attack by the thiol sulphur atom of HSCoA, leading to a simple thiol exchange reaction, producing a new thioester, **acetyl-CoA**, and the dithiol form of enzyme-bound LA, which is subsequently recycled to its cyclic disulfide form by FAD oxidation. This complex sequence of events, mediated by a multi-enzyme complex called *pyruvate dehydrogenase complex*, is not only important because of the key role acetyl-CoA plays in biosynthesis, it is especially important in the basic biochemistry of living systems, as it links glucose metabolism (glycolysis, the main source of pyruvic acid) to the citric

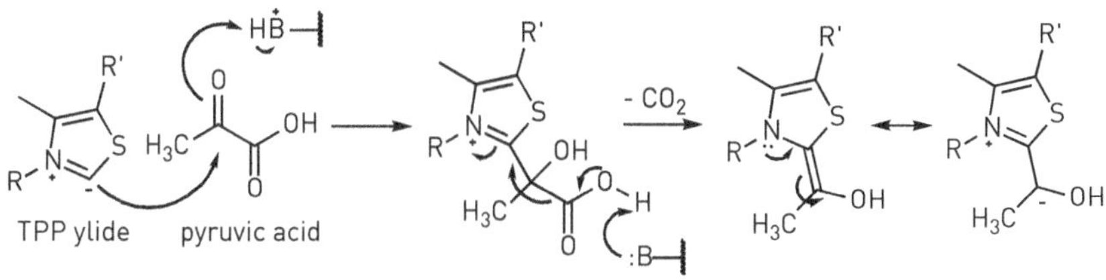

FIGURE 2.30
Carbanion intermediate from decarboxylation of pyruvic acid by TPP ylide.

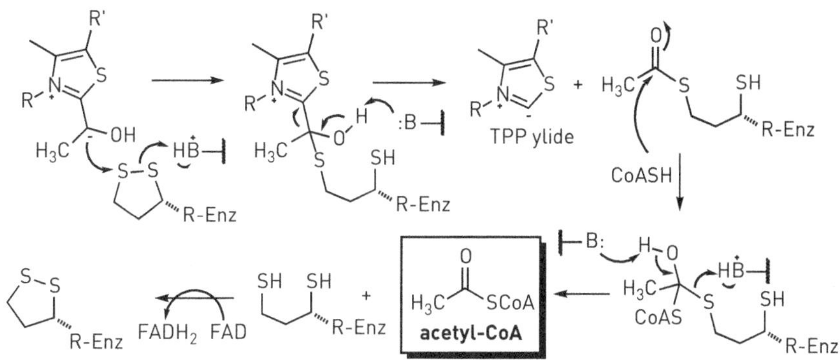

FIGURE 2.31
Action of enzyme-bound lipoic acid in formation of acetyl-CoA and regeneration of TPP ylide.

acid cycle (or Krebs cycle), a process that combines acetyl-CoA and oxaloacetic acid (the ketoacid of aspartic acid) as part of the many biochemical reactions involved in cellular respiration.

As we will see, acetyl-CoA has many other important uses in biosynthesis, functioning as an acyl transfer agent and participating as a **two-carbon fragment** for the building up of complex structures via a variety of familiar transformations, not only as an electrophilic carbonyl compound, but also as a nucleophile via its enolate form. Some of these transformations, summarized in Fig. 2.32, take advantage of the higher reactivity of thioesters relative to simple esters in both nucleophilic acyl substitution reactions (acyl transfer) and the ease with which their α-protons may be removed to form enolate ions (pK_a ~20 for thioesters vs. pK_a ~25 for esters).

BIOTIN: THE CO_2 CARRIER, TRANSPORT, AND TRANSFER AGENT

The product derived from α-carboxylation of acetyl-CoA is called **malonyl-CoA**, an especially important derivative of acetyl-CoA for generation of its thioester enolate ion in certain Claisen condensation reactions. The incorporation of CO_2 in this fashion requires a cofactor called **biotin** which acts as a CO_2 carrier, transport, and transfer agent in several biosynthetic sequences, working in conjunction

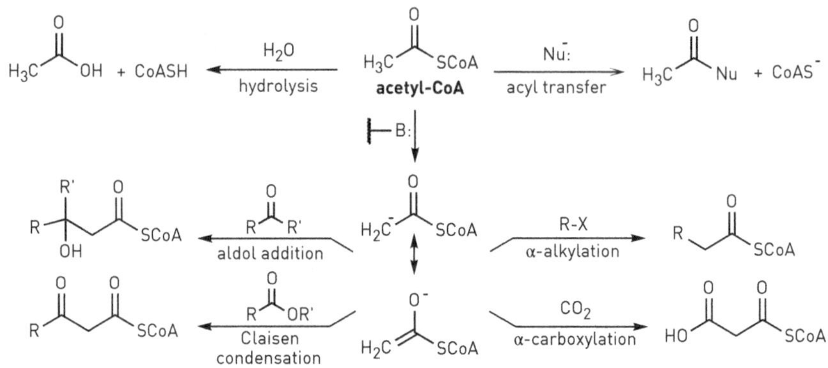

FIGURE 2.32
Some important reactions in biosynthesis involving acetyl-CoA as a key intermediate.

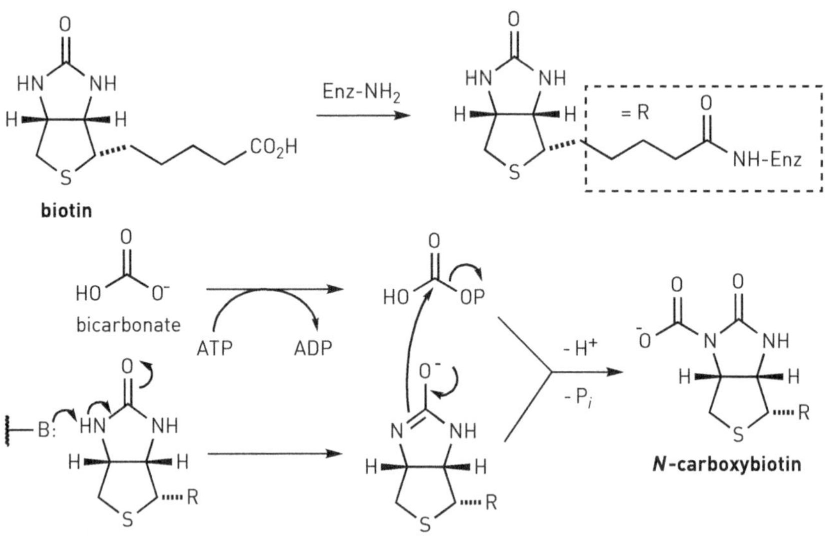

FIGURE 2.33
Structure of biotin and mechanism of formation of *N*-carboxybiotin.

with specific carboxylase enzymes (Fig. 2.33). Biotin is first converted to an enzyme-bound form which is then deprotonated to give an enolate-like anion which attacks a carboxyphosphate derived from phosphorylation of bicarbonate (HCO_3^-) with ATP. This results in formation of **N-carboxybiotin**, formally the carrier of CO_2 for subsequent carboxylation reactions.

A number of different mechanisms have been proposed and studied regarding how *N*-carboxybiotin delivers CO_2 to nucleophilic agents. We will consider what is regarded by some as the most likely mechanism for carboxylation of acetyl-CoA enolate ion, namely initial release of CO_2, with concurrent deprotonation of acetyl-CoA by the biotin enolate ion. This regenerates biotin (enzyme-bound) with the acetyl-CoA enolate ion and CO_2 in close proximity, resulting in nucleophilic carboxylation to give **malonyl-CoA** as shown in Fig. 2.34. Later on we'll see how malonyl-CoA functions as

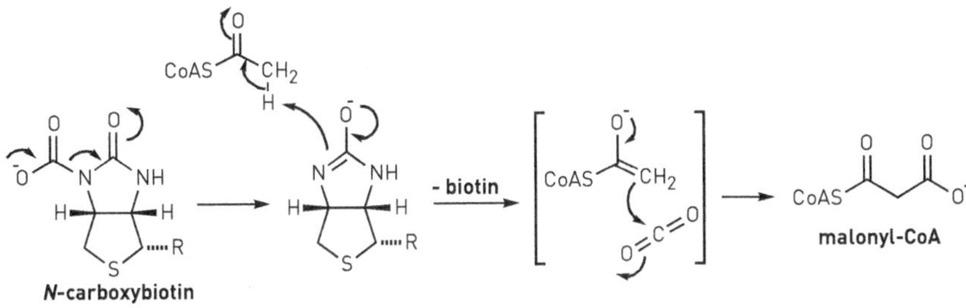

FIGURE 2.34
N-carboxybiotin and carboxylation of acetyl-CoA to give malonyl-CoA.

a source of the enolate ion of acetyl-CoA via simple decarboxylation, forming an essential component in the biosynthesis of fatty acids and polyketide products in Chapter 5.

SAM: A C_1 FRAGMENT FOR METHYL GROUPS

CO_2 can be considered a **one-carbon fragment** for the building up of larger structures from smaller ones by nucleophilic carboxylations; S_N2-type alkylations can also accomplish this by simple nucleophilic methylation using the methyl donor **S-adenosylmethionine** or **SAM** (Fig. 2.35). This important alkylating agent may be thought of as the biological equivalent of methyl iodide for such reactions, as it takes advantage of the leaving group ability of a positively charged sulphur atom. Generally speaking, CH_3–O–, CH_3–N– or CH_3–S– groups found in natural product structures usually originate via SAM alkylation of the heteroatom involved. For simplicity's sake, when SAM methylation is depicted mechanistically, we will employ the abbreviated representation shown.

Certain methyl groups attached to carbon chains or rings may also be derived from SAM alkylation in systems where carbon is available in a suitable nucleophilic form, such as in enolate ions or phenolate ions (Fig. 2.36). Such ions may alkylate either on carbon (*C*-alkylation) or oxygen (*O*-alkylation) with site selectivity mainly controlled by the enzyme mediating the process. The conversion of tyramine to *N*-methyltyramine via SAM represents a typical *N*-methylation process. Both compounds are naturally occurring phenethylamine alkaloids (Chapter 7) isolated from a number of different sources and able to act as CNS stimulants.

DMAPP: AN ALLYLIC C_5 FRAGMENT FOR STRUCTURE BUILDING

We've seen CO_2 and SAM as C_1 fragments and acetyl-CoA as a C_2 fragment for the building up of more complex organic structures from simpler ones. Another important alkylating agent that will be employed in a variety of biosynthetic transformations is **dimethylallyl diphosphate** or **DMAPP**. Unlike SAM, which is limited

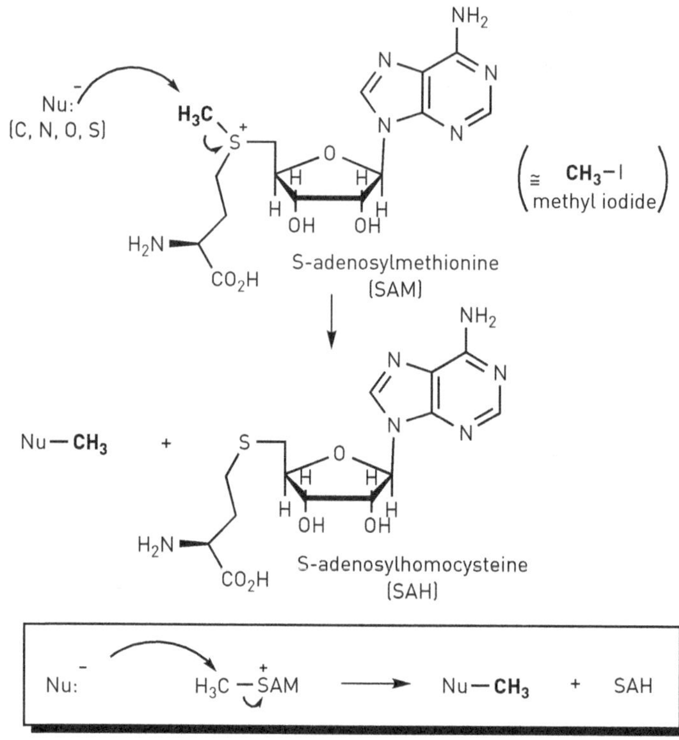

FIGURE 2.35
Structure and function of S-adenosylmethionine (SAM) for biosynthetic methylations.

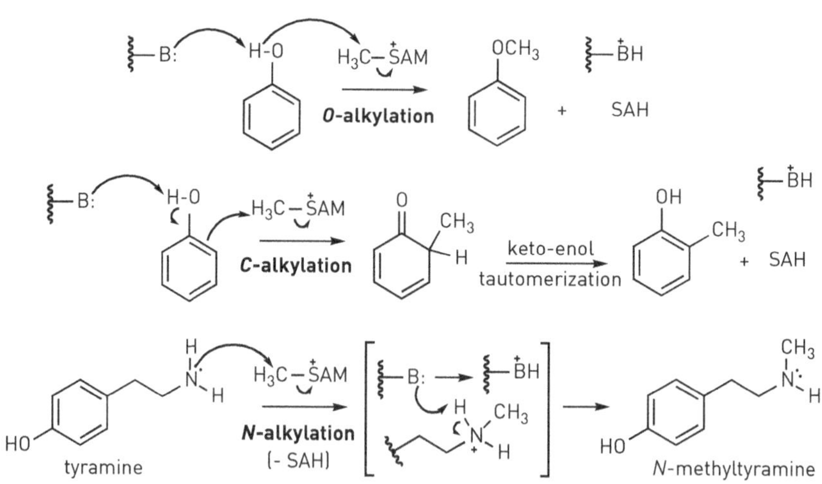

FIGURE 2.36
O-, *C*-, and *N*-alkylations using SAM as a methyl group donor.

to S_N2-type alkylations due to the inherent instability of the CH_3^+ that would be required for S_N1-type processes, DMAPP has its leaving group attached to a primary allylic carbon, so this important **five-carbon fragment** may be delivered via either S_N2 nucleophilic substitutions or via S_N1 substitutions, owing to the resonance stabilization of the allylic carbocation resulting from initial loss of the diphosphate leaving group (Fig. 2.37). In both cases, DMAPP may be thought of as the biological equivalent of allyl bromide, the laboratory reagent most commonly used for allylic alkylation reactions. Alternatively, alkyl diphosphates (R–OPP) may be thought of as biological equivalents of the more familiar tosylate or mesylate ester derivatives of alcohols which, like the diphosphate ester, transform a normally poor leaving group (–OH) into an excellent leaving group for nucleophilic substitution processes.

An example of one way DMAPP is used in biosynthesis is the late-stage *ortho*-alkylation of the phenolic ring of the compound shown in Fig. 2.38 (a *pterocarpene*), producing the alkylated isoflavonoid derivative **erypoegin H**, a potent natural antibiotic isolated from the roots of the ornamental plant *Erythrina poeppigiana* and found to be active against vancomycin-resistant strains of bacteria.

We will also see that DMAPP and its closely related partner, **isopentenyl diphosphate (IPP)** constitute the basic so-called **isoprene** building blocks of the biosynthetic pathway that produces the large and structurally diverse group of natural products known as the **terpenes**. This terpenoid pathway produces many compounds of great significance to human health. **Retinal**, a key molecule involved in

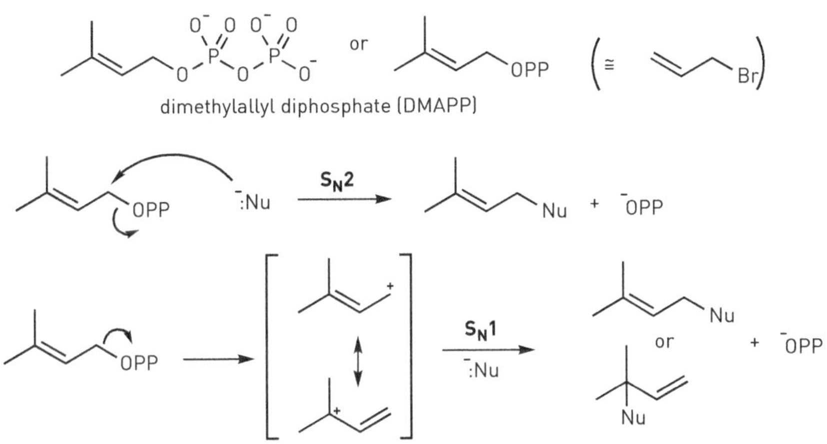

FIGURE 2.37
Structure of dimethylallyl diphosphate (DMAPP) and its use in S_N2 and S_N1 alkylations.

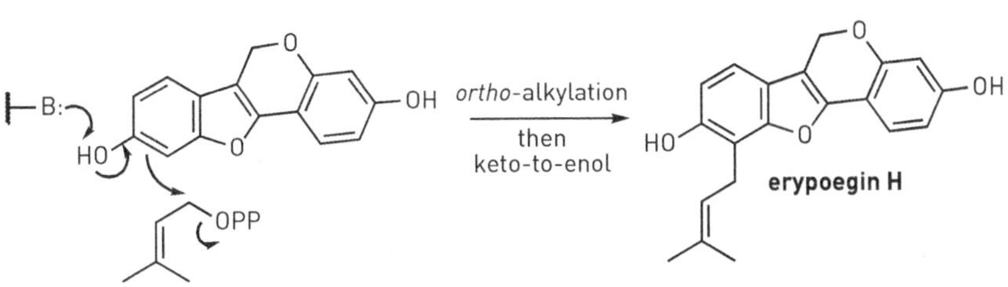

FIGURE 2.38
Ortho-alkylation of a phenolic ring with DMAPP to produce the antibiotic erypoegin H.

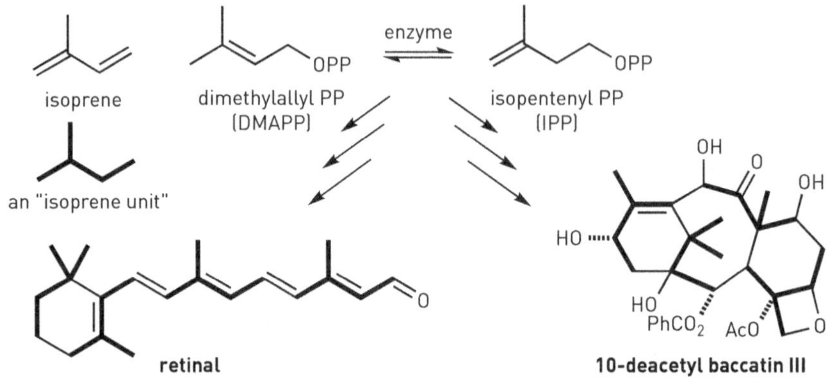

FIGURE 2.39
Isoprene units highlighted in the structures of two important natural products.

human vision and **taxol**, a powerful clinical chemotherapy drug derived from the intermediate compound **10-deacetylbaccatin III**, are just two examples; their five-carbon isoprene structural units are highlighted in Fig. 2.39. In Chapter 4, we will also see how this pathway assembles all the steroid structures, from cholesterol to the sex hormones.

OTHER ESSENTIAL STRUCTURAL FRAGMENTS: PUTTING IT ALL TOGETHER

Two key amino acids, **ornithine** and **lysine**, often provide basic $\mathbf{C_4N}$ and $\mathbf{C_5N}$ fragments that are essential components in biosynthetic pathways leading to different types of alkaloids that we will examine later in Chapter 7. The pyrrolizidine alkaloids, which include compounds such as **retronecine** (Fig. 2.40) are examples of frequently toxic C_4N-containing alkaloids produced by plants as a defense against herbivores. **Swainsonine**, a C_5N-containing alkaloid toxin found in locoweed, causes significant annual economic losses in grazing livestock but has also shown potential as an anticancer agent in a number of studies.

Many structures containing $\mathbf{PhC_1}$, $\mathbf{PhC_2}$, $\mathbf{PhC_3}$, as well as $\mathbf{PhC_2N}$ fragments are derived from two important amino acids: phenylalanine (Phe) and tyrosine (Tyr) (Fig. 2.41). Starting with either structure, decarboxylation can provide PhC_2N fragments which may then be further modified by transamination or amine oxidations to give PhC_2 fragments for further elaboration. If Phe or Tyr undergo 1,2-elimination of ammonia, a PhC_3 fragment is produced which can also undergo various subsequent biosynthetic transformations. 1,2-elimination of NH_3 followed by oxidative C=C bond cleavage can also provide access to PhC_1 fragments such as benzaldehydes or benzoic acids which may then be modified further. We can recognize the tyrosine origin of the PhC_3 units of **etoposide**, an important chemotherapy agent in the treatment of lymphoma, lung, and testicular cancer. This nonalkaloid lignan

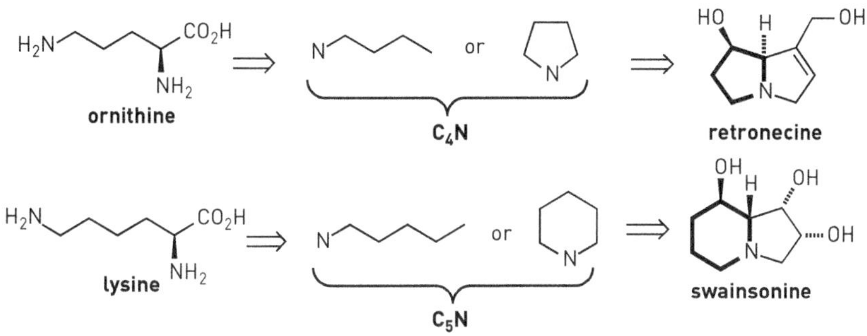

FIGURE 2.40
Lysine and ornithine as sources of C_4N and C_5N fragments in biosynthesis.

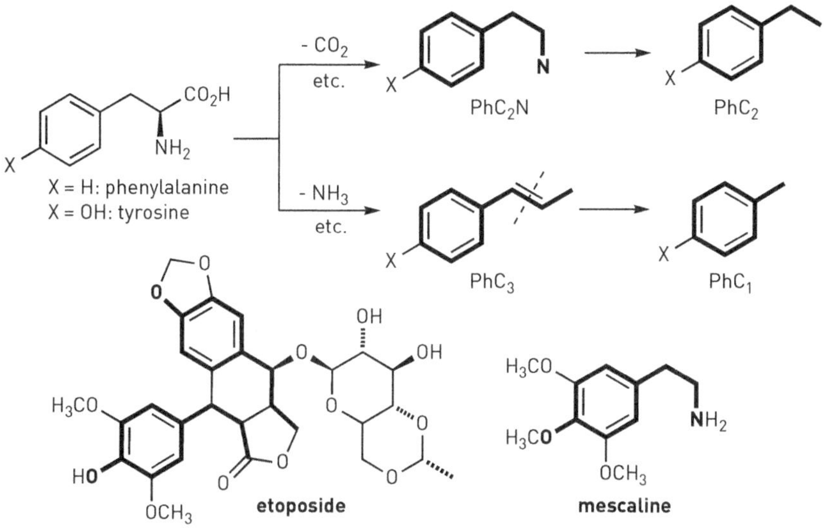

FIGURE 2.41
Phenylalanine and tyrosine as PhC_2N, PhC_2, PhC_3, and PhC_1 structural fragments.

is derived from a compound isolated from a common forest plant, the American Mayapple (*Podophyllum peltatum*), which produces it via the shikimic acid biosynthetic pathway we will examine in detail in Chapter 6. We can also see the tyrosine origins of the PhC_2N component of **mescaline**, a notorious psychedelic alkaloid isolated from peyote cactus which, though used in religious rites and other contexts by native American populations for several millennia, is nevertheless illegal in most countries. Note the SAM-derived methyl groups on the aromatic ring oxygens in both these compounds. SAM is also the origin of the single C atom bridging the two *ortho*-oxygen atoms on the second aromatic ring of etoposide. This so-called **methylenedioxy group** is a common function in many natural product structures, the formation of which we will address later on.

A final example of an essential fragment in biosynthesis finds its origin in the amino acid tryptophan (Trp) to give the so-called **indole-C_2N** fragment (Fig. 2.42)

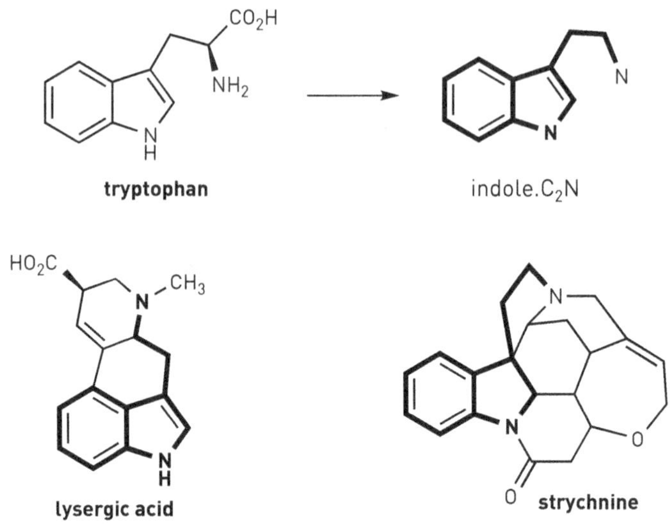

FIGURE 2.42
Tryptophan as an indole.C_2N fragment source in biosynthesis.

which forms portions of the structures of a number of familiar alkaloids including **lysergic acid**, a member of the ergot alkaloid family whose diethyl amide derivative is the potent hallucinogen LSD (lysergic acid diethylamide). **Strychnine**, a highly toxic terpene alkaloid which contains the indole-C_2N fragment from Trp as well as isoprene units from the terpenoid pathway, has been known for thousands of years, most famously as the poison that may have found its way into the wine goblet of the young Alexander the Great in 323 BC, taking his life at the tender age of 32.

LOOKING AHEAD

New and remarkably complex natural products continue to be discovered almost on a daily basis, some of which not only show important beneficial effects for human health, but may even provide new perspectives for historical analysis. For example, a recent theory suggests that rather than being killed by strychnine poisoning, Alexander the Great might actually have been poisoned by drinking water from the river Styx, which is now known to be contaminated with a highly toxic compound called **calicheamicin** (Fig. 2.43). Discovered in 1981, this nonalkaloid natural product is produced by certain soil bacteria utilizing the polyketide biosynthetic pathway (Chapter 5) with acetyl-CoA as the principal building block. Because of its highly specific toxicity to DNA in virtually all kinds of cells, many derivatives of calicheamicin have now been prepared, examined, and used in carefully targeted cancer chemotherapy regimens. Two of these derivatives, *calicheamicin γ1* and *esperamicin*, are considered to be the most powerful anticancer agents known to science.

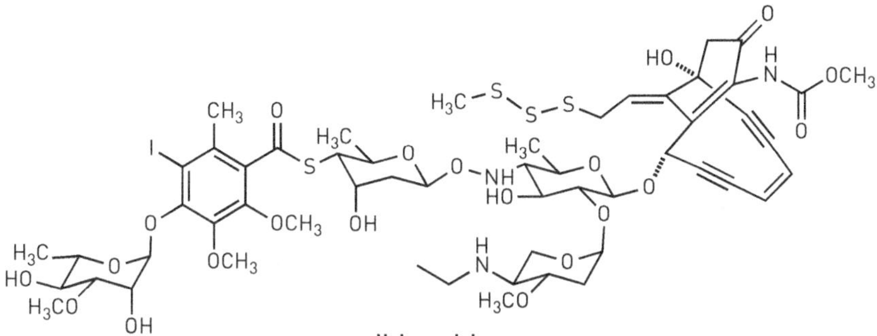

calicheamicin

FIGURE 2.43
Structure of calicheamicin, another poison possibly responsible for the death of Alexander the Great.

You may recognize a number of the components of calecheamicin's structure as being carbohydrate (sugar) molecules. Clearly, we will need a good grasp of carbohydrate structure and chemistry, one of the topics of our next chapter, before moving further into the study of biosynthetic pathways. We have already seen the significance of familiar amino acids such as phenylalanine, tyrosine, and tryptophan as well as some less familiar nonpeptidic amino acids such a ornithine as basic biosynthetic building blocks, so our next chapter will also include a brief overview of amino acid structure and biosynthesis. Once we understand how amino acids and carbohydrates are constructed, how and where they are often added to the core structures of more complex molecules, and how they are used to assemble some of the key starting materials for biosynthesis, we will then be in a position to systematically build up our knowledge of how specific natural products are produced by a set of fundamental pathways in living systems. In doing so, we will eventually be able to dissect even a remarkably complex compound like calicheamicin to see what kinds of reactions and components are likely to have been used by organisms to assemble such a marvel of molecular architecture and how such knowledge can be used to solve problems in structure identification and to even generate new ideas for the assembly of complex organic compounds of our own making.

STUDY PROBLEMS

1. Provide the missing structures expected for **A** and **B**, then provide a reasonable mechanism for the NAD^+ oxidation that would produce the thioester.

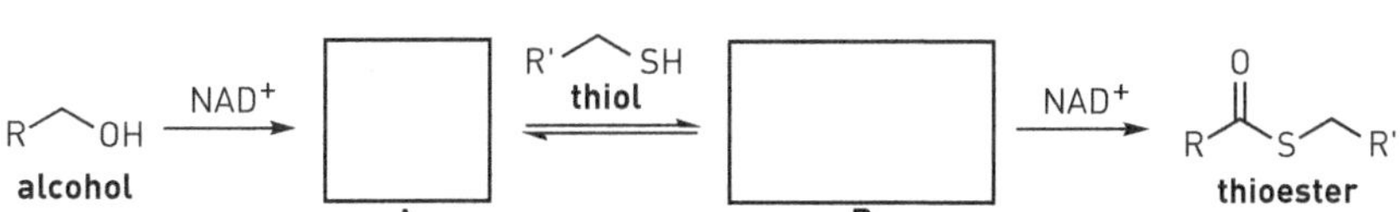

2. Propose a reasonable sequence of three different biological oxidation reactions that could transform **A** into **B**. Show the intermediate product from each step and provide mechanistic details of the third oxidation step.

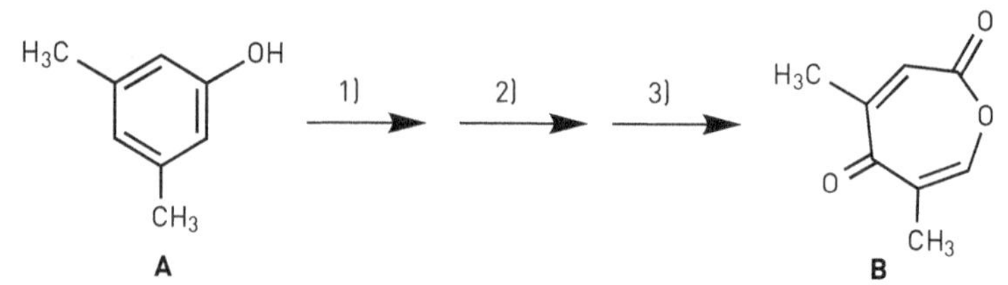

3. Pummerer's Ketone **2** is formed by oxidation of *para*-cresol **1** using potassium ferricyanide under basic conditions and has also been formed by biological oxidation of **1** in the presence of a specific enzyme. Propose reasonable mechanisms for the sequence of steps leading to **2**. To begin, you may assume that **1** initially oxidizes to the usual phenoxy radical intermediate.

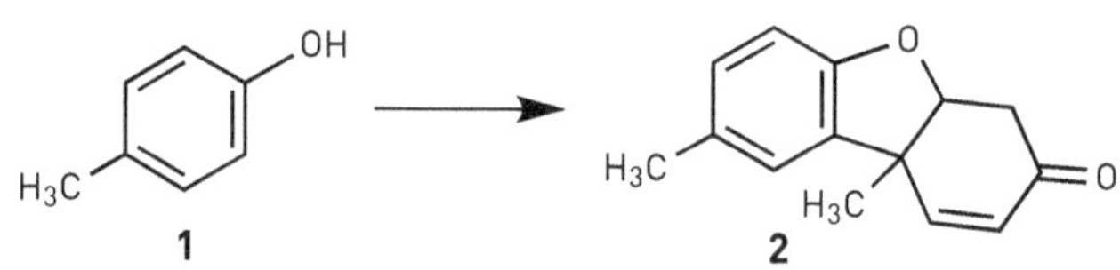

4. Give a brief and concise description of the role that each of the following enzyme cofactors plays in biosynthetic processes.
   a. DMAPP
   b. NADH/NAD$^+$
   c. SAM
   d. PLP
   e. TPP
   f. FAD/FADH$_2$
   g. ATP
   h. Biotin
5. In animals, phenylalanine is oxidized to tyrosine via a monooxygenase-mediated process. If deuterium-labeled phenylalanine (as shown) undergoes this oxidation, what happens to the deuterium? Provide a mechanistic analysis (keep in mind that a C–D bond is stronger than a C–H bond).

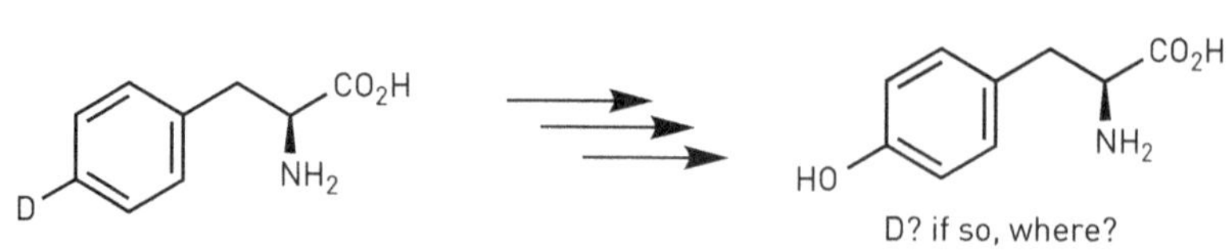

6. *Rifamycin* B, a potent inhibitor of *Escherichia coli* RNA polymerase, contains a number of different functional groups, some of which are labeled in the structure shown.

What is the likely biosynthetic origin (precursor molecule) of methyl group **A**?

What is the likely biosynthetic origin (precursor molecule) of methyl group **B**?

What is the likely biosynthetic origin (precursor molecule) of methyl group **C**?

Name the functional group involving the atoms directly attached to (and including) carbon **1**.

Name the functional group involving the atoms directly attached to (and including) carbon **2**.

Name the functional group involving the atoms directly attached to (and including) carbon **3**.

For the H atoms labeled **X**, **Y**, and **Z**, what is their correct order of increasing acidity (least acidic first)?

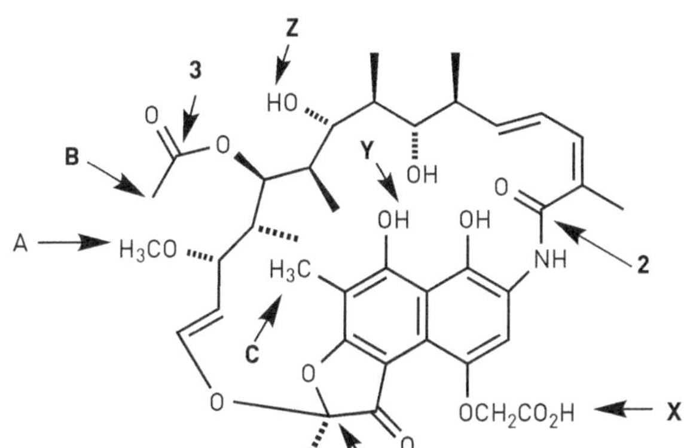

rifamycin B

7. Compound **A** is a substituted pyrrole intermediate used in porphyrin biosynthesis. Propose a detailed mechanism to account for the formation of **A** from 5-aminolevulinic acid as shown (hint: the mechanism requires no co-factors or reagents other than water and a proton. You need only use imine formation, imine-enamine tautomerism, and other intramolecular steps including a final dehydration).

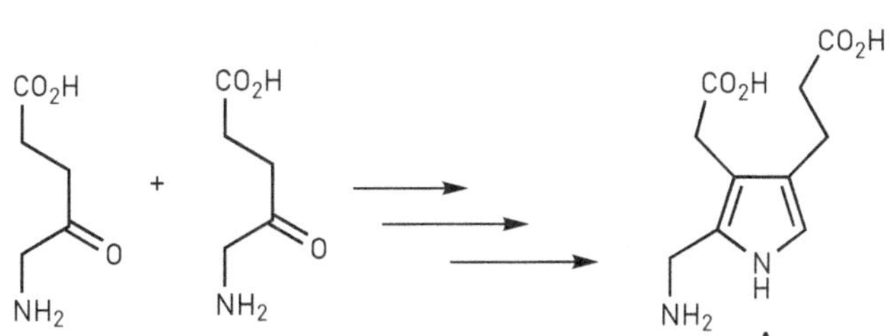

5-aminolevulinic acid

8. *Frontalin* is the sex attractant of the southern pine beetle (*Dendroctonus frontalis*), a remarkably destructive pest of pine forests of the American south. The attractant is biosynthesized from 6-methyl-6-heptene-2-one by the action of a monooxygenase-mediated epoxidation followed by a simple acid-catalyzed cyclization. Propose a mechanism for the epoxide cyclization leading to *frontalin*.

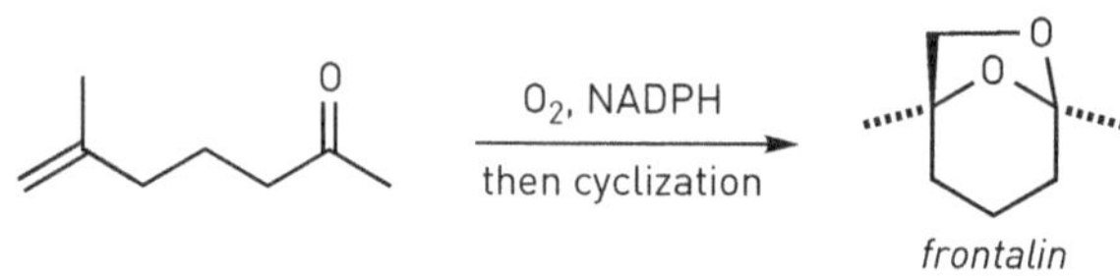

9. *Fenchone* undergoes a monooxygenase-mediated oxidation and hydrolysis to give a mixture of hydroxyacids as shown.Provide a reasonable mechanism to account for the formation of the two products shown from *fenchone* and flavin–OOH.

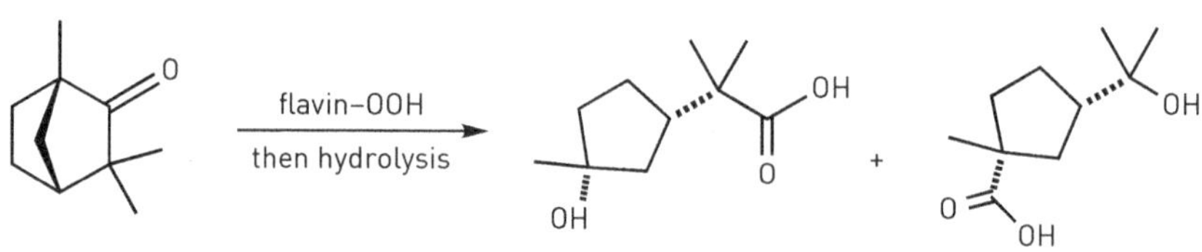

10. When the antibiotic emycin F is treated with aqueous acid, a molecular rearrangement occurs, producing a constitutional isomer known as emycin E. Write a reasonable mechanism that accounts for this acid-catalyzed transformation.

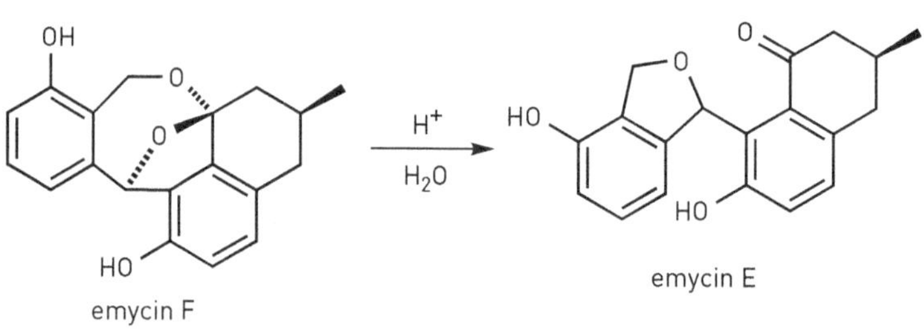

11. The enzyme-catalyzed reaction of TPP ylide with 2 moles of pyruvic acid leads to the product shown plus CO_2. Propose a reasonable mechanism for this transformation.

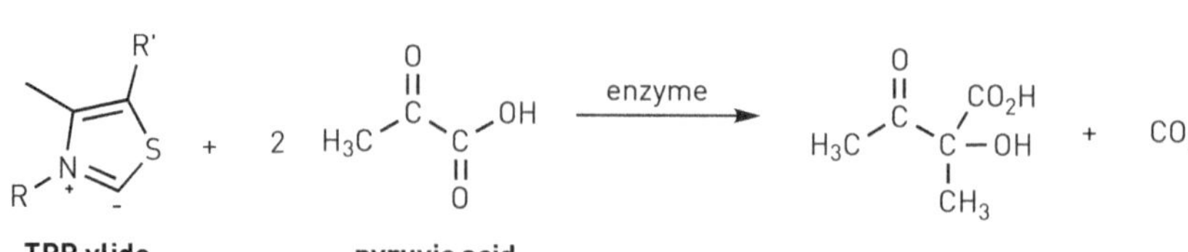

3 Biosynthesis of Carbohydrates and Amino Acids

I called it ignose, not knowing which carbohydrate it was. This name was turned down by my editor. "God-nose" was not more successful, so in the end "hexuronic acid" was agreed upon. Today the substance is called "ascorbic acid" and I will use this name.
—Albert Szent-Gyorgyi (*Nobel Prize in Medicine, 1937*)

We have already seen that some of the basic building blocks used in the biosynthesis of natural products are amino acids such as phenylalanine, tyrosine, and others. These and other crucial construction materials such as the acyl group in acetyl-CoA are all ultimately derived from carbohydrates. In this chapter, we will present an abbreviated overview of the components of carbohydrate structure and metabolism sufficient for our purposes going forward, with a schematic flowchart showing how carbohydrates and amino acids are modified, combined, and branched off in various ways to yield the distinct set of biosynthetic pathways that will form the core of the remainder of the text. We will finish the chapter with a brief, general review of amino acid nomenclature and structure with emphasis on the key amino acids that will be used throughout the remainder of the text.

WHAT MAKES A CARBOHYDRATE?

We know that plants make *glucose* ($C_6H_{12}O_6$) by photosynthesis using light, water (H_2O), and carbon dioxide (CO_2). Another way of looking at the formula for glucose is $C_6(H_2O)_6$, that is, six carbon atoms and six water molecules. Thus, glucose was originally referred to as a hydrated form of carbon—a **carbohydrate**. But this is a very general term since there are many different types of carbohydrate compounds. One way to broadly classify carbohydrates is to identify them as either mono- (one), di- (two), oligo- (a few) or poly- (many) **saccharides.** For example, *glucose* ($C_6H_{12}O_6$) cannot be broken down into simpler carbohydrates by simple hydrolysis, so it is classified as a **monosaccharide**, that is, a single, discrete carbohydrate compound. On the other hand, the carbohydrate *sucrose* ($C_{12}H_{22}O_{11}$) is classified as a **disaccharide** since when it is subjected to aqueous hydrolysis, it yields two different monosaccharide carbohydrates, namely *glucose* ($C_6H_{12}O_6$) and *fructose* ($C_6H_{12}O_6$). Noting that glucose and fructose are different compounds but with the same molecular formula, they must be related to one another either as stereoisomers or as constitutional isomers, so further refinement of classification is needed.

Structurally speaking, most monosaccharide carbohydrates are simply polyhydroxyaldehydes (*aldoses*) or polyhydroxyketones (*ketoses*) which can be further classified using a combination of aldo- or keto- prefixes along with suffixes such as triose, tetrose, pentose, or hexose to designate the number of carbon atoms. Thus, a ketopentose is a five-carbon ketone carbohydrate, while an aldotetrose is a four-carbon aldehyde carbohydrate. In the hydrolysis of the disaccharide sucrose above, the resulting monosaccharide carbohydrates produced (Fig. 3.1) turn out to be an aldohexose (glucose) and a ketohexose (fructose).

Note that carbohydrates are further classified as D or L sugars according to whether the last (lowermost) chiral –OH group in a Fischer projection of the monosaccharide is pointing to the right (D) or to the left (L). D-sugars are by far the predominant naturally occurring forms. Remember that a D or L designation is related to *configuration* at a specific chirality center (like *R* or *S*) and *not* to the direction of optical rotation as referred to by the familiar lower case *d* (+) or *l* (–).

The smallest carbohydrate possible is (+)-D-glyceraldehyde, the reduced form of which is glycerine, the triol component from which triacylglycerol esters (triglycerides) are formed from condensation with fatty acids (more on this later). Beyond this simple D-aldotriose, we see in Fig. 3.2 the names and configurational relationships of most of the important aldotetrose, aldopentose, and aldohexose carbohydrates.

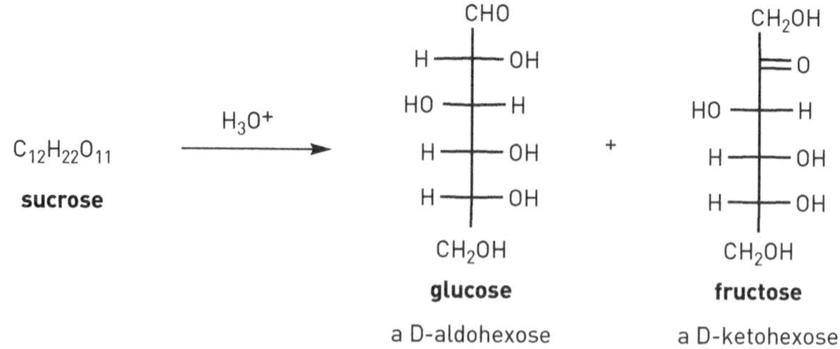

FIGURE 3.1
Fischer projections of D-glucose and D-fructose obtained from sucrose hydrolysis.

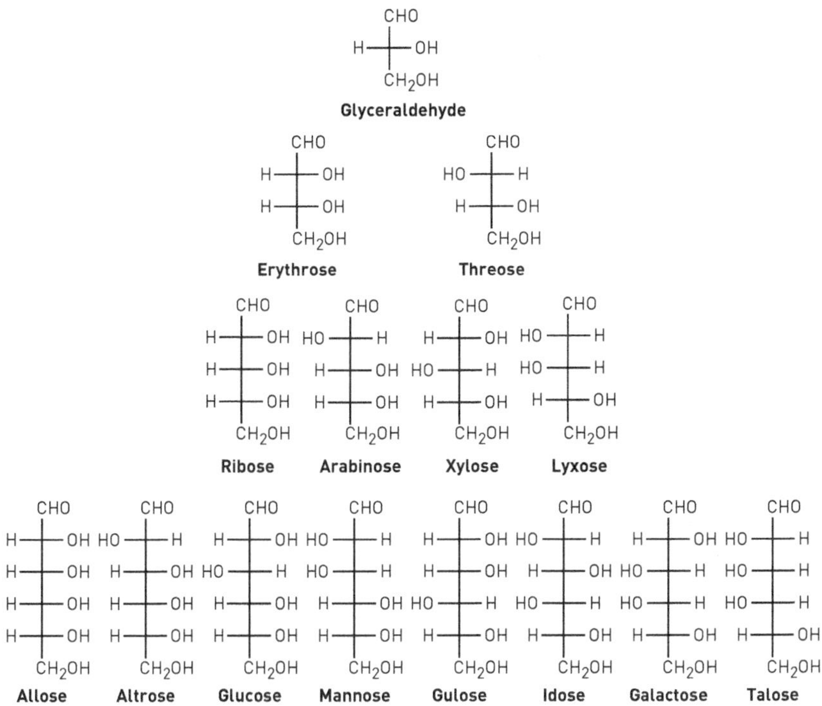

FIGURE 3.2
The configurational relationships of the aldose carbohydrates.

CYCLIC HEMIACETALS AND ANOMERS

We noted previously (Chapter 1, Fig. 1.49) that aldehydes and ketones containing one or more –OH groups elsewhere in the molecule can undergo intramolecular hemiacetal formation and that the equilibrium favors the cyclic hemiacetal if the ring is 5- or 6-membered. This is especially important in carbohydrates; most any sugar molecule with a hydroxyl O atom four or five atoms distant from the carbonyl carbon will prefer to exist in cyclic hemiacetal form. The resulting 5-membered ring hemicacetals are called **furanose** forms (from the O-heterocycle *furan*), while the 6-membered rings are called **pyranose** forms (from the O-heterocycle *pyran*); when multiple hydroxyl groups are located relative to the carbonyl in such a way as to allow formation of either type, the predominant form will depend on the relative isomeric relationships of the hydroxyls to one another on the resulting ring system. The relationship of open-chain versus furanose and pyranose forms for D-glucose in aqueous solution is illustrated in Fig. 3.3.

Note that in open-chain form, glucose has only four chirality centers, but in cyclic form, an additional chirality center is present due to formation of the hemiacetal functional group. Since the carbonyl carbon can present either of its two prochiral faces to the attacking nucleophilic –OH group, the resulting hemiacetal is produced as a mixture of stereoisomers at the hemiacetal carbon; this unique chiral carbon in the

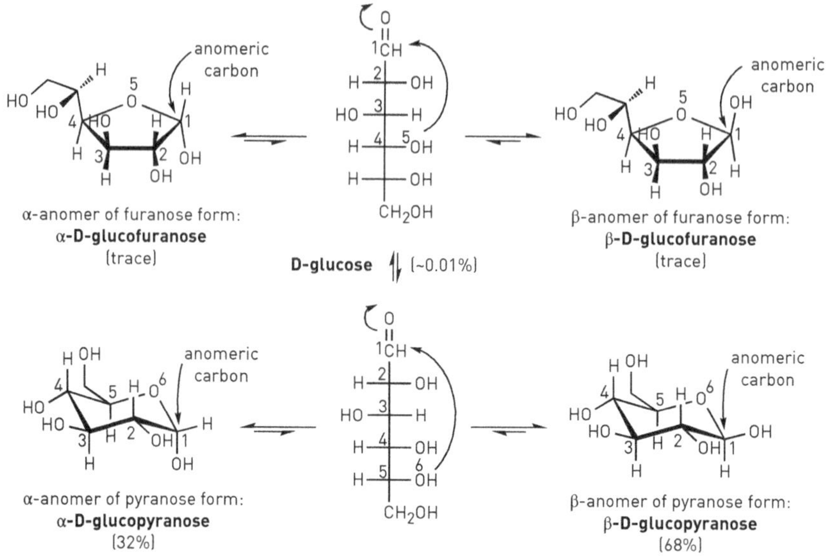

FIGURE 3.3

Interconversion and equilibrium composition of glucose forms in aqueous solution: open chain vs. furanose and pyranose α and β anomers.

cyclic carbohydrate is known as the **anomeric carbon** and the resulting stereoisomers are a special class of diastereomers called **anomers.** The anomeric carbon is always an easy one to spot in cyclic carbohydrates; it will be the only carbon atom bonded to two oxygens; all other carbons will be bonded to only one oxygen. In chair representations of glucose, the so-called **α-anomer** will have the anomeric hydroxyl group in an axial position on the ring (**a** for **a**xial as well as **a** for **a**lpha is an easy way to remember this) when all other groups on the ring are equatorial, while the **β-anomer** will have the anomeric hydroxyl in an equatorial position on the ring when all other groups are equatorial. In the case of glucose, the predominant equilibrium form in aqueous solution is the pyranose form, with the β-anomer (68%) favored over the α-anomer (32%); the furanose and open-chain polyhydroxyaldehyde forms are present in only trace amounts. The process of interconversion of anomers to their equilibrium composition in solution is known as **mutarotation** and occurs for all aldopentoses and aldohexoses.

Technically, α and β anomeric assignments are based on comparison of the *R/S* configurational assignments at the anomeric carbon and the corresponding Fischer projection's lowermost chiral carbon. If the two assignments are the same, the anomer is β; if different, the anomer is designated α. Most D-sugars have the *R* configuration at the lowermost chiral carbon, and an anomeric equatorial –OH will correspond to the *R* configuration for most aldopyranose forms drawn in their most stable chair conformation; thus, the equatorial anomer will usually be β for D-aldopyranose sugars, while the axial anomer will be α. But keep in mind that the axial and equatorial positions of chair conformations are interconvertible via chair-chair "ring flipping," so the above analysis only holds for single conformations of the most stable chair form where only

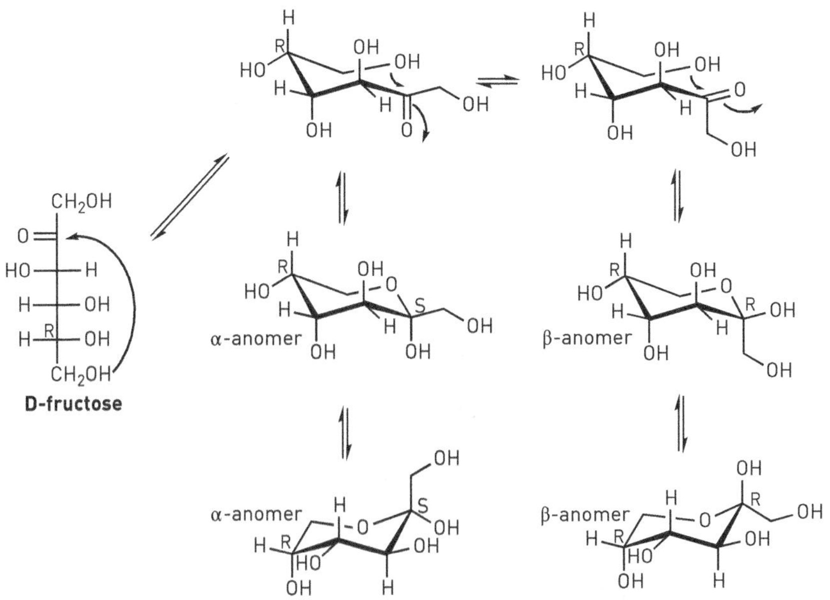

FIGURE 3.4
Chair representations of α and β-D-fructopyranose.

the position of the pyranose anomeric –OH is changeable relative to all the others. To see how this applies, consider the pyranose forms of fructose, by far the most significant ketohexose carbohydrate, as shown in Fig. 3.4.

As we can see, the β-anomer (*R*,*R*) may have the anomeric –OH either axial or equatorial, depending on which chair form is represented. The more stable form of most pyranose carbohydrates will correspond to the chair form that maximizes the number of equatorial substituents versus axial substituents. That's an easy call with D-glucopyranose, since the conformation that has all groups equatorial (including the anomeric –OH) equatorial is the β-anomer. It's a closer call in the case of D-fructopyranose; for the β-anomer, the conformation with the anomeric –OH group axial has two axial and three equatorial substituents; with the anomeric –OH equatorial, there are three axial and two equatorial substituents, so the former is probably slightly more stable. It is exactly the opposite case for the α-anomer. Thus, unlike D-glucopyranose, the most stable chair form of the β-anomer will have the anomeric –OH axial rather than equatorial, while the α-anomer will have its anomeric –OH equatorial in the more stable chair form.

The above discussion is less important for furanose forms since they lack the obvious axial and equatorial chair positions characteristic of pyranose forms. **Haworth projections** can be useful for representation of both furanose and pyranose forms, as shown for D-fructose and D-glucose in Fig. 3.5, though their use is more common for furanose forms. For D-sugars, converting a Fischer projection into a Haworth projection is straightforward. With the ring O-atom projecting toward the back of the structure and the anomeric carbon on the right side, any atom or group pointing

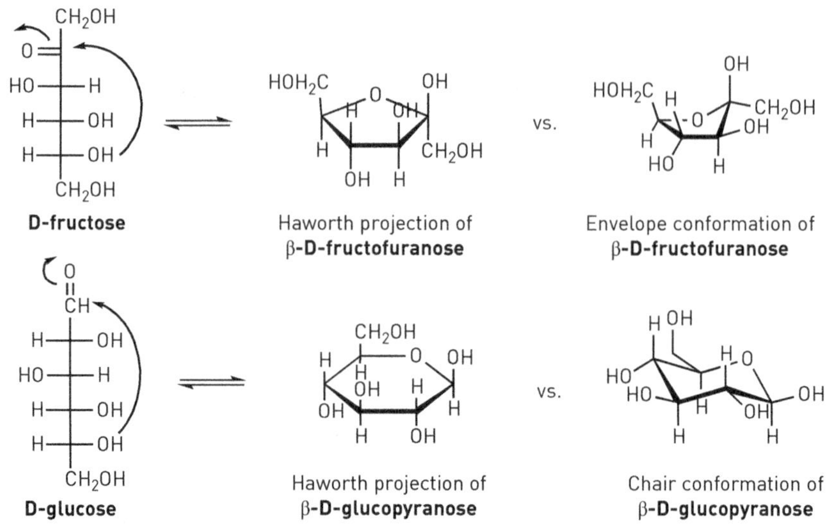

FIGURE 3.5
Haworth vs. conformational representations of β-D-fructofuranose and β-D-glucopyranose.

to the right in the open-chain Fischer projection points down in the corresponding Haworth projection; if pointing to the left in the Fischer projection, it points up in the Haworth projection.

Significantly, the terminal $-CH_2OH$ group of D-carbohydrates (located at the bottom of the Fischer projection) is always pointing up in the Haworth projection (unless this group's –OH is used to form the hemiacetal). Using the terminal $-CH_2OH$ group as a reference point (Fig. 3.5), for most Haworth projections of D-furanose or D-pyranose forms, the α-anomer will have the anomeric –OH group pointing down (*trans* to the terminal $-CH_2OH$), while the β-anomer will have the anomeric –OH pointing up (*cis* to the terminal $-CH_2OH$). Note how the all-*trans* relationship of the ring substituents in β-D-glucopyranose is emphasized by the Haworth projection compared to the chair representation. The latter is more correct from a conformational point of view, while the former helps to clarify the stereochemical relationships.

C-2 EPIMERS AND ENEDIOLS—SIMPLE CONVERSION OF ONE CARBOHYDRATE INTO ANOTHER

We know that protons attached to the α-carbon of carbonyl compounds are especially acidic and may be removed by bases to form enolate ions. When such α-protons are located at a chirality center, proton removal followed by reprotonation of the resulting enolate leads to racemization of the chirality center, since the intermediate enolate is only prochiral and so will produce one enantiomer from protonation at its *Si* face and the other from protonation at its *Re* face. If this process occurs in the open-chain form of aldose sugars, the chirality center at C-2 becomes scrambled, resulting in the formation of diastereomeric sugars. In systems such as

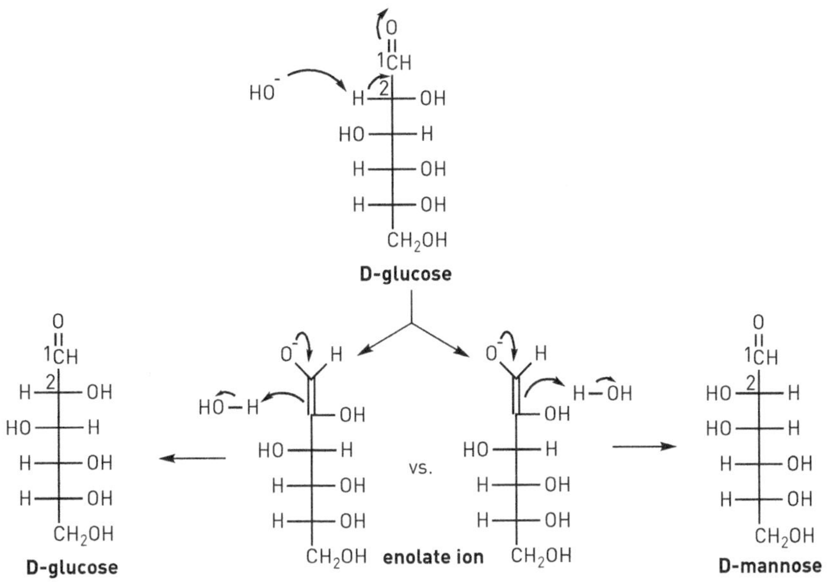

FIGURE 3.6
Base-catalyzed conversion of D-glucose into C-2 epimers: D-glucose and D-mannose.

this, the process is referred to as **epimerization**, and in the case of aldose carbohydrates, the two disastereomers formed are related to one another as so-called **C-2 epimers** as illustrated in Fig. 3.6. Thus, under base-catalyzed conditions, D-glucose is **epimerized** to a mixture of D-glucose and D-mannose.

If the intermediate enolate ion protonates at oxygen instead of carbon, the resulting enol form of the carbonyl group has an –OH group at each end of the double bond, making it a so-called **enediol**. These unusual enol forms, when reverting back to carbonyl form, can produce either an aldehyde or a ketone, depending only on which of the two enediol carbons acts as the proton acceptor (Fig. 3.7).

In practice, treatment of glucose with aqueous base leads to a mixture that consists of about 65% glucose, 32% fructose, and 3% mannose. A mixture with the same composition will also be produced starting from mannose, since mannose and glucose both form the same enediol intermediate under these conditions. Similarly, other pairs of D-aldohexoses may be epimerized and/or converted to give the three remaining D-ketose sugars: *psicose, sorbose,* and *tagatose*, while two D-aldopentose pairs each yield the two possible corresponding D-ketopentose sugars. Thus, either ribose or arabinose may yield the ketopentose *ribulose*, while either xylose or lyxose will give the ketopentose *xylulose* as in Fig. 3.8.

OTHER IMPORTANT MONOSACCHARIDES: DEOXY AND AMINO SUGARS

Not all carbohydrates are formulated as shown for the aldose sugars. Two important examples are D-2-deoxyribose, a **deoxy sugar**, and D-glucosamine, an **amino**

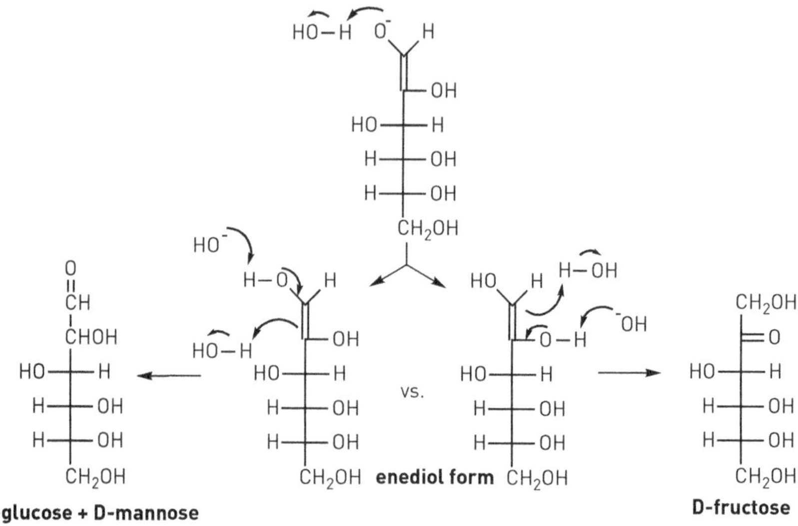

FIGURE 3.7
Base-catalyzed epimerization of D-glucose vs. conversion to D-fructose via the enediol.

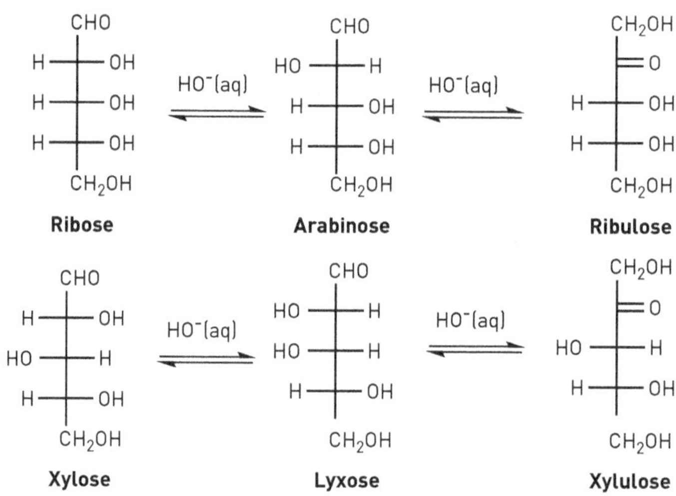

FIGURE 3.8
Base-catalyzed interconversion of D-aldopentose and D-ketopentose carbohydrates.

sugar (Fig. 3.9). The prefix "deoxy" is related to the missing –OH group at C-2 in what would otherwise be the structure of D-ribose. This simple deoxy designation is widely used in natural products chemistry, providing a consistent and useful nomenclature that uses familiar names (and structures) of known compounds rather than deriving yet another new common name. Similarly, D-glucosamine, also known as D-2-deoxy-2-aminoglucose, relates the name and structure of a known compound to a new one.

Deoxyribose may be familiar as an essential component in **DNA (deoxyribonucleic acid)** and as one of the features that distinguish DNA from **RNA (ribonucleic acid**, which employs ribose rather than 2-deoxyribose). **Glucosamine** is a

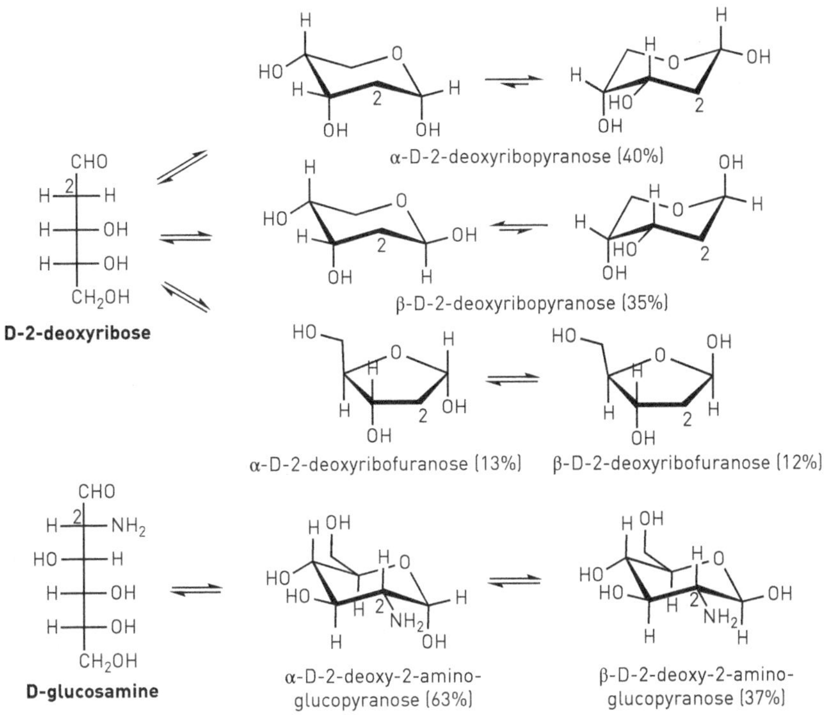

FIGURE 3.9
D-2-deoxyribose, D-glucosamine, and their cyclic hemiacetal forms.

common dietary supplement (usually as the hydrochloride salt or sulphate ester) used by many who suffer from osteoarthritis. More importantly, glucosamine is the biosynthetic precursor of all other nitrogen-containing sugars. For dietary purposes, glucosamine is obtained commercially from the hydrolysis of a crustacean exoskeleton material known as **chitin**, a biopolymer in which glucosamine, as its N-acetyl derivative, is the principal component.

THE SIGNIFICANCE OF THE ANOMERIC CARBON: GLYCOSIDE FORMATION

Recall that any **hemiacetal** may be considered an intermediate in the formation of a corresponding **acetal** under acid catalyzed conditions if the protonated –OH group of the hemiacetal is lost as water and the resulting resonance-stabilized carbocation is trapped by an available nucleophilic alcohol –OH group. Such acetals, when derived from the cyclic hemiacetals of carbohydrates, are known as **O-glycosides.** The nucleophilic alcohol component, ROH, may be any simple alcohol such as methanol or ethanol or a more complex alkanol or cycloalkanol. The general process is illustrated in Fig. 3.10.

This is an example of a very general and extremely important process: namely, the linking of carbohydrate structures to other molecules by nucleophilic trapping at the anomeric carbon of the carbohydrate. This is how **disaccharides** and

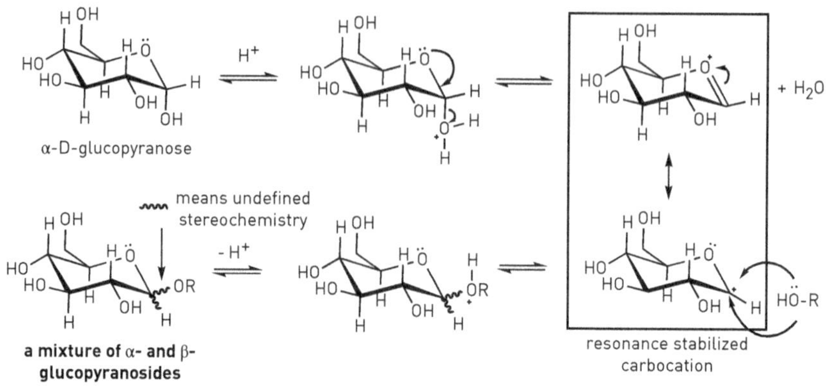

FIGURE 3.10
Acid-catalyzed formation of a glucopyranoside mixture from glucose and an alcohol, ROH.

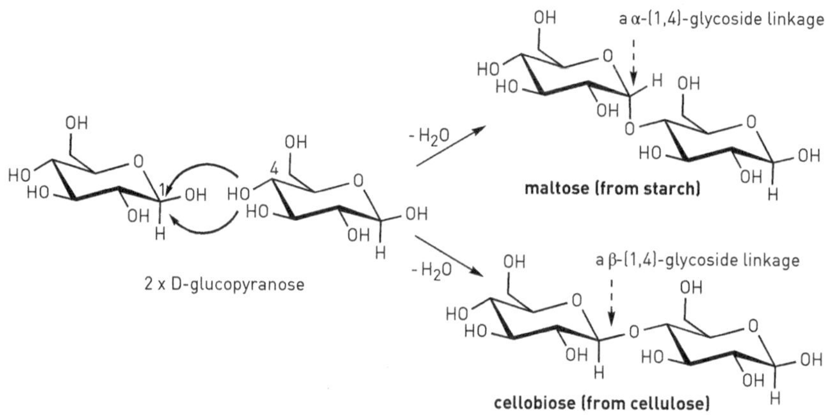

FIGURE 3.11
Structures of disaccharides derived from glucose: maltose and cellobiose.

polysaccharides are formed from monosaccharides. Since the resulting products are acetals, we can also now see how and why di- and polysaccharides are hydrolyzed to monosaccharides: the reverse process is simply **acetal hydrolysis.**

Unfortunately, the simple scheme in Fig. 3.10 is not one that is stereospecific. Glycosides, like their hemiacetal precursors, can be formed as either α or β anomers, and under such conditions we would also expect the carbohydrate to exist as a mixture of pyranose and furanose forms, thus leading to complex mixtures. For laboratory synthesis of glycosides, there are a number of ways to overcome these limitations, though we will not address those methods here. We will be more interested in seeing how biological systems overcome these limitations to produce different glycosides stereospecifically. But before we look into how nature manages this process, let's learn a bit more about some important examples of glycoside structures.

Figure 3.11 illustrates in general terms how two glucose molecules come together by linking the –OH group at C-4 of one glucose to C-1 (the anomeric carbon) of a

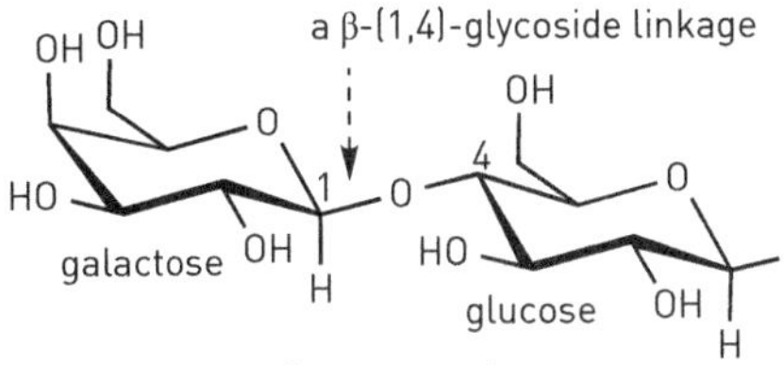

lactose (milk sugar)

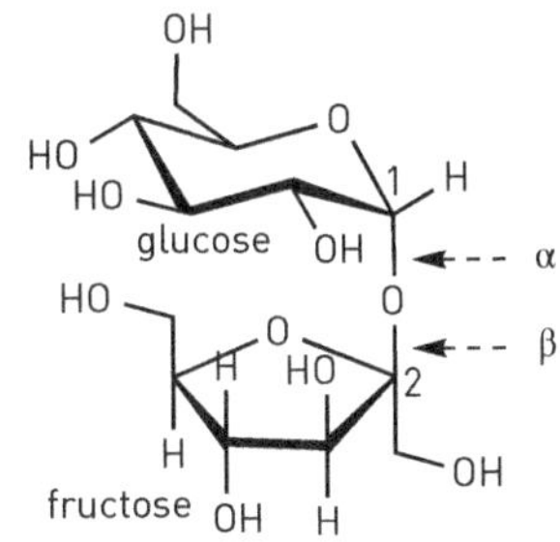

sucrose (table sugar)

FIGURE 3.12
Structures of the disaccharides lactose and sucrose.

second glucose. Since the C-4 –OH can attack at either the top face or the bottom face at C-1, both α and β glycoside linkages are possible. When the bond to C-1 is axial, the disaccharide formed is **maltose,** a product obtained by hydrolysis of starch and whose glycoside bond is designated an α-(1,4)-glycoside linkage, making maltose an α-glycoside. Alternatively, if the resulting glycoside bond at C-1 is equatorial, the disaccharide product is **cellobiose**, a hydrolysis product obtained from cellulose. Here the glycoside bond is designated as a β-(1,4)-glycoside linkage, making cellobiose a β-glycoside.

Other O-glycosides of note are the disaccharides **lactose** and **sucrose** (Fig. 3.12). These disaccharides both contain a single glucose component joined to a second monosaccharide that is different in each case. Thus, lactose, also known as *milk sugar*, is composed of a galactose unit linked at its anomeric carbon (C-1) to the C-4 hydroxyl of a glucose unit via a β-(1,4) linkage. In humans, a deficiency in the enzyme *lactase*, which facilitates the hydrolysis and breakdown of lactose, leads to a condition known as **lactose intolerance**, a genetic trait that leads to gastric upset when milk products are ingested, though the condition is easily treated with over-the-counter *lactase* formulations. Sucrose, better known as *table sugar*, is composed of glucose and fructose, but it differs from other disaccharides in that the glycoside linkage between the two individual sugar components involves the anomeric carbon of both units. Thus, the unique glycoside linkage of sucrose is designated a 1α, 2β glycoside linkage, since it is α with respect to the glucose anomeric carbon (at C-1) and β with respect to the fructose anomeric carbon (at C-2).

Polysaccharides are examples of high molecular weight carbohydrate biopolymers. Unlike some disaccharides, most polysaccharides contain only a single type of carbohydrate as a repeating unit. Two important examples are **amylose**, one of two glucose-derived components of starch, and **cellulose**, the main structural material of many plants and also a polymer of glucose (cotton is essentially pure cellulose). These two very different biomaterials (one is a key food source while the

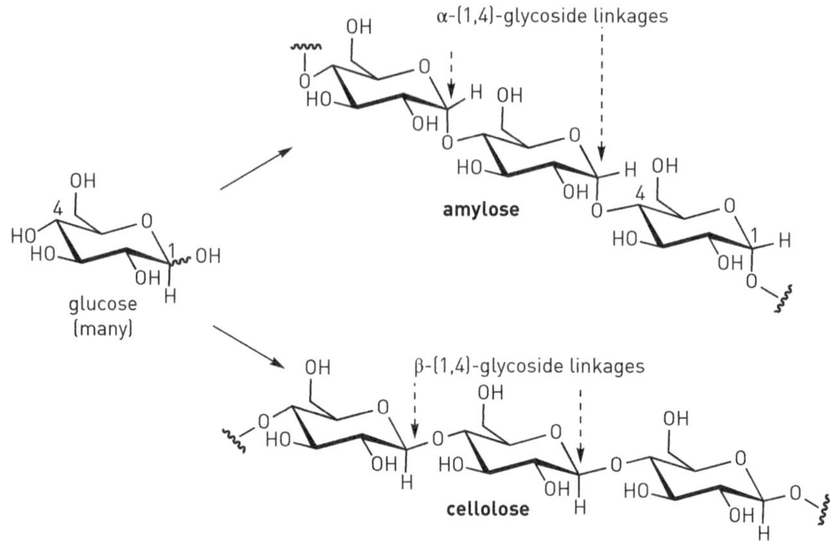

FIGURE 3.13
Structures of the glucose polysaccharides amylose and cellulose.

other is an indigestible yet valuable source of tough, durable fibers) differ mainly in the type of O-glycoside bond linking the individual glucose units. Thus, amylose utilizes α-(1,4)-glycoside linkages while cellulose utilizes β-(1,4)-glycoside linkages (Fig. 3.13). That is a relatively small difference, chemically speaking, for two such very dissimilar materials. The polysaccharide **chitin,** mentioned earlier as a component of crustacean exoskeletons, also uses β-(1,4)-glycoside linkages, but the repeating unit is N-acetylglucosamine rather than glucose.

Other glycosides in which a carbohydrate is linked to various types of organic alcohols or amines are quite abundant in nature, and many are important biological or medicinal compounds. When the carbohydrate component of such compounds is released from the glycoside linkage by hydrolysis, the residual organic structure is known as the **aglycone** of the original glycoside structure. For example, **salicin** is a component of willow bark, whose use in the treatment of fever, pain, and inflammation dates to as far back as 400 BC. When ingested, hydrolysis of the glycoside linkage in the stomach affords the aglycone, **salicyl alcohol** (Fig. 3.14). Metabolic oxidation of this benzylic alcohol affords salicylic acid, the active principal in acetyl salicylic acid (aspirin).

The O-glycoside **daunorubicin**, a powerful cancer chemotherapy agent isolated from *Streptomyces peucetius*, contains a unique amino sugar called **daunosamine** (Fig. 3.15) which is essential for the bioactivity of the compound. In other glycosides, the sugar anomeric carbon may be linked to N- or C-atoms of the organic aglycone. **Adenosine**, a familiar nucleic acid component of RNA, is an N-glycoside of β-D-ribfuranose and the heterocyclic base **adenine. Mangiferin**, an antioxidant found principally in mangoes but also widely distributed in other higher plants, is a

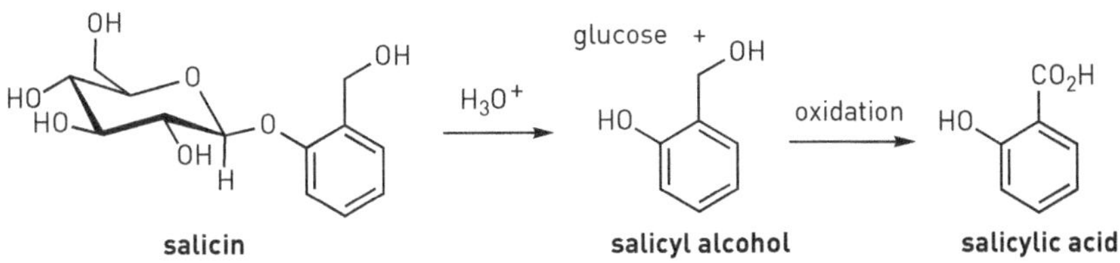

FIGURE 3.14
The O-glycoside salicin, its aglycone salicyl alcohol, and the oxidation product salicylic acid.

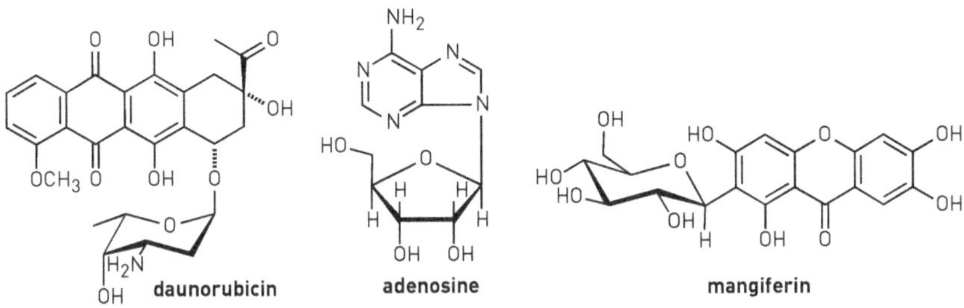

FIGURE 3.15
An O-glycoside (daunorubicin), an N-glycoside (adenosine), and a C-glycoside (mangiferin).

glucose C-glycoside connected to a xanthone-based aglycone and is a constituent of some folk medicines.

A large and important class of antibiotics, the **aminoglycosides**, are so-called not because they are necessarily N-glycosides, but rather because they contain one or more amino sugars. Examples include **streptomycin**, the first antibiotic found to be effective against tuberculosis, and **neomycin**, widely used in many over-the-counter topical antibiotic ointments and creams. By contrast, the relatively rare **thioglycosides** are specifically S-glycosides, such as the white mustard seed component **sinalbin,** a well-known example found in many wild plants. All these are shown in Fig. 3.16 using conventional organic stereochemical structural representations rather than conformational or Haworth diagrams.

In many glycosidic natural products, we will be mainly interested in how the aglycone portion is biosynthesized, since the carbohydrate component is usually added to the aglycone by glycosylation at a late stage. But before moving on, let's take a look at how glycoside linkages are usually formed biosynthetically.

UDP-SUGARS AND GLYCOSIDE FORMATION: S_N2 CHEMISTRY AT WORK

As mentioned earlier, acid-catalyzed reaction of alcohols with carbohydrate hemiacetals to give glycoside (acetal) derivatives can be difficult to control in terms of pyranose versus furanose and α versus β anomeric forms. Nature takes a different approach to glycoside formation by making use of carbohydrate derivatives known

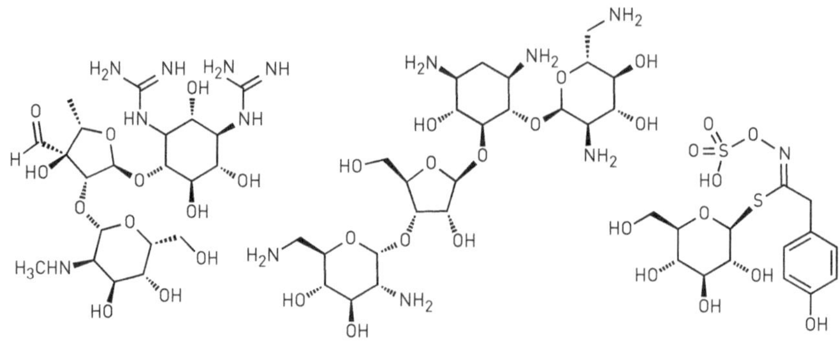

streptomycin neomycin sinalbin

FIGURE 3.16
The aminoglycoside antibiotics streptomycin and neomycin and the S-glycoside sinalbin.

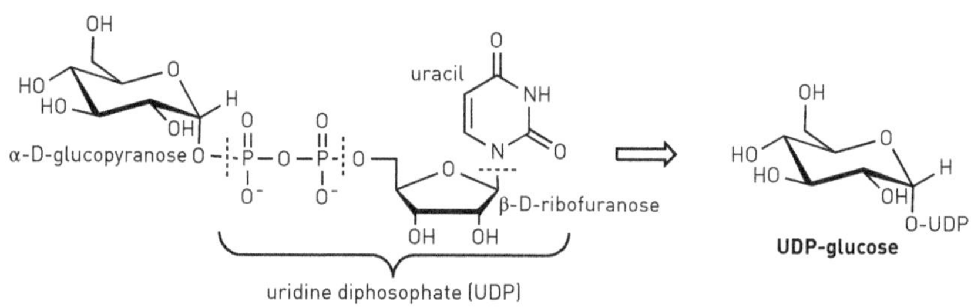

FIGURE 3.17
Structural components of UDP-glucose.

as **uridine diphosphosugars** or simply **UDP-sugars**. The most important of these is UDP-glucose (Fig. 3.17). This seemingly complex structure is relatively easy to dissect into its various components: an α-D-glucopyranose linked by a diphosphate bridge to the C-5 hydroxyl group of D-ribofuranose which in turn is connected via a β-N-glycoside linkage to a molecule of uracil (uracil + β-D-ribofuranose = **uridine**, an RNA base).

The diphosphate ester linkage is employed here to activate the anomeric O-atom of the glucose component toward nucleophilic displacement in much the same way that an alcohol -OH group can be turned into a good leaving group by conversion to a tosylate or mesylate ester. When a nucleophile (Nu–H) attacks at the activated anomeric center, an S_N2 reaction displaces the UDP component as a leaving group, leading to inversion of configuration at the anomeric carbon and formation of a β-glycoside derivative (Fig. 3.18). Many O-, N-, and C- glycosides found in nature are β-anomers for this reason.

When α-linkages are required, a nucleophilic functional group in an enzyme mediating the process may form a β-linkage by initial nucleophilic S_N2 attack; this group may then be displaced by an external nucleophile (Nu–H) in a second S_N2 process (Fig. 3.19). Inversion followed by inversion leads to a net retention of

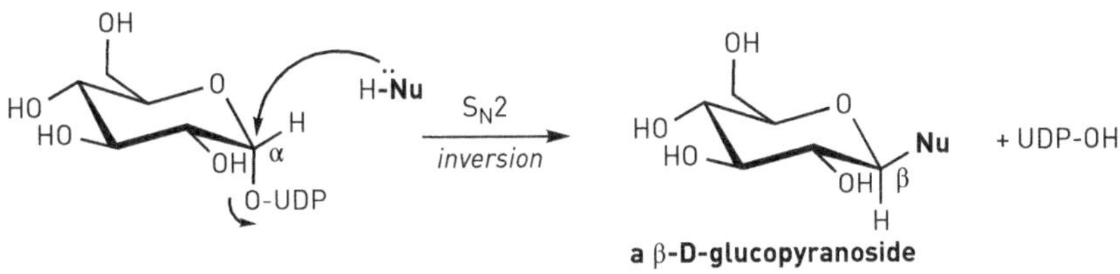

FIGURE 3.18
Conversion of UDP-glucose to a β-D-glucopyranoside via S_N2 reaction with a nucleophile, Nu–H.

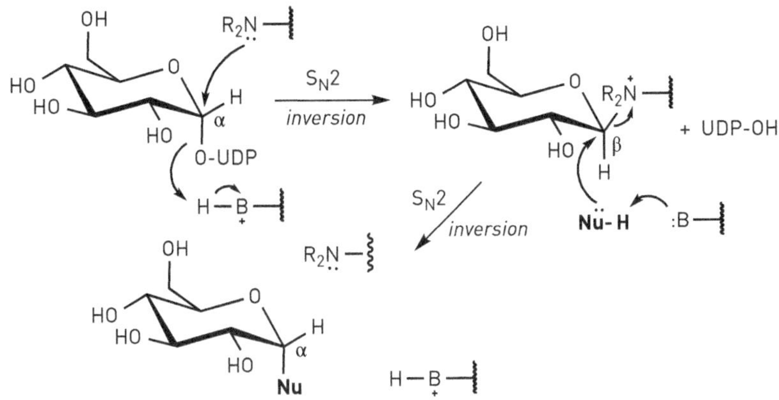

FIGURE 3.19
α-D-glycoside formation via enzyme-mediated sequential S_N2 reactions.

configuration and formation of an α-linkage at the anomeric center of the resulting glycoside.

In some instances, certain important UDP-carbohydrates are biosynthesized from other UDP-carbohydrates. For example, the UDP derivative of the aldopentose **xylose**, an important building block component of plant cell wall structure, is synthesized from UDP-glucose using some of the biological transformations introduced in Chapter 2. The steps involved are NAD^+ oxidation of the C-6 hydroxyl group of UDP-glucose to give the corresponding **UDP-glucaronic acid**. Further oxidation at C-4 gives a β-keto acid which decarboxylates to yield the corresponding ketone. The sequence is completed by NADH reduction of the ketone to give the pyranose form of UDP-xylose. Hydrolysis of the phosphate linkage will yield free xylose, as shown in Fig. 3.20. The name xylose is derived from *xylos*, the Greek word for wood, from which xylose was first isolated.

Another example is found in the synthesis of the O-methyl sugar **cladinose**, one of two unusual carbohydrate O-glycoside components connected to the core structure of **erythromycin**, a macrolide antibiotic whose aglycone biosynthesis we will explore further in Chapter 5. The sequence of reactions involved in cladinose synthesis (Fig. 3.21) again begins with glucose, but linked not to a uridine diphosphate (UDP) group, but rather to thymidine diphosphate group (TDP, a β-N-glycoside of thymine and 2-deoxyribose). TDP-glucose is first oxidized at C-4 by NAD^+; the

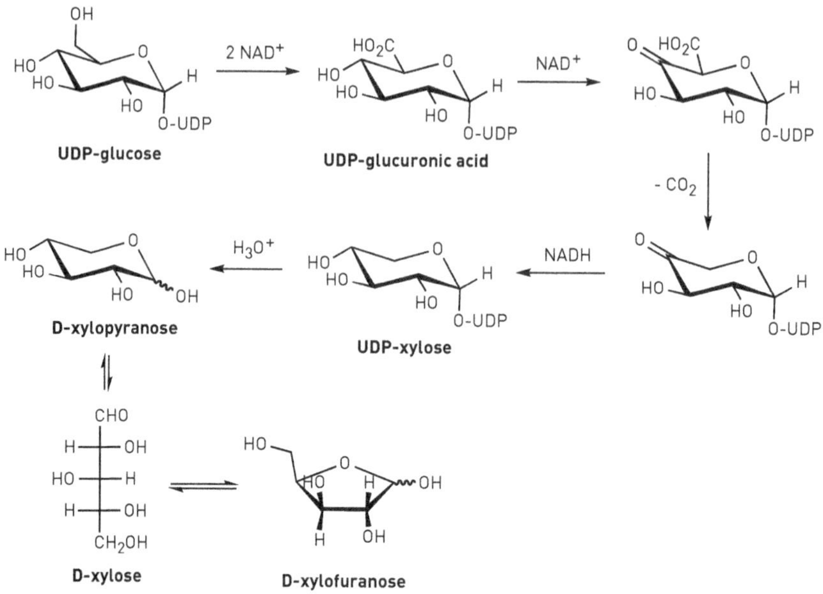

FIGURE 3.20
Biosynthesis of UDP-xylose from UDP-glucose.

resulting β-hydroxyketone then undergoes a base-induced dehydration to give the corresponding α,β-unsaturated ketone (similar to dehydration of aldol products); conjugate (1,4) reduction by NADH and keto-enol tautomerization gives the α-methylketone **TDP-6-deoxy-D-xylo-4-hexulose**. Subsequent dehydration (loss of the C-2 hydroxyl) gives a second α,β-unsaturated ketone which then undergoes epimerization at C-5 (α to the ketone carbonyl) followed by conjugate addition of hydride from NADH with concurrent methylation of the intermediate enolate by S-adenosylmethione (SAM). NADH reduction of the ketone, SAM methylation of the C-3 hydroxyl, and hydrolysis completes the sequence to give cladinose.

Now that we have a reasonable grasp of carbohydrate nomenclature and structure, let's take a bit of time to look at the chemistry involved in the breaking down of glucose into various organic fragments, some of which constitute the essential building blocks for the biosynthesis of all other primary and secondary metabolite structures.

ORGANIC REACTIONS IN CARBOHYDRATE CHEMISTRY: OVERVIEW OF GLUCOSE METABOLISM

Glucose is broken down as a fuel source for energy production in living systems in two important ways, both of which will be of interest to us, though mainly from the point of view of the mechanistic aspects of the organic chemistry of the processes as well as the use of the products produced by these pathways. The more important of these two processes is known as **glycolysis**, but we will also be interested in a second

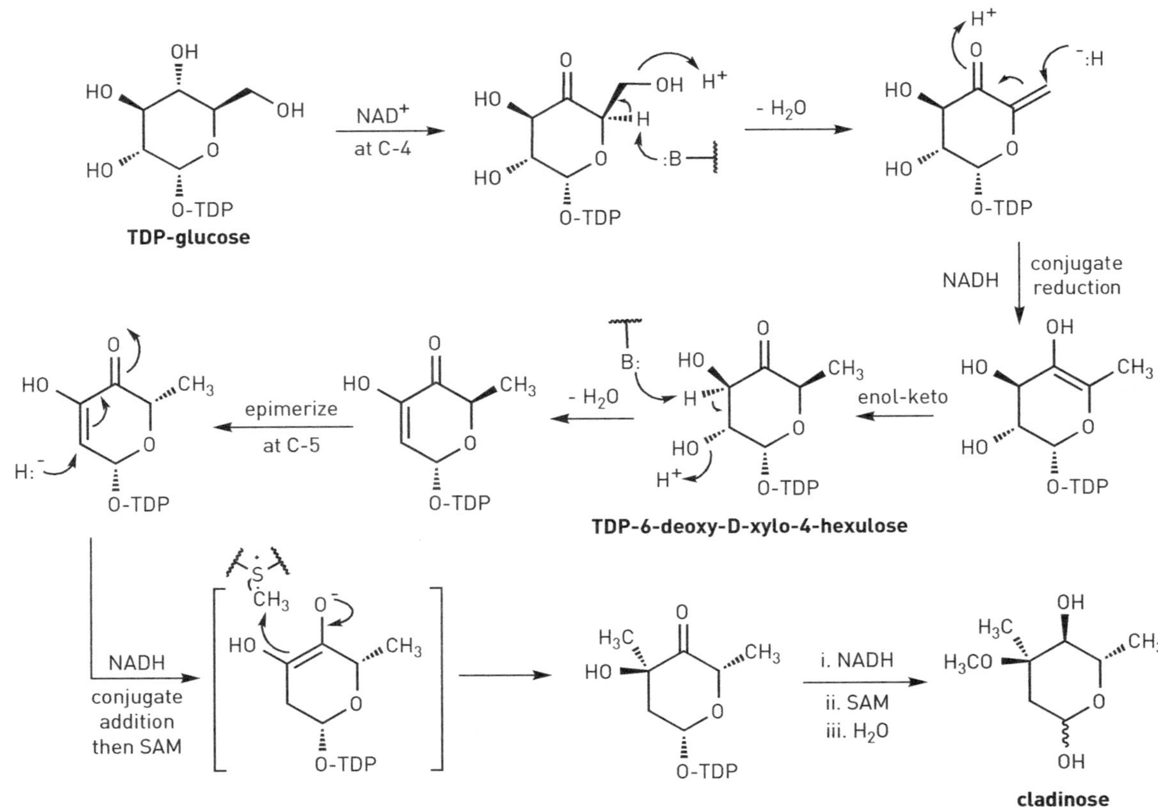

FIGURE 3.21
Biosynthesis of cladinose, a carbohydrate component of the antibiotic erythromycin.

route to glucose metabolism known as the **pentose phosphate pathway**. To keep our studies focused on the organic chemistry of these processes and the uses of their products, we will defer to specialists in biochemistry for detailed treatments of their enzymology, pathway regulation, and thermochemical aspects. To begin to sort all this out, let's take it one pathway at a time, starting with **glycolysis**.

GLYCOLYSIS: A 10-STEP PROGRAM

While glycolysis ultimately converts glucose to pyruvic acid, the overall transformation occurs through a complex series of individual enzyme-mediated reactions which we can now examine one by one. The first step in glycolysis is essentially an esterification reaction that converts α-glucopyranose to **glucose-6-phosphate** (Fig. 3.22) by reaction with ATP. A basic site of the meditating enzyme *hexokinase* is involved in the deprotonation of the alcohol function, while Mg^{+2} (not shown) coordinates with ATP to facilitate the phosphorylation. We will continue to use our previous abbreviations of ROPPP for triphosphates, ROPP for diphosphates, and ROP for monophosphate derivatives of alcohols for simplicity. This reaction is very similar to the conversion of an alcohol to an acetate ester by reaction with acetic anhydride.

The next step involves conversion of glucose-6-phosphate to fructose-6-phosphate. This *phosphoglucose isomerase*-catalyzed process is basically a sequence that involves ring opening of the pyranose form to the open-chain aldohexose form, followed by keto-enol tautomerization from aldohexose to enediol to the ketohexose form, giving **fructose-6-phosphate** which then cyclizes to its α-furanose hemiacetal form (Fig. 3.23).

The third step, catalyzed by *phosphofructokinase*, is similar to the first in that it again involves phosphorylation of a primary alcohol by ATP, in this case converting fructose-6-phosphate to **fructose-1,6-diphosphate**, as shown in Fig. 3.24. This represents the standard formulation of this step, though some studies indicate that the actual product is the β-anomer rather than the α-anomer, though this distinction has little bearing on the very important chemistry that follows, since either

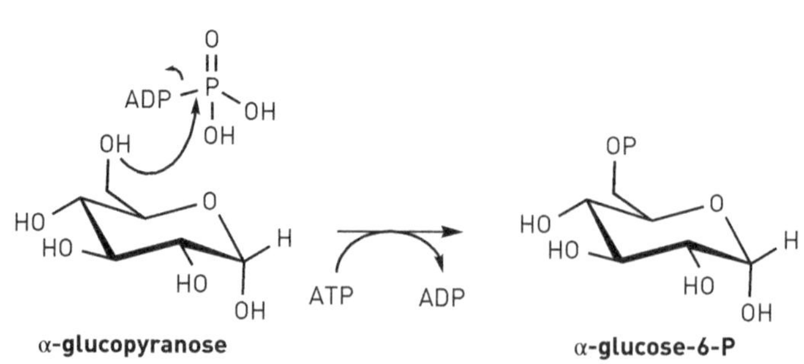

FIGURE 3.22
Glycolysis Step 1: Conversion of α-glucopyranose to α-glucose-6-phosphate.

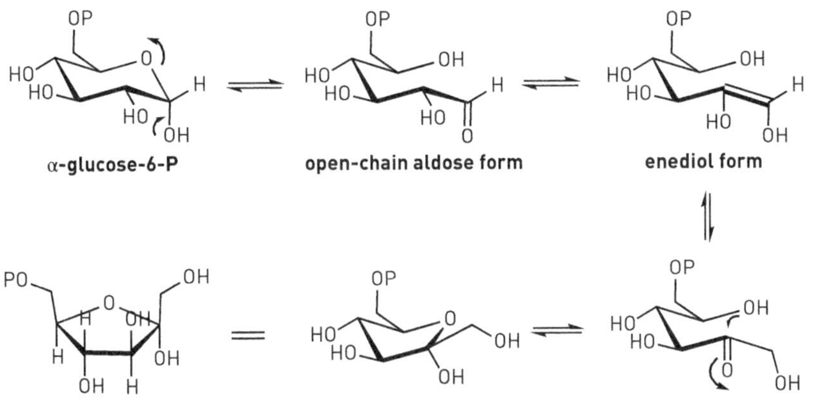

FIGURE 3.23
Glycolysis Step 2: Isomerization of α-glucose-6-phosphate to α-fructose-6-phosphate.

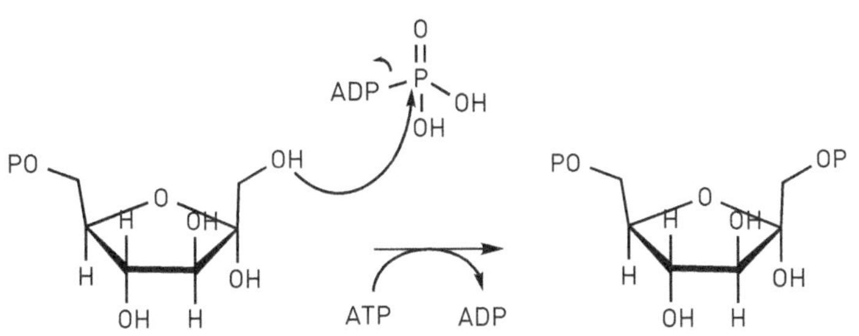

FIGURE 3.24
Glycolysis Step 3: Phosphorylation of α-fructose-6-phosphate to α-fructose-1,6-diphosphate.

anomeric form may be considered to be in equilibrium with the open-chain ketose form which participates in the next step.

For the fourth step in glycolysis, the open-chain ketohexose form of fructose 1,6-diphosphate undergoes what is essentially a **retroaldol addition** reaction catalyzed by *fructose diphosphate aldolase*. This key step is the only one in which a C–C bond is broken in the glycolysis pathway and therefore may be considered the most important step in the breakdown of glucose from a structural degradation perspective, since all previous and subsequent steps mainly involve simple condensations, isomerizations, or oxidations. The retroaldol addition reaction involves initial conversion of the fructose 1,6-diphosphate carbonyl group into an imine link using an enzyme side chain amino group, thereby holding the carbohydrate chain in a favorable position relative to other basic and acidic enzyme functions that facilitate the subsequent cleavage reaction, as shown in Fig. 3.25. The retroaldol process releases the first three-carbon carbonyl product, **glyceraldehyde-3-phosphate**, as well as an enzyme-bound three-carbon enamine component which undergoes subsequent isomerization and hydrolysis of the resulting imine linkage to regenerate

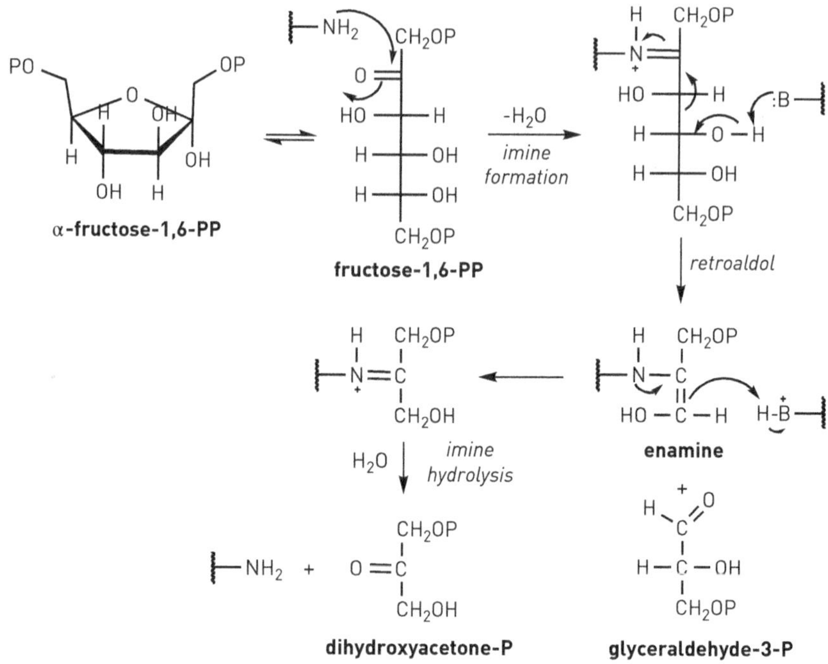

FIGURE 3.25
Glycolysis Step 4: Retro aldol cleavage of fructose-1,6-diphosphate to glyceraldehyde-3-phosphate and dihydroxyacetone phosphate.

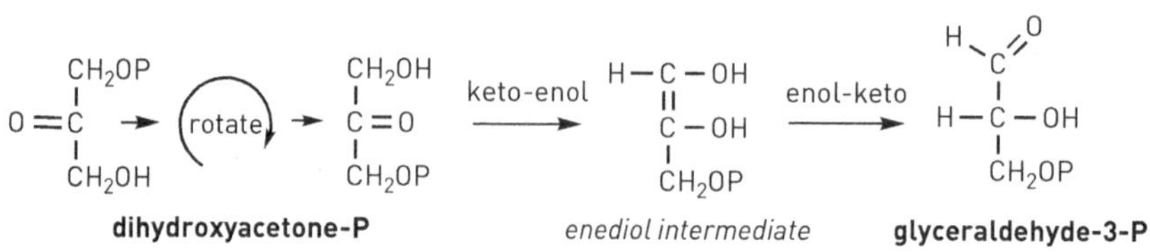

FIGURE 3.26
Glycolysis Step 5: Conversion of dihydroxyacetone phosphate to glyceraldehyde-3-phosphate.

the enzymatic amino group and release the second three-carbon carbonyl product, **dihydroxyacetone phosphate**.

The fifth step in the sequence, catalyzed by *triose phosphate isomerase*, converts **dihydroxyacetone phosphate** to **glyceraldehyde 3-phosphate** via the corresponding intermediate **enediol**, similar to the process seen in step 2. Thus, at this point in glycolysis, we note that one molecule of glucose has been broken down into two molecules of glyceraldehyde 3-phosphate (Fig. 3.26).

In the sixth step, glyceraldehyde-3-phosphate undergoes nucleophilic attack by a *glyceraldehyde phosphate dehydrogenase* thiol group to give a **hemithioacetal** intermediate which then undergoes oxidation by NAD^+ to give to the corresponding **thioester** (Fig. 3.27). Recall that thioesters are more reactive than

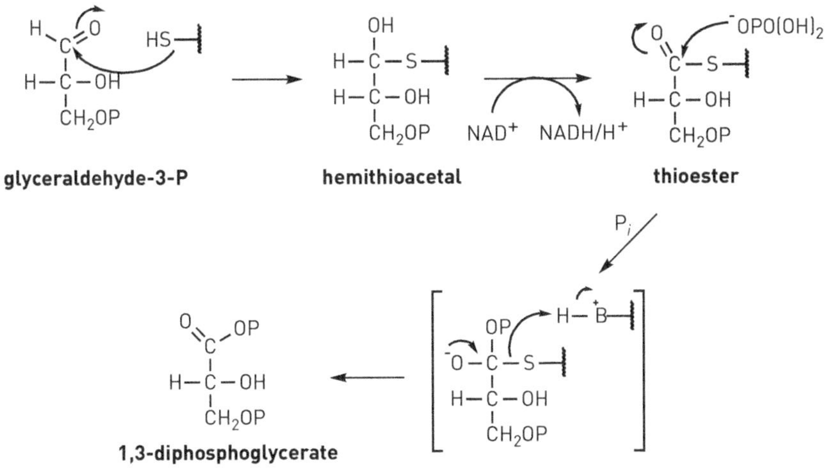

FIGURE 3.27
Glycolysis Step 6: Oxidation and phosphorylation of glyceraldehyde-3-phosphate.

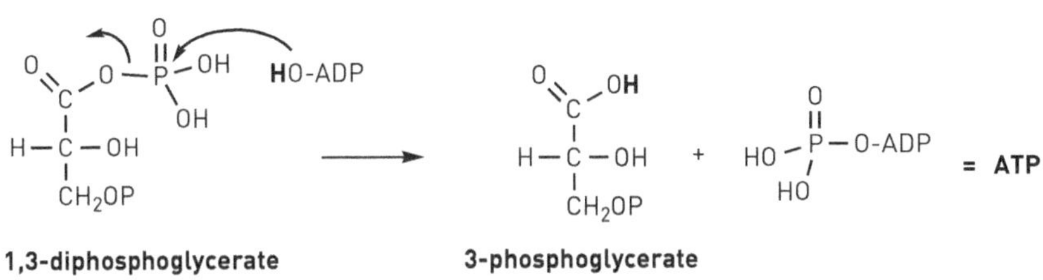

FIGURE 3.28
Glycolysis Step 7: Conversion of 1,3-diphosphoglycerate to 3-phosphoglycerate and ATP.

simple esters toward nucleophilic acyl substitution. Thus, we see inorganic phosphate ion (PO_4^{3-} or P_i) acting as a nucleophile to attack the thioester carbonyl, leading to a tetrahedral intermediate that expels the enzymatic thiol component, giving **1,3-diphosphoglycerate**: a phosphate ester at C-3 and an **acylphosphate** at C-1. The acylphosphate linkage is actually a mixed anhydride of a carboxylic acid and phosphoric acid, and as we saw earlier in Chapter 2, such mixed anhydrides are similar in reactivity to acid chlorides in terms of nucleophilic acyl substitutions reactions.

This enhanced reactivity of the acyl phosphate is then exploited in the seventh step of glycolysis for the transfer of the phosphate group from 1,3-diphosphoglycerate to ADP via a simple nucleophilic substitution catalyzed by *phosphoglycerate kinase*. This step generates **3-phosphoglycerate** as well as giving back ATP in the process (Fig. 3.28). This important step in glycolysis is an example of **substrate-level phosphorylation**, as it directly produces energy for the cell by generating ATP through transfer of phosphate from a phosphorylated substrate intermediate directly to ADP (again, note that all acids in these schemes are shown in nonionized form for simplicity's sake).

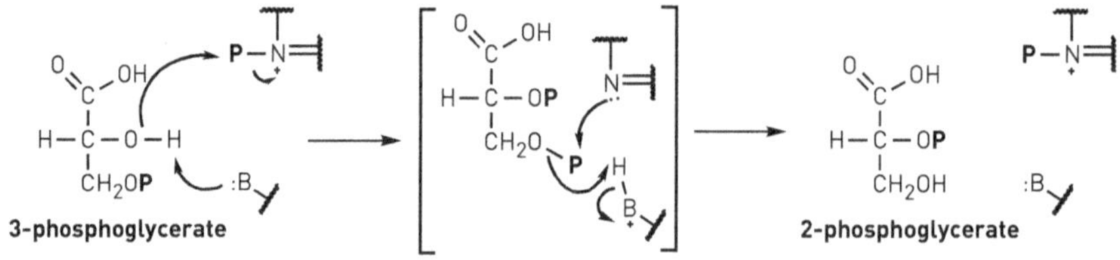

FIGURE 3.29
Glycolysis Step 8: Isomerization of 3-phosphoglycerate to 2-phosphoglycerate.

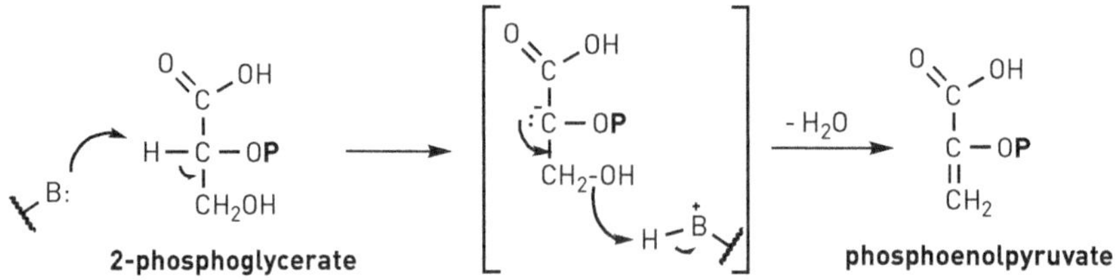

FIGURE 3.30
Glycolysis Step 9: Dehydration of 2-phosphoglycerate to phosphoenolpyruvate.

The eighth step in glycolysis is essentially an isomerization reaction catalyzed by *phosphoglycerate mutase* that converts 3-phosphoglycerate to **2-phosphoglycerate.** This transformation actually proceeds via formation of the intermediate **2,3-diphosphoglycerate** from which the C-3 phosphate group is removed. This is accomplished by a phosphorylated enzyme function which initially gives up its phosphate group to the C-2 hydroxyl group and then recovers it from the phosphate ester at C-3 to regenerate the active enzyme, as shown in Fig. 3.29.

The ninth step is an *enolase*-mediated dehydration reaction somewhat analogous to the dehydration of β-hydroxycarbonyl compounds from aldol-type processes to give α,β-unsaturated carbonyl compounds. The acidic α-proton of 2-phosphoglycerate is removed by a basic enzyme side chain (Fig. 3.30) to give the corresponding enolate ion in what is formally an E1cB-type mechanism. Loss of HO^- to a proton donor gives H_2O and the α,β-unsaturated product, **phosphoenolpyruvate.** This is essentially a phosphate ester of the enol form of pyruvic acid, as the name implies.

In the tenth and final step of glycolysis, the phosphate group of phosphoenolpyruvate is transferred to ADP by *pyruvate kinase*-catalyzed nucleophilic substitution, thereby regenerating ATP. The resulting enol form then tautomerizes to the keto form as expected, yielding the final product of glycolysis, **pyruvic acid**, as shown in Fig. 3.31. The reactivity of the enol phosphate ester here is greater than that of a simple alcohol phosphate due to the final driving force of tautomerization from enol to keto form; note also that this is another example of substrate-level phosphorylation.

If we add up all the balanced equations from each step, we find the overall process of glycolysis boils down to: $C_6H_{12}O_6$ + 2 NAD^+ + 2 ADP + 2 P_i ¾→ 2 $C_3H_4O_3$ + 2 NADH + 2 H^+ + 2 ATP + 2 H_2O.

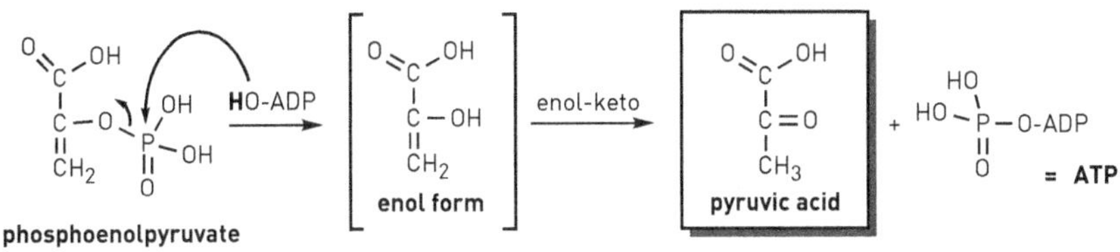

FIGURE 3.31
Glycolysis Step 10: Conversion of phosphoenolpyruvate to pyruvic acid and ATP.

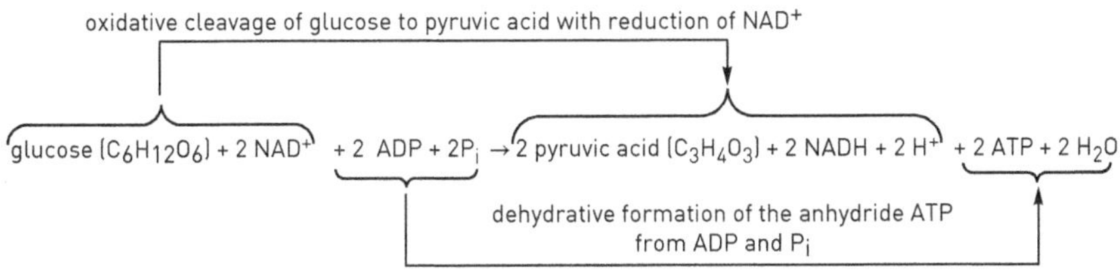

FIGURE 3.32
Overall transformations of glycolysis separated into redox and anhydride formation components.

Another way of viewing this chemically is to consider the overall glycolysis process as being one that essentially couples the oxidative breakdown of 1 mole of glucose into 2 moles of pyruvic acid (with concurrent reduction of 2 $NAD^+ \rightarrow$ 2 NADH + 2 H^+) with the dehydrative conversion of 2 moles of ADP and 2 moles of P_i (inorganic phosphate) into 2 moles of the more reactive "anhydride" ATP (with concurrent removal of 2 moles of H_2O). The two balanced processes are linked on each side of the equation in Fig. 3.32.

If we assume that ample glucose, ADP, and P_i are available, the limiting reagent in the process will be the cofactor NAD^+. This means that for glycolysis to continue, more NAD^+ will be required. In principle, this could be accomplished under aerobic conditions by regeneration of NAD^+ from NADH via direct reaction with molecular oxygen (2 NADH + 2 H^+ + $O_2 \rightarrow$ 2 NAD^+ + 2 H_2O); in practice, this is accomplished using protons and a complex, multistep sequence known as the **electron transport chain** that regenerates NAD^+ from NADH and eventually passes the hydrogen atoms and associated electrons along to O_2 which is discharged in reduced form as H_2O. The details are very complicated, so for now let's just briefly consider how NAD^+ is regenerated under conditions of limited O_2 supply or under anaerobic conditions.

WHAT HAPPENS TO THE PYRUVIC ACID FROM GLYCOLYSIS

The whole point of the oxidative breakdown of glucose is to supply energy in the form of ATP to power other cellular processes. In some situations, an oxygen supply may be limited for a period of time; one example is when the energy demands of vigorous exercise exceed the ability to supply oxygen to muscle cells for regenerating the NAD^+ needed to keep glycolysis going. There is an alternative way to recycle

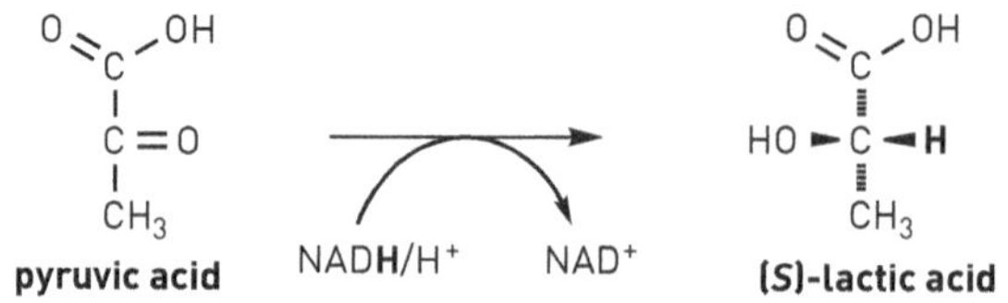

FIGURE 3.33
Stereospecific reduction of pyruvic acid to (*S*)-lactic acid and regeneration of NAD^+ from $NADH/H^+$.

NADH back to NAD^+ when O_2 availability as an electron acceptor (via the electron transport chain) is limited. What is needed is another electron acceptor and this constitutes another use for the pyruvic acid formed from glycolysis. Direct NADH/H^+ reduction of the ketone function of pyruvic acid to the corresponding alcohol produces **lactic acid** (Fig. 3.33), but also recycles NADH back to its oxidized form, NAD^+. In this way, ATP and energy production from glycolysis can continue for a while—at least until lactic acid build-up leads to the eventual aches and pains associated with muscle exhaustion. Note the reduction is *stereospecific*, with the hydride from NADH being delivered to the *Re* face of pyruvic acid, giving only the *S* enantiomer of lactic acid.

Of course, lactic acid has fates of its own, mainly to be transported to the liver where it is oxidized back to pyruvic acid to serve as a building block for the local production of more glucose by a pathway called **gluconeogenesis** ("new glucose formation"). While this nonphotosynthetic process is not precisely the reverse of glycolysis, it is fair to say that the two processes share many of the same steps and intermediates, so we need not examine it in further detail for our current purposes.

Under **anaerobic** conditions, some microorganisms like yeast can break down glucose by glycolysis, then convert the resulting pyruvic acid to ethanol and CO_2 in the familiar process of **fermentation** that forms the basis for production of beverage alcohol in beer and wine. The conversion of pyruvic acid is driven by TPP ylide, as shown in Fig. 3.34. After decarboxylation and protonation, **acetaldehyde** is released which is then reduced by $NADH/H^+$ to give **ethanol**. Retention of the by-product CO_2 in beverage fermentation is responsible for the carbonation of beer and sparkling wines, while removal of the residual CO_2 by degassing leads to "still" wines. When yeasts are used in baking, CO_2 production causes dough to expand or "rise" while the volatile ethanol produced is driven off during the baking process.

As we saw previously in Chapter 2, another fate of pyruvic acid closely related to the nonoxidative decarboxylation process shown above is the *oxidative* decarboxylation and conversion of pyruvic acid to **acetyl-CoA** by the action of TPP ylide and enzyme-bound lipoic acid. The process shown in Fig. 3.35 is *nonoxidative*, since the ketone carbonyl carbon in pyruvic acid ends up in the same oxidation state in the product acetaldehyde, while in the conversion to acetyl-CoA, its oxidation state increases to that of a thioester carbonyl carbon.

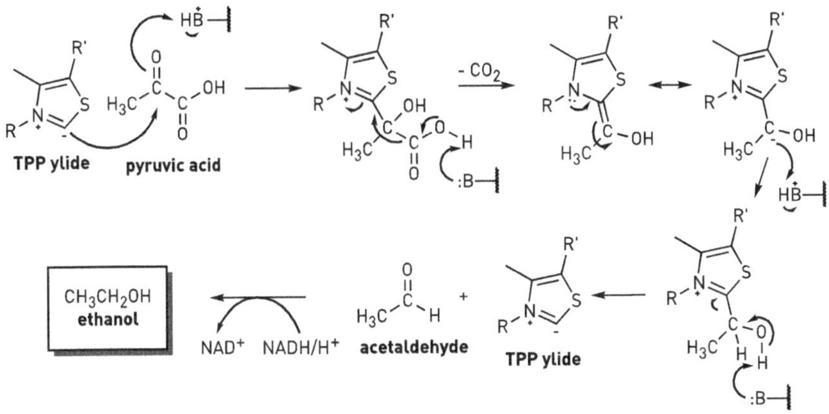

FIGURE 3.34
Decarboxylation of pyruvic acid by TPP ylide, generation of acetaldehyde and its subsequent reduction to ethanol by NADH/H^+ in anaerobic fermentation.

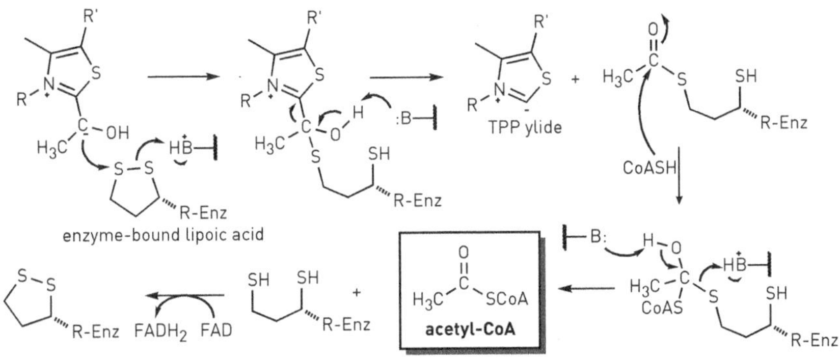

FIGURE 3.35
Formation of acetyl-CoA via decarboxylated TTP-bound intermediate from pyruvic acid.

THE CITRIC ACID CYCLE: ANOTHER 10-STEP PROGRAM

While production of acetyl-CoA represents one of the more important uses of pyruvic acid, another essential sequence that utilizes both pyruvic acid and acetyl-CoA is the so-called **citric acid cycle**. This indispensable closed loop of reactions is used by all aerobic organisms for the oxidative conversion of the acetate in acetyl-CoA (derived not only from metabolism of carbohydrates, but also from fats or proteins) into CO_2. The cycle also generates important precursors for the biosynthesis of specific amino acids as well as some of the NADH required by numerous other biochemical processes. The pivotal nature of the cycle in terms of cellular metabolism indicates that it was surely one of the earliest biochemical mechanisms associated with the evolution of living systems.

To gain an appreciation for the subsequent workings of the citric acid cycle, we will begin by looking briefly at how pyruvic acid and acetyl-CoA can come together to set things into motion. This starts by a carboxylation reaction of pyruvic

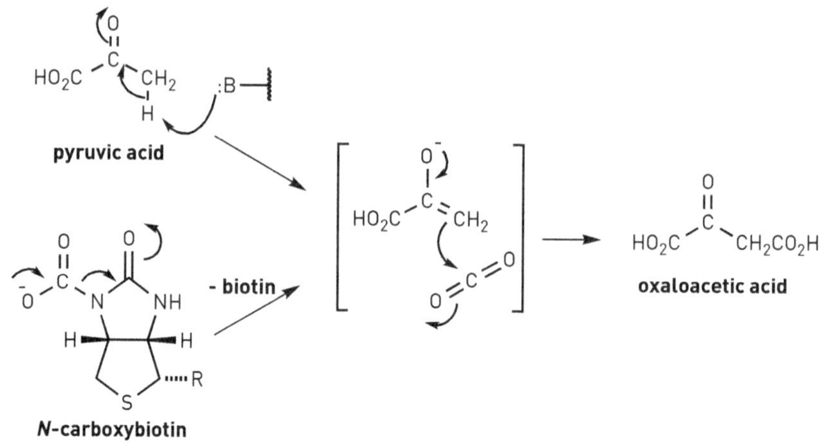

FIGURE 3.36
N-carboxybiotin and carboxylation of pyruvic acid enolate ion to give oxaloacetic acid.

acid catalyzed by *pyruvate carboxylase* (for the purposes of depicting the chemistry and intermediate products of the citric acid cycle, we will stick with showing all carboxylic acids and their names in nonionized form). Here, the universal carboxyl carrier **N-carboxybiotin** releases CO_2 to be trapped by the ketone enolate ion of pyruvic acid, leading to the carboxylated product known as **oxaloacetic acid** (Fig. 3.36). This important intermediate is the starting point for the process of gluconeogenesis referred to earlier and as we will see, is also produced by the citric acid cycle itself, making this process a closed loop of reactions, since this starting reactant is also the final product of the process. Once we have seen how these starting materials set the process in motion, we can examine the overall cycle in a single diagram. It should be pointed out that seven of the ten intermediates involved in the cycle, including oxaloacetic acid, are also available from other metabolic processes such as the degradation of amino acids and may enter the cycle at any point.

Next comes the crucial step in which acetyl-CoA reacts with oxaloacetate. This reaction is essentially a *citrate synthase*-catalyzed aldol addition process in which the enolate ion of acetyl-CoA attacks the ketone carbonyl of oxaloacetic acid, yielding the intermediate product **(S)-citryl-CoA** which is then hydrolyzed to liberate **citric acid**, the product from which the cycle takes its name, and HSCoA (Fig. 3.37). Note that while (*S*)-citryl-CoA is chiral, its hydrolysis product, citric acid, is an achiral compound.

While we are not emphasizing the role of the enzymes here, this step reminds us of the important specificity of such enzyme-mediated processes. Since oxaloacetic acid is a β-dicarbonyl compound, we would ordinarily expect its carbonyl α-protons to be much more acidic than those of acetyl-CoA, which is a simple thioester. This means that this addition reaction could never take place under ordinary base-catalyzed conditions, since the oxaloacetic acid enolate ion would form much more readily. Here, the enzyme holds both reactants in a specific spatial relationship that

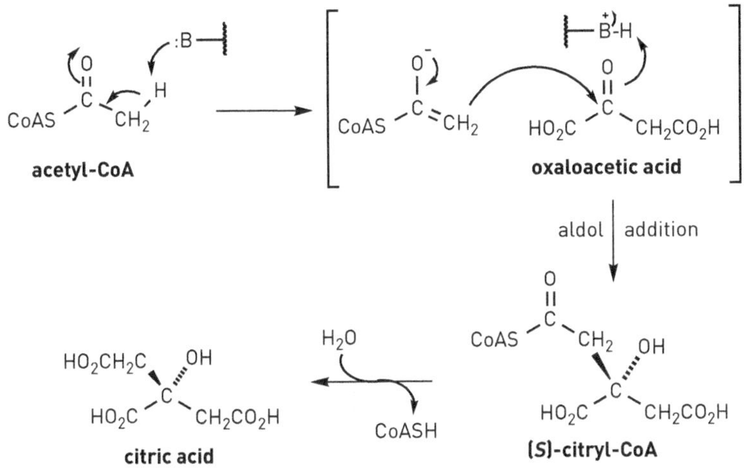

FIGURE 3.37
Aldol addition reaction of acetyl-CoA and oxaloacetate to produce citrate.

brings the thioester into intimate contact with the basic site needed for its deprotonation while simultaneously providing the acidic site necessary to protonate the oxygen of the oxaloacetic acid ketone carbonyl as the nucleophilic attack proceeds.

Now that we have initiated the citric acid cycle by the introduction of oxaloacetic acid and acetyl-CoA, we can take a look at the overall process to see how it consumes acetyl-CoA, discharging its carbons as CO_2 and its hydrogens to NADH and $FADH_2$ as shown in Fig. 3.38. It is convenient to think of the cycle as a sort of engine that uses acetyl-CoA for fuel, while discharging CO_2 and water. As long as there is fuel available, the engine will keep running, though this engine can also run on any of the intermediate products for fuel, most of which are available through the degradation of amino acids and other sources.

Referring to Fig. 3.39, we can see that step 2 in the process is simply a dehydration reaction, as often follows an aldol-type addition process. This reaction, catalyzed by the enzyme *aconitase*, is a stereospecific *anti* elimination, producing the α,β-unsaturated dicarbonyl product, ***cis*-aconitic acid**. This is followed by step 3, which is a nucleophilic conjugate addition of water at C-2 of the α,β-unsaturated dicarbonyl catalyzed by the action of the same enzyme as in step 2. This *anti* addition is also stereospecific, with the OH of water being delivered to the *Re* face at C-2 and the proton at the *Re* face of C-3 in *cis*-aconitic acid, leading to the isomeric product **(2*R*,3*S*)-isocitric acid**. Thus, the stereospecificity of both the elimination and the addition processes further emphasizes the careful regio- and stereochemical control exerted by the mediating enzyme.

Step 4 is the simple *isocitrate dehydrogenase*-catalyzed NAD^+ oxidation of the secondary alcohol function at C-2 of isocitric acid to give the corresponding ketone, **oxalosuccinic acid**, while also producing the first of three equivalents of NADH from the cycle (Fig. 3.40). This β-ketoacid product then undergoes the expected

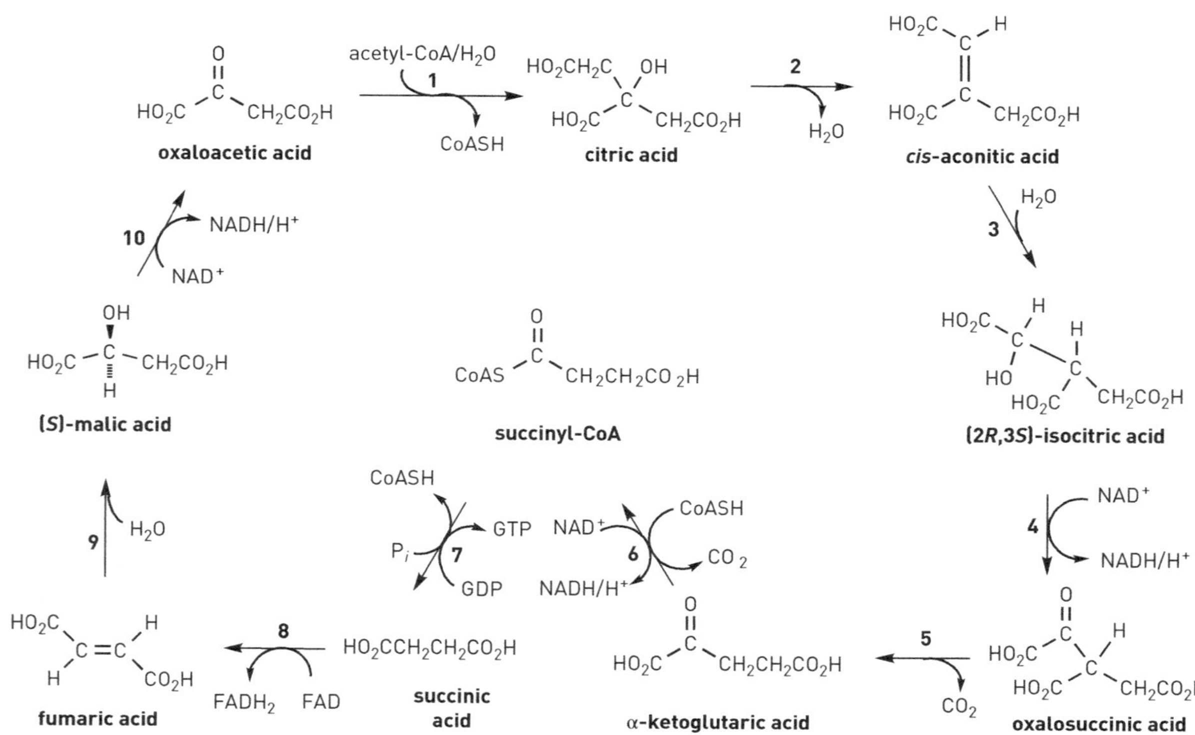

FIGURE 3.38

The 10 steps of the citric acid cycle.

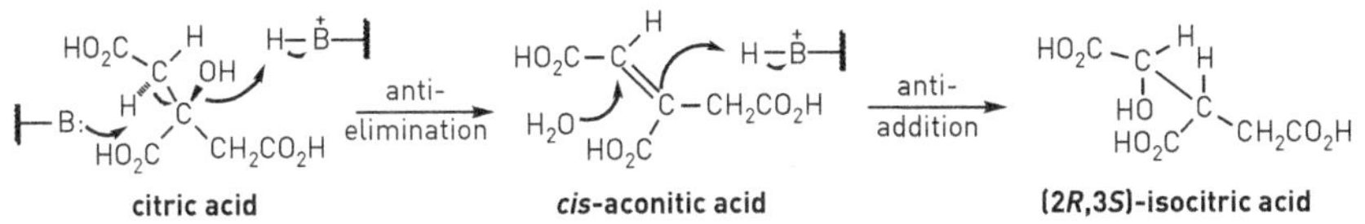

FIGURE 3.39
Stereospecific dehydration of citric acid and hydration of *cis*-aconitic acid to give isocitric acid.

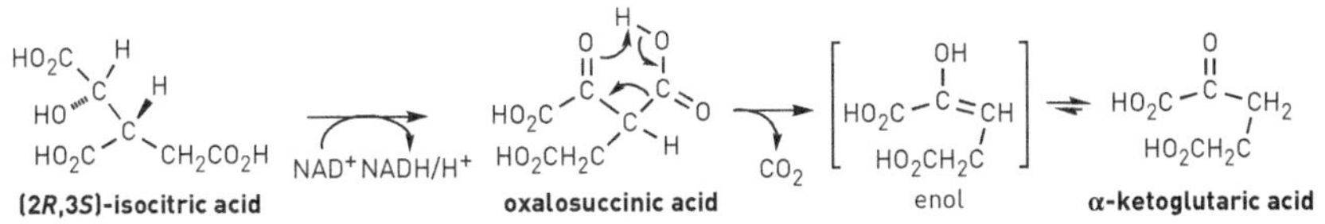

FIGURE 3.40
Oxidation of isocitric acid to oxalosuccinic acid and decarboxylation to α–ketoglutaric acid.

decarboxylation in step 5. While enzyme-assisted, this decarboxylation would probably occur spontaneously as it does for most β-keto acids; in this instance, the decarboxylation generates the first of the two equivalents of CO_2 produced in each turn of the cycle. We also note that the product, **α-ketoglutaric acid**, is a common biomolecule that could also be introduced into the cycle from other sources; it could also be diverted to the production of the amino acid **glutamic acid** via transamination.

The next two steps in the cycle are somewhat more complex and require some additional clarification. Step 6, catalyzed by a multienzyme complex, represents the second decarboxylative step in the cycle, and while not as simple as the decarboxylation in the previous step, is actually just another example of the oxidative decarboxylation of α-ketoacids to acylCoA derivatives facilitated by TPP ylide, CoASH, and enzyme-bound lipoic acid, as seen previously in the conversion of pyruvic acid to acetyl-CoA (see Figs. 3.35 and 3.36). Thus, this sequence converts α-ketoglutaric acid to **succinyl-CoA** with concurrent loss of CO_2. The role of NAD^+ in this step is to regenerate FAD from the $FADH_2$ produced during oxidative regeneration of the lipoic acid cyclic disulfide ring (Fig. 3.41).

Step 7 in this sequence, catalyzed by *succinyl-CoA synthetase*, is very similar to the seventh and eighth steps of the glycolysis pathway (see Figs. 3.28 and 3.29) which involved conversion of a thioester to an intermediate acylphosphate by nucleophilic acyl substitution by inorganic phosphate (P_i) which in turn transferred the reactive phosphate group to ADP, converting it to ATP and releasing the corresponding carboxylic acid. Here, the thioester is **succinyl-CoA** rather than an enzyme-bound thioester, and the resulting intermediate acylphosphate undergoes nucleophilic attack by GDP (guanosine diphosphate) rather than by ADP, producing GTP and **succinic acid** (Fig. 3.42). While this is also an example of substrate-level phosphorylation as seen previously in glycolysis, there is no obvious reason why GDP is used here rather than ADP.

In the final steps of the citric acid cycle (Fig. 3.43), we see first the oxidative dehydrogenation of succinic acid to **fumaric acid** by the action of FAD and catalyzed

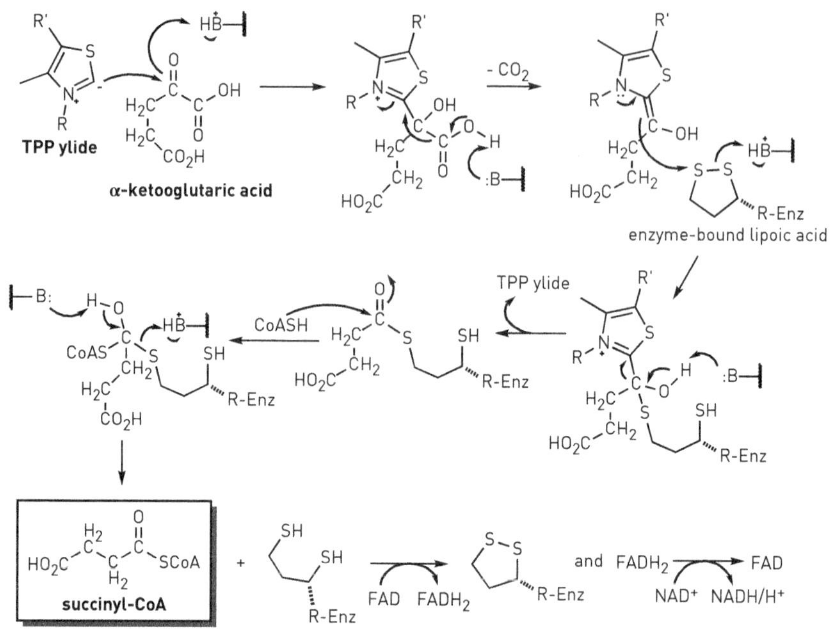

FIGURE 3.41
Formation of succinyl-CoA via decarboxylated TTP-bound intermediate from α-ketoglutaric acid.

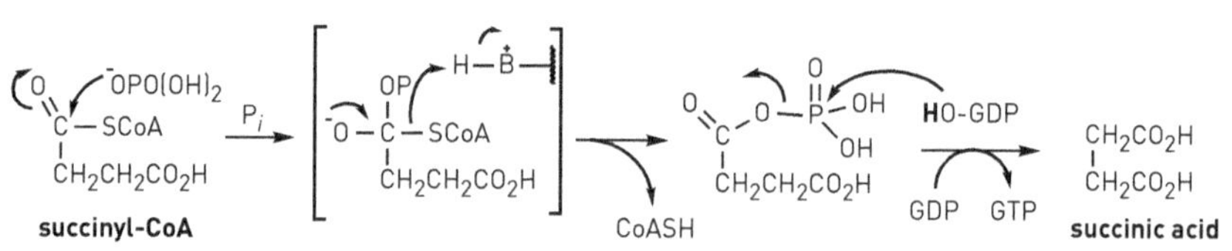

FIGURE 3.42
Conversion of succinyl-CoA to succinic acid coupled to conversion of GDP to GTP.

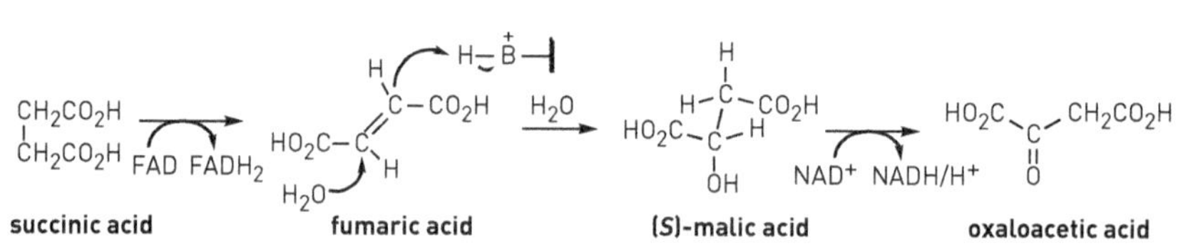

FIGURE 3.43
Oxidation of succinic to fumaric acid, hydration to (*S*)-malic acid, and final oxidation to oxaloacetic acid.

by *succinate dehydrogenase* in step 8. Once again, this is a stereospecific *anti* elimination, removing the elements of H_2. This is followed by a stereospecific conjugate addition of water to the α,β-unsaturated dicarboxylic acid catalyzed by *fumarase* in step 9 to produce **(S)-malic acid** in a process quite similar to the *anti* addition of water in step 4 that produced isocitric acid. We complete one full turn of the cycle in *malate dehydrogenase*-catalyzed step 10 by regenerating the **oxaloacetic acid** we originally started with in step 1 by NAD^+ oxidation of the secondary alcohol function of (*S*)-malic acid to the corresponding ketone.

So, one turn of the citric acid cycle accomplishes the degradation of a single acetyl-CoA into 2 CO_2 while also producing one each of HSCoA, GTP, and $FADH_2$, along with 3 NADH and 3 H^+. Now that we have a better appreciation for glycolysis and one of the important fates of its principal product, pyruvic acid, let's turn to another pathway for metabolism of glucose that also provides some useful intermediate products.

THE PENTOSE PHOSPHATE PATHWAY: SEVEN ALTERNATIVE STEPS TO SOME FAMILIAR INTERMEDIATES

While not as significant as glycolysis for the degradation of glucose, the pentose phosphate pathway is nevertheless still important for a number of different reasons. In animals, liver and mammary gland tissues use this pathway to metabolize a significant fraction of the glucose they process, though other tissues, such as skeletal or heart muscle, rely almost completely on glycolysis. The main reasons for use of the pentose phosphate cycle are to generate ribose-5-phosphate needed for the production of ribonucleic acids, to produce erythrose-4-phosphate for the production of aromatic amino acids in plants, and to convert $NADP^+$ to NADPH, a reducing agent whose function and usage are similar to those of NADH. For our purposes, we mainly want to gain an appreciation for the structures and nomenclature associated with the intermediates produced as well as an understanding of the stoichiometry, which at first glance appears unusual, as can be seen in the overall equation for metabolism of one mole of glucose via this pathway:

glucose-6-P + 2 $NADP^+$ + H_2O → 2/3 fructose-6-P + 1/3 glyceraldehyde-3-P
+ CO_2 + 2 NADPH + 2 H

We can think of this as a sort of partial degradation pathway which produces some CO_2 but also produces intermediate products that may be diverted to other uses as well as products that may ultimately be passed along to glycolysis for further degradation. An overview of the entire pathway is shown in Fig. 3.44.

The first step in the pathway is the *glucose-6-phosphate dehydrogenase*-catalyzed $NADP^+$ oxidation of the β-anomeric hydroxyl of glucose-6-phosphate at C-1 to give the corresponding **6-phosphogluconolactone** which is hydrolyzed in the second step to give the open-chain form, **6-phosphogluconic acid**. The third step, (Fig. 3.45), is catalyzed by *phosphogluconate dehydrogenase* and involves the regiospecific oxidation of the C-3 hydroxyl by $NADP^+$, giving an intermediate β-ketoacid which then decarboxylates as expected to give **ribulose-5-phosphate**. Note that steps 1 and 3 are responsible for the production of the two equivalents of NADPH produced per equivalent of glucose metabolized by this pathway.

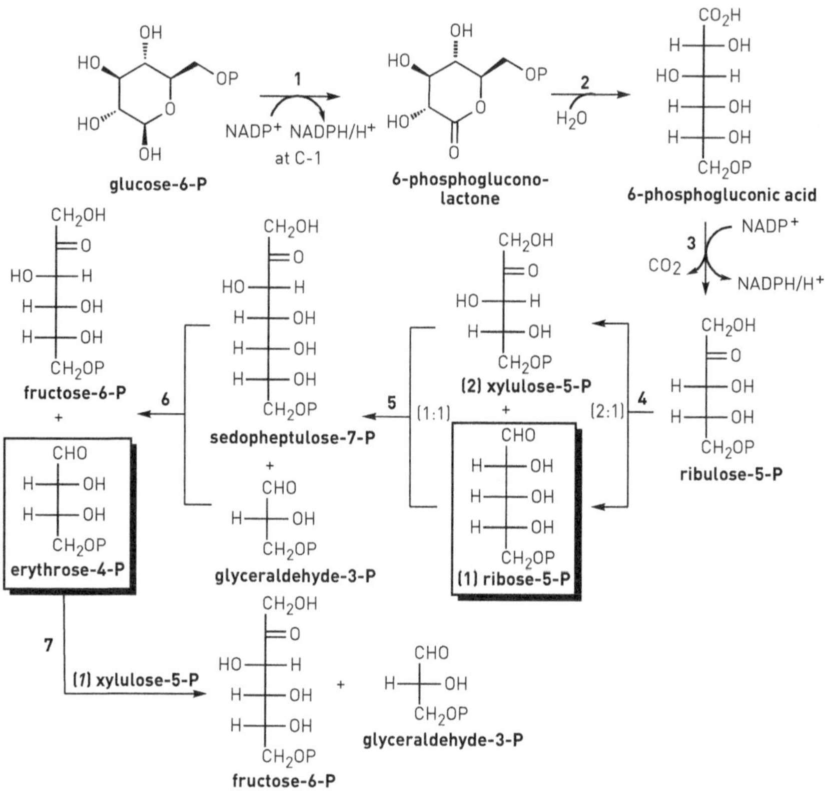

FIGURE 3.44
The seven steps of the pentose phosphate pathway.

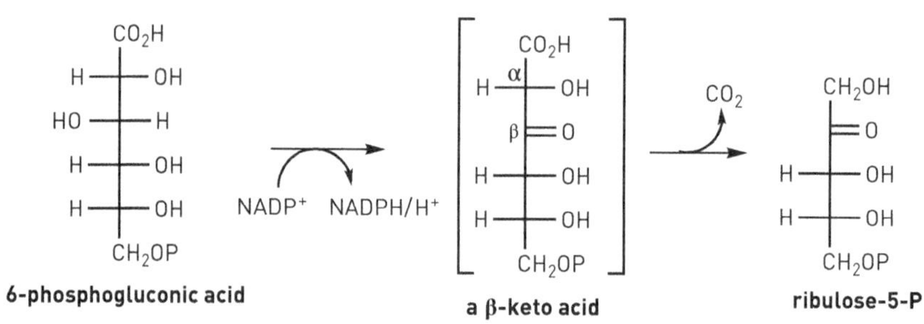

FIGURE 3.45
Step 3 of the pentose phosphate pathway.

The fourth step in the sequence involves a series of isomerizations of ribulose-5-phosphate, each involving the generation of isomeric products derived from two different enediol forms of ribose-5-phosphate, namely the corresponding 1,2-enediol and 2,3-enediol forms, as shown in Fig. 3.46. These isomerizations, each catalyzed by a specific *epimerase* enzyme, lead to the corresponding carbohydrates **xylulose-5-phosphate** and **ribose-5-phosphate** in a relative ratio of 2:1, respectively, for bookkeeping purposes (or, 3 moles of ribulose-3-phosphate will be isomerized to 2 moles of xylulose-3-phosphate and 1 mole of ribose-3-phosphate). The actual amount and

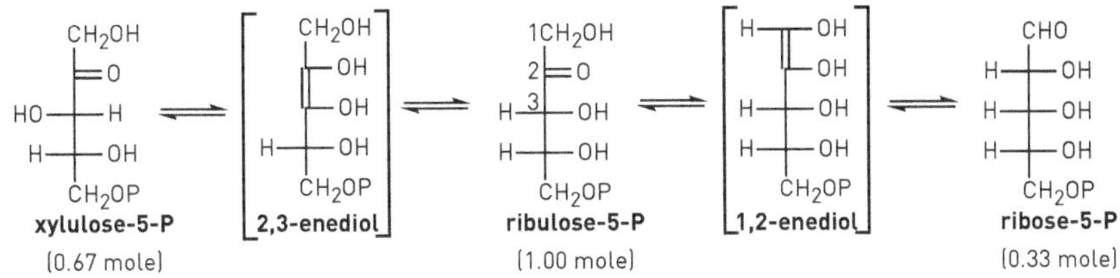

FIGURE 3.46
Step 4 of the pentose phosphate pathway: isomerizations of ribulose-5-P.

fate of the products will depend in part on local demands for ribose-5-phosphate production in connection with ribonucleic acid synthesis.

Excess ribose-5-phosphate can then be converted to other products by the reactions in the subsequent steps. In the fifth step, catalyzed by a *transketolase* enzyme, TPP ylide adds to the ketone carbonyl of half of the xylulose-5-phosphate from step 4 in a manner similar to that seen in the TPP reaction with α-ketoacids (Fig. 3.47, i)); the resulting intermediate undergoes a retroaldol reaction which leads to formation of **glyceraldehyde-3-phosphate**; the TPP-bound two-carbon fragment originally from xylulose-5-P then acts as a nucleophile, attacking the aldehyde carbonyl of ribose-5-phosphate, leading to an intermediate that undergoes deprotonation, followed by expulsion of TPP ylide and simultaneous production of the ketoheptose product, **sedoheptulose-7-phosphate** (Fig. 3.47, ii)). Thus we have the reaction of one equivalent of ribose-5-P and one equivalent of xylulose-5-P producing one equivalent of glyceraldehyde-3-P and sedoheptulose-7-P. In this process, we see an aldotriose produced from a ketopentose by removal of a two-carbon fragment which is then added to an aldopentose to produce a ketoheptose. Also note that half of the xylulose-5-P produced in step 4 remains available for use in step 7, but for now we have to carry out step 6 to produce a product that will ultimately react with this remaining xylulose-5-P.

In the sixth step of the sequence, a *transaldolase* enzyme amino group condenses with the ketone carbonyl of sedoheptulose-7-phosphate, producing an iminium ion derivative (a protonated imine in this case) which then undergoes a retroaldol cleavage (nearly identical to that seen in the fourth step of the glycolysis pathway). This produces a three-carbon enamine intermediate as well as the important product **erythrose-4-phosphate** (Fig. 3.48, i)) which will ultimately play a key role in the plant biosynthesis of aromatic amino acids and other natural products from the *shikimic acid pathway* (Chapter 6); the erythrose-4-P will also serve as a key reactant in the seventh and final step of the pentose phosphate pathway by combining with unused xylulose-5-P produced in the fourth step.

In the next part of step 6 (Fig. 3.48, ii), the three-carbon enamine intermediate acts as a nucleophile, attacking the aldehyde carbonyl of glyceraldehyde-3-phosphate (from step 5) in an aldol-type addition process. This step knits together the two three-carbon fragments to produce a six-carbon carbohydrate imine derivative which upon hydrolysis yields **fructose-6-phosphate.** For further metabolic degradation, this may be passed along to the glycolysis pathway, along with the

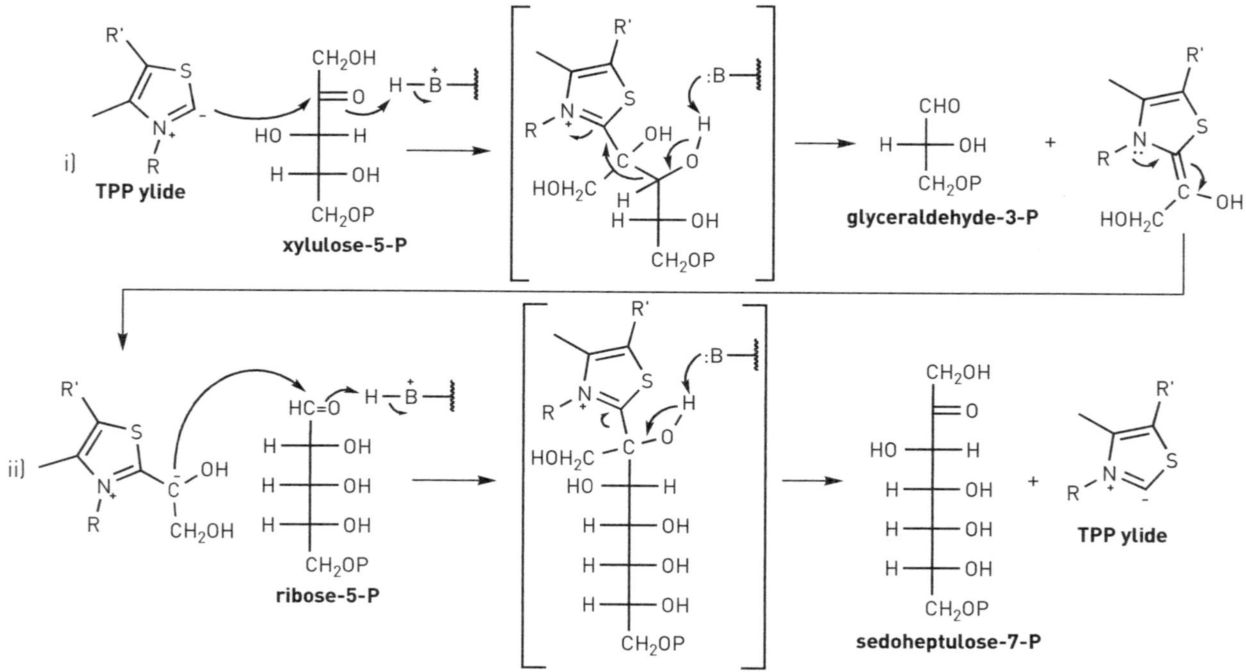

FIGURE 3.47
Step 5 of the pentose phosphate pathway: production of glyceraldehyde-3-P and sedoheptulose-7-P.

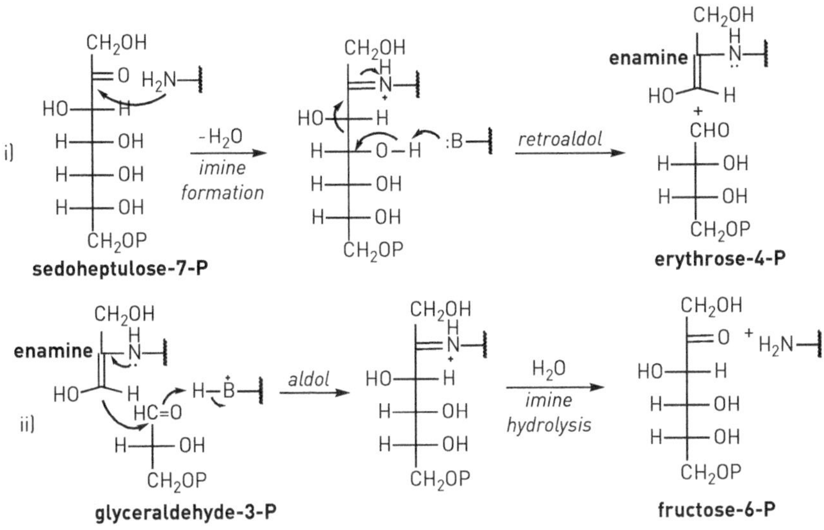

FIGURE 3.48
Step 6 of the pentose phosphate pathway: production of erythrose-4-P and fructose-6-P.

fructose-6-P and glyceraldehyde-3-P that are the products in the seventh and final step, which we can now examine.

As mentioned above, the pathway is completed in the seventh and final step in which the remaining unused xylulose-5-phosphate from step 4 reacts with TPP ylide in a sequence that is identical mechanistically to that in step 5 (Fig. 3.49).

Thus, the *transketolase*-mediated retroaldol process again releases the first final product of the sequence, **glyceraldehyde-3-phosphate**, along with the usual TPP-bound nucleophilic two-carbon fragment. This fragment then attacks the aldehyde carbonyl of the erythrose-3-phosphate derived from the previous step. After knitting together these two-carbon and four-carbon fragments, a subsequent deprotonation and elimination of TPP ylide yields the second final product of the step, **fructose-6-phosphate**. Again, these two products are familiar ones which now may enter the glycolysis pathway for final degradation to pyruvic acid.

THE BIG PICTURE

As we have seen, the breakdown of glucose via glycolysis ultimately leads to the production of pyruvic acid, and our interest in the process stemmed in part from the various fates of pyruvic acid itself which we have looked at in some detail. To gain a better perspective of where all of this is leading, a general overview of the relationships of the various pathways we have examined is presented in Fig. 3.50. While this may look hopelessly complex at first glance, if we look more closely we can see how this flow chart actually organizes and simplifies a lot of information while pointing the way to where we will be going later in the text in Chapters 4–7.

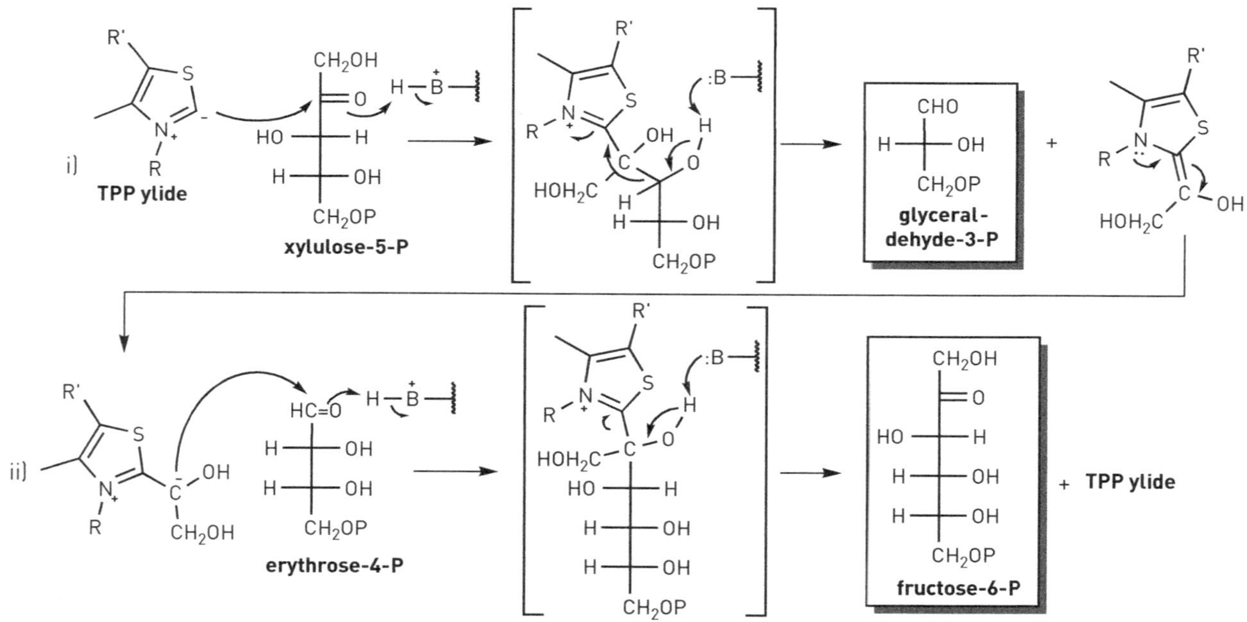

FIGURE 3.49
Step 7 of the pentose phosphate pathway: final production of glyceraldehyde-3-P and fructose-6-P.

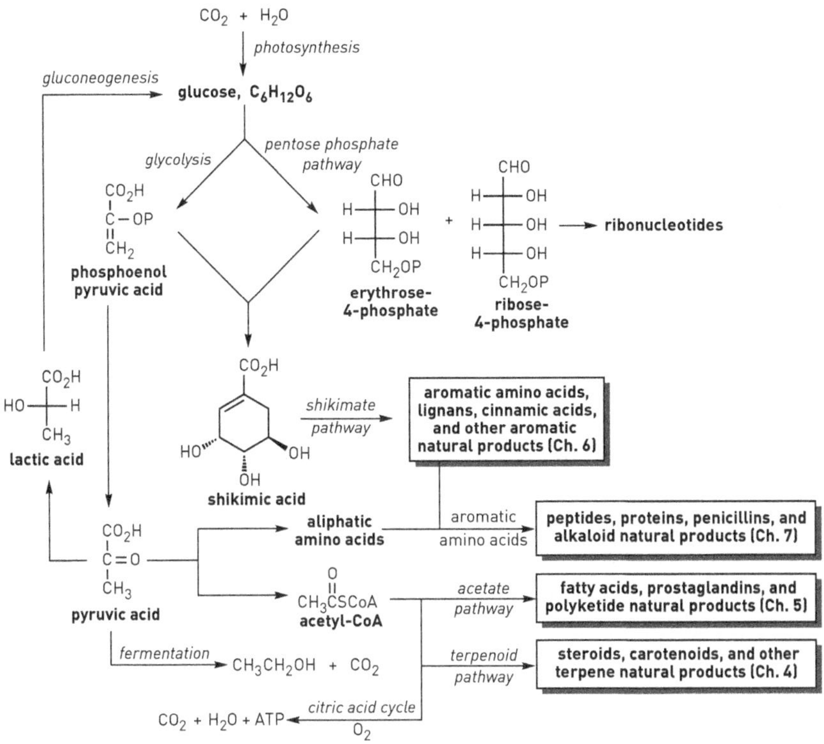

FIGURE 3.50
Breakdown of glucose and use of its components in various biosynthetic pathways.

As we saw, reduction of the ketone carbonyl of pyruvic acid gives **lactic acid,** which can then be used to resynthesize glucose in **gluconeogenesis,** a process very nearly the reverse of glycolysis in terms of the chemistry of its individual steps. Alternatively, under anaerobic conditions, such as in yeasts, pyruvic acid underwent reduction and decarboxylation, producing ethanol and CO_2 in the familiar process of **fermentation.** We then saw how **acetyl-CoA** (prepared from pyruvic acid by the action of TPP ylide and enzyme-bound lipoic acid) moved into the citric acid cycle where its two carbons were subsequently degraded to CO_2. Later on we'll see how acetyl-CoA is used as one of the fundamental building blocks for the biosynthesis of **terpene** natural products (The Terpenoid Pathway, Chapter 4) as well as fatty acids, prostaglandins and a wide variety of **polyketide** natural products (The Acetate Pathway, Chapter 5). The alternative route for glucose breakdown, the **pentose phosphate pathway,** was also significant for a number of reasons, but mainly for its production of ribose-5-phosphate, needed for the synthesis of ribonucleotides and erythrose-4-phosphate used by plants for the production of **shikimic acid,** from which they then produce the aromatic amino acids **phenylalanine** and **tyrosine,** two fundamental building blocks of the so-called **shikimate** natural products (The Shikimic Acid Pathway, Chapter 6). The aromatic amino acids, when combined with aliphatic amino acids derived from pyruvic acid and intermediates from the citric acid cycle, lead not only to familiar peptides, proteins, and important modified peptides such as the **penicillins,** but also to the vast array of nitrogen-containing natural products known as the **alkaloids** (Biosynthesis of Alkaloids and Related Compounds, Chapter 7).

Since amino acids will clearly play a prominent role in much of what comes later in our text, we will devote the remainder of this chapter to an overview of their structure and nomenclature, followed by an abbreviated look at how a number of them are produced from some of the carbohydrate-derived intermediate products produced in the pathways we have just studied. The biosynthesis of a number of others, especially the aromatic amino acids, will be deferred to later chapters as topics of special significance in terms of their relationship to specific pathways or classes of natural products.

AMINO ACIDS: MORE IMPORTANT PRIMARY METABOLITE BUILDING BLOCKS FOR BIOSYNTHESIS

Amino acids are, of course, among the most important components of all biological systems, since they combine to make peptides and proteins, the biopolymers that constitute hair, nails, connective tissue, tendons, antibodies, and so on. More importantly, all enzymes are specialized proteins with very specific catalytic properties essential to the functioning and integrated operation of all cellular processes. Clearly, life goes nowhere without amino acids. But our limited focus here will be on their organic chemistry: their structure and functions as *individual* amino acids, both from the standpoint of their biosynthesis as well as their later use as important raw materials for the construction of a variety of important *nonpeptide* natural products in subsequent chapters. The acid–base properties of amino acids, their analysis, purification, and catabolism, along with the intimate and seemingly infinite details of peptide, protein, and enzyme structure and function are all core components of the study of biochemistry which, except for a very brief review, we must leave to more advanced texts in the field.

To begin our brief tour, we turn to Fig. 3.51 which lists the names, abbreviations, and structures of the 20 **standard α-amino acids**, in alphabetical order. The names of the **essential amino acids**, those which humans cannot biosynthesize and therefore must obtain in the diet, are indicated with an asterisk (*). This is by no means an exhaustive list of all amino acids (well over 100 have been isolated from various sources), but rather a list of the so-called **proteinogenic amino acids,** those which are genetically coded for and produced by cellular machinery for peptide and protein synthesis. Some lists also include *pyrrolysine* and *selenocysteine*, though the former is rather rare and the latter is not truly proteinogenic.

With the exception of glycine, all the α-amino acids are produced in nature as the corresponding (*S*)-enantiomers at the amino-bearing carbon (note that (2*S*,3*R*)-threonine and (2*S*,3*S*)-isoleucine have a second chirality center). When drawn in Fischer projection form with their carboxyl group at the top of the vertical chain, the α-amino group at the chirality center of the standard amino acids will be found pointing to the left, leading to their being designated as **L-amino acids** in the same way that

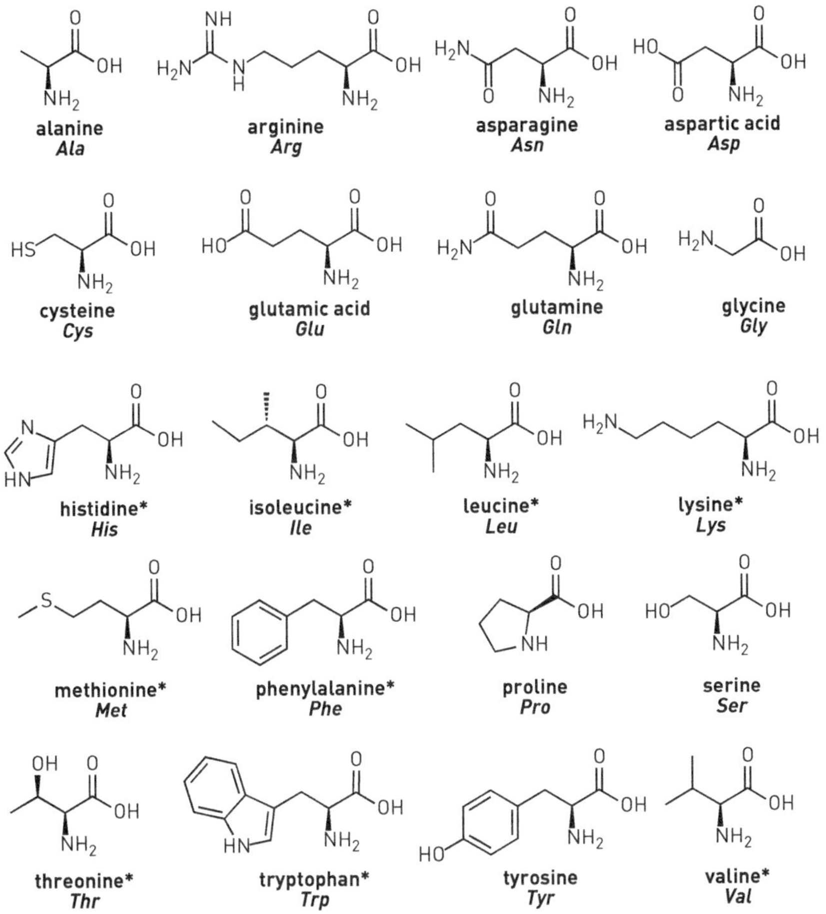

FIGURE 3.51
Names, abbreviations, and structures of the 20 standard α-amino acids (*essential).

D-carbohydrates are designated as such when their lowermost chiral carbon hydroxyl group points to the right in a Fischer projection (Fig. 3.52). We will ordinarily represent these structures in "amino acid" form as shown here, rather than in their more appropriate "zwitterionic" form, for simplicity's sake, when used in mechanistic schemes.

To consider the biosynthesis of amino acids, we can use the flow chart in Fig. 3.53 which makes it clear that all trace their origin back to one or more of the basic carbohydrate intermediates that we encountered in glycolysis, the citric acid cycle, or the pentose phosphate pathway. Using this chart as a starting point, we will spend a bit of time looking at some of the reactions involved in the biosynthesis of several of the amino acids, though as mentioned above, we will defer detailed treatment of the aromatic amino acids (His, Trp, Phe, and Tyr) and some of the others to later chapters in the text.

As can be gleaned from the flow chart, all nonessential amino acids are synthesized from common intermediates from glycolysis and the citric acid cycle: pyruvic acid, 3-phosphoglycerate, oxaloacetic acid, and α-ketoglutaric acid. Similarly,

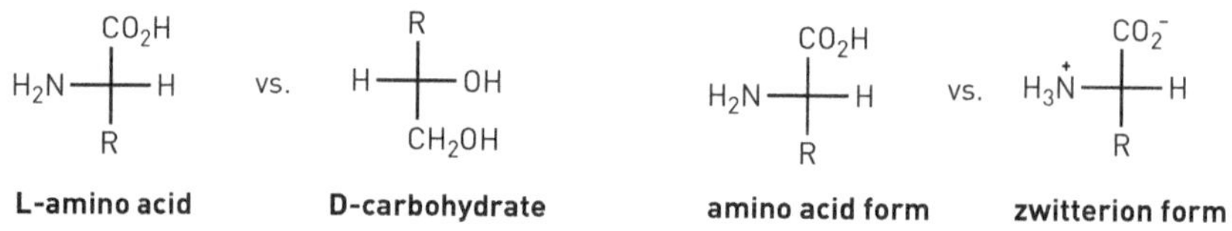

FIGURE 3.52
Stereoisomeric forms and zwitterionic forms of α-amino acids.

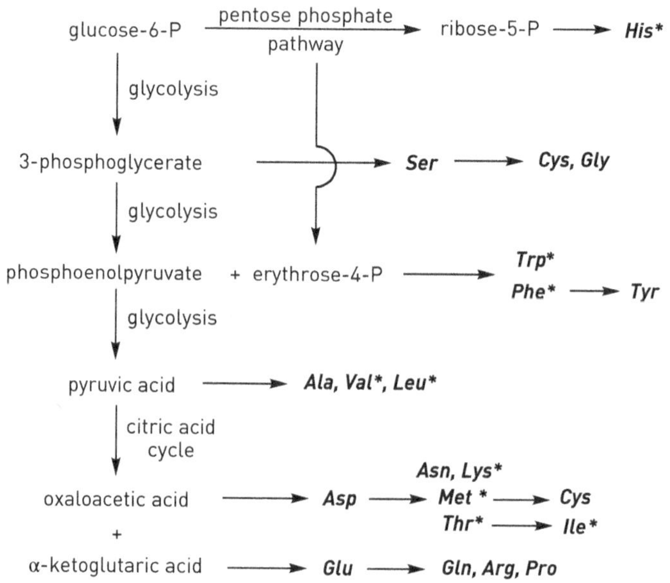

FIGURE 3.53
Flow chart for amino acid biosynthesis from carbohydrate metabolism components (*essential).

some of the essential amino acids come from pyruvic acid or phosphoenolpyruvate, while others may come from the pentose phosphate pathway products erythrose-4-phosphate or ribose-5-phosphate, and still others come from subsequent transformations of other amino acids. For the essential amino acids, biosynthesis is limited to plants or microorganisms and may occur by different pathways in different species, in stark contrast to the pathways associated with carbohydrate or lipid metabolism, which are nearly universal. It is presumed that the loss of these pathways in mammals is a result of various evolutionary mutations over time.

BIOSYNTHESIS OF SERINE: A GOOD PLACE TO START

We can start by taking a look at how 3-phosphoglycerate is converted to **serine**, from which cysteine and glycine are both subsequently derived. The process begins with NAD^+ oxidation of the secondary alcohol of 3-phosphoglycerate to give the corresponding ketone (Fig. 3.54). This α-ketoacid then undergoes a PLP-mediated transamination reaction with glutamic acid acting as the amine donor, giving

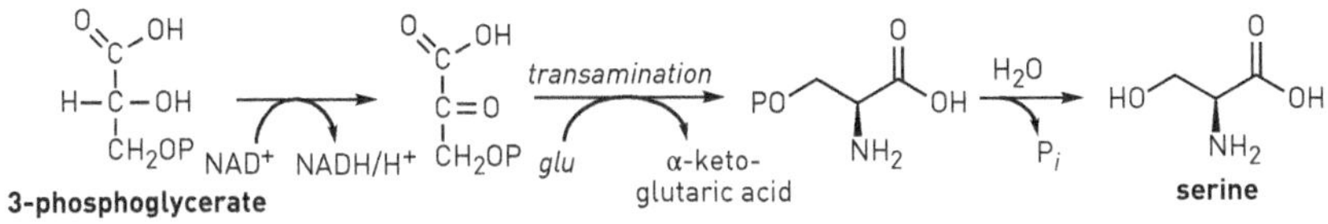

FIGURE 3.54
Biosynthesis of serine from 3-phosphoglycerate.

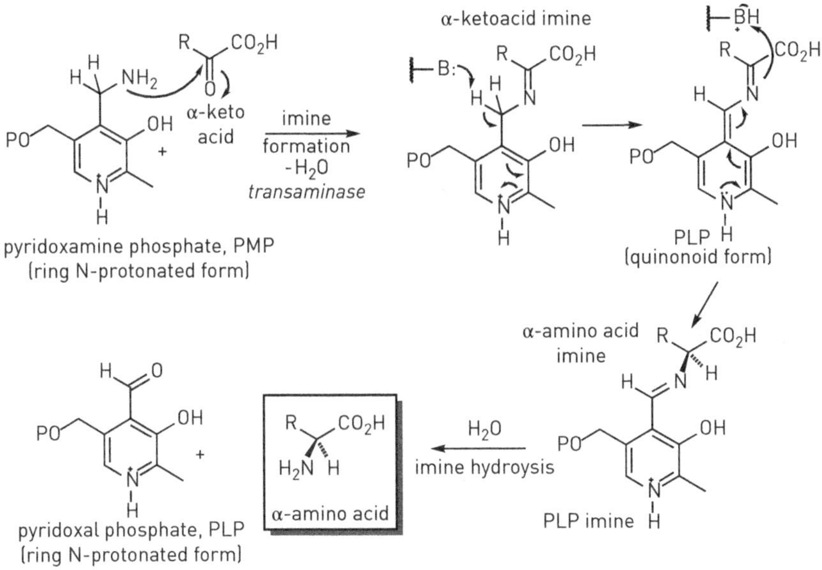

FIGURE 3.55
Mechanism for transamination of α-ketoacids by the action of pyridoxamine phosphate (PMP).

α-ketoglutaric acid and the intermediate product, 3-phosphoserine. Simple hydrolysis of the phosphate ester to the alcohol and inorganic phosphate leads directly to serine.

We'll take this as an opportunity to briefly revisit the mechanism for transamination that was discussed previously in Chapter 2. In Fig. 3.55, we see the general scheme for converting an α-ketoacid to an α-amino acid by reaction with pyridoxamine phosphate (PMP) which in turn is converted to pyridoxal phosphate (PLP). To see how the overall process works to deliver the amino group of glutamic acid to 3-phosphoserine in Fig. 3.54, work the process in Fig. 3.55 in the reverse direction using PLP and glutamic acid as the amino acid (or see Fig. 2.22); this will produce α-ketoglutaric acid and PMP. Subsequent reaction of PMP with the ketoacid from Fig. 3.54 will then produce 3-phosphoserine.

This basic transamination scheme is one we will see used for the synthesis of several other amino acids and is also an important and useful process for a variety of other transformations in biosynthetic schemes to come later. One way to gain a further appreciation for the complexity and versatility of PLP-mediated transformations is to take a detailed look at the 10-step conversion of serine into cysteine, as shown in Fig. 3.56. This remarkably complex process stands in stark contrast to the relative simplicity of the formation of serine itself, and while we will not look at all subsequent biosynthetic schemes in such detail, this one should give us a good sense of how to push the arrows in other PLP processes and to develop the intuition needed to predict likely mechanistic details in similar or related transformations.

In step 1, imine formation between PLP and serine takes places, followed by a deprotonation to give the usual quinonoid form of the PLP imine in step 2. A rearomatization of the PLP ring with loss of H_2O from the serine chain in step 3 is followed by nucleophilic addition of the thiol S atom of homocysteine (derived from methionine), then PLP ring aromatization and protonation to yield an imine in steps 4 and 5. Temporary loss of PLP from the imine to produce the key intermediate **cystathionine**

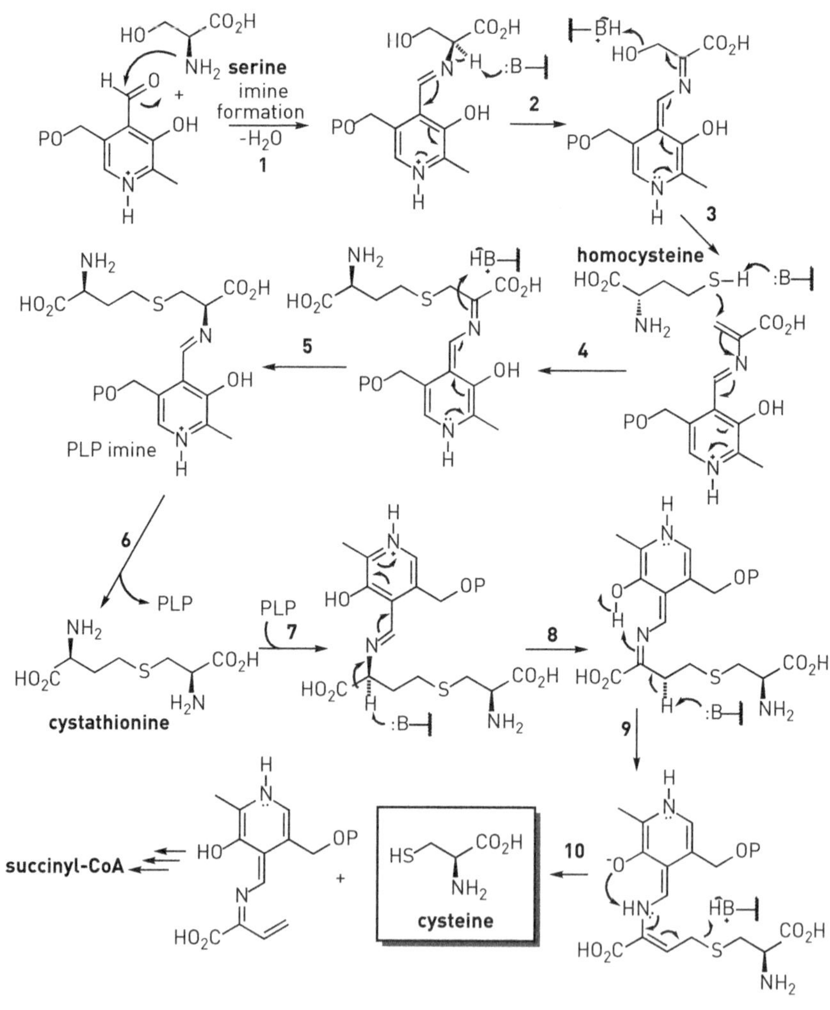

FIGURE 3.56

Ten-step PLP-mediated conversion of serine to cysteine by way of intermediate cystathionine.

in step 6 is then followed by PLP imine formation at the second amino group of cystathionine in step 7. Sequential deprotonations in steps 8 and 9 are followed by a final elimination and protonation in step 10 that produces cysteine as well as a PLP-bound intermediate which is subsequently transformed in a number of steps into succinyl-CoA. While perhaps intimidating at first, the sequence makes good chemical sense and nicely illustrates a number of commonly encountered transformations that are made possible via PLP intermediates.

The same PLP imine derivative of serine from step 1 in the cysteine biosynthesis is also involved in conversion of serine to the achiral amino acid **glycine**, further illustrating the versatility of PLP-mediated processes (Fig. 3.57). The reaction begins with a simple elimination reaction that produces formaldehyde and the usual PLP-imine quinonoid form. PLP ring rearomatization and subsequent hydrolysis gives glycine directly.

As shown in Fig. 3.58, transamination plays a direct role in the conversion of pyruvic acid to **alanine**, oxaloacetic acid to **aspartic acid**, and α-ketoglutaric acid to

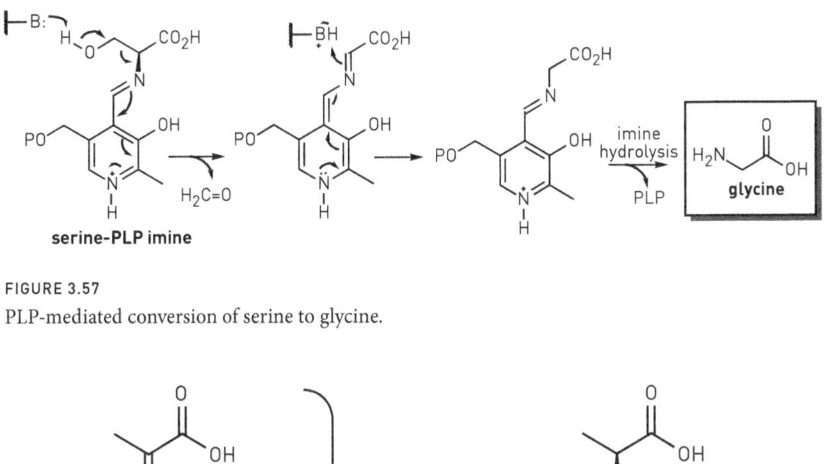

FIGURE 3.57
PLP-mediated conversion of serine to glycine.

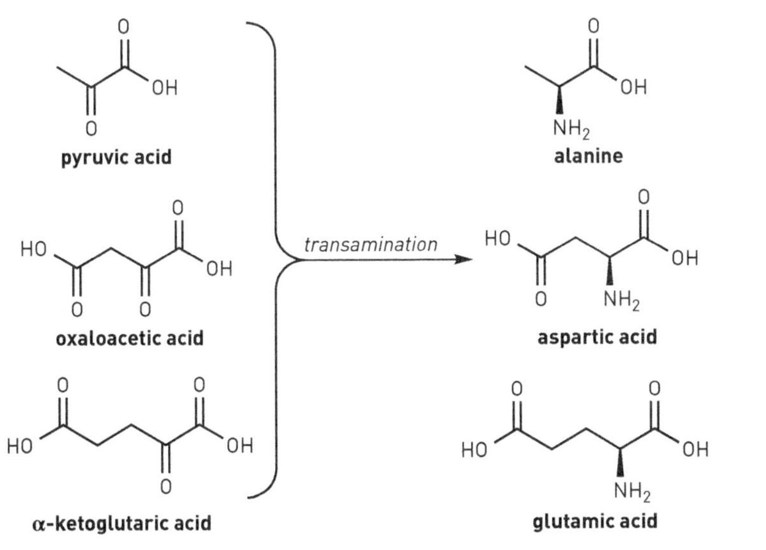

FIGURE 3.58
Transaminations for the direct biosynthesis of alanine, aspartic acid, and glutamic acid.

glutamic acid, three other nonessential amino acids for which PLP plays a pivotal role in their biosynthesis.

Both aspartic acid and glutamic acid are subsequently converted to the corresponding amides **asparagine** and **glutamine** by acyl phosphate substitution reactions using ammonia as the nucleophile, as shown in Fig. 3.59. The acyl phosphates are derived from ATP but differ from one another in that the acyl phosphate intermediate from aspartic acid is connected to adenosine by a monophosphate bridge, while the glutamic acid phosphate is an acyl monophosphate. The amide function of glutamine is as an amino group donor in numerous other biosynthetic processes and provides the nitrogen used here as well as serving as a general storehouse for NH_3 derived from other processes; *glutamine synthetase*, the enzyme responsible for glutamic acid amidation, clearly plays a key role in nitrogen metabolism regulation.

In addition to alanine, pyruvic acid is also a direct precursor of **valine** and **leucine** and provides two of the six carbons of **isoleucine**. For both valine and isoleucine, the usual TPP-bound intermediate from pyruvic acid decarboxylation serves as a nucleophilic 2-carbon fragment. In the case of valine, the electrophile is a second equivalent of pyruvic acid while for isoleucine, this role is played by a degradation product of threonine, **α-ketobutyric acid** (Fig. 3.60). The key step in both cases is an α-ketol-type 1,2-alkyl migration, a common base-catalyzed rearrangement reaction of α-hydroxyketones. Reduction to the respective 1,2-diols followed by dehydration gives enols which tautomerize to the corresponding α-ketoacids which are then converted to the α-amino acids by the usual transamination process.

Interestingly, the same α-ketoacid intermediate that undergoes transamination to **valine** may also participate in an aldol condensation via the enolate of acetyl-CoA. Hydrolysis and dehydration gives an α,β-unsaturated 1,4-dicarboxylic acid which then undergoes conjugate addition of water to give a β-hydroxy-1,4-dicarboxylic acid. NAD^+ oxidation of the hydroxyl group gives a β-keto-1,4-dicarboxylic acid which decarboxylates to the α-ketoacid. A final transamination then affords **leucine**. Much of the chemistry involved in biosynthesis of the remaining amino acids is more specialized and will be encountered later in Chapters 6 and 7 when we

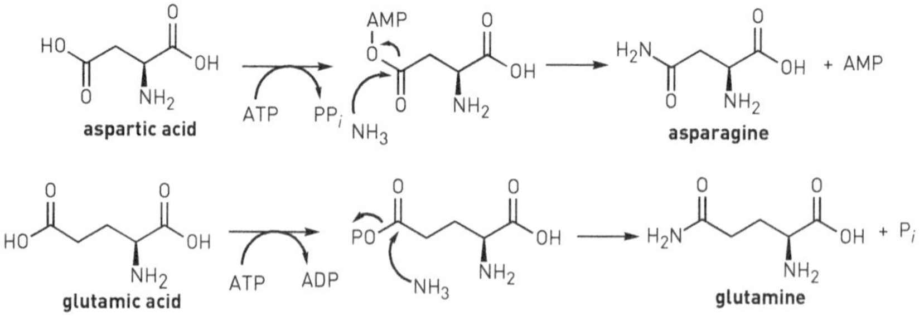

FIGURE 3.59
Conversion of aspartic and glutamic acids to asparagine and glutamine.

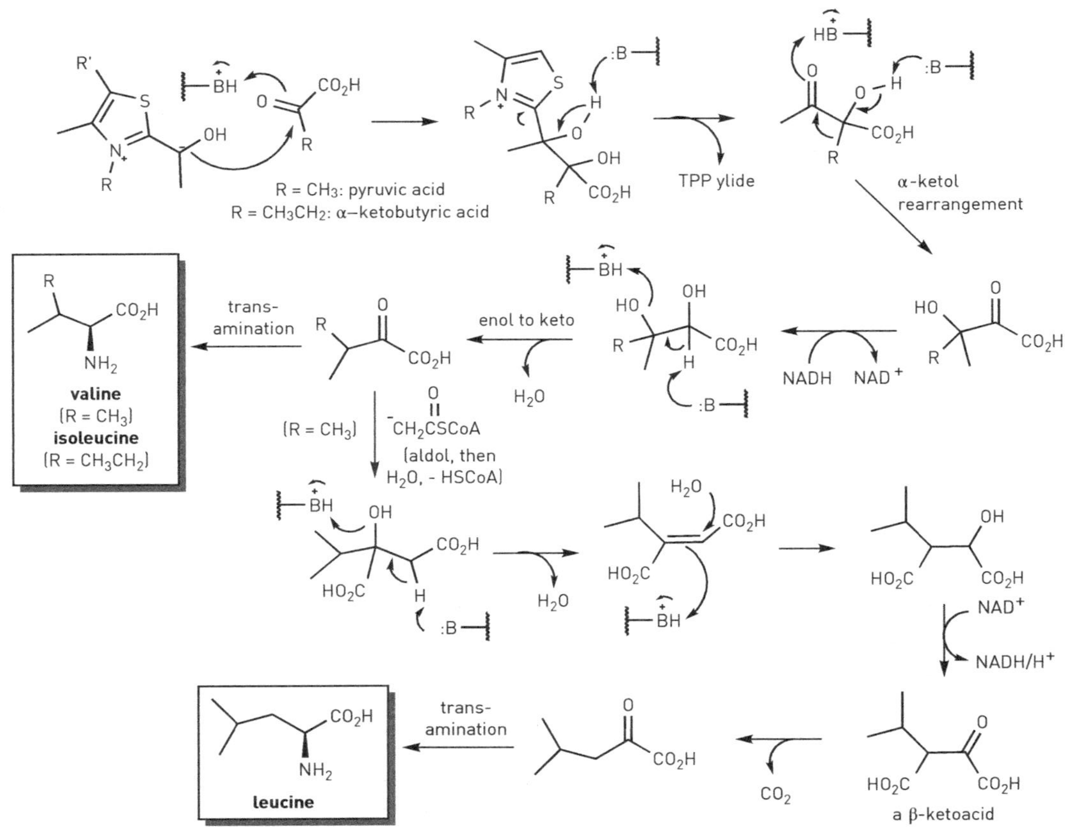

FIGURE 3.60
Conversion of pyruvic acid and α-ketobutyric acid to valine, leucine, and isoleucine.

look at both the shikimic acid pathway and the fascinating chemistry involved in the biosynthesis of some of the major plant alkaloids.

PEPTIDES AND PROTEINS: A VERY BRIEF REVIEW

While we need not labor over peptides and proteins for the purposes of this text, it will be helpful to at least briefly review how amino acids bond with one another to produce these important biomolecules, not only to better appreciate the essential connection between protein shape and enzyme catalytic action, but to also recognize that peptides and proteins may bond with other organic structures such as carbohydrates to produce biologically significant "hybrid" biopolymer structures.

Like saccharides, peptides may be classified according to the number of monomer units linked together to form them by using the usual prefixes (di-, tri-, tetra-, etc.). Thus, a tripeptide is made up of three amino acids linked together by **peptide bonds** which, of course, are simply amide linkages derived from the amino group of one amino acid and the carboxyl group of another. However, unlike many polysaccharides such as cellulose or glycogen, polypeptides are not usually made up of a single monomer as the repeating unit, making peptide structures and properties more complex and dependent on the specific sequence of amino acids present. For example, using the three simple amino acids alanine (Ala), glycine (Gly), and phenylalanine (Phe), a total of six distinctly different tripeptides may be assembled: Ala-Gly-Phe, Ala-Phe-Gly, Gly-Ala-Phe, Gly-Phe-Ala, Phe-Ala-Gly, and Phe-Gly-Ala (Fig. 3.61). Taking into account the possibility of using any one of the three amino acids more than once in a sequence leads to even more possible tripeptide structures. In all peptides, there will be one amino acid whose amino group is

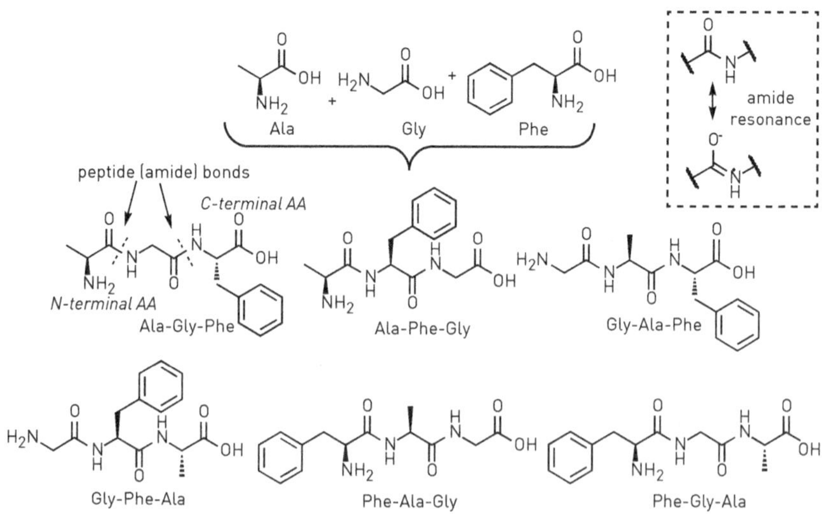

FIGURE 3.61
Amino acid sequences of six different tripeptides, each derived from Gly, Ala, and Phe.

unbound, known as the **N-terminal amino acid** and another whose carboxyl group is unbound, known as the **C-terminal amino acid**. Peptide amino acid sequences are usually read and drawn from left-to-right, with the N-terminal AA at the left end and the C-terminal AA at the right end. It is also useful to recall that since peptide bonds are actually amide linkages, **amide resonance** plays an important role in the 3-D structure of most polypeptides and proteins due to the restricted rotation around peptide C–N bonds, all of which have a substantial degree of π-bond character. The resulting staggered or "zigzag" structure of peptide chains, as represented in the tripeptides shown, has a specific geometry that places the N–H and C=O bonds of peptide linkages in a "trans" orientation relative to one another, again the result of the partial π-bond character of amide C–N bonds.

Peptides and proteins are generally linear polymers, though some cyclic peptides are of particular interest and will be seen in later chapters. Various polypeptides can have a variety of different functions, acting as enzymatic catalysts, neurotransmitters, hormones, cellular structural components, or even sometimes toxins. This wide variation in biological function is always dependent on the overall 3-D shape of the protein, which in turn is determined exclusively by the original amino acid sequence, often referred to as the **primary structure** of the protein, a structural designation which may also include any **disulfide linkages** (or **disulfide bridges**) which may be present. Such linkages can form when **cysteine thiol groups** located at various positions within a peptide chain become bonded to one another (or to thiol groups in another chain) by the oxidative transformation given by RSH + RSH → RS–SR. Since the reaction is reversible, disulfide linkages may be broken reductively to regenerate the original thiol groups. As shown in Fig. 3.62, the connections possible are nicely illustrated by the peptide hormone *insulin* which consists of two separate peptide chains A (with 21 amino acids) and B (with 30 amino acids) linked together by disulfide bridges at two different points. Note also that the A-chain contains an internal disulfide link between two cysteine residues fairly close together

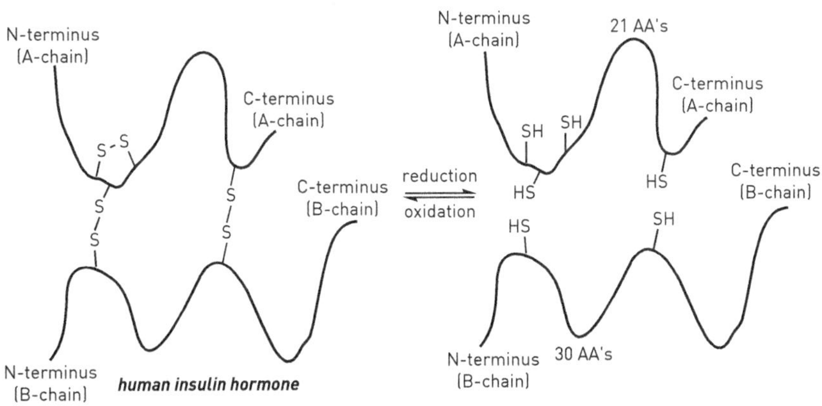

FIGURE 3.62
Insulin molecule composed of two polypeptide chains A and B, cross-linked by disulfide bridges.

in the chain. We thus see that in order to correctly determine the primary structure (amino acid sequence) of any peptide that contains disulfide linkages, it will first be necessary to reductively break the disulfide linkages; only then can the actual sequence of the chain be established. It is for this reason that disulfide linkages are often included as components of primary peptide structure, since these are true covalent bonding interactions, though these links may also be considered as part of the tertiary structure of proteins, as we will see shortly.

Further refinement of protein shape is determined by **secondary** structure, a result of **H-bonding** interactions between peptide bond N–H and C=O functions fairly close to one another within the chain. The zigzag structure of peptide chains, as represented earlier in Fig. 3.61, minimizes steric interactions between amino acid side chains but also allows for extensive H-bonding interactions. The two principal types of secondary structure in proteins that result from such H-bonding are the **α-helix** and the **β-pleated sheet**, either of which will be important in determining the final overall shape of a given protein, which in turn will determine its biological properties and function. The α-helix illustrated in Fig. 3.63 shows the preference for a "right-handed" twist in the helical arrangement in proceeding from the N-terminus to the C-terminus (the mirror image of such a helix would be "left-handed") for proteins derived from naturally occurring L-amino acids (unnatural D-amino acid-derived proteins would prefer a left-handed helix). For natural proteins, this "telephone cord" helical arrangement maximizes H-bonding interactions between the amino acid N–H and C=O functions; note also that the amino acid side chains (R groups) always project outward from the backbone of the helix (Fig. 3.63).

As shown in Fig. 3.64, the β-pleated sheet form of secondary protein structure can involve C=O and N–H hydrogen bonding interactions in either a **parallel** or an **antiparallel** fashion and may involve intermolecular interactions between separate peptide chains or the formation of pleated sheet regions within a large polypeptide (and therefore a protein) which has folded back upon itself.

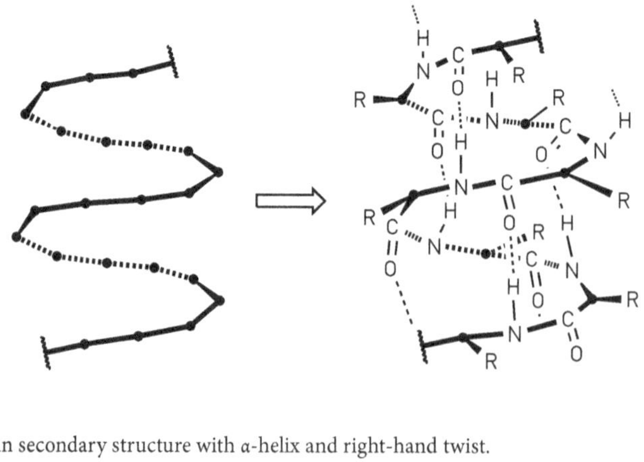

FIGURE 3.63
Peptide chain secondary structure with α-helix and right-hand twist.

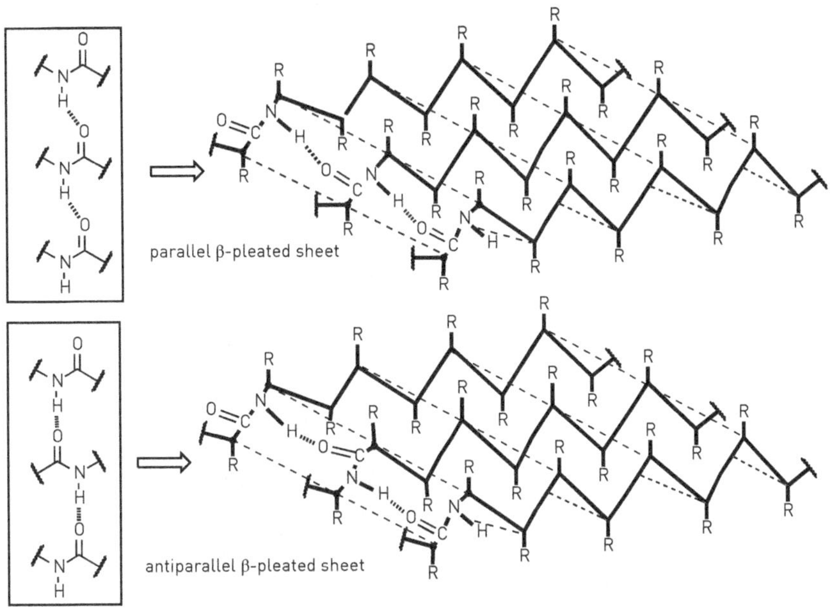

FIGURE 3.64
Comparison of C=O and N–H hydrogen bonding interactions in parallel vs. *anti* parallel β-pleated sheets.

Some proteins are **fibrous** in nature and may be elastic, due to the spring-like behavior of helical structure H-bonding, or inelastic pleated sheets which have H-bonding essentially perpendicular to the directionality of the chain backbone. However, most proteins are termed **globular** in shape; such a spherical 3-D conformation is due to multiple levels of structural interaction.

The third level of protein structure, known as **tertiary structure**, is associated with the unique overall 3-D shape of a protein. This shape results from the characteristic **folding** of a given protein due to **noncovalent** interactions between amino acid side chain groups often far removed from one another in the amino acid sequence. These interactions may include: i) **salt bridges** which are pH-dependent attractive interactions between positively and negatively charged ionic functions contained within certain amino acid side chains; ii) **H-bonding** interactions between neutral hydrophilic side chains containing C=O, N–H, O–H, or S–H groups; and iii) weak **van der Waals interactions** between hydrophobic hydrocarbon side chain groups (Fig. 3.65). Disulfide bridges are also sometimes included in tertiary structure since they involve interactions between amino acid side chains that significantly affect overall protein shape, though these are covalent rather than noncovalent interactions and are therefore much more robust.

Quaternary structure represents an additional level of protein complexity that can arise in certain cases when two or more separate chains interact to form an overall aggregate protein unit with specific functionality that is present only when this higher order of assembly is employed. For all proteins, their unique nooks, crannies,

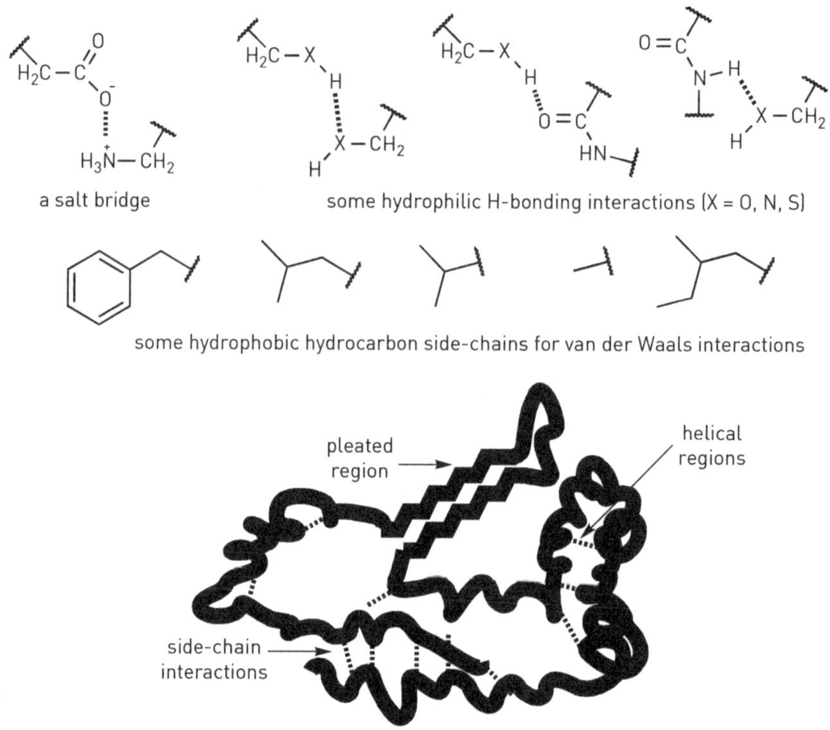

FIGURE 3.65
Examples of different amino acid side chain interactions within an overall protein structure.

and surfaces can allow binding or interaction with other molecules, giving enzymes their characteristic catalytic activity as a result of the subtle interplay of side chain inter- and intramolecular forces that ultimately produce the three-dimensional shape of these biopolymers.

The foregoing represents a bare minimum of some of the essentials of an extraordinarily complex area of study. For now, we'll leave the treatment of the organic mechanisms associated with the biosynthesis of peptides to a later chapter dealing with some specific examples of peptide-based natural products and turn finally to some concluding examples of how carbohydrates, amino acids, peptides and proteins can further interact with one another in biologically significant ways.

PUTTING PROTEINS AND CARBOHYDRATES TOGETHER: GLYCOPROTEINS VERSUS PROTEIN GLYCOSYLATION

We know that carbohydrates can have electrophilic components (an aldehyde carbonyl or an activated anomeric leaving group) while amino acids, peptides, and proteins have many potential nucleophilic components (either N-terminal amino groups or side chain nitrogens or oxygens), so it should not be surprising that these two separate groups of biomolecules can interact with one another to form hybrid structures in some instances. We finish off this chapter by taking a brief look at some examples of these complex and important biopolymers.

When linkages between carbohydrates and proteins are formed in an enzyme-controlled process, a hybrid structure known as a **glycoprotein** is formed. The linkage is usually one in which a nucleophilic atom from a protein is attached to the anomeric center of the carbohydrate component via a UDP-glucose-type nucleophilic substitution. The final carbohydrate component is usually an oligosaccharide and can play a critical role in protein folding and function. These interesting structures are normally divided into three groups: 1) **N-linked**, in which the carbohydrate component is linked to the nitrogen of the amide side chain of an asparagine residue in the protein; 2) **O-linked**, in which the carbohydrate is linked to the hydroxyl side chain of a serine or threonine residue in the protein; and 3) **C-linked**, in which the carbohydrate is linked to an indole ring carbon of a protein tryptophan residue (C-linked glycoproteins are rare compared to N- and O-linked). The basic components of the three are outlined in Fig. 3.66.

In membranes, glycoproteins can play an important role in interactions between cells. They also appear as structural features in connective tissues and are especially important components of the immune system. One interesting example is the group of glycoproteins involved in designation of blood types, as shown in Fig. 3.67. The disaccharide shown is connected to a carbohydrate hydroxyl component of a glycoprotein, with **A, B, and O blood types** being distinguished by the substitution on the C-3 oxygen of the galactose component. Connection at this position to an N-acetyl galactosamine (Type A), a galactose (Type B), or simple termination in a –OH group (Type O) distinguishes the three blood group antigens.

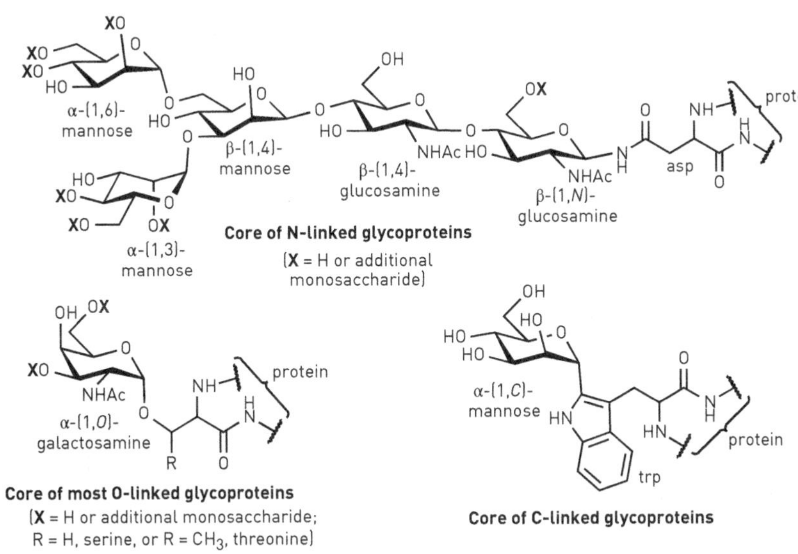

FIGURE 3.66
Core components of N-, O-, and C-linked glycoproteins.

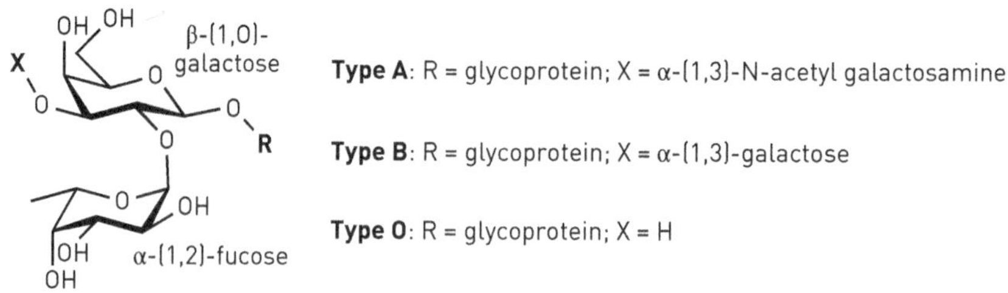

FIGURE 3.67
Glycoproteins and core structural components of type A, B, and O blood group antigens.

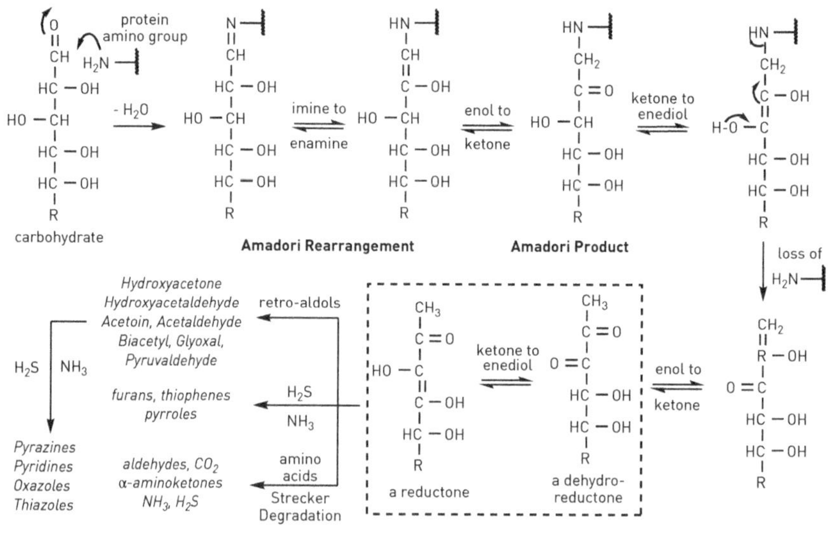

FIGURE 3.68
Maillard reaction of carbohydrates and proteins to produce food flavor and fragrance compounds.

The relatively recent interest in glycoproteins was partly inspired by early 20th century studies by Louis Camille Maillard (1878–1936) of the reactions between carbohydrates and proteins that occur during the cooking of foods. The **Maillard reaction**, as it is now known, is actually a complex series of reactions between protein N-terminal or side chain amino groups and reducing sugars that results in a form of "browning" that produces a vast array of different flavor and fragrance compounds associated with baking, frying, brewing, and other high-temperature cooking processes (Fig. 3.68).

Unlike glycoprotein formation, the Maillard reaction is an example of **nonenzymatic** protein **glycosylation** or **glycation**, a process that begins with simple imine (Schiff's base) formation followed by an **Amadori rearrangement** which is akin to enediol formation but involves an α-hydroxyimine rather than an α-hydroxy aldehyde or ketone. The resulting **Amadori compounds** may then break down further via loss of the amino function as shown in Fig. 3.68 to form so-called **reductones** and **dehydroreductones** which can then undergo a variety of further rearrangements, retroaldol reactions, condensation reactions with simple amino acids and various cyclizations to produce more complex flavor and fragrance products that may be characteristic of the type of food or beverage involved, the cooking process or other factors such as temperature and pH.

In biological systems, nonenzymatic Maillard-type processes have been implicated in protein–carbohydrate reactions that result in the formation of so-called **advanced glycation end-products** or **AGEs**. It turns out that many cells in the body bear a receptor (commonly referred to as RAGE) for AGEs that, upon binding such products, is believed to contribute to various age- and diabetes-related chronic inflammatory conditions including atherosclerosis, asthma, and arthritis, some of which may result from loss or modification of normal protein function due to Maillard nonenzymatic glycation processes. Investigation of the health-related implications of the biological Maillard reaction is bound to be an area of active research interest in the years ahead.

LOOKING AHEAD

For now, we have accumulated enough organic review and foundational material in bioorganic reactions, carbohydrates, and amino acids to take the next step in our journey through the organic chemistry of the major biosynthetic pathways, starting with the vast array of significant terpenoid natural products covered in the chapter that follows.

STUDY PROBLEMS

1. Two D-aldopentose carbohydrates give optically inactive (achiral) products when treated with sodium borohydride in ethanol. Which ones and why?
2. Most aldose monosaccharides and disaccharides will react with certain mild oxidizing agents in aqueous solution and are thus known as "reducing sugars."
   a) At which carbon would such oxidations most likely take place in β-D-glucopyranose? In maltose?
   b) Neither methyl-α-D-glucopyranoside nor sucrose are reducing sugars. Explain why not.
3. Draw Haworth and conformation diagrams for each of the following:
   a) methyl-β-D-lyxofuranoside
   b) L-cellobiose
   c) β-D-arabinofuranosyl-α-L-arabinofuranoside.
4. A proposed mechanism for conversion of UDP-1-glucose to UDP-1-galactose involves a 3-keto-UDP-1-glucose derivative. Suggest an overall mechanism for the conversion.

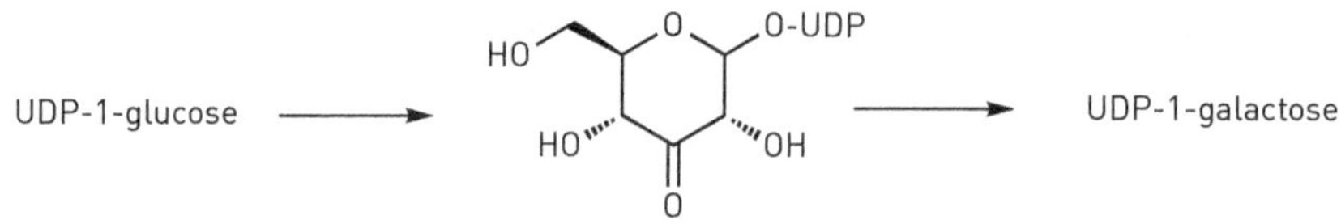

5. Provide a likely mechanism for the laboratory reactions shown:

a)

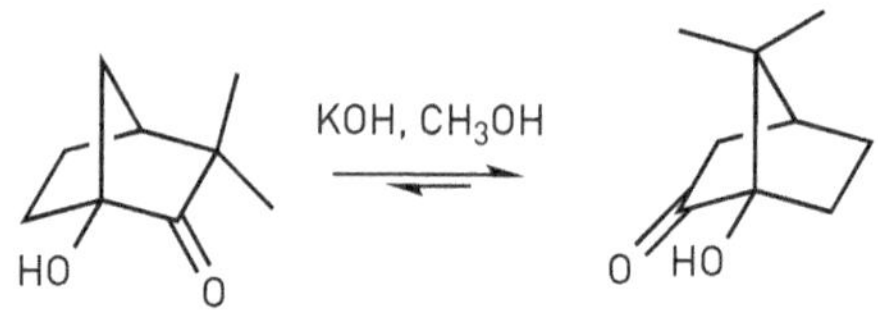

b)

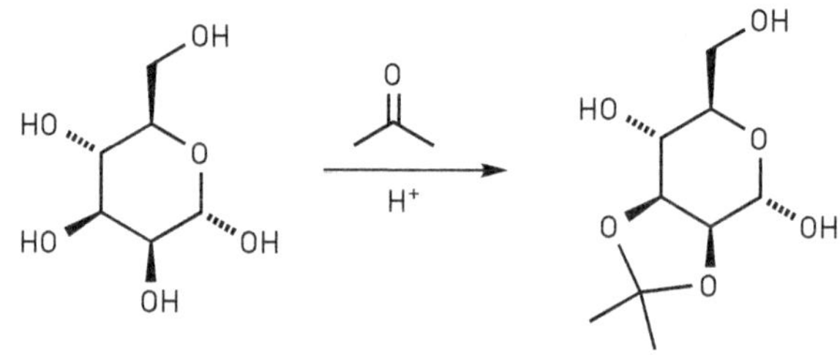

c)

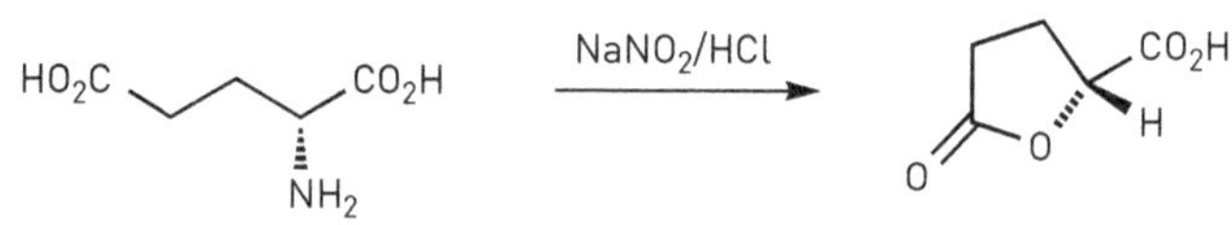

6. While achiral, citric acid is a prochiral compound, since a specific chemical change applied to one of the "wings" attached to the central carbon of the structure would produce an *R* enantiomer while the same change applied to other wing would produce the *S* enantiomer.

   a) Label the *pro-S* and *pro-R* wings of citric acid.

   b) When citric acid is prepared by reaction of oxaloacetic acid with ^{13}C-labeled acetyl-CoA (labeled at its carbonyl carbon), subsequent reactions of the citric acid cycle convert it to only one of the two ^{13}C-labeled α-ketoglutaric acid structures shown. If the enzyme *aconitase* which catalyzes the *anti* elimination dehydration step is completely selective for the *pro-R* wing of citric acid, which of the two labeled α-ketoglutaric acid structures **A** or **B** would be produced?

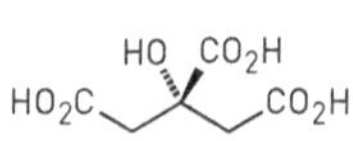

citric acid

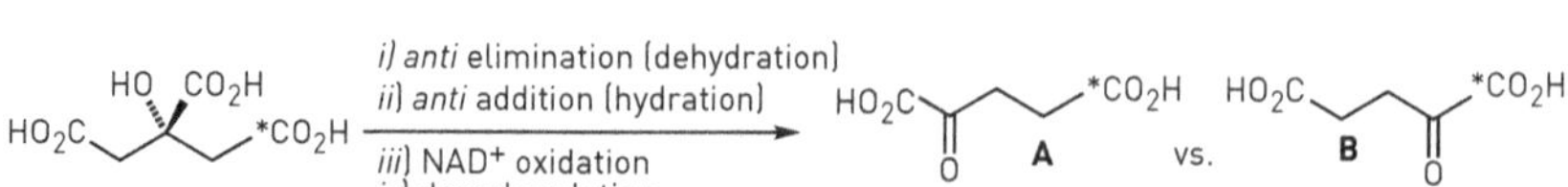

^{13}C-labeled citric acid **^{13}C-labeled α-ketoglutaric acid**

c) The methylene carbon of the *pro-R* wing of citric acid has two H atoms; one is *pro-R* and one is *pro-S*. When the *pro-R* wing of citric acid undergoes *anti* elimination of water to produce *cis*-aconitic acid, which H atom was removed in the elimination, the *pro-R* or the *pro-S* proton? Explain.

7. Cleavage of the aminoacid **A** shown employs a pyridoxal-5'-phosphate (PLP) dependent enzyme that catalyzes the hydrolytic cleavage of the β,γ-carbon-carbon bond in the amino acid to give anthranilic acid **B** and *L*-alanine **C**. Propose a chemical mechanism for this transformation using your knowledge of other PLP dependent reactions.

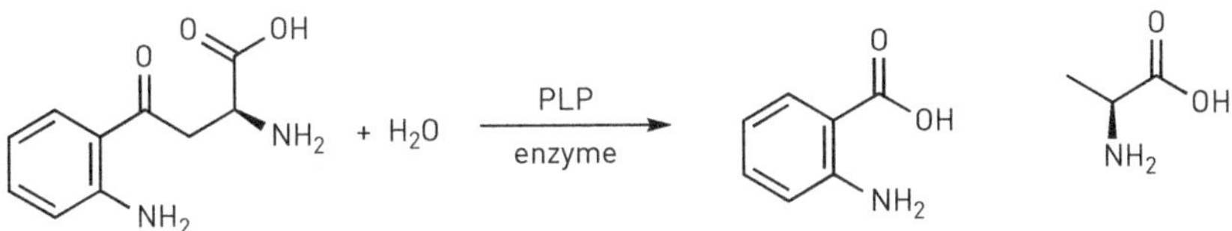

A B C

8. Glucosamine 6-phosphate is normally produced by enzyme-mediated NH_3 transfer from glutamine to fructose-6-phosphate. Propose a reasonable mechanism.

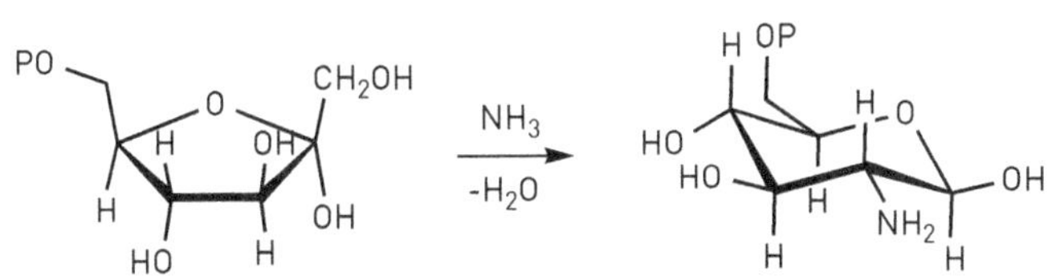

9. In a polypeptide with an α-helical structure, the α-helix would be considered a part of which of the polypeptide's structures?
   a) primary structure
   b) secondary structure
   c) tertiary structure
   d) quaternary structure

10. The planar peptide linkage in proteins and peptides has restricted rotation about the C–N bond. The reason for this restricted rotation is mainly due to which of the following?
   a) the presence of disulfide linkages in the polypeptide
   b) the presence of H-bonding to the carbonyl oxygens
   c) the presence of amide resonance which gives peptide bonds some partial π-bond character
   d) the presence of bulky amino acid side chains which sterically interact with one another

11. The major form of D-glucose present in aqueous solution is:
   a) the open-chain hydroxyaldehyde form

b) α-D-glucopyranose
c) β-D-glucopyranose
d) β-D-glucofuranose

12. Explain the principal structural difference between amylose (from starch) and cellulose. Use appropriate structural diagrams to illustrate your explanation.
13. Acid-catalyzed reaction of D-maltose with excess methanol yields a mixture of products. Draw the structures of those products.
14. Side chains of proteins or peptides are frequently involved in H-bonding or ionic bonding with other side chains. What level of protein structure do these interactions involve?
    a) primary structure
    b) tertiary structure
    c) secondary structure
    d) quaternary structure

4 The Terpenoid Pathway

Products from Mevalonic Acid and Deoxyxylulose Phosphate

But behind all this there looms a vast new problem, in comparison to which the one already solved seems quite small: the problem, what kind of chemical processes in the plant organism cause the formation of essential oils. How can we explain their infinite variety?
—Otto Wallach (*Nobel Prize in Chemistry, 1910*)

It was Otto Wallach (1847–1931) who first coined the term "terpene" and made the observation that many plant-derived essential oils had chemical structures whose composition was based on multiples of a basic five-carbon unit. His work with turpentine and the organic products derived from it was consistent with earlier studies of natural rubber which had shown that its thermal decomposition released "isoprene" (2-methyl-1,3-butadiene) as the principal product (Fig. 4.1). This led eventually to the formulation of the so-called biogenetic isoprene rule of Leopold Ruzicka (1887–1976) in 1953 which stated that "the carbon skeleton of the terpenes is composed of isoprene units linked in regular or irregular arrangement."

As it turns out, biosynthetic pathways to terpenes are found in nearly all organisms, producing a remarkable variety of different structural types, as we will soon see. In fact, something in excess of over 25,000 different terpenes with a wide variety of biological functions have been isolated from the plant kingdom over the years. Interestingly, while many terpenes are simple achiral compounds, others are chiral as can be seen in the case of α-pinene in Fig. 4.1. But unlike the naturally occurring L-amino acids and D-carbohydrates, different organisms may produce the same terpene product but in different enantiomeric forms. For example, limonene is formed by more than 300 plants, with the (+)-(*R*) enantiomer being the most

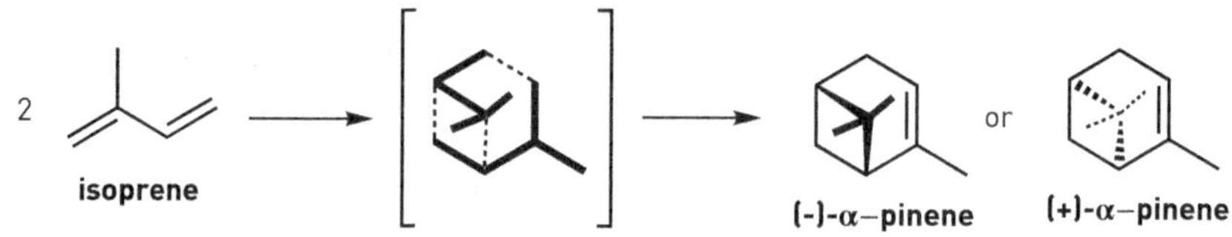

FIGURE 4.1
Isoprene units in α-pinene, a component of "terpentine" obtained from pine tree resin distillation.

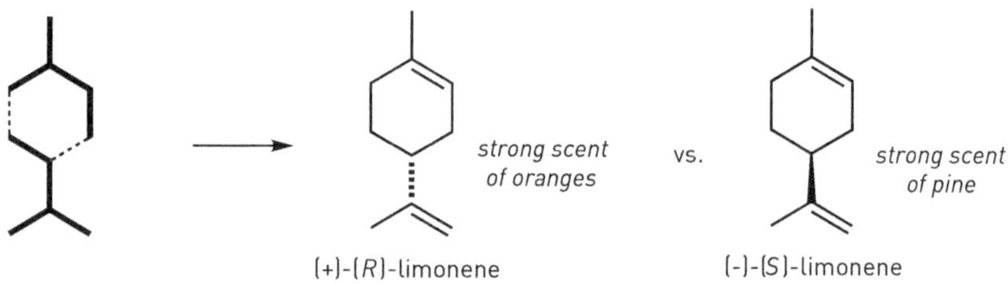

FIGURE 4.2
Enantiomers of limonene, the principal component of orange oil.

widespread form as the major constituent of citrus peel essential oils (orange oil). As the most abundant of all terpenes, its pleasant citrus fragrance and flavor have led to its worldwide use in the food and fragrance industries and also as a botanical insecticide. A number of plants produce both enantiomers of limonene, while others produce only the (–)-(*S*)-enantiomer which possesses a strong pine smell reminiscent of turpentine. This obviously speaks to the chirality and enantioselectivity of our own olfactory receptor sites which can readily distinguish between the two enantiomers, thus signaling a different odor response in each case (Fig. 4.2).

As with the carbohydrates, any detailed examination of a large and widely variable family of compounds such as the terpenes must begin with a systematic classification of the principal structural types that are encountered in nature. From there, we may then move forward to begin to understand how members of each group are constructed, modified, and elaborated into a vast array of organic natural products of extraordinary structural complexity, diversity, and biological significance.

CLASSIFICATION OF TERPENES: HOW MANY ISOPRENE UNITS?

Terpenes are usually classified into one of seven basic families according to the number of individual isoprene units employed in their basic construction (Table 4.1). Polymeric terpenes such as natural rubber or gutta-percha are of variable chain length and so do not truly represent a distinct terpene family.

Within these families, the vast majority of compounds are found to be so-called **"regular" terpenes**, those in which the isoprene units are linked in a **"head-to-tail"** fashion (Fig. 4.3). As the name implies, the far less abundant **"irregular" terpenes**

Table 4.1 General Classifications for Terpenes

Terpene family name	*# of isoprene units*	*# of carbon atoms*	*# of rings possible*
Hemiterpenes	1	5	0
Monoterpenes	2	10	0,1,2
Sesquiterpenes	3	15	0,1,2,3
Diterpenes	4	20	0,1,2,3,4
Sesterterpenes	5	25	0,1,2,3,4,5
Triterpenes	6	30	0,1,2,3,4,5,6
Tetraterpenes	8	40	Various
Rubber/gutta-percha	>100	>500	—

are those which have their isoprene units linked in some fashion other than head-to-tail. Beyond this, terpenes may be further subdivided into alicyclic, cyclic, bicyclic, tricyclic, and so on, according to the number of rings that may be present in a given compound from a particular family. The term **"terpenoid"** is sometimes used interchangeably with terpene, though the latter usually implies a hydrocarbon structure while the former ordinarily refers to modified terpenes containing one or more oxygen-containing functional groups such as alcohols, aldehydes, esters, or ethers.

Among the examples of **monoterpenes** shown in Fig. 4.3 are **menthol** and **camphor**, familiar and highly fragrant terpenes derived from mint oils with mild anesthetic properties that make them both common ingredients in throat lozenges, cooling gels, and vaporizer additives. The **sesquiterpene** examples include **farnesol**, a natural antibacterial agent and pesticide, and **zingiberene**, a fragrant component of ginger oil. The **diterpene** alcohol **vitamin A** (retinol) is essential to the eye for low-light and color vision in its aldehyde form (retinal) and is also a cellular growth factor as the corresponding carboxylic acid (retinoic acid). **Taxadiene** is the important terpene precursor to the terpenoid anticancer drug **taxol**, widely used in the treatment of breast, ovarian, lung, head, and neck cancers as well as the AIDS-related Kaposi's sarcoma. The pentacyclic diterpene **cafestol**, an aroma compound found in Arabica coffee beans, has also been shown to possess some anticancer activity and may also have the ability to inhibit the progress of Parkinson's symptoms.

Some examples from the remaining families of terpenes are shown in Fig. 4.4. Of these, the **sesterterpenes** such as the plant toxin **ophiobolin A** are relatively rare, being produced mainly by fungi and marine organisms. Conversely, the **triterpenes** are widely distributed in nature, with **squalene** serving as the essential terpene precursor to all the **steroids**, including **cholesterol** and the **sex hormones** (note that there is a single nonhead-to-tail isoprene linkage in squalene). The **tetraterpenes**, also known as **carotenoids**, include a vast array of biologically significant pigments such as β-carotene and related structures, but unlike the triterpenes, these compounds are not biosynthesized in animals and so must be obtained in the diet. The polyterpenes are represented by the two isoprene polymers, **natural rubber** (from rubber trees) and **gutta-percha** (dental rubber), differing from one another only in

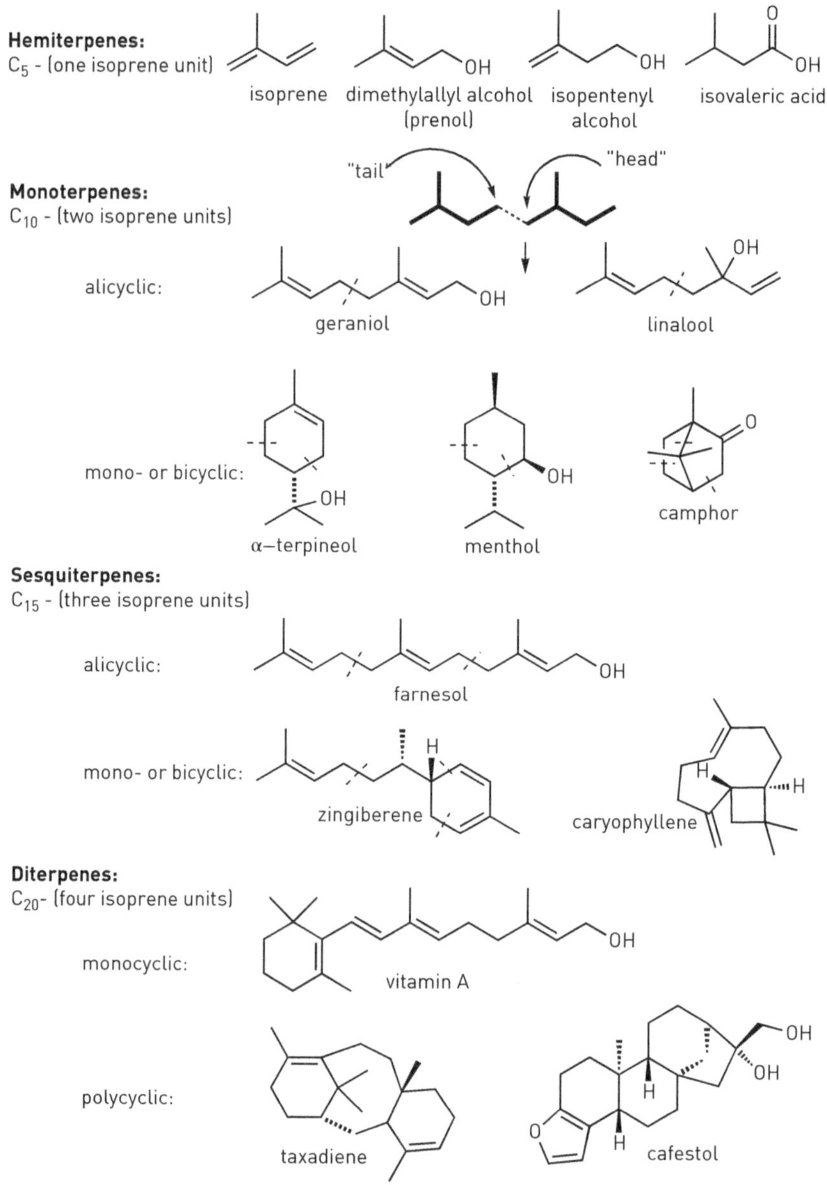

FIGURE 4.3
Head-to-tail isoprene linkage and some hemi-, mono-, sesqui-, and diterpene natural products.

the geometry of their isoprene linkages, that is, *cis*-1,4 for latex versus *trans*-1,4 for gutta-percha. Both are natural forms of polyisoprene with similar physical properties; gutta-percha is less common but notable as a "bioinert" material compared to natural rubber, which can elicit a latex allergy response in some individuals. It is for this reason that gutta-percha is used for certain kinds of dental fillings and in some surgical equipment.

Now that we have a better idea of what terpenes look like and how they are classified, we can now take a closer look at exactly how these important compounds

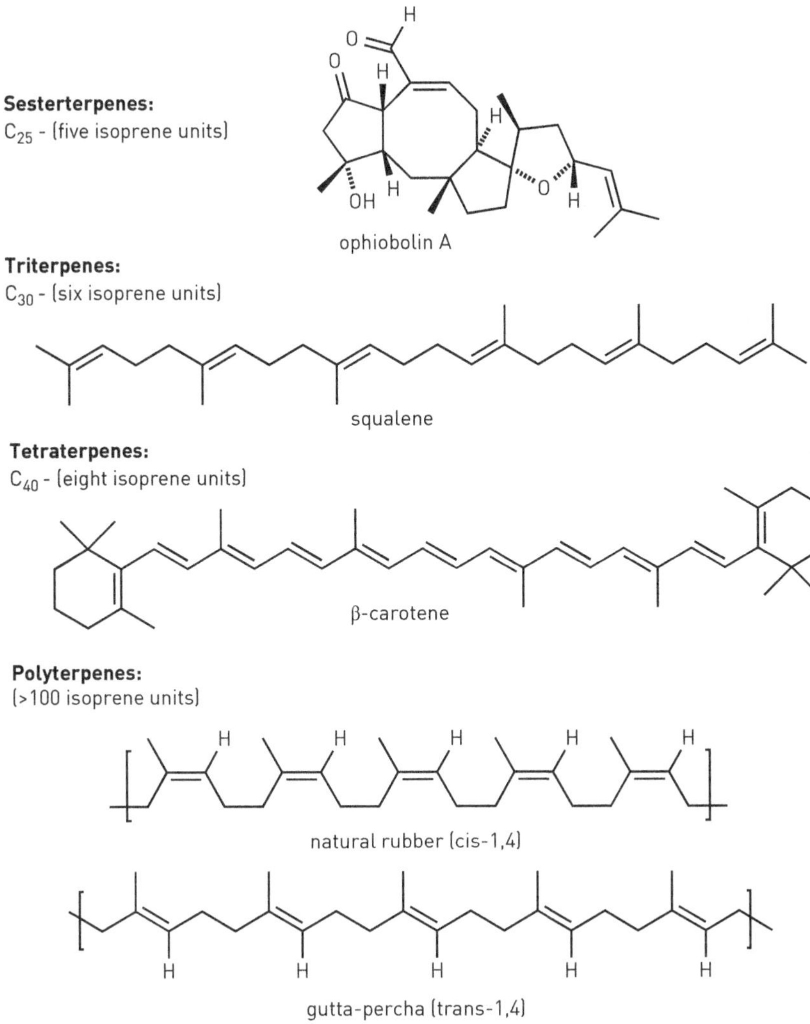

FIGURE 4.4
Examples of some sester-, tri-, tetra-, and polyterpene natural products.

are constructed. In fact, we will see that products of the terpene pathway are not actually constructed from individual units of isoprene, but rather from two closely related building blocks, **dimethylallyl diphosphate (DMAPP)** and **isopentenyl diphosphate (IPP)**, both of which are produced by two distinctly different routes in different organisms.

THE MEVALONIC ACID ROUTE TO DMAPP AND IPP

The mevalonic acid route begins with intermediates we have encountered before, as shown in Fig. 4.5. The process begins in step 1 with a Claisen condensation between **malonyl-CoA** and **acetyl-CoA** to produce **acetoacetyl-CoA**. This is followed by step 2 in which acetoacetyl-CoA, converted to an enzyme-bound thioester, undergoes attack

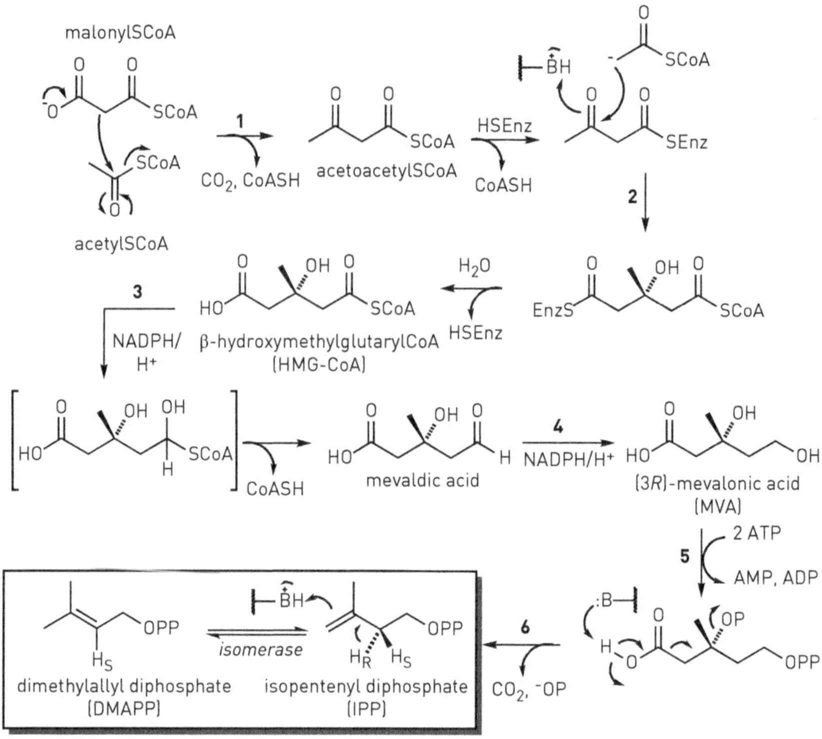

FIGURE 4.5
Biosynthesis of the terpene precursors IPP and DMAPP via mevalonic acid.

by acetyl-CoA enolate at the ketone carbonyl in a simple aldol addition process. After hydrolysis and loss of EnzSH, the resulting **β-hydroxymethylglutarylCoA** then undergoes NADPH reduction of the thioester to a hemithioacetal in step 3 which suffers loss of HSCoA to yield the corresponding aldehyde, **mevaldic acid**. A second NADPH reduction of the aldehyde to the primary alcohol in step 4 yields the important intermediate product, **(3*R*)-mevalonic acid (MVA)** from which this particular route to DMAPP and IPP takes its name. In step 5, the tertiary and primary alcohols are converted to mono- and diphosphate derivatives, respectively, by the action of 2 ATP. A final decarboxylation-driven elimination of the monophosphate as a leaving group in step 6 produces **isopentenyl diphosphate (IPP)** which is then converted to **dimethylallyl diphosphate (DMAPP)** by the action of an isomerase enzyme via what is believed to be an intermediate carbocation. Note the stereospecificity of this interconversion in which only the pro-*R* proton of IPP is lost in the formation of DMAPP.

THE DEOXYXYLULOSE PHOSPHATE ROUTE TO IPP AND DMAPP

Unlike the mevalonic acid route, which Ruzicka first described in the early 1950s and which operates in animals, plants, bacteria, algae, fungi, and protozoa, the **deoxyxylulose phosphate route** to IPP and DMAPP, discovered only in the early 1990s,

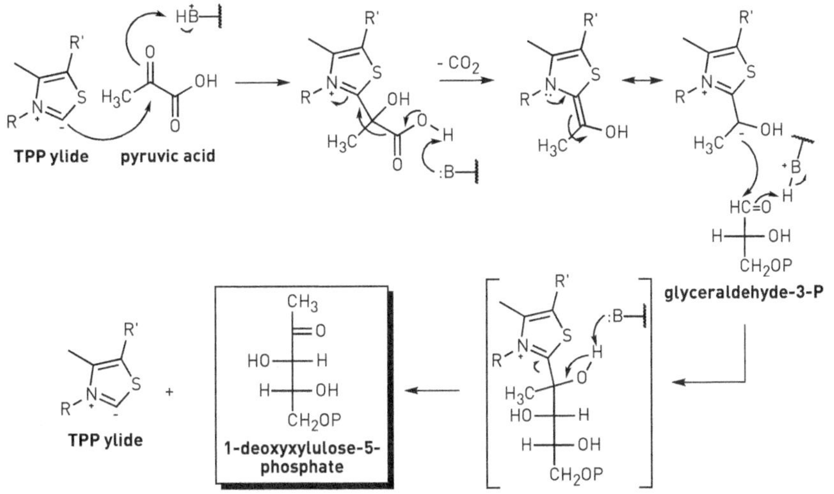

FIGURE 4.6
Biosynthesis of 1-deoxyxylulose-5-phosphate from pyruvic acid, glyceraldehyde-3-P, and TPP ylide.

has been found to operate widely in bacteria and in some algae, plants, and protozoa but not in animals or fungi, though some organisms have been found to utilize both pathways for terpene biosynthesis. Also unlike the mevalonate route, whose details have been extensively studied over many years, a number of the late steps associated with the deoxyxylulose phosphate route remain obscure or are not entirely understood. What is known for certain is that the process begins with biosynthesis of **1-deoxyxylulose-5-phosphate** from TPP ylide, pyruvic acid, and glyceraldehyde-3-phosphate, as shown in Fig. 4.6.

As shown, the nucleophilic TPP-bound dipolar intermediate from decarboxylation of pyruvic acid attacks the aldehyde carbonyl of **glyceraldehyde-3-P** derived from glycolysis. The resulting tetrahedral intermediate is then deprotonated with subsequent expulsion of TPP ylide and generation of **1-deoxyxylulose-5-phosphate** in a process similar to that seen earlier in the generation of sedoheptulose-7-phosphate from ribose-5-phosphate in the pentose phosphate pathway (see Fig. 3.47).

For the subsequent conversion of 1-deoxyxylulose-5-phosphate to IPP and DMAPP, much is known of the early steps, as shown in Fig. 4.7. An **α-ketol rearrangement** similar to one seen previously (see Fig. 3.59) in the conversion of pyruvic acid to valine begins the sequence (the starred carbons emphasize the 1,2-carbon migration that occurs in this step). NADPH reduction of the resulting aldehyde gives **2C-methylerythritol-4-phosphate**. The details of the next transformation involve nucleophilic displacement of diphosphate (PP_i) from cytidinetriphosphate (CTP) by an oxygen of the C4-phosphate group of 2C-methylerythritol-4-phosphate to form a CMP derivative. Conversion of the C2 tertiary –OH to a monophosphate by reaction with ATP is then followed by an intramolecular nucleophilic displacement of CMP by an oxygen from the C-2 monophosphate group to yield an unusual cyclic derivative, **2C-methylerythritol-2,3-cyclodiphosphate**. Beyond this key

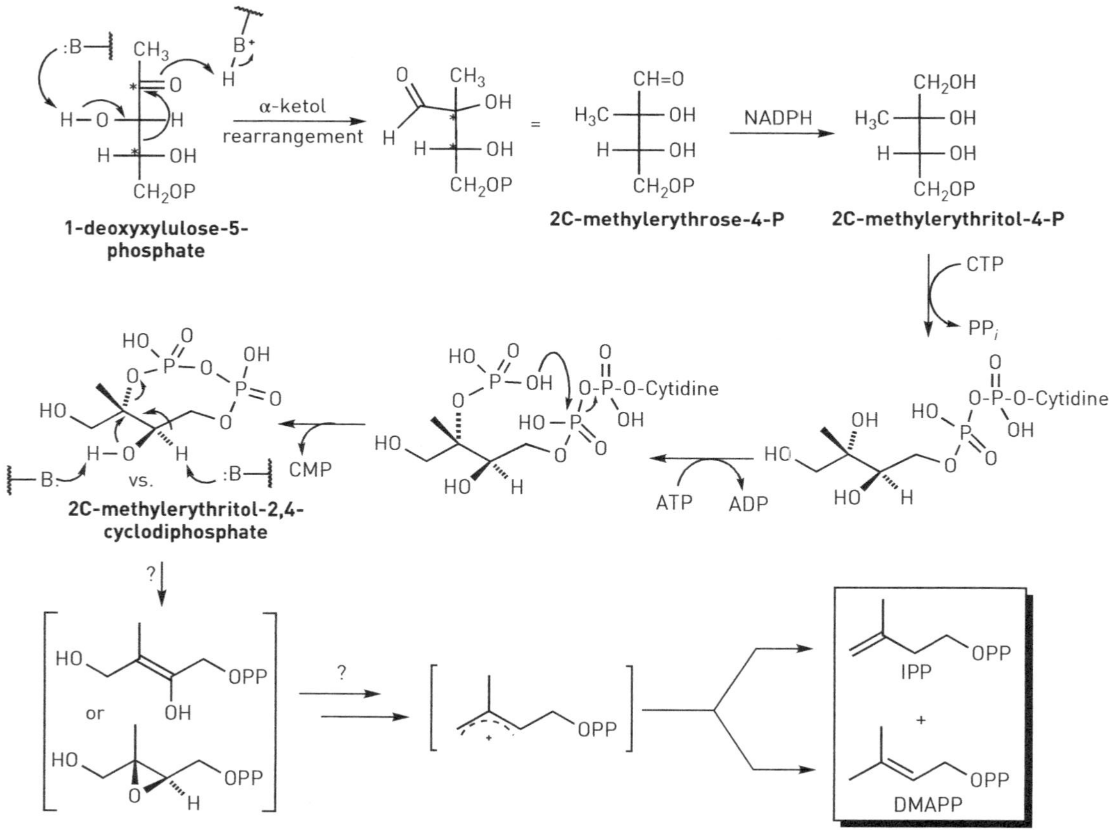

FIGURE 4.7
Known and speculative steps in the conversion of 1-deoxyxylulose-5-P to IPP and DMAPP.

intermediate, not much is known regarding the remaining chemistry involved in its conversion to IPP and DMAPP. Some speculation centers on two alternative elimination pathways to yield the terminal diphosphate group, one involving formation of an enol, the other an epoxide, as shown. In either case, both oxygen atoms must eventually be lost, perhaps giving rise to the final products via reductive transformation of an allylic cation, though it is not clear if the products are formed separately or if IPP is formed first and then isomerized to DMAPP as in the mevalonic acid route.

Now that we have seen how the two basic building blocks of the terpene biosynthetic pathway are formed, we next take a systematic look at how IPP and DMAPP are combined with one another and subsequently modified in various ways to create each of the families of the vast array of abundant terpene and steroid natural products that are found in nature, starting with the hemiterpenes.

HEMITERPENES: JUST ONE ISOPRENE UNIT

While isopentenyl alcohol would appear to be a simple hydrolysis product from IPP, dimethylallyl alcohol probably arises from addition of water to the intermediate carbocation after loss of diphosphate from DMAPP. Isoprene is similarly produced by *isoprene synthase*-mediated proton loss from the same carbocation (Fig. 4.8).

Aside from its commercial production, an enormous amount of isoprene, the monomer of natural rubber, is emitted into the atmosphere by many species of trees and shrubs including oaks and poplars, among others, in a process that is believed to help moderate heat stress in these plants. Global annual emissions of isoprene from natural sources rival those of methane, accounting for approximately one-third of the hydrocarbon content of the atmosphere. When combined with water and oxygen, isoprene may be converted into a variety of haze-producing derivatives, some of which are partly responsible for such phenomena as the so-called smoke of the Smoky Mountains portion of the Appalachian chain. Interestingly, the average human may produce as much as 17 mg of isoprene daily.

In the United States, the annual market for synthetic isoprene is about two billion pounds, of which roughly 60% is used in the manufacture of **synthetic rubber** for

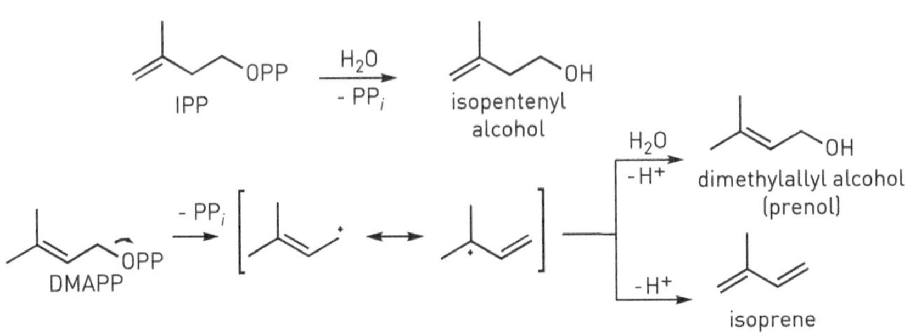

FIGURE 4.8
Hemiterpenes from IPP and DMAPP.

automobile tires, with the remainder used in adhesives or specialty chemicals. While most of this isoprene currently comes from petroleum feedstock, the production of biological isoprene or **bioprene** via fermentation technologies is rapidly maturing and may eventually supplant petroleum sources for the manufacture of polyisoprene; fermentation may also eventually prove to be a viable source of hydrocarbon building blocks for the manufacture of isoprene-based **biofuels**.

MONOTERPENES (C_{10}) AND ISOPRENE LINKAGE: HEADS, IPP WINS; TAILS, DMAPP LOSES

As was mentioned earlier, when isoprene units are linked together, it is not actually isoprene itself that is involved, but rather DMAPP and IPP units which are linked together in a specific way, that is, in a head-to-tail fashion. To see how that works, it helps to recognize that these two components are complementary to one another in their chemical behavior, with DMAPP serving as an electrophilic component while IPP serves as a nucleophilic component (Fig. 4.9). Thus, the diphosphate group is lost as a leaving group from the "tail" of DMAPP, yielding an electrophilic allylic carbocation which then undergoes nucleophilic attack by the electron pair from the π-bond at the "head" of the IPP component. The resulting C–C bond-forming reaction produces a tertiary carbocation that is discharged by simple loss of a proton to yield a new derivative called **geranyl diphosphate** or **geranyl PP** or just **GPP**. This process is completely analogous to the simple acid-catalyzed cationic dimerization of isobutylene to yield diisobutylene.

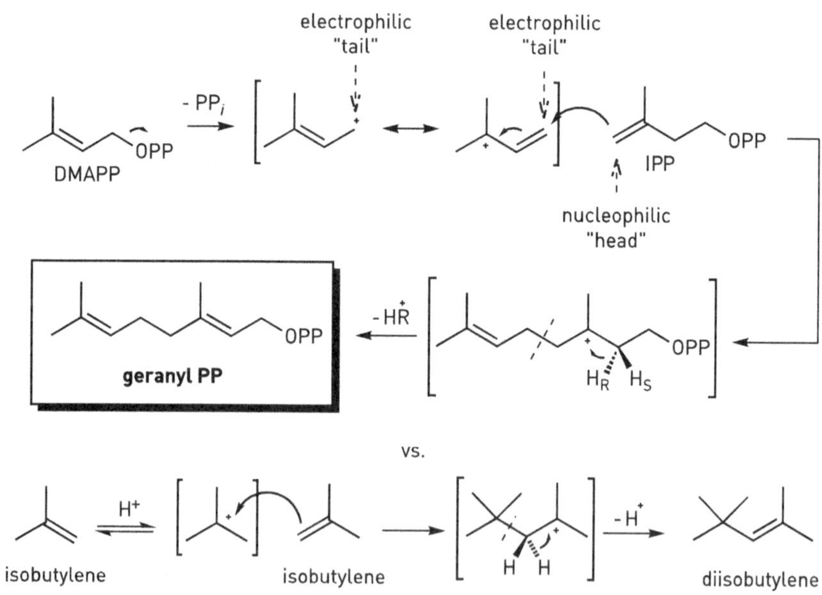

FIGURE 4.9

Linkage of IPP and DMAPP to produce geranyl PP compared to isobutylene dimerization.

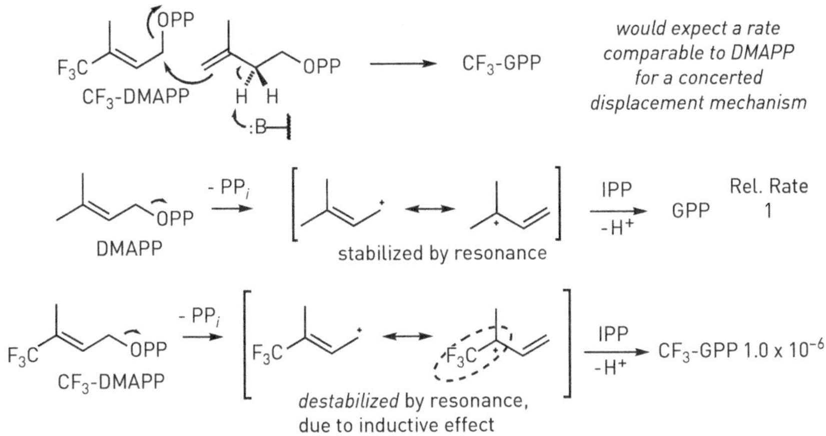

FIGURE 4.10
Evidence for carbocation intermediate in DMAPP/IPP linkage via prenyl transferase.

Note that in the linkage of DMAPP and IPP units catalyzed by enzymes called *prenyl transferases*, final proton loss to discharge the intermediate cation is specific for removal of the H_R proton and is analogous to the proton loss in isomerization of IPP to DMAPP. Note also the (*E*) configuration for the double bond produced in the GPP product.

The importance of carbocation intermediates in the organic chemistry of terpenes cannot be overemphasized, and the linkage of DMAPP and IPP is just one example, though we might reasonably ask for evidence supporting such an intermediate here, since a direct nucleophilic displacement process to produce GPP might seem just as reasonable (Fig. 4.10). As it turns out, if trifluoromethyl-substituted DMAPP is used, the rate of conversion to GPP is over a million times slower. This result makes sense only if a carbocationic intermediate is involved, since no effect on the rate of GPP production would be expected from a concerted nucelophilic displacement mechanism. The validity of such an experiment rests in part on the similarity in size of a CF_3 group to a CH_3 group which would lead to similar enzyme binding rates for the two substrates, that is, both would be expected to "fit" in the active site. But the electronic properties of the two substrates would be vastly different; the rate of formation of a CF_3-substituted allylic carbocation should be very slow due to a highly unfavorable inductive effect.

GERANYL PP TO NERYL PP VIA LINALYL PP: THE IMPORTANCE OF ALKENE STEREOCHEMISTRY

We noted that DMAPP can readily ionize to an allylic carbocation via loss of diphosphate as a leaving group. As we will see, GPP can do the same thing, and for this reason GPP is an important precursor to a variety of other terpene products. However, the (*E*) geometry of its double bond limits what can happen to the

GPP-derived carbocation, since such allylic carbocations tend to preserve their geometry. At first glance, this might seem unreasonable, since the allylic carbocation has a resonance form that would appear capable of isomerization via a simple rotation around a C–C single bond. But we must remember that the actual carbocation is a resonance hybrid of the two structures we ordinarily draw, meaning that there is actually some double bond character between all three carbons involved, preventing a simple C–C bond rotation, as shown in Fig. 4.11.

Note, however, that if after initial loss of ⁻OPP, the tertiary carbon of GPP cation recaptures the ⁻OPP as a nucleophile, the structurally different product **linalyl diphosphate (LPP)** is produced (Fig. 4.12). Since this tertiary diphosphate is free to rotate about the C–C bond adjacent to the tertiary center, subsequent ionization of the less sterically congested conformer leads to an allylic carbocation which, if

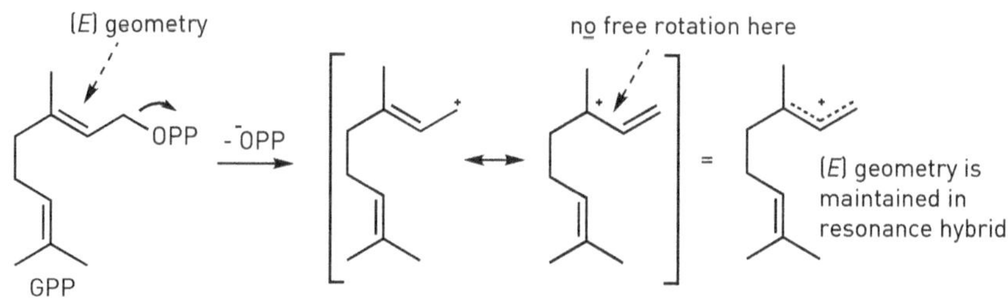

FIGURE 4.11
Retention of double-bond geometry in allylic carbocation from GPP.

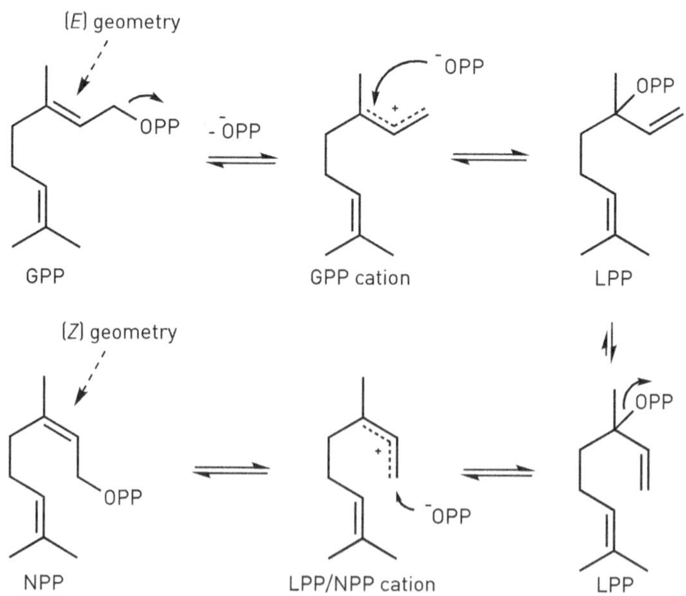

FIGURE 4.12
Isomerization of geranyl PP (GPP) to neryl PP (NPP) via linalyl PP (LPP).

trapped again by $^{-}$OPP at its *primary* carbon, affords an isomer of GPP with (*Z*) stereochemistry at its double bond. This isomer is known as **neryl diphosphate (NPP)**. The significance of this isomerization of GPP to NPP by way of LPP will be evident shortly when we examine the various transformations possible for each of these derivatives via their corresponding carbocations.

SOME ACYCLIC MONOTERPENES AND THEIR USES

From GPP, a number of different products are possible, as shown in Fig. 4.13. Many of these are significant flavor or fragrance compounds found in volatile aroma plant extracts known as **essential oils**. Few are found in pure form, but rather appear as complex mixtures which can be challenging to separate. **Geraniol** is a component of many different essential oils such as geranium oil, rose oil, and lemon oil, among others. **Citronellol** is obtained from geraniol by reduction of the allylic alcohol double bond and is an example from this group with a chirality center. Both enantiomeric forms are found in nature; the (+) isomer (shown) is a component of lemon grass oil and is more common, while the (–) isomer is found in rose oil.

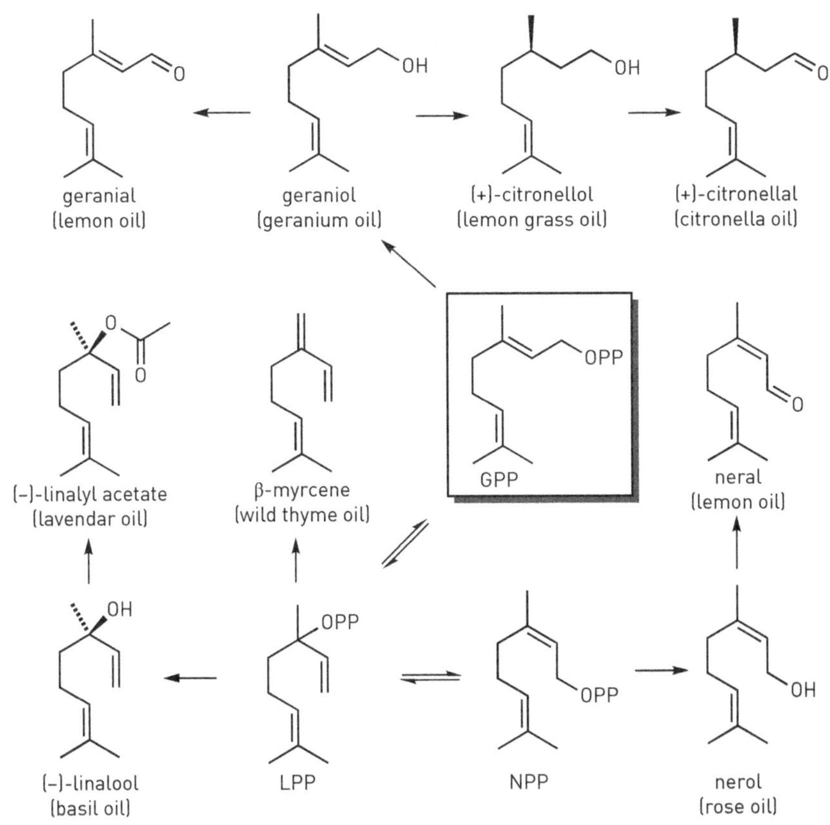

FIGURE 4.13

Representative structural relationships among some acyclic monoterpenes derived from GPP.

Oxidation of citronellol produces the corresponding aldehyde **citronellal**, a component of citronella oil which has been shown to have significant mosquito repellent activity, hence the use of citronella candles for summer evenings out of doors. Oxidation of geraniol itself leads to the α,β-unsaturated **geranial** (also known as (*E*)-citral), a component of lemon, orange, and many other essential oils which is widely used in the flavor and perfumery industries due to its intense lemon fragrance. Isomerization of GPP to LPP and NPP ultimately leads to similar derivatives. Loss of diphosphate and a proton from LPP can give different triene hydrocarbons, one of which is **β-myrcene**, a fragrant component from wild thyme and hops. The alcohol corresponding to LPP is known as **linalool** which, like citronellol, possesses a chirality center. The (+) form predominates in many plants producing it, such as basil and thyme, while a near racemic mixture is isolated from lemon and grapefruit. The corresponding **linalyl acetate** ester is found in lavender oil, among others. **Nerol**, the alcohol derived directly from NPP, is found in hops, lemon grass, and neroli oil from the bitter orange tree. It has a rose fragrance similar to that of geraniol. Its aldehyde derivative is **neral** (also known as (*Z*)-citral), another important flavor and fragrance compound with a sweet lemon aroma that is less intense than geranial.

MONO- AND BICYCLIC MONOTERPENES VIA CATIONIC CYCLIZATIONS AND WAGNER–MEERWEIN SHIFTS

As we have seen above, much of the chemistry of the terpenes is dominated by typical carbocation reactions such as nucleophilic additions, eliminations, or rearrangements. Furthermore, we have seen that *bimolecular* C–C bond formation between carbocations and alkenes, as in the linking of IPP and DMAP units, is also commonly encountered. Not surprisingly, ring-forming *unimolecular* C–C bond-forming reactions are also quite common for appropriate substrates and rings sizes. Such simple transformations can lead to a remarkably diverse array of monocyclic and bicyclic monoterpenoid structures.

Unlike the GPP cation which is geometrically constrained, intramolecular C–C bond formation by cationic cyclization is readily achievable via the NPP/LPP cation with its (*Z*)-configuration, as shown in Fig. 4.14. This cascade of reactions illustrates some typical transformations possible in the series, though the compounds shown are by no means an exhaustive catalogue of the many products that can be produced from this same starting material.

Initial cyclization of the NPP/LPP cation leads to **α-terpinyl cation** which may then be quenched by the usual sequences of carbocation quenching reactions. For many of these compounds, different enantiomers may be produced by different sources, so for our purposes the more commonly encountered form is usually shown. Simple proton loss from α-terpinyl cation affords **limonene**, a terpene from citrus fruits with a strong lemon smell that makes it useful as a food additive, as a fragrance for household products, and as a hydrocarbon from renewable sources

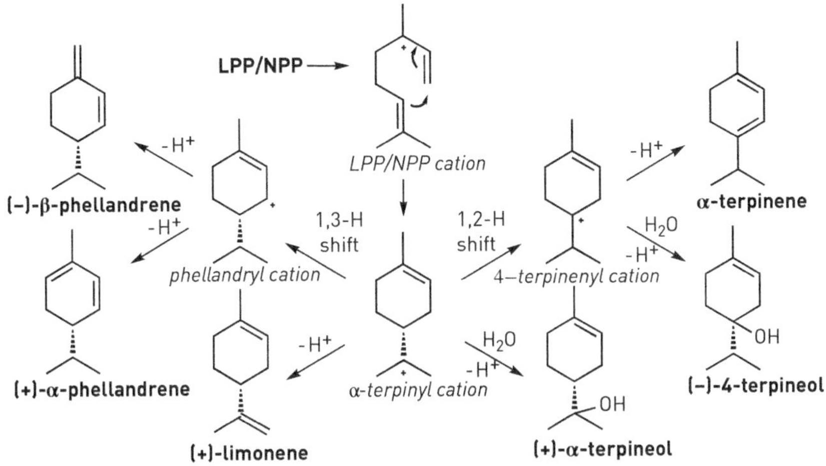

FIGURE 4.14
Cationic cyclization and some representative examples of monocyclic monoterpenoids.

that is sometimes used to replace petroleum-based solvents. An unusual 1,3-hydride shift in α-terpinyl cation may seem unlikely at first, but note that the resulting carbocation is allylic and therefore more stable. Proton loss from different resonance contributors affords **α-** and **β-phellandrenes**, both useful fragrance compounds with peppery-minty aromas. Nucleophilic trapping of α-terpinyl cation by water leads directly to **α-terpineol**, which has a heavy floral aroma similar to lilac. Products arising from a 1,2-H shift of α-terpinyl cation followed by either nucleophilic trapping of water or simple proton loss are **4-terpineol**, with a fragrance similar to α-terpineol, and **α-terpenine** with an odor that has been described as "woody."

Further intramolecular cyclizations of α-terpinyl cation are also possible, leading to various bicyclic monoterpene derivatives, as shown in Fig. 4.15 using a more conformationally useful representation of the carbocation.

Initial cyclization occurs by trapping of the isopropyl cation side chain by the ring π-bond, leading to **pinyl cation**. For each of the two representations of the pinyl cation shown, two pathways a and b are possible: path a involves alternate 1,2-C shift possibilities leading to either **fenchyl cation** or **bornyl cation**, while path b involves alternate modes of β-proton loss for pinyl cation, leading to **α-** or **β-pinene** products. Like isoprene, α-pinene is a significant source of atmospheric hydrocarbon, being emitted in substantial quantities by pine trees. Both enantiomeric forms are produced in nature, with the (–)-form predominant from European pines, and the (+)-form predominant in North America. The isomeric β-pinene is found in rosemary, parsley, rose, and hop; both pinenes are major constituents of turpentine.

Nucleophilic trapping of **fenchyl cation** by water affords the terpenoid alcohol **fenchol**, which has a bittersweet, lime-like flavor and is found in lime oil, grape, nutmeg, and rosemary. Fenchol undergoes oxidation to **fenchone** which possesses a distinct camphor-like aroma and is one of the components of absinthe, a potent,

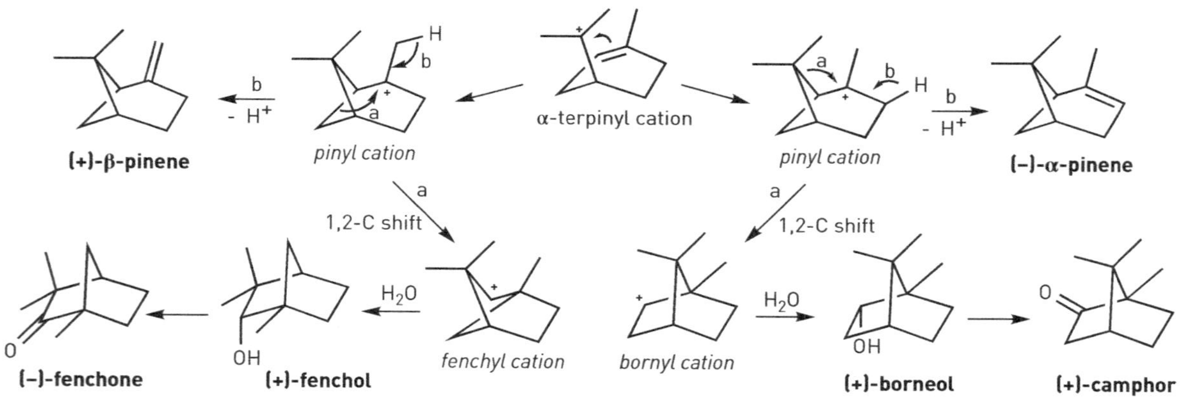

FIGURE 4.15

1,2-C shifts in α-terpinyl cation and some representative bicyclic monoterpenoids.

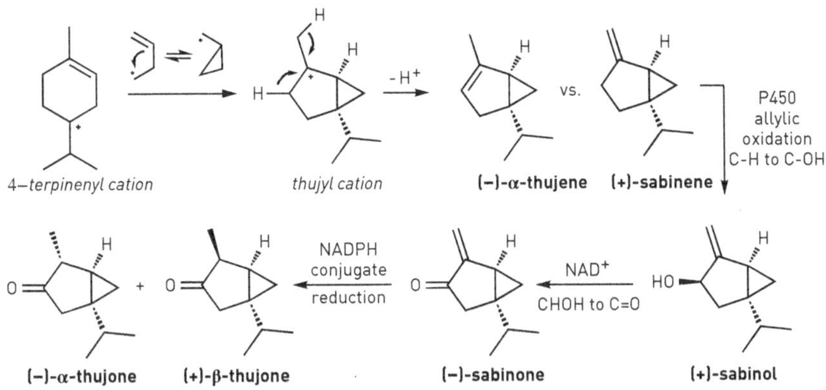

FIGURE 4.16
Cyclization of 4-terpinenyl cation to thujyl cation and some resulting bicyclic monoterpenoids.

light-green alcoholic spirit made from distilled wormwood and anise (also known as the "Green Fury") which only recently has returned to popularity after being outlawed in most countries for nearly 100 years. **Camphor** itself is a ketone similar to fenchone and is derived from oxidation of **borneol**, the alcohol obtained from nucleophilic trapping of **bornyl cation** by water. Camphor, available from many sources, can comprise as much as 20% by weight of the dried leaves of rosemary and has many medicinal and culinary uses similar to those of **menthol**, as mentioned earlier.

The **4-terpinenyl cation** (Fig. 4.14) can also undergo a further intramolecular cyclization as shown in Fig. 4.16. The initial cyclization is reminiscent of the homoallyl-cyclopropyl carbinyl cation system we discussed briefly in Chapter 1 and leads to the so-called **thujyl cation.** Two fates for thujyl cation involve simple proton loss to afford either **α-thujene**, a pungent flavor component of the herb summer savory, or **sabinene**, a component of carrot seed oil and black pepper among other sources. A P450-type allylic oxidation of sabinene leads to **sabinol** which after NAD^+ oxidation to **sabinone** finally undergoes conjugate addition of hydride to the α,β-unsaturated carbonyl system to produce **α-** or **β-thujone**, both of which are readily epimerized due to the acidic proton at the chirality center α- to the ketone carbonyl. As another terpene component of **absinthe**, thujone was partly responsible for the formerly illegal status of this spirit due to its supposed dangerous psychedelic properties. While that role has now largely been discredited, thujone levels in absinthe are still controlled by law.

WHAT'S THAT SMELL? LIMONENE DERIVATIVES AS FLAVOR AND FRAGRANCE COMPOUNDS

As shown in Fig. 4.17, limonene undergoes oxidation at two different allylic positions to give the corresponding alcohol derivatives ***trans*-carveol**, a component of both

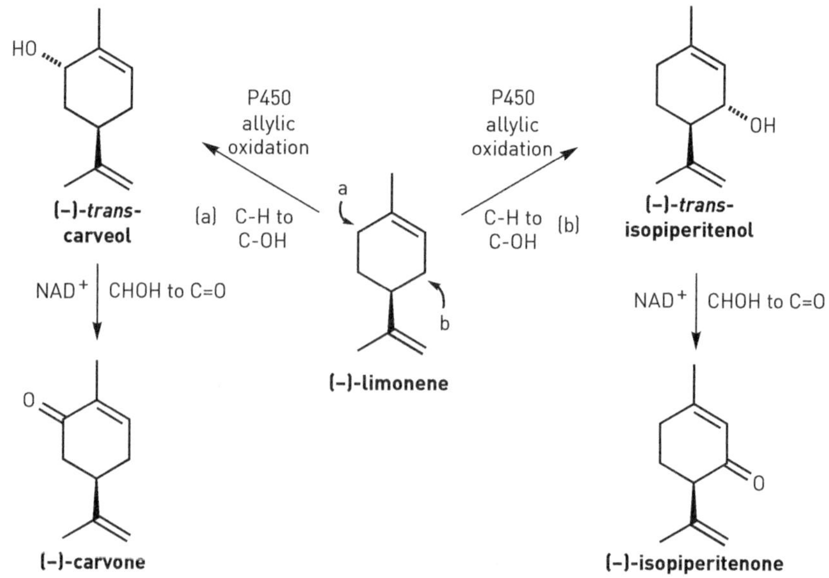

FIGURE 4.17
Biosynthesis of some oxygenated terpenoid derivatives of (–)-limonene.

spearmint and caraway seed oil among others, and also ***trans*-isopiperitenol**; both undergo NAD^+ oxidations to the corresponding ketones **carvone** and **isopiperitenone** as shown. The (–)-enantiomer of carvone (shown) has a strong spearmint odor, while the (+)-enantiomer has the aroma of caraway seeds; these are frequently invoked as prime examples of the ability of the olfactory senses to distinguish between enantiomeric forms. Isopiperitenone has a herbal fragrance described as sweet or fruity. All these are useful as flavor or fragrance additives in various commercial applications.

Isopiperitenone undergoes a series of further transformations that ultimately lead to the widely distributed peppermint oil component **menthol** (Fig. 4.18). Initial conjugate reduction of the α,β-unsaturated ketone function yields ***cis*-isopulegone** which then undergoes an allylic isomerization via its acidic tertiary α-hydrogen to yield the α,β-unsaturated ketone isomer, **pulegone**. Conjugate addition of hydride may lead to the *trans*-stereoisomer, **menthone**, while alternate face reduction (or epimerization of menthone at the tertiary center alpha to the carbonyl) produces the corresponding *cis*-isomer, **isomenthone**. Final addition of hydride to the *Si* face of the menthone carbonyl gives **menthol**, while addition to the *Re* face gives **neomenthol**. Similar face-selective reductions of isomenthone yield **isomenthol** and **neoisomenthol**. Production of the different stereoisomers is controlled by different enzymes in different plants at each step, but of all the stereoisomers, **(–)-menthol** is the most abundant and it is also the most important, owing to its mild analgesic, anesthetic, and cooling properties as well as its pleasing aroma.

There are other mono- and bicyclic monoterpenes structures far too numerous to deal with in a brief overview, and new ones are isolated on a regular basis.

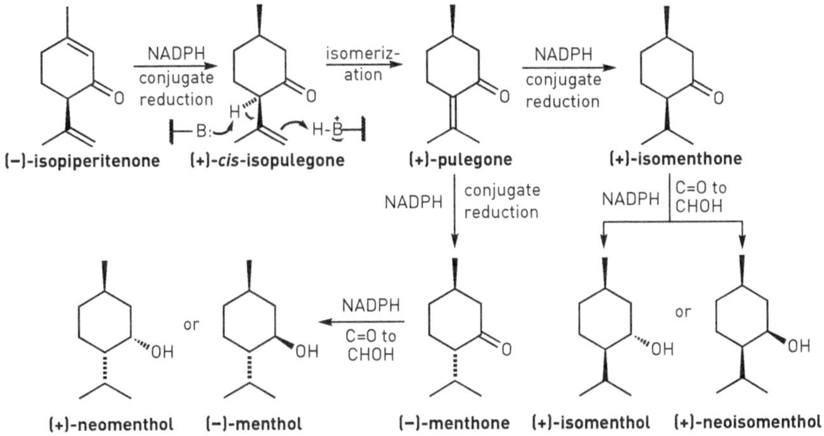

FIGURE 4.18
Biosynthetic transformations in the conversion of isopiperitenone to menthol and its disastereomers.

Nevertheless, we have seen some representative groups which have given us a fairly good idea of some of the basic principles involved in their synthesis and subsequent modifications. But before we move on from the monoterpenes to the C_{15} sesquiterpenes, let's take a brief look at some rather unusual monoterpene structures in which the individual C_5 units are *not* connected in a head-to-tail fashion.

IRREGULAR MONOTERPENES: IF NOT HEAD-TO-TAIL, THEN HOW?

As the name suggests, the irregular monoterpenes are compounds which violate the biogenetic isoprene rule. The four structural groups shown in Fig. 4.19 are chosen to illustrate basic differences among some of the irregular monoterpenes, with the connectivity between C_5 units clearly not head-to-tail, but nevertheless variable from group to group. In the first three, that is, the **artemisyl, chrysanthemyl**, and **lavandulyl** groups, the individual isoprene units are intact, while in the fourth **santolinyl** group, one of the isoprene units appears to have been disconnected and reassembled by a skeletal rearrangement. Neither GPP nor LPP appear to be involved in the biosynthesis of these monoterpenes; overall there are relatively few members in each family with the exception of the chrysanthemyl group whose members appear as various esters derived from the corresponding parent carboxylic acids. We'll take a brief look at how one of these systems is assembled by an alternate method for the linking of isoprene units and which we will encounter again later in the chapter when we look at the earliest stages of steroid biosynthesis and the assembly of the carotenoid terpenes.

The generally accepted route to the chrysanthemyl family involves linkage of two DMAPP units rather than one DMAPP and one IPP unit as is the case for regular terpenes (Fig. 4.20). Esterification of the resulting **chrysanthemic acid** or **pyrethric**

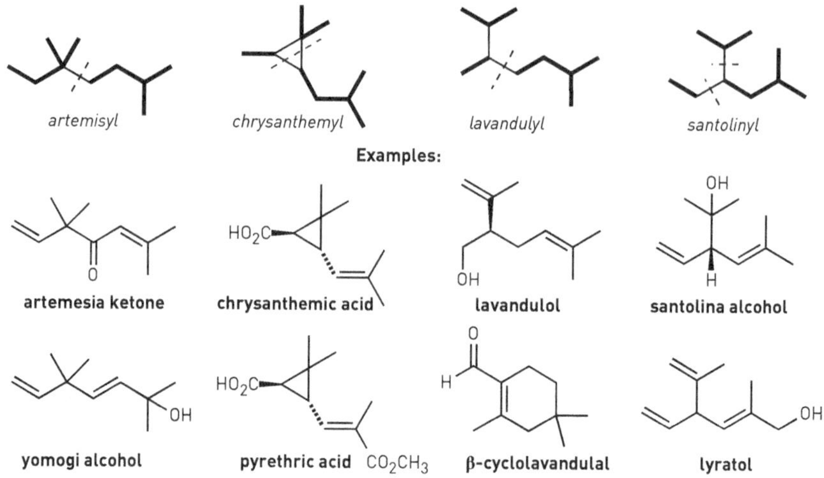

FIGURE 4.19
Four families of irregular monoterpenes and some representative members.

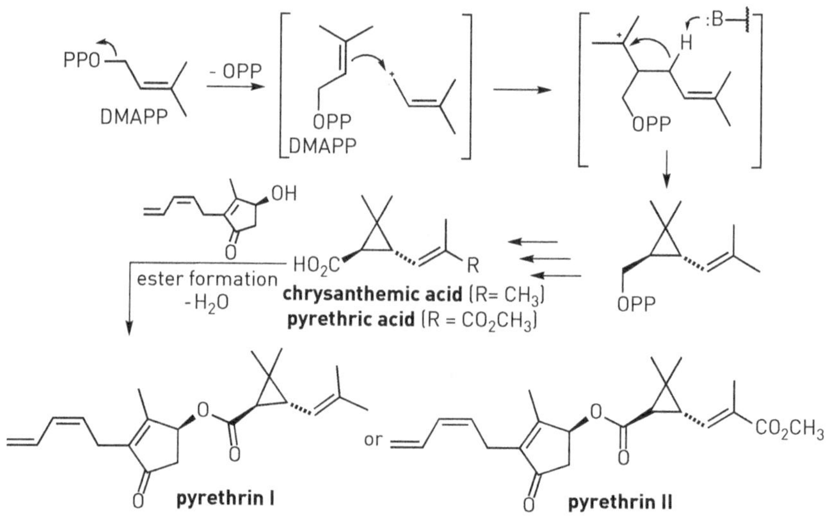

FIGURE 4.20
Biosynthesis of chrysanthemic and pyrethric acids and conversion to pyrethrins I and II.

acid with the cyclopentyl alcohol **(Z)-pyrethrelone** leads to **pyrethrins I** and **II**, two members of a group of six potent, structurally related insect neurotoxins that are widely used as natural insecticides.

A good deal of experimental work has been done to try to determine the pathways involved in biosynthesis of the other systems. Since the chrysanthemyl, artemisyl, and santolinyl are the most commonly encountered systems, it has been proposed that the chrysanthemyl PP may be a precursor to the others via ionization to a cyclopropylmethyl carbocation which would be expected to rearrange by different paths to yield the santolinyl or artemisyl cation systems, both of which would readily produce some of the known products of these families by simple additions of water, and so on (Fig. 4.21).

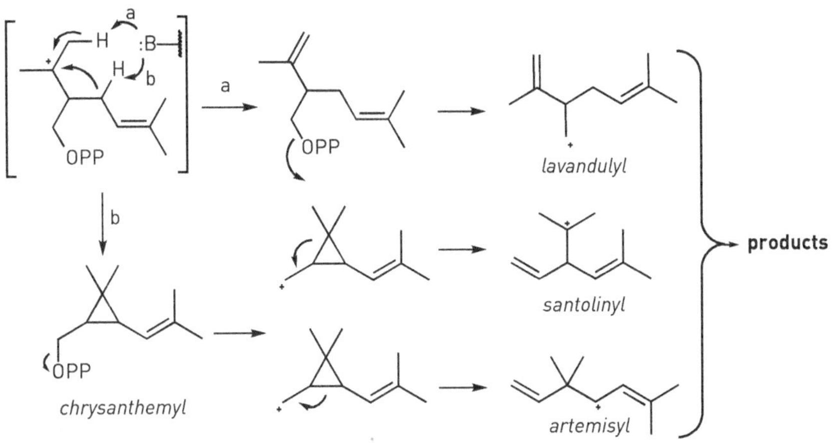

FIGURE 4.21
Possible relationships among the chrysanthemyl, santolinly, and artemisyl systems.

Similarly, the chrysanthemyl PP precursor may follow an alternate pathway of deprotonation rather than cyclization which would afford the lavandulyl system.

IRIDOIDS: FROM CATNIP TO ALKALOIDS

Iridoids are a class of monoterpene-derived compounds characterized by a cyclopentane ring fused to a six-membered cyclic ether. They have their biosynthetic origin in geraniol and are found in a wide variety of plants, usually appearing as glycosides bound to a glucose molecule. Most iridoids are produced as a defense against infection by microorganisms or as a bitter-tasting deterrent to herbivores and can exhibit a wide range of biological activities including cardiovascular, analgesic, anti-inflammatory, antimutagenic, antitumor, and antiviral activities among others. Figure 4.22 traces a typical sequence for iridoids starting from **geraniol** in the biosynthesis of **nepetalactone**, from the catnip plant. Roughly 50% of all cats react to nepetalactone, suggesting a genetic susceptibility to its aroma; it is also an active repellent for insects such as cockroaches and mosquitoes. As can be seen, oxidation of geraniol to **8-hydroxygeraniol** is followed by oxidation of the two primary alcohols to the corresponding dialdehyde, **8-oxogeranial**. This is the key intermediate in formation of the cyclopentane ring system, formulated as occurring via intramolecular conjugate addition by the aldehyde enolate moiety initially derived from conjugate hydride addition to the β-methyl-α,β-unsaturated aldehyde end of the dial system as shown. Intramolecular trapping of the resulting enol form of the cyclic **irododial** by the proximate aldehyde function leads to a hemiacetal which is then oxidized to give **nepetalactone**.

A small number of iridoids are also found in the insect world. Figure 4.23 shows the structures of **chrysomelidial** from leaf beetle larvae *Chrysomelidae*, **anisomorphal** from the defensive glands of the large walking stick *Anisomorpha buprestoide*, and **dolichodial** from ants of the genera *Dolichoderus*, along with two important

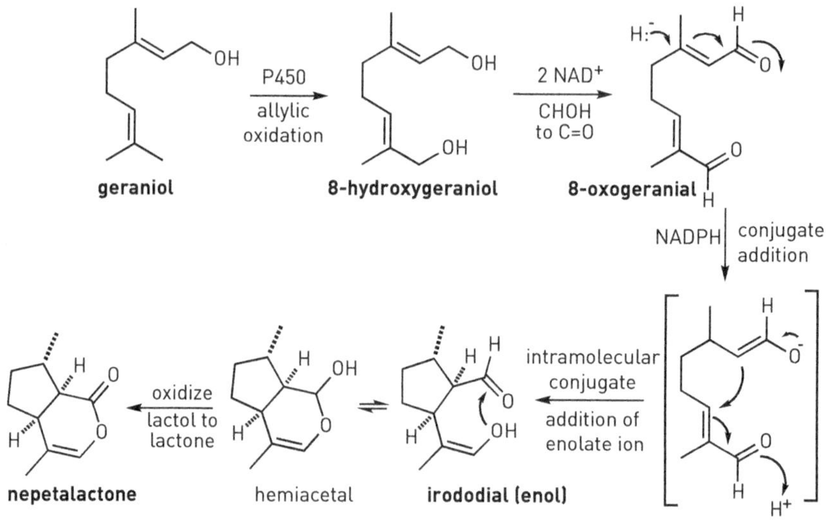

FIGURE 4.22
Biosynthesis of the iridoid nepetalactone from geraniol by way of irododial.

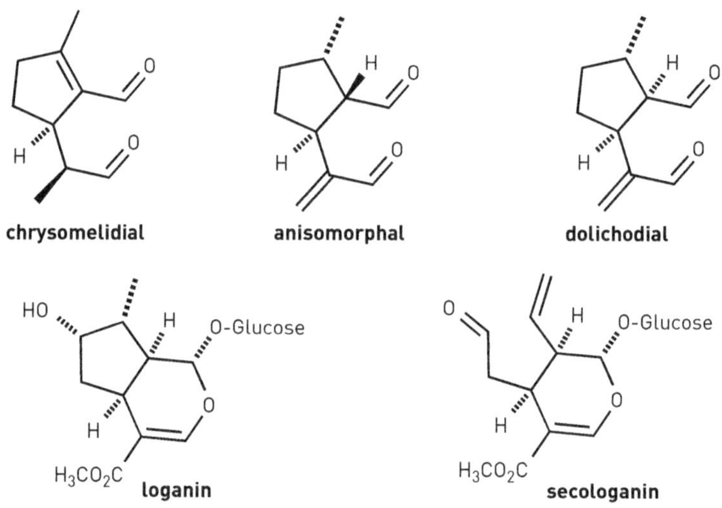

FIGURE 4.23
Three insect-derived iridoids and plant-derived iridoid glucosides loganin and secologanin.

plant-derived glucoside iridoids, **loganin** and **secologanin**, the latter of which plays an important role in the biosynthesis of terpene indole alkaloids which we will address later on in Chapter 7.

SESQUITERPENES (C_{15}): LINKING OF DIFFERENT STARTER UNITS

Given that geranyl diphosphate (GPP) is a primary allylic diphosphate ester, it is able to participate as a substrate for the same kind of head-to-tail linking with IPP units

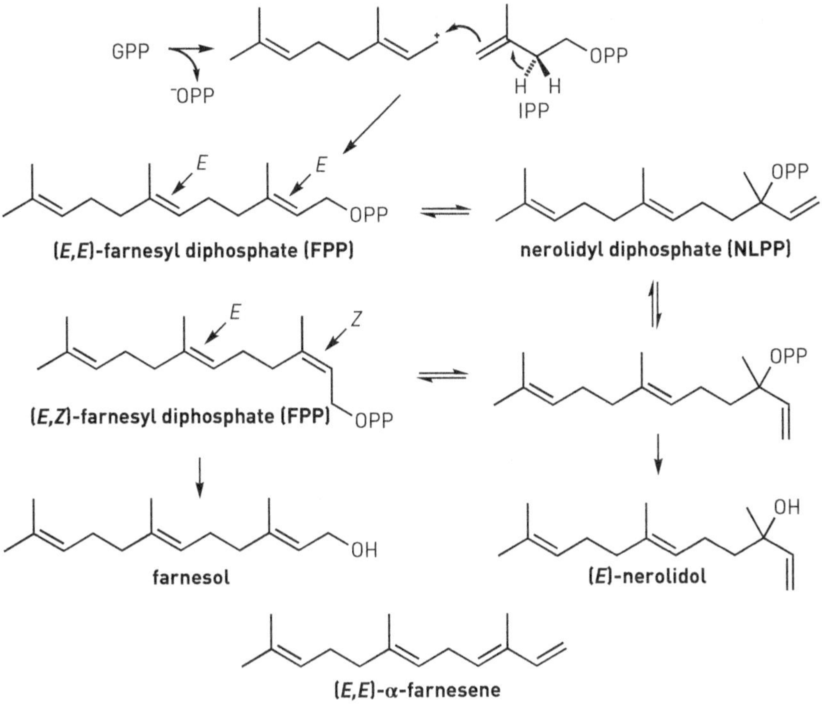

FIGURE 4.24
Linkage of GPP and IPP to give FPP, NLPP, and some sesquiterpene products.

that we previously saw for DMAPP. The result of this linkage is a C_{15} sesquiterpene diphosphate, **(*E,E*)-farnesyl diphosphate** or **FPP.** Like the cationic isomerizations associated with GPP, we find that FPP undergoes ionization and allylic isomerization to give the corresponding tertiary derivative, **nerolidyl diphosphate (NLPP)**, which after bond rotation may then undergo allylic isomerization to give the stereoisomeric derivative, **(*E,Z*)-FPP** as shown in Fig. 4.24.

Farnesol, the alcohol derivative of (*E,E*)-FPP, is present in citronella, lemongrass, and many other essential oils and is widely used in perfumery as a **fixative** to enhance the scent and moderate the volatility of other fragrance compounds. It has also been shown to induce apoptosis (cell suicide) in a number of different cancer cell types and to inhibit the initiation of cancer cell development in some animal models. **Nerolidol** is a fragrant tertiary alcohol from NLPP that is found in ginger, jasmine, lavender, and lemongrass; its ability to penetrate the skin has led to testing for potential use in transdermal drug delivery. As would be expected, elimination of diphosphate and a proton from FPP leads to the common acyclic sesquiterpene **(*E,E*)-α-farnesene**, which is partly responsible for the scent of gardenia and green apples. The biosynthesis of this compound has been extensively studied due to its key role in a condition of stored fruit deterioration known as **scalding** which has been found to involve oxidative transformations of α-farnesene in the skins of affected fruits.

SOME FPP CYCLIZATIONS IN SESQUITERPENE BIOSYNTHESIS

In the formation of cyclic sesquiterpenes, various enzymes in different organisms may establish specific folding patterns for FPP that lead to specific modes of cyclization after initial cation formation via loss of diphosphate, as seen in Fig. 4.25.

The cyclization of **farnesyl cation** leads initially to **germacryl cation** which may then lose a proton to produce **germacrene A** or **B.** For germacrene B, sequential oxidation at the indicated allylic position produces **germacrone**, a bioactive sesquiterpene ketone from *Rhizoma curcuma* which has been shown to inhibit proliferation of breast cancer cell lines via an apoptosis pathway. Alternatively, germacrene A undergoes a more complex sequence of enzyme-controlled, site-specific oxidative transformations, ultimately leading to the sesquiterpene lactone **parthenolide**, the active principal found in the herbal remedy plant **feverfew** (*Tanacetum parthenium*)

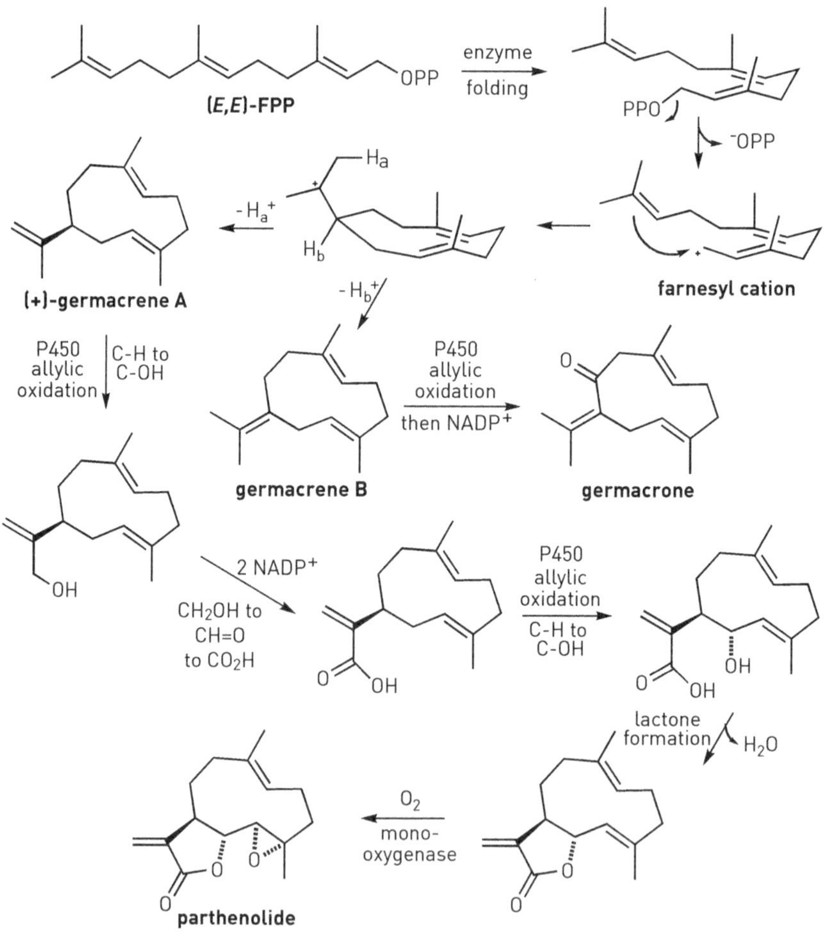

FIGURE 4.25

Cyclization of farnesyl cation to germacrene A or B and conversion to germacrone or parthenolide.

used for reduction of fevers (hence the name of the plant) and arthritis; it has also been shown to be particularly effective for the treatment of migraine headaches.

As might be expected, the stereochemistry of **nerolidyl PP** allows folding patterns and cyclization modes that are not possible via farnesyl PP. A significant example of cyclization of nerolidyl cation is shown in Fig. 4.26, resulting in the formation of **bisabolyl cation**, a key intermediate which has been studied extensively due to its role in the biosynthesis of a variety of different bicyclic and tricyclic sesquiterpenoids.

Simple proton loss from bisabolyl cation as shown gives **(Z)-alpha-bisabolene**, one of the main constituents of opopanax oil. Alternatively, a 1,3-hydride shift (favorable due to the formation of a resonance-stabilized allylic carbocation) affords a second opportunity for cationic cyclization leading to a 6,6-bicyclic ring system which loses a proton to afford **amorpha-4,11-diene**, the terpene precursor to the globally important antimalarial drug **artemisinin**, a sesquiterpenoid lactone isolated from the plant *Artemisia annua*, long used as a medicinal herb in Chinese folk remedies.

In Fig. 4.27, we can see an alternative 1,2-hydride shift for bisabolyl cation which leads to another tertiary carbocation; this cation subsequently cyclizes to give the spirocyclic **acoryl cation** which then may either lose a proton (via pathway a) to afford the spiro[6.5]decane derivative, **α-acoradiene**, a flour beetle aggregation

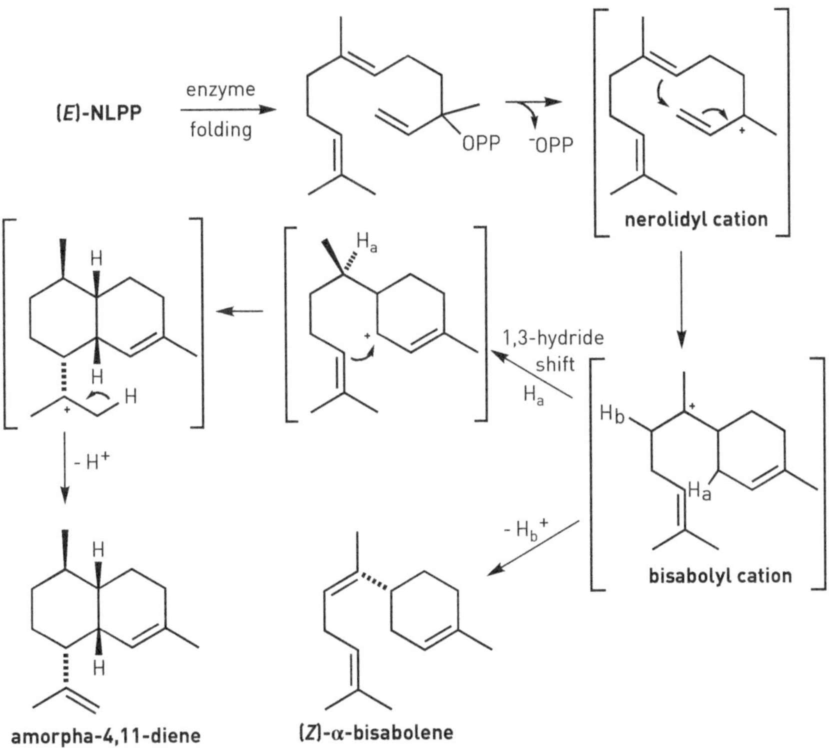

FIGURE 4.26
Nerolidyl to bisabolyl cation and its conversion to mono- and bicyclic sesquiterpenes.

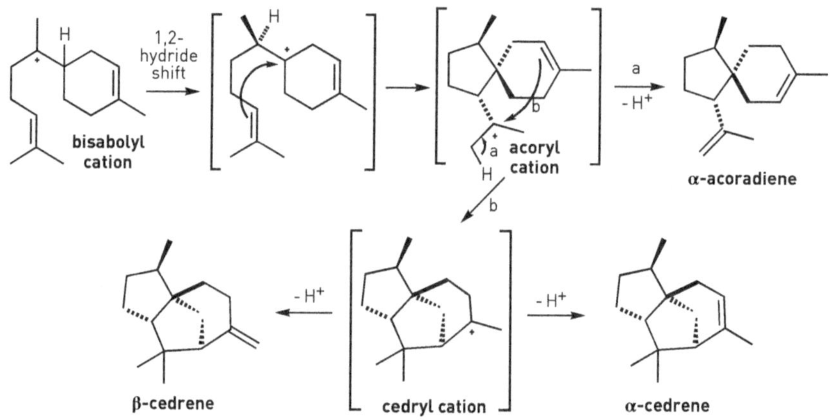

FIGURE 4.27
Bisabolyl cation and its conversion to other bi- and tricyclic sesquiterpenes.

pheromone, or it may undergo intramolecular cyclization (pathway b) to give the tricyclic **cedryl cation**. Loss of an available β-proton yields either **α-cedrene** or **β-cedrene**, two of several fragrant sesquiterpene cedar oil components found in pencil cedar which are responsible for the familiar smell of our favorite writing instrument.

TRICHODIENE AND THE TRICHOTHECENES: HOW TO TRACE A REARRANGEMENT PATHWAY

Another very important aspect of the chemistry of bisabolyl cation is its role in the biosynthesis of **trichodiene**, the fundamental precursor to an especially important family of mold toxins known collectively as the **trichothecenes**, a large family of related toxins including **deoxynivalenol (DON)** and **T-2 toxin** which are produced by various mold species including *Fusarium* and *Acremonium* among others (Fig. 4.28). The T-2 mycotoxin in particular is a significantly toxic trichothecene that has been implicated in a number of historic outbreaks of toxin-related illness and death as the result of the ingestion of moldy grains, including a massive case of toxin poisoning in post-WWII Russia and the famous Plague of Athens in 430 B.C.

The unusual structure of trichodiene relative to bisabolyl cation indicates that converting one into the other is a cationic process involving a number of different specific hydride and methyl shifts. This provides us with an opportunity to see how organic chemists have used isotopic labeling studies to determine the likely sequence of events that produce various substitution patterns in such complex rearrangements.

The process outlined in Fig. 4.29 begins with the use of **^{13}C-labeled DMAPP** as shown. Note that providing labeled DMAPP precursors to an organism in "feeding studies" would also lead to formation of some **^{13}C-labeled IPP**; therefore we would expect the eventual formation of **^{13}C-labeled NLPP** and thence **bisabolyl cation** with enhanced ^{13}C content at specific positions within the structure.

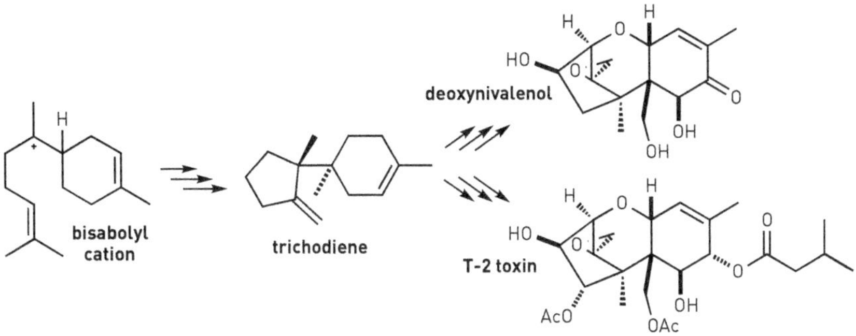

FIGURE 4.28
Rearrangement of bisabolyl cation to trichodiene and conversion to trichothecene toxins.

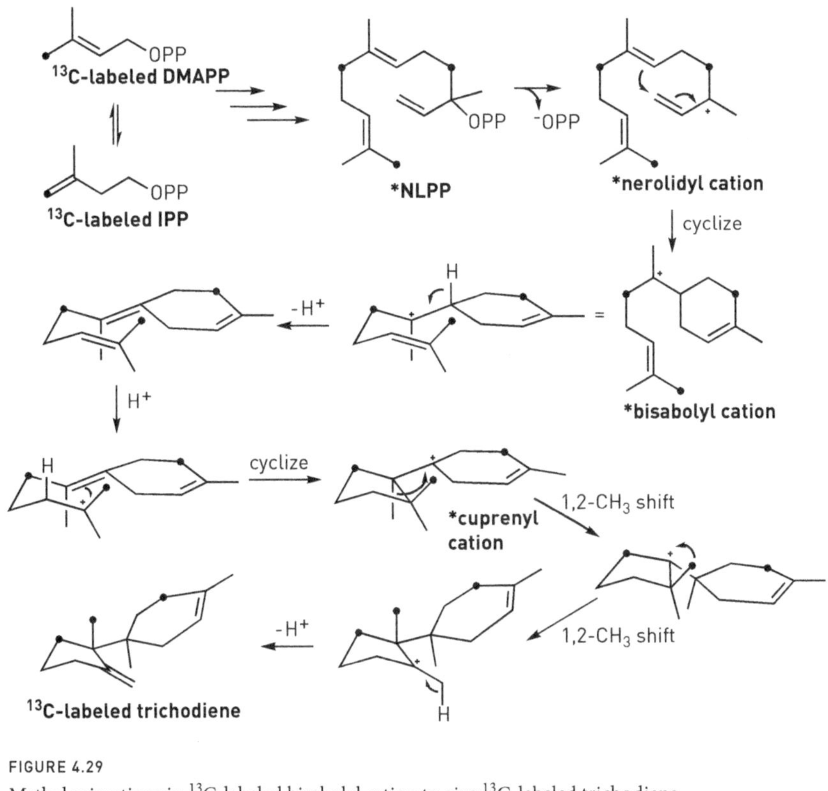

FIGURE 4.29
Methyl migrations in ^{13}C-labeled bisabolyl cation to give ^{13}C-labeled trichodiene.

By using ^{13}C NMR to determine where the labeled carbon atoms eventually reside in the final product **trichodiene**, one may infer the sequence of events involved in the various hydride and methyl migrations that must have taken place during the rearrangement. Thus, rearrangement of labeled bisabolyl cation may be formulated as undergoing initial loss of a proton followed by protonation of the remote alkene function shown to give a carbocation that can cyclize to generate the 5-membered ring, thereby producing labeled **cuprenyl cation**, a known intermediate in the process (a more traditional alternative formulation of this sequence involves initial cyclization of bisabolyl cation to produce the five-membered ring, followed by a 1,5-hydride shift to produce cuprenyl cation). From here, a sequence of 1,2-methyl shifts produces the final carbocation which then undergoes loss of a proton to produce **trichodiene** with enhanced ^{13}C-content at the indicated positions. While broken down here into a series of individual steps to aid in visualization of the overall sequence, it should be recognized that this is probably a concerted process, with cyclizations, methyl migrations, and proton loss occurring simultaneously. Labeling studies such as this one have proven to be invaluable in helping to determine the complex mechanistic sequences likely to be involved in the biosynthetic conversion of many different intermediates into specific products in diverse enzyme-mediated processes, and we will be seeing other uses of this technique throughout the remaining chapters.

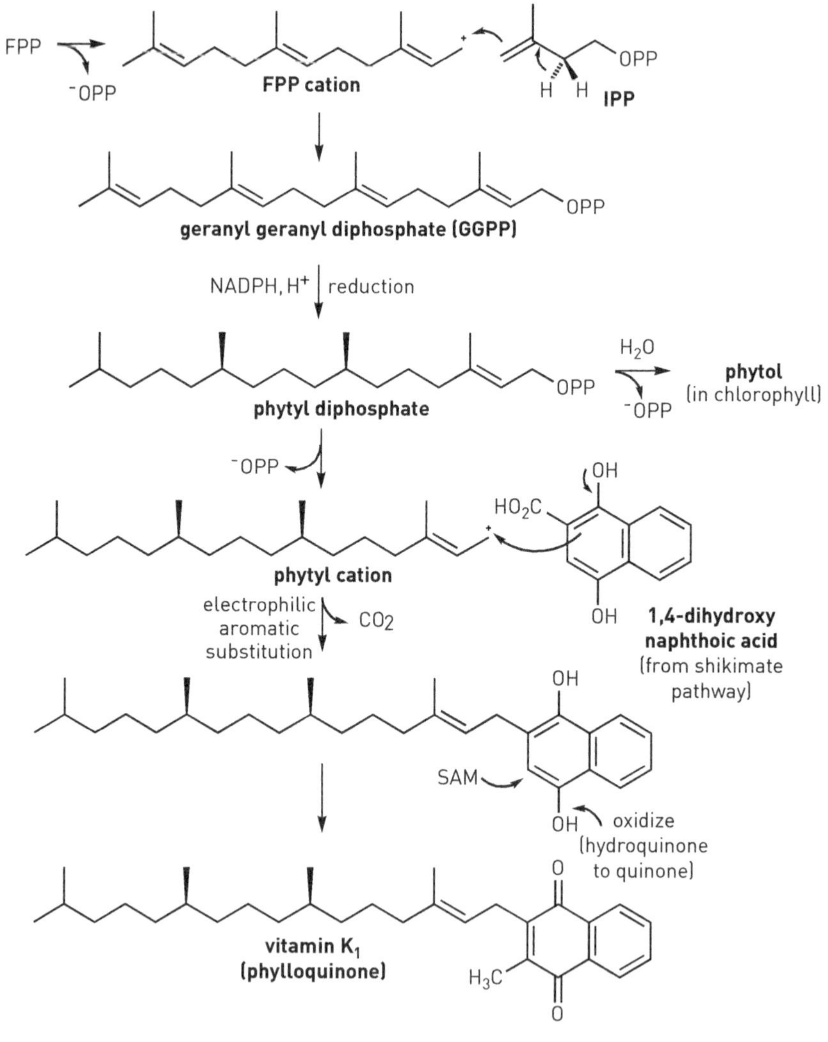

FIGURE 4.30

Reaction of FPP and IPP to give GGPP, precursor to C20 diterpenes and vitamin K_1.

What has been shown here may be thought of as only the tip of the iceberg with regard to the number of different possible sesquiterpene compounds produced in nature, with many more being discovered on a regular basis and all sharing their origin in farnesyl diphosphate. The isolation, characterization, and elucidation of biosynthetic pathways associated with these compounds represent some of the most significant achievements of the organic chemistry community over the last 50 years. And the wealth of information obtained in these studies finds applicability not only in many other related fields, but also serves as the foundation upon which further understanding of more complex terpene structures is built. So, while we still have far to go in exploring the vast world of terpenoid organic structures, we are now armed with a depth of knowledge that will make what remains much easier to grasp. So now it's time to move on.

DITERPENES (C_{20}): TAKING IT TO THE NEXT LEVEL OF MOLECULAR COMPLEXITY AND DIVERSITY

Just as we used GPP and IPP as the building blocks to produce the FPP from which the sesquiterpenes were all derived, the same sequence may be applied to the utilization of FPP and IPP in combination to produce the basic starting material for the production of all diterpenes, **geranyl geranyl diphosphate** or **GGPP**, as shown in Fig. 4.30. Thus, in spite of its name, this C_{20} derivative is *not* produced from dimerization of two C_{10} geranyl PP units. Among the important acyclic diterpene derivatives is a reduced form of GGPP known as **phytyl PP** which is used in conjunction with the shikimic acid pathway product **1,4-dihydroxynaphthoic acid** in the biosynthesis of the mixed pathway product **vitamin K$_1$**, an essential human blood-clotting factor. The hydrolysis product of phytyl PP is **phytol**, an alcohol constituent incorporated into the structure of chlorophyll that is also used in the manufacture of synthetic vitamin E.

CYCLIC DITERPENES: FROM BASEBALL AND PLANT HORMONES TO ANTICANCER DRUGS

Cyclization of GGPP is involved in the biosynthesis of **abietic acid**, the most abundant of a group of related compounds found in the rosin of coniferous trees, a substance which acts to help seal wounds in such trees and has long been used as a watertight caulking for ships and also as an agent to improve the grip on baseballs and baseball bats or the tackiness of the bow used with stringed instruments (Fig. 4.31). Anyone who has gotten pine tree resin on their hands is familiar with its uniquely dry and tacky sensation.

Gibberellic acid, a potent diterpenoid plant hormone first identified in Japan as a byproduct of the plant pathogen *Gibberella fujikuroi* has been found to regulate

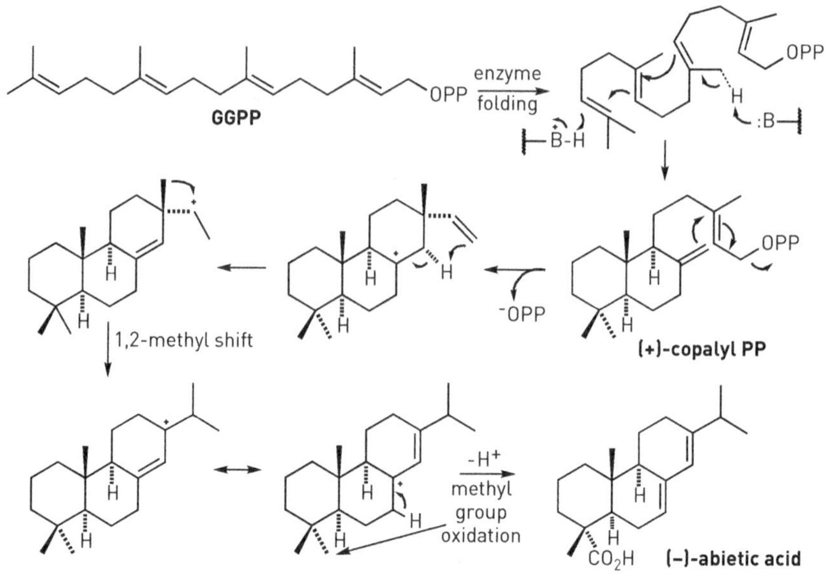

FIGURE 4.31
Cyclization of GGPP and transformations to produce abietic acid.

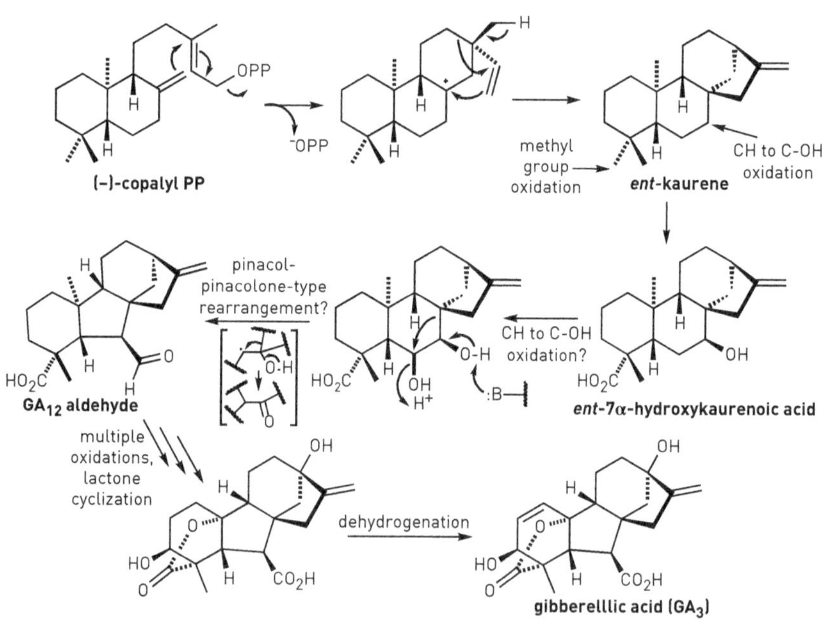

FIGURE 4.32
Cyclization of (−)-copalyl PP and subsequent transformations to produce gibberellic acid.

stem and root growth as well as seed germination and dormancy. Its proposed biosynthesis from (−)-copalyl PP is shown in Fig. 4.32. It begins with a cyclization and 1,2-carbon migration to produce the intermediate compound ***ent*-kaurene**. Oxidation of one geminal dimethyl group to the carboxylic acid and C–H to CH–OH oxidation leads to ***ent*-7α-hydroxykaurenoic acid** which is then oxidatively converted to the key ring-contracted intermediate, **GA_{12}-aldehyde**. The ring contraction mechanism shown is speculative, being formulated as a pinacol-pinacolone-type rearrangement from a vicinal diol (pinacol) intermediate, though alternative radical-type mechanisms are also possible. Additional oxidations, a decarboxylation, lactonization, and dehydrogenation with introduction of a ring double bond finally completes this complex sequence of biotransformations to give gibberellic acid or **GA_3**.

A final example of diterpene cyclizations is a very important one, as it outlines the basic biosynthetic pathway to the important cancer chemotherapy agent **taxol**, mentioned earlier as a compound widely used in the treatment of breast, ovarian, lung, head, and neck cancers as well as AIDS-related Kaposi's sarcoma. Originally isolated from the bark of the **Pacific yew tree**, *Taxus brevifolia*, intense interest in taxol in the late 1970s stemmed in part from the discovery that the compound manifested a previously unknown mechanism of anticancer action that involved stabilization of cell microtubules in a way that interfered with their normal breakdown during the process of cell division. While its effectiveness was clearly demonstrated, a major drawback to clinical use of taxol was its relative scarcity, after it was determined that more than 350,000 Pacific yew trees would have to be stripped annually to yield sufficient amounts of taxol for projected needs. This drawback was initially addressed by semisynthetic routes to taxol, but more recently, plant cell fermentation technology has been developed that now supplies most of the taxol produced worldwide. Thus, the biosynthetic pathway to this compound, shown in abbreviated form in Fig. 4.33, is not only interesting from an organic chemistry standpoint but also represents a highly valuable commercial process for taxol production.

Initial carbocationic cyclization of GGPP, initiated by loss of $^{-}$OPP, yields a tertiary carbocation that loses a proton to give a cyclic diene; this diene then captures the same proton in an enzyme-mediated 1,5-proton transfer process to initiate a second carbocationic cyclization; subsequent proton loss from the resulting cation yields **4,11-taxadiene,** the basic diterpene precursor to taxol. Allylic oxidation of 4,11-taxadiene at C-5 with rearrangement yields an intermediate allylic alcohol which is esterified by the action of acetyl-CoA to produce the intermediate product **4,11-taxadien-5α,10β-diol monoacetate**. The remaining steps involve a complex series of oxidations and esterifications to produce the terpenoid alcohol precursor **10-deacetylbaccatin III**; sequential steps for construction and installation at the C-5 hydroxyl of an ester side chain derived from β-phenylalanine and phenylisoserine finally yields **taxol**. As mentioned earlier, semi-synthetic routes to taxol developed in the 1990s helped to relieve potential stress on the Pacific yew population by synthesizing and installing the required ester side chain onto 10-deacetylbaccatin III harvested from the more abundant European yew. So, while the plant cell fermentation technology referred to earlier continues to provide taxol, efforts are still underway to find other economical and ecologically

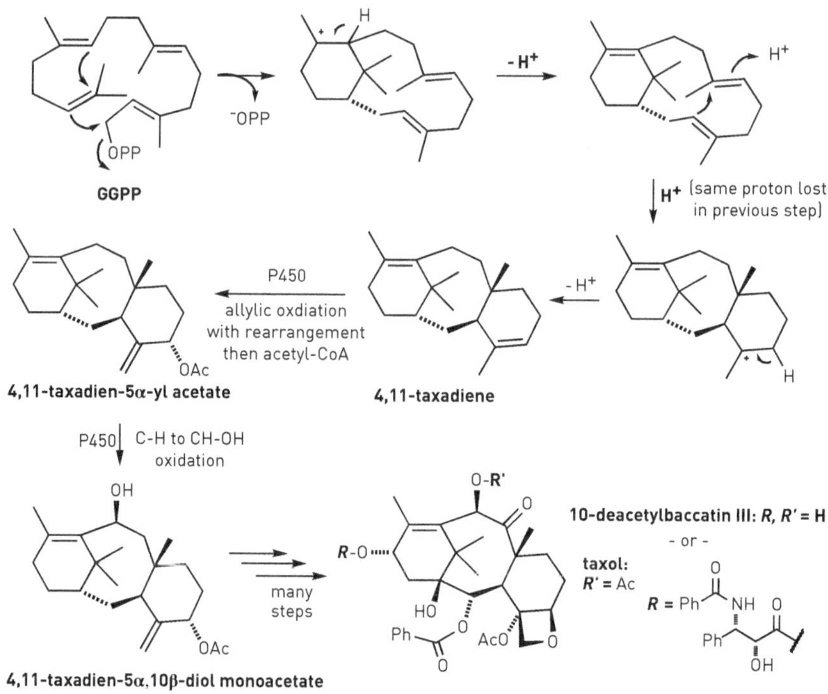

FIGURE 4.33

Cyclization of GGPP to yield 4,11-taxadiene and its subsequent conversion to taxol.

favorable natural sources of these compounds from plant waste materials like nut shells or seed hulls.

SESTERTERPENES (C_{25}): LESS COMMON, MORE COMPLEX

The C_{25} group of terpenes is derived from initial linkage of IPP and GGPP units to produce **geranylfarnesyl PP (GFPP)** as shown in Fig. 4.34. Representative members of this somewhat rare family of natural products, derived mainly either from fungi, as with **ophiobolin F**, or from marine organisms, as with **cheilanthadiol**, are shown.

The **ophiobolin F** cationic cyclization is initiated by loss of ⁻OPP, with a GFPP folding pattern enforced by the mediating enzyme which presumably holds the reacting π-bonds in the necessary alignment to facilitate the resulting cascade of C–C bond and ring formation. The intermediate tertiary carbocation formed then undergoes a concerted, stereospecific 1,5-hydride shift leading to further cyclization and discharge of the final stage carbocation by nucleophilic attack by water. Several other members of the ophiobolin group are produced by further oxidative transformations of this precursor, including **ophiobolin A**, a phytotoxic calmodulin inhibitor

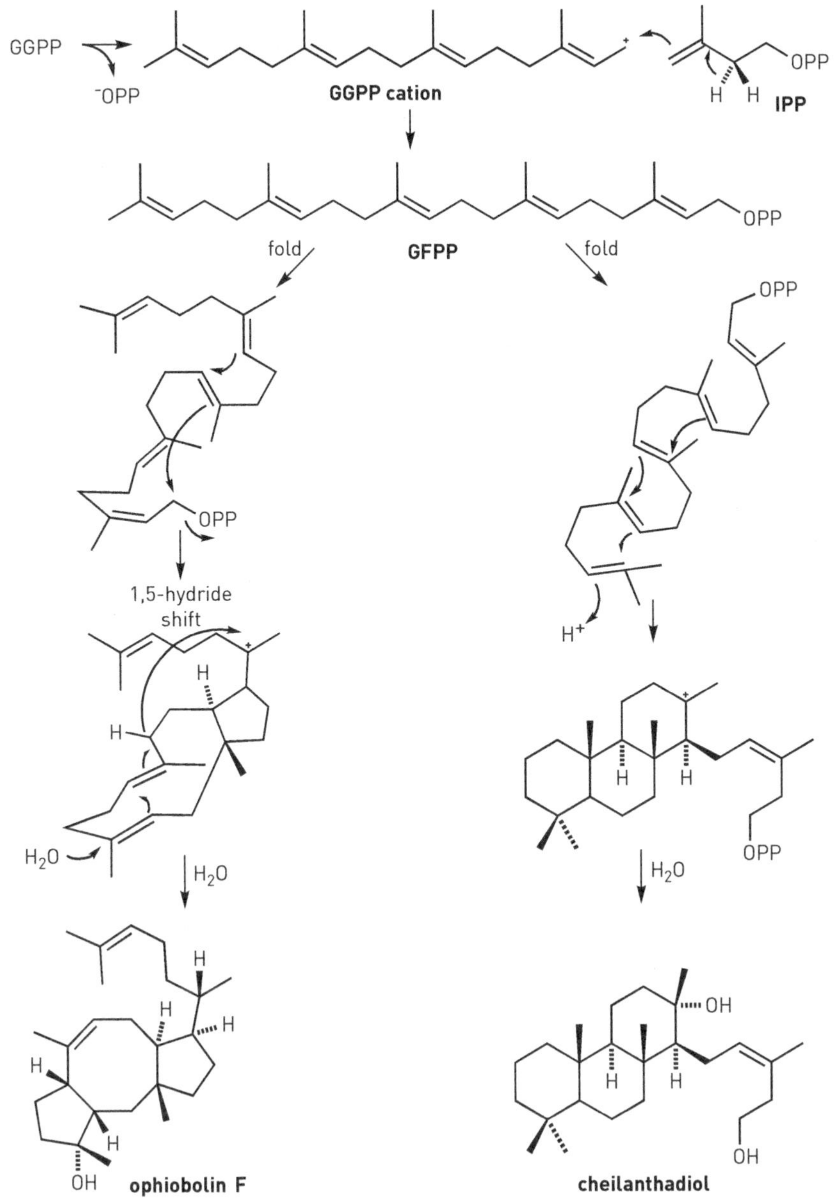

FIGURE 4.34

Formation of GFPP and cyclization to some representative sesterterpenoid products.

that has also shown antitumor and antibacterial activity. An alternate folding pattern and cyclization mode for GFPP is shown in the biosynthetic sequence leading to **cheilanthadiol**. Note that cyclization here is initiated by π-bond protonation rather than initial loss of $^{-}$OPP. Again, following the cyclization cascade, the final carbocation is discharged by water which presumably also hydrolyzes the diphosphate linkage to produce the diol product as shown.

TRITERPENES AND STEROIDS: ANOTHER CASE OF IRREGULAR LINKAGE OF TERPENE UNITS

For the generation of C_{30} triterpenes and triterpenoids, we might expect head-to-tail linkage of a C_{25} GFPP unit and a C_5 IPP unit for the usual terpene chain extension process. Instead, two C_{15} units of FPP are linked together in what is essentially a "tail-to-tail" fashion, mediated by an enzyme known as *squalene synthase*, leading to production of the fundamental C_{30} triterpene hydrocarbon, **squalene**, from which all other triterpenes, steroids, sterols, and related compounds are ultimately derived (Fig. 4.35). Squalene was originally isolated from shark liver oil (hence its name, which comes from that of the shark, *Squalus* spp.), though many plant oils such as olive, palm, and wheat-germ oil also contain significant amounts of squalene.

Upon careful consideration of the structure of squalene, we have to wonder exactly how this tail-to-tail linkage of FPP units actually takes place mechanistically. After all, instead of pairing a nucleophilic carbon with an electrophilic carbon to form a C–C bond in the usual fashion, this process somehow involves the joining of two electrophilic carbons (each with a leaving group, OPP), to make a single C–C bond. Fig. 4.36 provides the standard sequence of transformations believed to be involved in this unusual C–C bond-forming process.

The sequence of events here is quite similar to the nonstandard linkage of DMAPP units seen in the formation of the irregular monoterpenes (see Figs. 4.20 and 4.21). Thus, ionization of the first FPP unit to its carbocation is followed by attack from the π-electrons of the second FPP unit, leading to C–C bond formation and an enzyme-stabilized carbocation. Discharge of this carbocation by subsequent deprotonation-cyclopropane ring closure leads to the formation of **presqualene diphosphate**, a key intermediate in this process which can be isolated and characterized in the absence of NADPH. Loss of $^{-}$OPP from presqualene PP leads to a

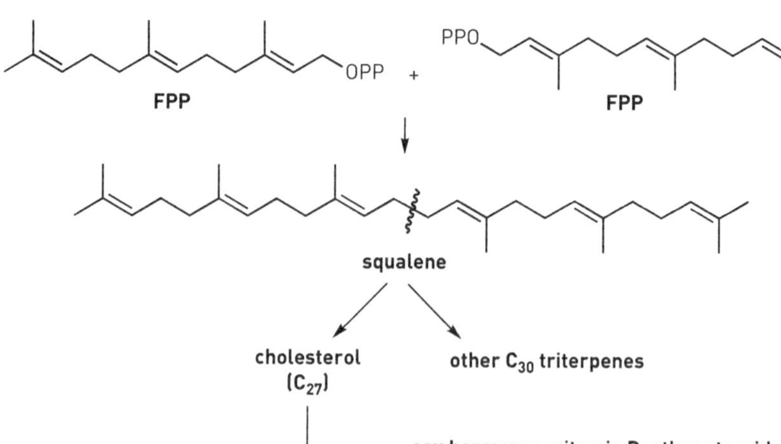

FIGURE 4.35
Linkage of two FPP units to form squalene for biosynthesis of other triterpenes and steroids.

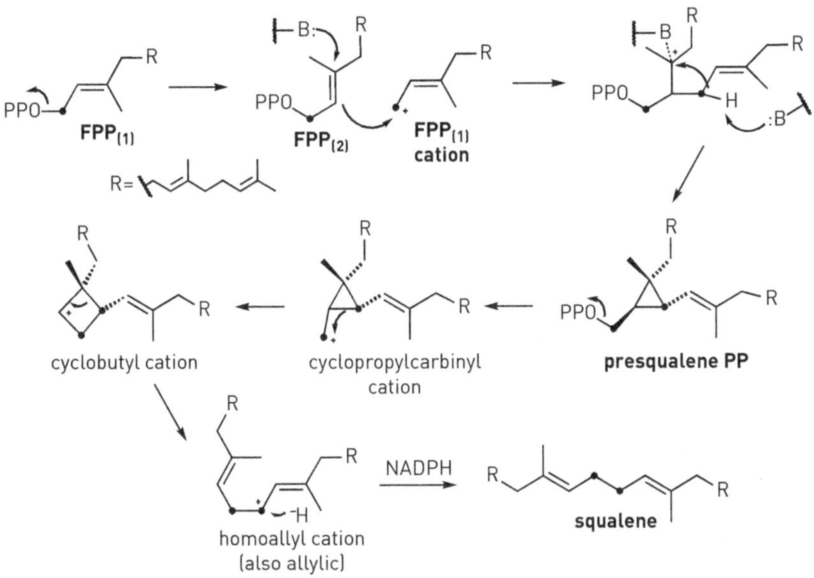

FIGURE 4.36
Tail-to-tail mechanism for squalene formation from two FPP units.

cyclopropylcarbinyl cation which, as we have seen previously, can lead to products derived from the *cyclopropylcarbinyl-cyclobutyl-homoallyl cation* system. In this instance, the homoallyl carbocation component of this system is also allylic and therefore especially stable; discharge of this carbocation by addition of hydride ion from NADPH completes the sequence, affording **squalene,** shown with the linked terminal carbons from each FPP unit highlighted for emphasis.

OXIDOSQUALENE AND STEROID BIOSYNTHESIS: CYCLIZATION TO LANOSTEROL AND BEYOND

The significance of squalene cannot be overstated as it is the basic template upon which all steroidal compounds are ultimately built. The fundamental biosynthetic transformation that sets all this into motion is the initial conversion of squalene to the epoxide derivative (*S*)-2,3-oxidosqualene (or (*S*)-2,3-epoxysqualene), catalyzed by squalene monooxygenase (Fig. 4.37).

The cationic cyclization of 2,3-oxidosqualene is of enormous significance, and different organisms deal with this process in different ways. For instance in some plants, 2,3-oxidosqualene is converted to an important plant sterol known as **cycloartenol**, while in animals and fungi it is initially converted to **lanosterol**, the essential precursor to all other steroids produced by this path. Both compounds are derived from a common **protosteryl carbocation** intermediate which arises from a "chair-boat-chair" cyclization conformation, as shown in Fig. 4.38, and the divergence of the fates of this cation are ultimately enzyme-dependent. The same set of 1,2-hydride and methyl migrations takes place in both systems but ultimately

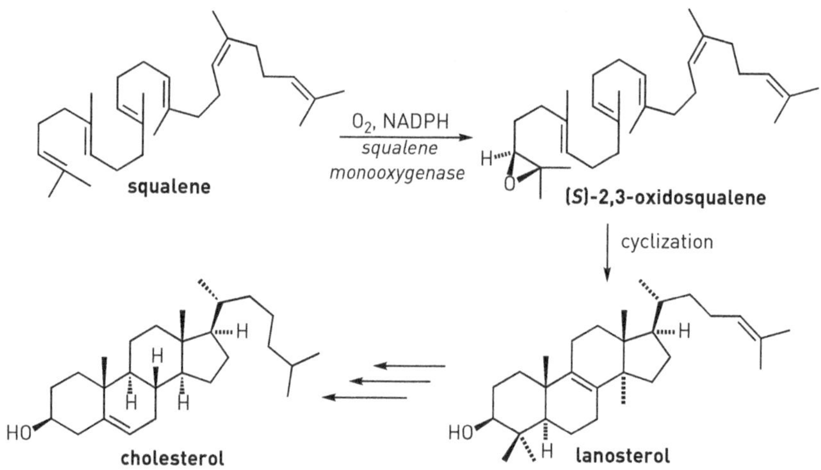

FIGURE 4.37
Conversion of squalene to (*S*)-2,3-oxidosqualene, precursor to lanosterol and cholesterol.

terminates in carbocation discharge by cyclopropane ring formation in plants while proton loss to yield an alkene is the final act in animals and fungi.

The sequence of stereospecific 1,2-migrations of hydride and methyl in the conversion of oxidosqualene to both products is probably concerted and involves precise orbital alignment in which each migrating group is *anti* to the one preceding it, as shown in Fig. 4.39 for the migrations involved in formation of **lanosterol.** Note the *cis* relationship between the last remaining hydrogen atom (ultimately lost as a proton) and the methyl group immediately adjacent to it: at this point, 1,2-migrations cease since the methyl group is not *anti* to the H-atom and π-bond formation results instead, leading to lanosterol.

In animals, lanosterol is the most significant product since it is ultimately converted to cholesterol and all the other essential steroid structures, as we will see shortly. In plants however, there is enormous variety in the various ways that triterpenoid cyclizations can take place and in the products ultimately produced. For instance, different enzymes can enforce alternate folding patterns for 2,3-oxidosqualene cyclization, such as the "chair-chair-chair-boat" folding pattern employed to yield **dammarenyl cation** (Fig. 4.40). This cation can then undergo a variety of different fates including a sequence of 1,2-shifts of hydride and carbon similar to those seen for protosteryl cation to give **euphol**; alternatively, simple capture of water by the cation gives the related product **dammerenediol.**

In some instances oxidosqualene cyclization can be interrupted and then extended, leading to unusual structural features. This is illustrated in Fig. 4.41 in the presumed biosynthetic origin of **malabaricol**, the main triterpene constituent of yellow pigments from the woody plant *Ailanthus malabarica*. Initial cyclization to the indicated tricyclic carbocation via the indicated folding pattern is then terminated by water trapping, after which a second epoxidation occurs, followed by

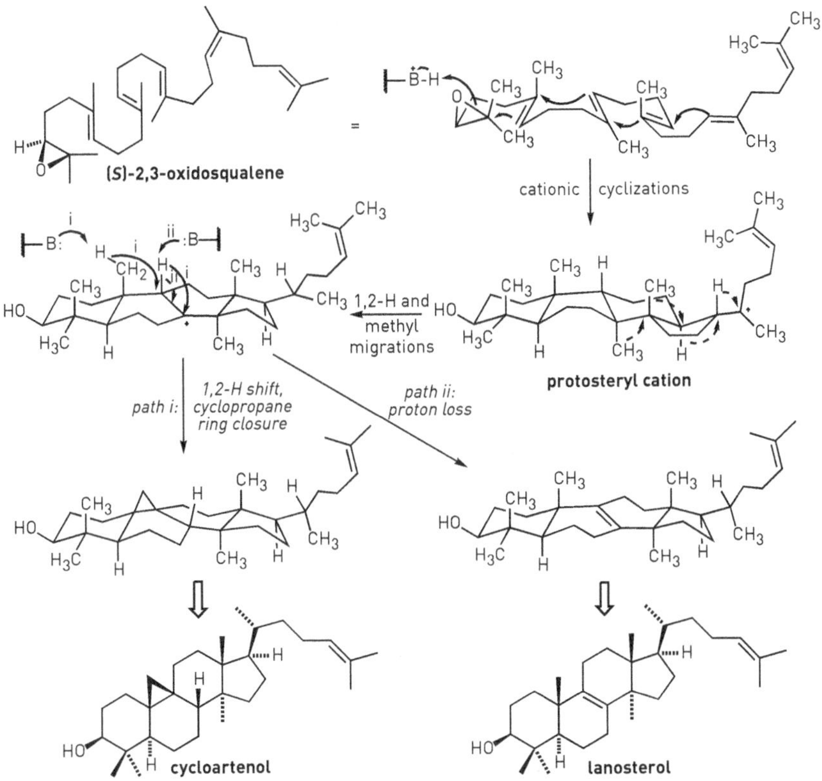

FIGURE 4.38
Cyclization of 2,3-oxidosqualene to protosteryl cation and final paths to cycloartenol and lanosterol.

another acid-catalyzed intramolecular nucleophilic ring opening of the epoxide which produces a five-membered cyclic ether. A final oxidation of alcohol to ketone produces the final product.

Before moving on to the complex chemistry involved in the conversion of lanosterol to cholesterol and beyond, we take a brief look at a small sampling of the remarkable structural variety possible in some of the medicinally significant plant-derived sterols shown in Fig. 4.42. Note that these compounds are highly modified structures relative to their precursor, cycloartenol; like cholesterol and other steroids, they contain fewer than the 30 carbons characteristic of standard triterpenes. **Stigmasterol**, a plant sterol believed to be beneficial in the prevention of certain cancers, is found in the oils of soybean, calabar bean, various nuts and seeds, and in a variety of medicinal herbs such as American ginseng. **Diosgenin**, the aglycone of a glycosidic sterol extracted from the tubers of wild yam, is of considerable economic significance as an abundant natural source of raw material used for the commercial semisynthesis of medicinal steroids such as cortisone, pregnenolone, progesterone, and other pharmaceutical steroids. Another useful example is **β-sitosterol**, a widely distributed phytosterol found in corn oil, soybean, saw palmetto, and avocado

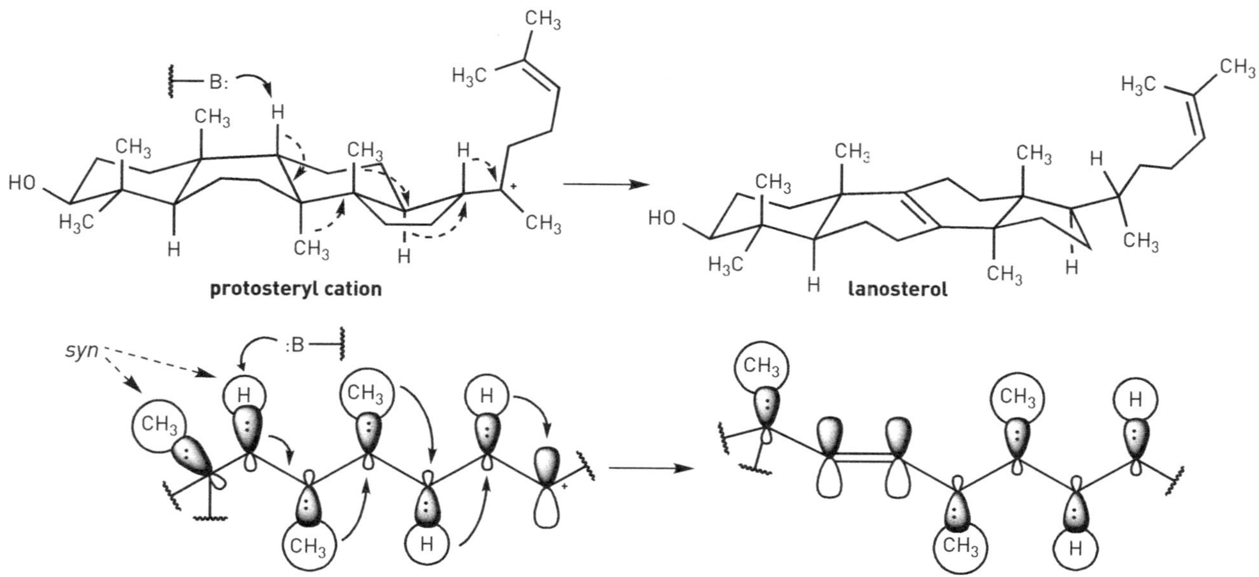

FIGURE 4.39

The *anti* relationship of migrating atoms or groups in conversion of protosteryl cation to lanosterol.

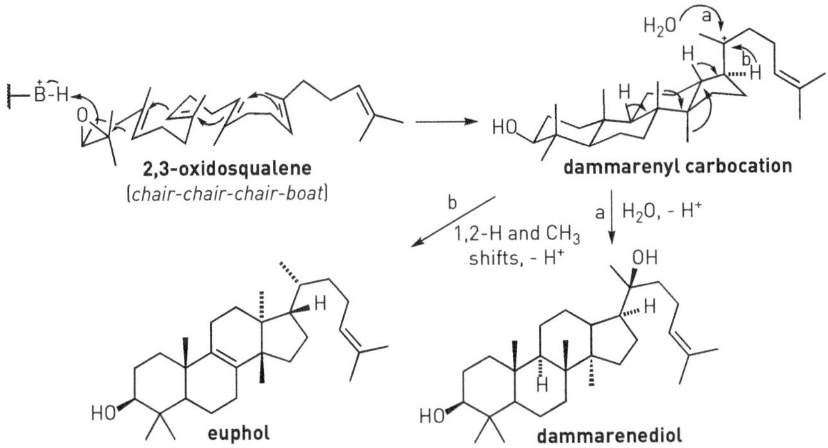

FIGURE 4.40
Chair-chair-chair-boat folding for oxidosqualene leading to dammarenediol or euphol formation.

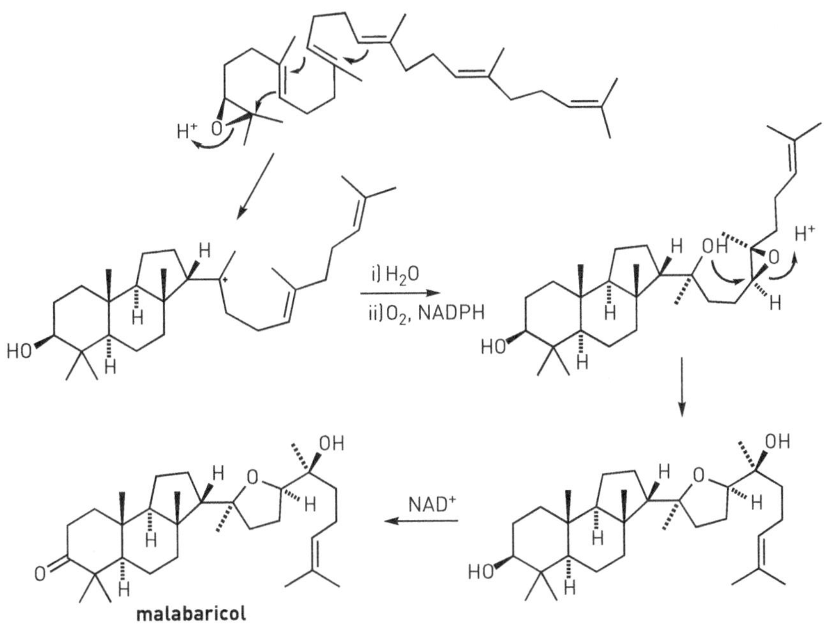

FIGURE 4.41
Alternative 2,3-oxidosqualene cyclization and subsequent biosynthetic steps to malabaricol.

which is often used to help treat elevated cholesterol levels by inhibiting cholesterol absorption in the intestine. The compounds **oubain** and **digoxin** are examples of the so-called cardiac glycosides (**digoxigenin** is the aglycone of digoxin), powerful compounds that act as inhibitors of the sodium pump (Na^+/K^+ ATPase pump) involved in heart electrical activity. While similar in action, digoxin, isolated from the foxglove plant, has replaced oubain for treatment of various heart conditions such as atrial fibrillation and flutter and congestive heart failure uncontrolled by other

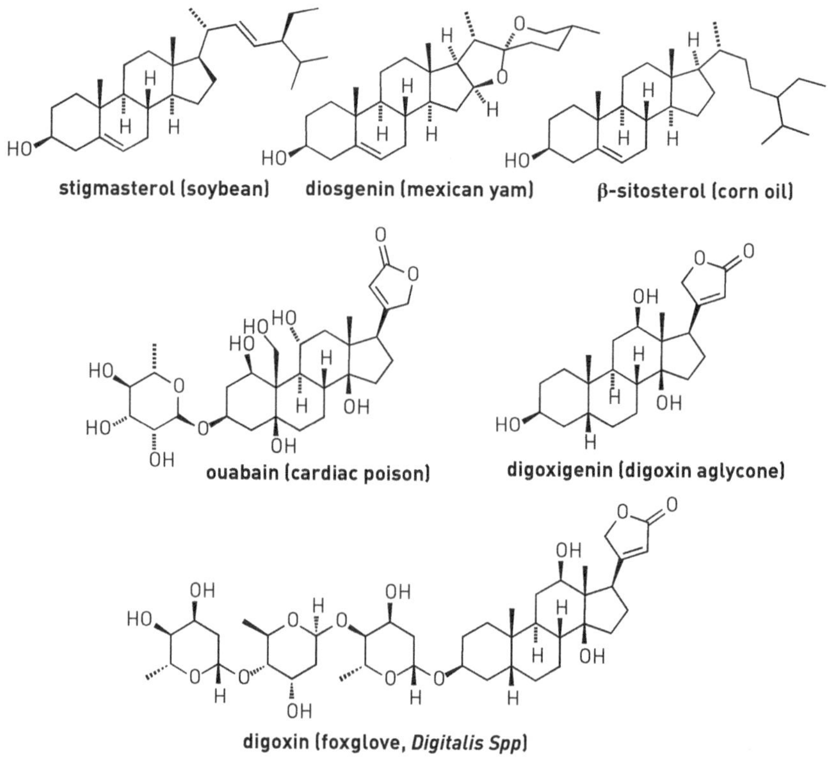

FIGURE 4.42
Examples of some significant plant sterols and cardiac aglycones and glycosides.

drugs. Oubain, derived from the seeds of the African plant *Strophanthus gratus*, was originally used in concentrated form by native tribesmen as an effective arrow poison for bringing down large game animals, presumably by inducing cardiac or respiratory failure.

CONVERSION OF LANOSTEROL (C_{30}) TO CHOLESTEROL (C_{27}): WHERE DID THE CARBONS GO?

Conversion of **lanosterol** to **cholesterol** is a complex, 19-step process of double-bond reductions, isomerizations, oxidations, reductions, decarboxylations, and eliminations occurring in four different regions of the tetracyclic ring system. The specifics of the process are summarized briefly in Fig. 4.43. As can be seen, the overall conversion of the lanosterol skeleton involves removal of a double bond (Region 1), removal of the C-14 methyl group (Region 2), sequential removal of the geminal dimethyl groups (Region 3) and a double-bond rearrangement (Region 4). While complex overall, many steps are fairly straightforward and some represent typical processes used for methyl group removal from other structures.

Step 1 is a typical reduction of the CH=CH to CH_2–CH_2 type using NADPH/H^+ and this is the only step in Region 1 of the tetracyclic structure. In Region 2,

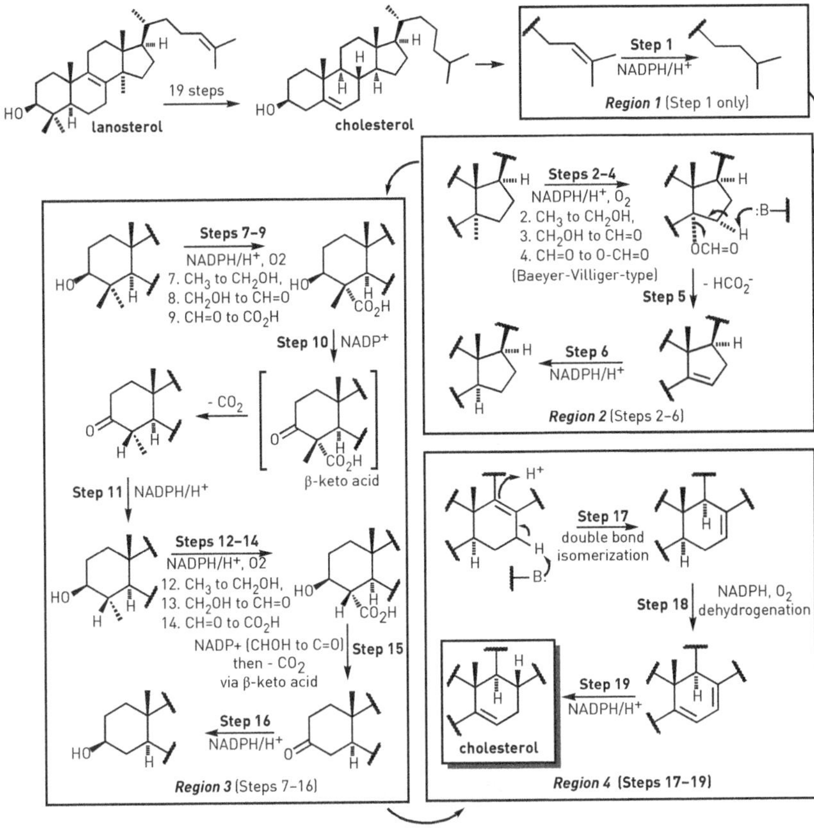

FIGURE 4.43
Conversion of lanosterol to cholesterol: 19 steps in four different regions.

steps 2–3 involve sequential C–H to C–OH oxidations of the C-14 methyl group of lanosterol, leading to the aldehyde. For simplicity's sake, step 4 is formulated here as Baeyer–Villiger-type oxidation to give a formate ester followed by a 1,2-elimination of formate to yield the alkene, though a more complex radical-type mechanism has been shown to be more likely for this step. A final CH=CH to CH_2–CH_2 reduction via NADPH/H^+ in step 6 gives the saturated cyclopentane ring with the indicated *trans* ring juncture. Moving to Region 3, we again see sequential oxidations of one of the geminal dimethyl groups, ending in formation of the corresponding carboxylic acid in steps 7–9. An $NADP^+$ oxidation of the cyclohexanol hydroxyl group in step 10 leads to the corresponding ketone after which the resulting intermediate β-keto acid undergoes decarboxylation as expected to give the monomethyl-substituted ketone. In step 11, the ketone function is reduced back to the alcohol oxidation state by NADPH/H^+, and the same oxidation sequence seen in steps 7–9 is repeated on the remaining methyl group in steps 12–14. Re-oxidation of the cyclohexanol hydroxyl in step 15 leads once more to a β-keto acid and subsequent decarboxylation. Final NADPH/H^+ reduction of the ketone function again affords the cyclohexanol component. To complete the conversion, the cyclohexene double bond in Region 4 of the tetracycle must undergo a change in position within the ring. This is accomplished by a simple isomerization in step 17, formulated as a simple protonation–deprotonation sequence. This is followed by an oxidative dehydrogenation to give a conjugated diene system in step 18 (the intermediate formed in this step, known as **7-dehydrocholesterol**, will play a role later on in the formation of vitamin D). The final double bond reduction in step 19 completes the process, affording **cholesterol**.

CONVERSIONS OF CHOLESTEROL: PRODUCTION OF THE SEX HORMONES

Cholesterol is the precursor to many other steroid structures, chief among these are the male and female sex hormones. As shown in Fig. 4.44, this process begins by removal of a six-carbon fragment from the acyclic portion of the structure. This is accomplished by sequential C–H to C–OH oxidations leading to a vicinal diol. Oxidative glycol cleavage produces the steroidal prohormone **pregnenelone.** Oxidation of the cyclohexanol hydroxyl to the corresponding ketone, accompanied by double bond isomerization, leads to **progesterone**, a hormone involved in the menstrual cycle and pregnancy in humans and which has a number of medicinal uses, including supplementation of declining natural levels of the hormone due to aging (the yam sterol **diosgenin** (see Fig. 4.42) is used for the manufacture of semisynthetic progesterone). The derivative **17-α-hydroxyprogesterone**, formed by a typical P450-type oxidation process, provides one of two possible pathways for biosynthesis of the important sex hormone precursor **androstenedione.** The pathway shown involves oxidative cleavage of the methyl ketone side chain of 17-α-hydroxyprogesterone, formulated here as another example of a Baeyer–Villiger type oxidation followed by loss of AcOH to give the cyclopentanone ring system of androstenedione as shown (the other pathway to androstenedione begins with pregnenolone instead of progesterone).

Androstenedione was taken as a supplement (known as "andro" prior to and during the time of the major league baseball doping scandals) to boost testosterone levels and to increase muscle mass, though these claims have not been substantiated over the years, and its use has now been banned in most sports. Its in vivo conversion to **testosterone** requires only NADPH reduction of the cyclopentanone function to the corresponding alcohol, while its conversion to the female sex hormones **estrone** and **estradiol** involves a complex sequence of C-10 methyl group oxidation and elimination that, for simplicity's sake, is formulated here as being essentially the same process used in removal of the C-14 methyl group of lanosterol (Fig. 4.43, Region 2), with elimination of formate being followed here by aromatization to the phenolic ring of estrone. NADPH reduction of the cyclopentanone function of estrone gives the corresponding alcohol, estradiol. Taken together, estrone, estradiol,

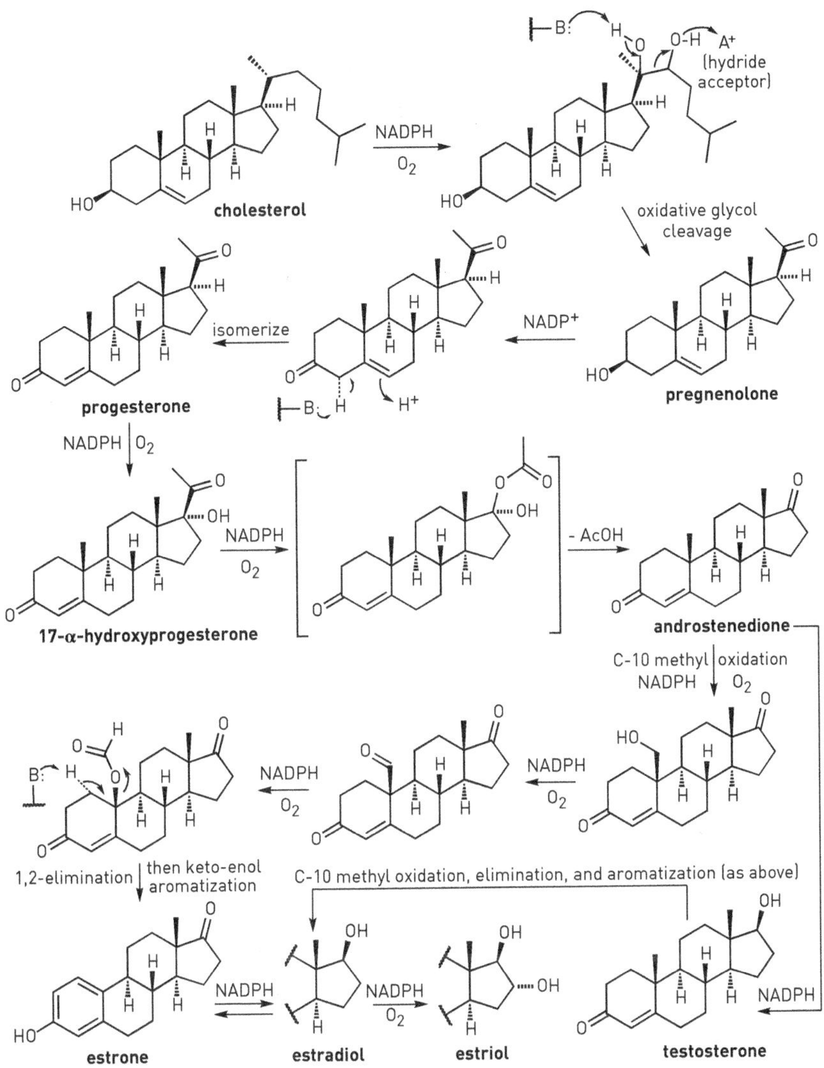

FIGURE 4.44
Pathways involved in conversion of cholesterol to various steroidal hormones.

and estriol form the fundamental group of natural estrogenic female hormones which have been used in both contraceptive formulations and in postmenopausal hormone replacement therapy for the prevention of osteoporosis.

Another group of steroid hormones of some significance is the **corticosteroids** (Fig. 4.45), so called due to their origin from the adrenal cortex in animals. They are ordinarily categorized into two main groups: the **glucocorticoids** and the **mineralocorticoids**, the former deriving their name from the role they play in glucose metabolism and the latter from their association with the management of sodium, potassium, and water levels in certain tissues. Glucocorticoids are usually typified by the presence of the primary/tertiary α-dihydroxyketone function at the

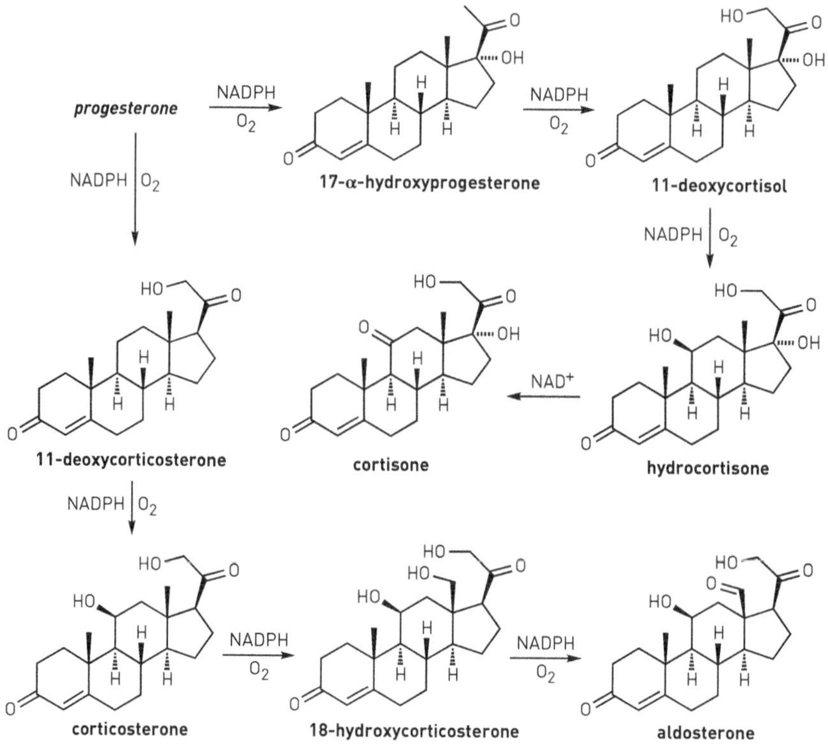

FIGURE 4.45
Conversion of progesterone into the corticosteroid hormones.

cyclopentane ring side chain, as seen in **hydrocortisone** (though **corticosterone** is an exception to this), while mineralocorticoids are usually limited to only a primary α-hydroxyketone function at this position, as in **aldosterone**. Of particular interest in this group is the corticosteroid known as **cortisone**. Proclaimed a "wonder drug" following its discovery in 1949 due to its dramatic anti-inflammatory effects in the treatment of patients suffering from debilitating rheumatoid arthritis, cortisone remains a widely used treatment for many different ailments, as it is remarkable in its ability to reduce pain and inflammation at sites of injury. Because of its suppression of the immune response, it is particularly effective in the reduction of swelling of the joints, hence its usefulness in the treatment of arthritic conditions. It is also used in therapies to help suppress organ rejection following transplants.

Since the long-term use of corticosteroids such as cortisone can have many undesirable side effects, considerable effort has been made over the years to develop synthetic analogues of cortisone designed to enhance anti-inflammatory properties at the expense of less desirable ones; many such steroidal drugs have been developed (Fig. 4.46) and are commonly prescribed for a variety of conditions, though most are still unsuitable for long-term use, hence the enduring interest in the development and marketing of synthetic **NSAIDs** (**n**onsteroidal **a**nti-**i**nflammatory **d**rugs) such as **ibuprofen**, **naproxen**, and **celecoxib** (celebrex).

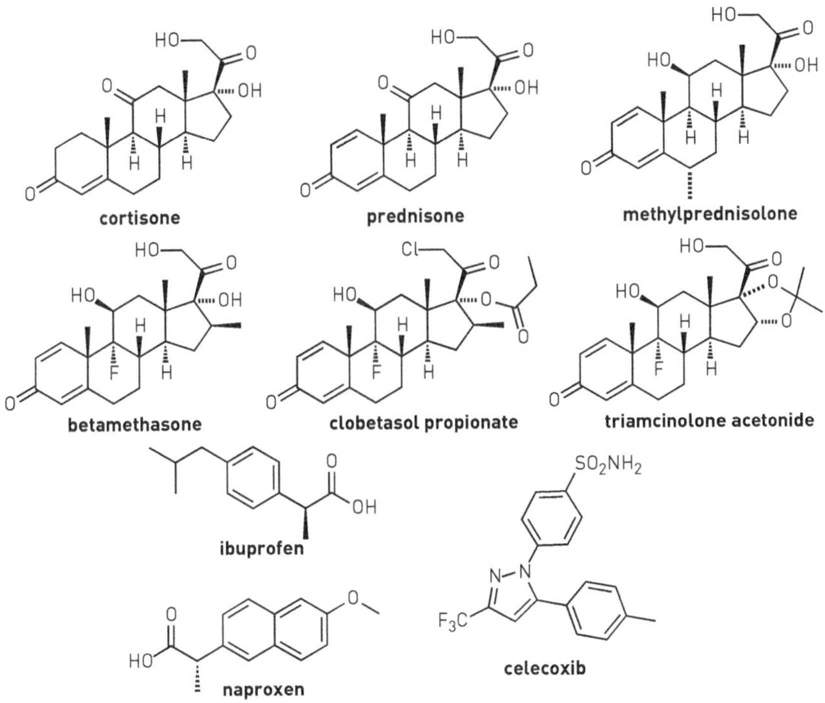

FIGURE 4.46
Some important steroidal and nonsteroidal anti-inflammatory drugs.

DEHYDROCHOLESTEROL, SUNSHINE, AND VITAMIN D_3 BIOSYNTHESIS

Before we leave our examination of steroidal structures, recall that we previously noted that in the double bond isomerization to complete the conversion of lanosterol to cholesterol, the system passes through an intermediate cyclohexadiene intermediate (Fig. 4.43, Region 4). This intermediate, known as **7-dehydrocholesterol**, also serves as the essential precursor to **vitamin D_3** (the so-called "sunshine vitamin") which plays a key role in bone health through regulation of absorption and metabolism of calcium and phosphorous. In this process, 7-dehydrocholesterol is photochemically converted to vitamin D_3 (or **cholecalciferol**) upon exposure of the skin to UV wavelengths between 270 and 300 nm in sunlight (Fig. 4.47).

The process involves a photochemical conrotatory cyclohexadiene ring-opening reaction to produce the corresponding hexatrienyl derivative **precholecalciferol**. This triene then undergoes a sigmatropic rearrangement or 1,7-hydride shift to produce the sterically crowded s-*cis* conformer of the isomerized hexatriene which relaxes to the more stable s-*trans* conformer of cholecalciferol (vitamin D_3). Our ability to synthesize this compound means that it is not a vitamin in the strictest sense, since these are usually considered to be compounds that must be obtained in the diet, but because deficient levels may lead to diseases such as rickets, vitamin D is commonly added to foods and is synthesized commercially by the same

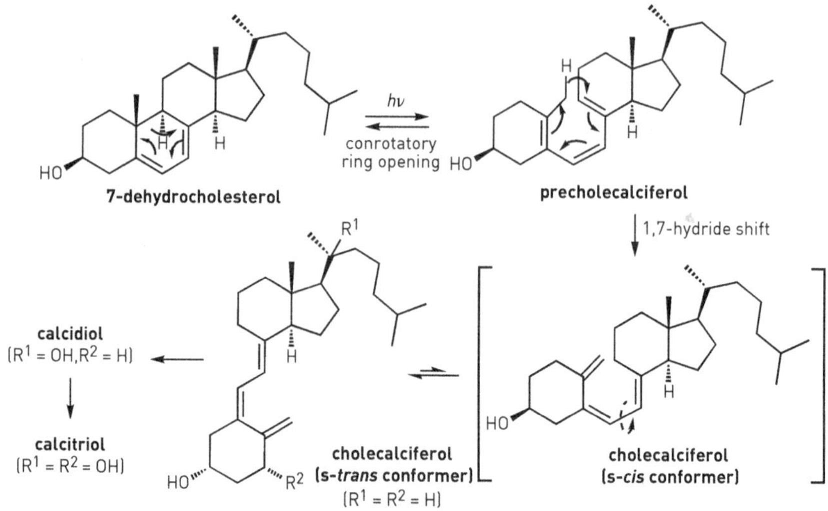

FIGURE 4.47
Photochemical conversion of cholesterol to cholecalciferol (vitamin D3), calcidiol, and calcitriol.

photochemical process. The vitamin D group includes two other components, **calcidiol** and **calcitriol**, both of which are derived from cholecalciferol by sequential hydroxylations occurring in the liver and the kidneys, with calcitriol being the most hormonally active component of the group.

TETRATERPENES AND CAROTENOIDS: TAIL-TO-TAIL LINKAGE OF C_{20} UNITS

You have probably noted that we are skipping over the possible C_{35} group of terpenes containing seven isoprene units. These are known as **sesquarterpenes** and are even less common than the sesterterpenes (C_{25}), being produced mainly in certain bacteria. Of far more interest are the C_{40} **tetraterpenes**, compounds which are produced by the initial tail-to-tail linkage of two C_{20} GGPP units in a fashion analogous to that used in the biosynthesis of squalene, but without the final NADPH reduction step. In plants or fungi, this process produces the basic precursor to tetraterpenes known as ***Z*-phytoene** (Fig. 4.48); in bacteria, the double bond isomerizes from *Z* to *E*.

From *Z*-phytoene, central bond isomerization and a sequence of four desaturation reactions ultimately leads to formation of **lycopene**, as shown in Fig. 4.49. This bright red pigment is found most commonly in tomatoes but is also present in many other red fruits and vegetables. The eleven conjugated double bonds in lycopene lead to an array of π-MOs with a HOMO–LUMO gap in the visible region of the spectrum, hence its deep red color. Though unproven, an association between lower levels of certain cancers and increased levels of lycopene consumption has been observed, due perhaps to its known antioxidant properties. Its lack of toxicity and bright red color make lycopene one of the most commonly used food colorings.

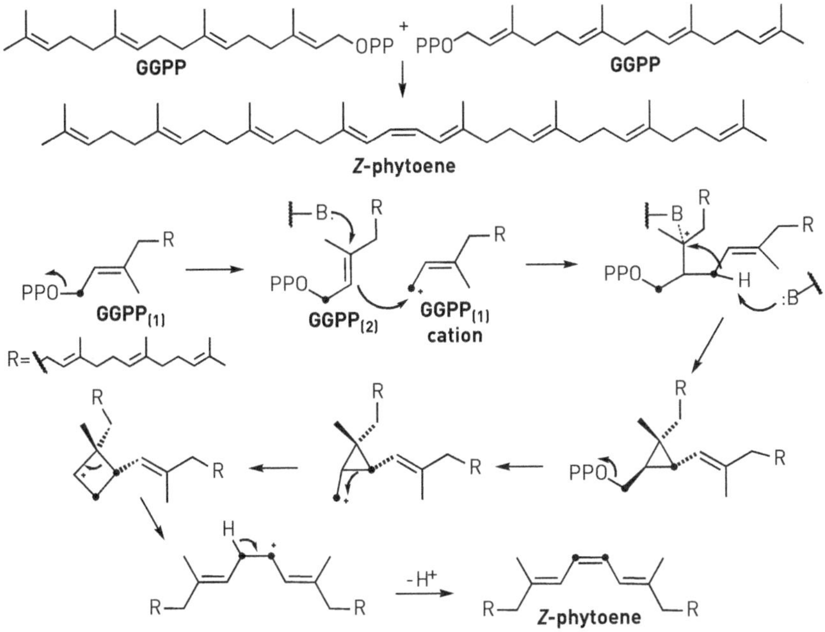

FIGURE 4.48
Linkage of GGPP units to form Z-phytoene for biosynthesis of tetraterpenes and carotenoids.

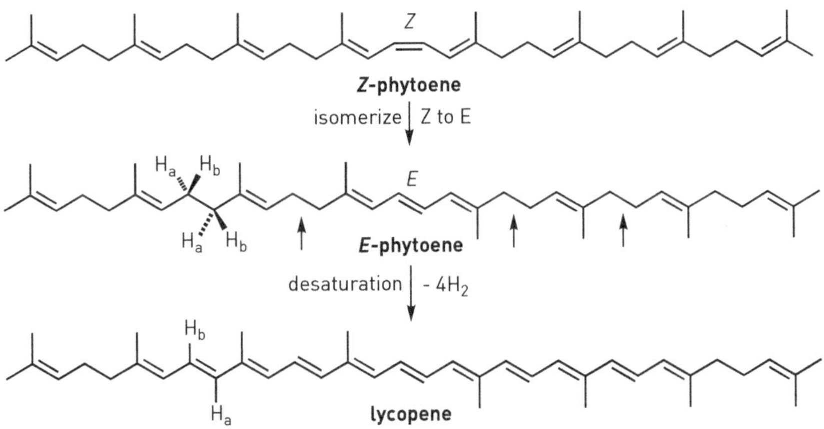

FIGURE 4.49
Z to *E* isomerization and stereospecific desaturation of phytoene to produce lycopene.

Phytoene and lycopene are the principal acyclic tetraterpenes of significance. Both acyclic and cyclic tetraterpenes are also known as **carotenoids**, plant pigments of special significance to human health and which must be obtained in the diet (animals do not biosynthesize tetraterpenes). The carotenoids are divided into two main groups, the hydrocarbon **carotenes** and the oxygenated **xanthophylls**, and typically result from cationic cyclization of one or both ends of the lycopene framework (end group cyclization). In the case of the xanthophylls, further oxidative transformations usually take place following initial cyclization. The combination of processes leads

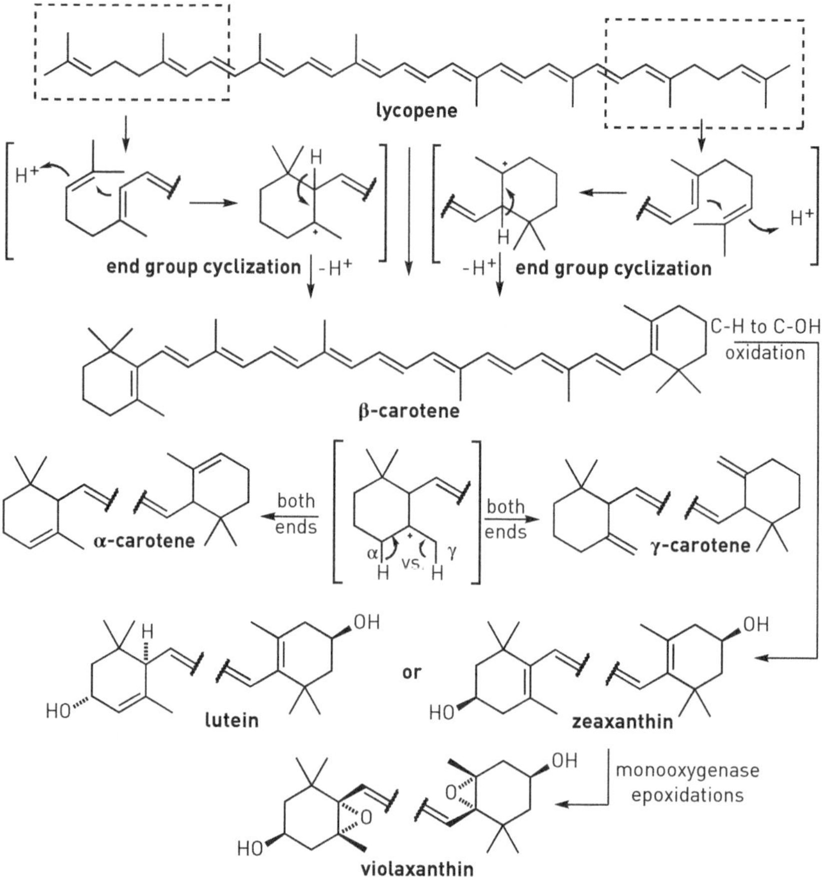

FIGURE 4.50

End group cyclization of lycopene and formation of α-, β-, and γ-carotenes and subsequent oxidative transformations to zeaxanthin, lutein, and violaxanthin.

to a wide variety of structurally related compounds; as seen in Fig. 4.50, cationic cyclization at both ends of the lycopene framework followed by simple proton loss leads to the widely distributed natural product **β-carotene**, a carotenoid responsible for the orange color of carrots and many other vegetables and fruits.

Oxidative transformations at one or both ends of the β-carotene ring system produce a number of important carotenoids. For example, both **zeazanthin** and **lutein** are found in many green vegetables such as spinach, peas, broccoli, romaine lettuce, and zucchini (note the double bond isomerization and epimerization at only one of the two end rings in lutein). These carotenoid alcohols are included in many ocular vitamin formulations since high levels of these have been associated with decreased incidence of age-related **macular degeneration**, possibly by protection of eye tissues from degenerative exposure to blue light, of which these compounds are strong absorbers. Further epoxidation of zeaxanthin also produces the common orange plant pigment **violaxanthin**.

The relationship between carotenoids and vision is also seen in the production and use of the essential terpenoid **vitamin A** which can be obtained in the diet but

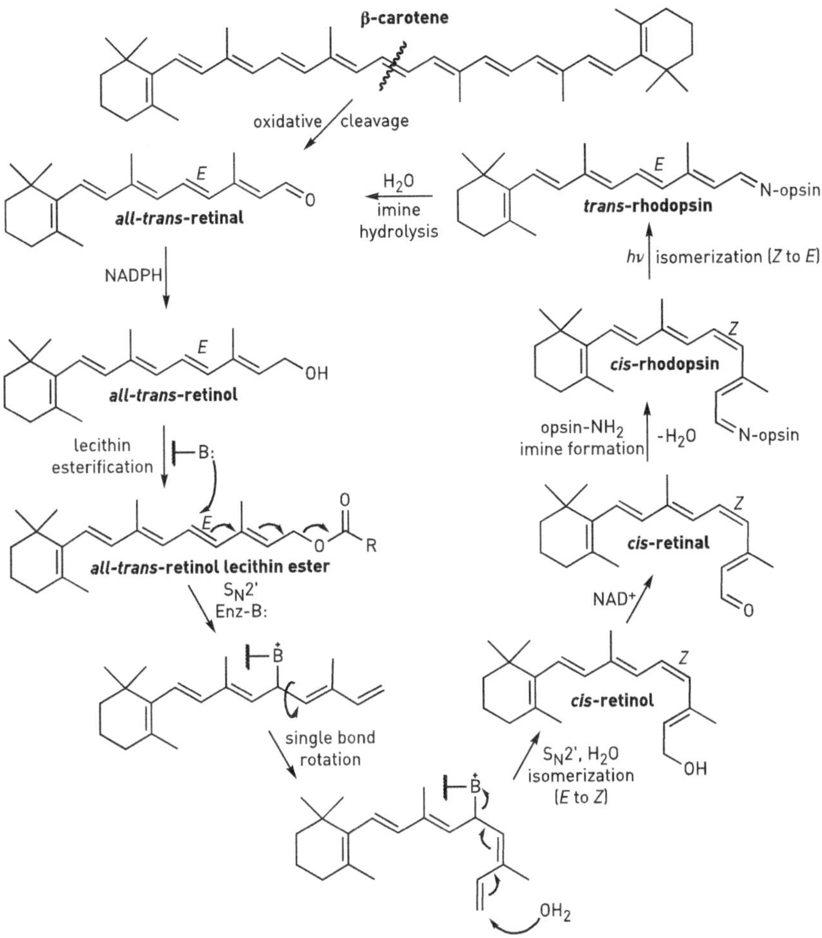

FIGURE 4.51
Cleavage of β-carotene to trans-retinal and retinol (vitamin A) as part of the visual cycle.

can also result from oxidative cleavage of the β-carotene central C=C bond in a process somewhat analogous to alkene ozonolysis, catalyzed by intestinal monoxygenase enzymes. This oxidative transformation leads initially to two molecules of the key vision-related aldehyde, **all-*trans*-retinal** or **vitamin A aldehyde**. Subsequent reduction by NADH then leads directly to **all-*trans*-retinol** (vitamin A). These important compounds are part of the **visual cycle**, as shown in Fig. 4.51.

Esterification of *trans*-retinol with lecithin yields an ester which can undergo an S_N2'-type nucleophilic substitution (nucleophilic attack at a π-bond of an allylic or a conjugated polyene system with a leaving group, leading to products with allylic isomerization) by an enzyme basic site, followed by a single bond rotation and a second S_N2' attack by nucleophilic water which displaces the enzyme base while producing the isomerized ***cis*-retinol.** An NAD^+ oxidation to ***cis*-retinal** is then followed by condensation with a lysine amino group within an **opsin** protein. The resulting imine linkage produces **rhodopsin**, the essential pigment associated with transmission of color-related visual signals in the retina of the eye. In this system, ***cis*-rhodopsin** captures a photon of light, leading to an electronically excited state that returns to the ground state with isomerization of the 11-position double bond (i.e., from 11-*cis*-rhodpsin to 11-*trans*-rhodopsin). This light-induced change in the overall molecular geometry of rhodopsin initiates a nerve impulse interpreted by the brain as a visual signal. Subsequent hydrolysis of the imine linkage releases the opsin protein and regenerates all-*trans*-retinal, which now can be cycled back into the system for subsequent visual events.

LOOKING AHEAD

We have not seen the last of DMAPP or terpenoid compounds by any means, as we will find that they can intersect with compounds from other pathways to produce a variety of mixed products. As we move on to the next chapter to further expand our knowledge of biosynthetic pathways, we will see some recurring themes from our study of terpenes: fundamental building blocks, repeating units, acyclic, monocyclic and polycyclic structures, alternate folding patterns, and pathways that can produce a seemingly endless variety of biologically significant organic structures, all with a logic behind them based on organic chemistry that we can understand, appreciate, and build upon.

STUDY PROBLEMS

1. From among all eight possible diastereomers of the basic structural system associated with menthol, neomenthol, isomenthol, and neoisomenthol, the menthol enantiomers would be expected to be the most stable. Explain using appropriate stereochemical diagrams.

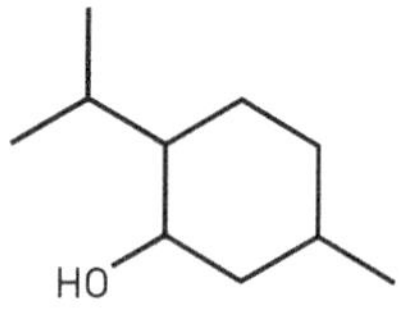

2. Eucalyptol, a simple bicyclic monoterpenoid ether, makes up about 90% of generic eucalyptus oil, but is also found in bay leaves, sweet basil, rosemary, and sage among other aromatic plants. Propose a biosynthesis from GPP.

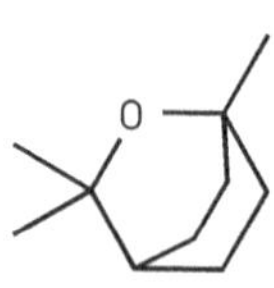

eucalyptol

3. When heated, *β-pinene* undergoes a retro-[2+2] cycloaddition reaction to give *myrcene* which may then be reacted with acrolein (propenal) to give a valuable commercial fragrance compound as a mixture of regioisomers (one major, one minor). Draw curved arrow mechanisms for both the formation of *myrcene* and also for the expected structure of the fragrance compound.

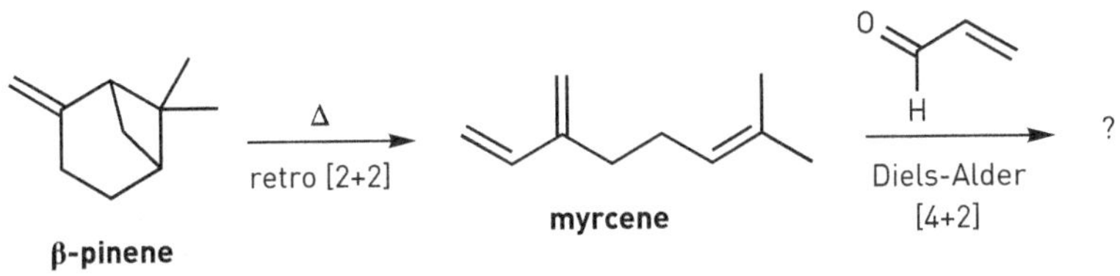

4. The boll weevil (*Anthonomus grandis*) has been among the most destructive of pests for the cotton-growing industry over the years. Many methods have been employed in attempts to eradicate or control the insect's spread, including the monitoring of populations using traps baited with its sex pheromone which is a mixture of four components, **grandlure I–IV** (grandlure I is also known as grandisol). It has been proposed that all four components of the pheromone mixture may be derived biosynthetically from geraniol. Propose a sequence of plausible reactions for the biosynthesis of these compounds from geraniol.

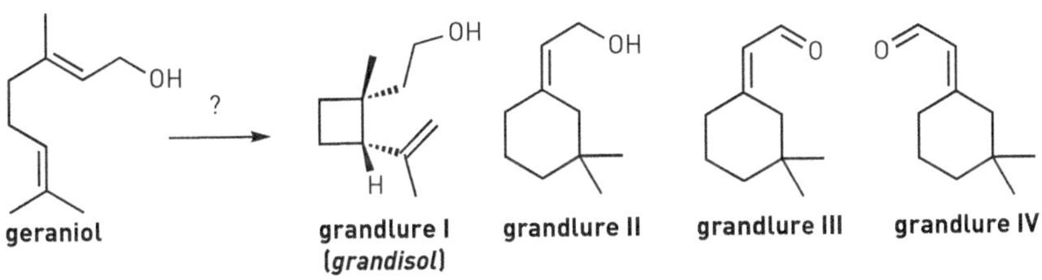

5. For a plant employing the DXP pathway for biosynthesis of geraniol, predict where ^{13}C labeling in geraniol would occur if the plant was provided with pyruvic acid labeled at the 2-position. Where would the labeled positions in geraniol occur from pyruvic acid labeled at the 3-position?

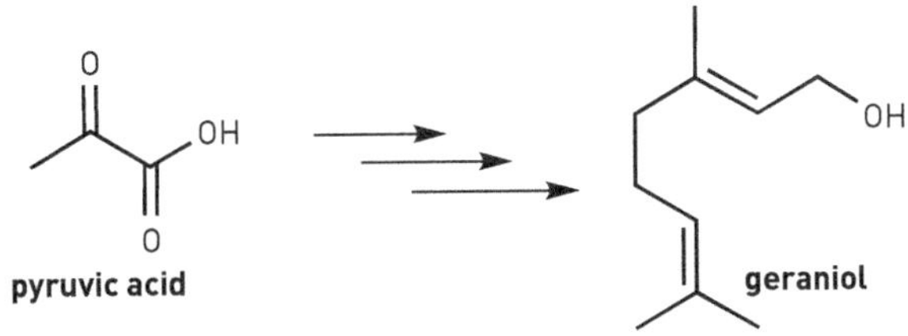

6. The enzyme *aristolochene synthase* catalyzes the cyclization of FPP to a large group of fungal toxins. The intermediate carbocation can undergo a variety of

simple shifts and deprotonations to give different products. Show how the two products shown are formed via a series of 1,2-shifts and a final proton loss.

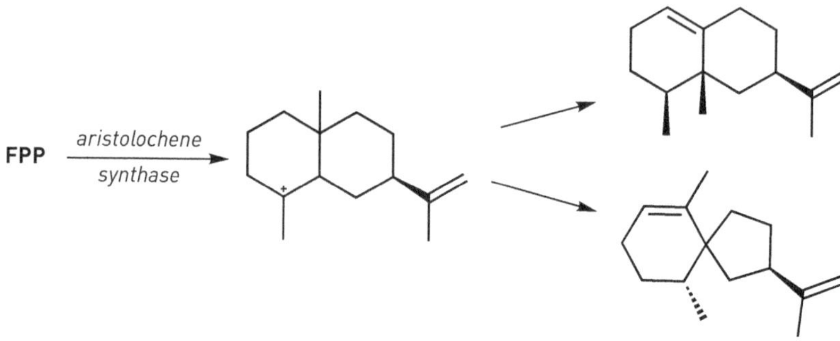

7. Mevalonic acid labeled with ^{13}C and ^{2}H has been used to demonstrate the sequence of cyclizations and 1,2-H shifts involved in biosynthesis of the hydrocarbon precursors to various natural products.
   a) Starting with the GGPP folding pattern provided, clearly demonstrate a likely sequence of cation-initiated cyclizations and accompanying 1,2-H shifts which would lead to the hydrocarbon precursor of Fusicoccin, a fungal metabolite responsible for plant wilting. Use only loss of OPP or a protonation to generate initial carbocations.

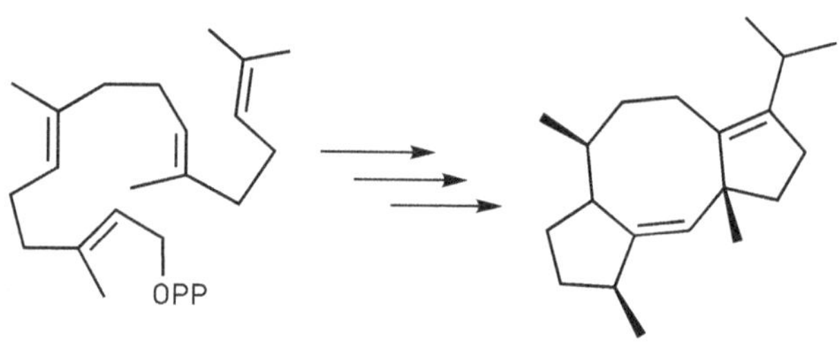

fusicoccin core

   b) Star (*) the carbons in the hydrocarbon core to indicate where labeled carbons would appear if acetate, ^{13}C-labeled at the methyl group, was introduced into the organism producing *fusicoccin* (assume generation of ^{13}C-labeled acetyl-CoA and a mevalonic acid pathway).
8. Propose a folding pattern and mechanism for cyclization of GGPP to produce *casbene*.

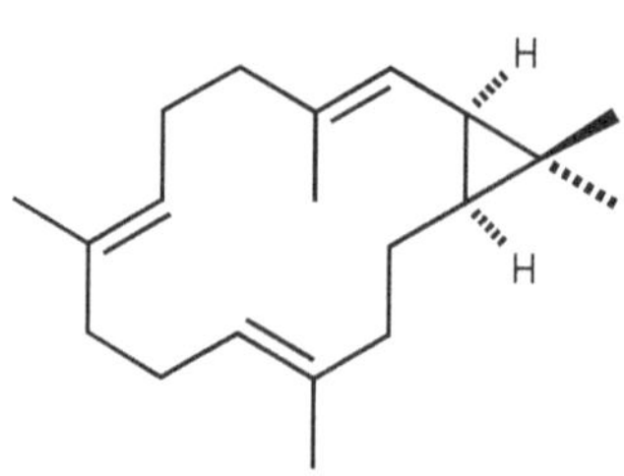

casbene

9. Vitamin E is an important group of structurally related phenolic antioxidants of which δ-tocopherol is the simplest member, possessing a monomethylated aryl ring (β- and γ-tocopherols are *para-* and *ortho*-dimethyl-substituted isomers, respectively, while α-tocopherol has a trimethylated aryl ring). Propose a biosynthetic sequence to produce δ-tocopherol from phytol PP and the shikimate-derived product, *homogentisic acid* (in the first step, both alkylation and decarboxylation occur). Indicate the reactions and cofactors most likely involved in producing the remaining tocopherol isomers (note that α-tocopherol is derived from γ-tocopherol).

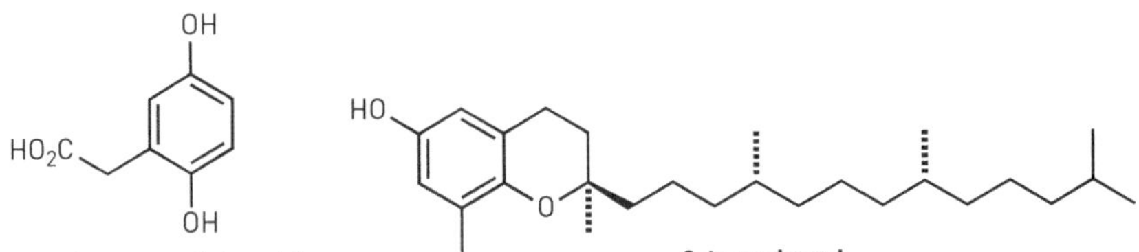

homogentisic acid **δ-tocopherol**

10. Propose a mechanism for the acid-catalyzed rearrangement of epoxidized derivatives of *zeaxanthin* to produce *capsanthin* and *capsorubin*, food colorings isolated from paprika oleoresin used as coloring agents for orange juice as well as various cheeses and spice mixtures.

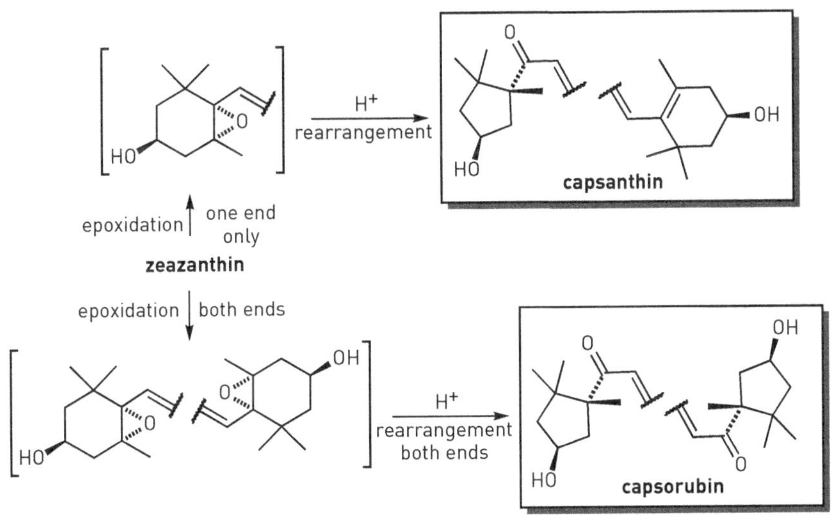

5 The Acetate Pathway

Biosynthesis of Polyketides and Related Compounds

The group $-CH_2CO-$ *(which the author proposes to call the "ketide" group) can be made to yield by means of the simplest reactions a very large number of interesting compounds, all belonging to classes which are largely represented amongst the compounds obtained from plants.*

—John Norman Collie (*1907 lecture*)

Within a few minutes these ideas reduced to order what had previously seemed an unrelated jumble of structures of many natural products. It was an immensely satisfying and emotional moment.

—Arthur J. Birch (*1952 lecture*)

We saw in the previous chapter how Otto Wallach's early proposal regarding the structural origin of terpenoid natural products was later refined by the insightful work of Leopold Rudzicka, leading to the biogenetic isoprene rule and all that it implies. In a nearly parallel fashion, we find in our present chapter a second, unrelated class of naturally occurring compounds whose characteristic structural features prompted an initial innovative hypothesis by J. N. Collie near the turn of the 20th century. Collie proposed that certain natural compounds might arise from precursors containing repeated "ketide" ($-CH_2CO-$) units which could then undergo subsequent condensations and other reactions typical of carbonyl compounds to produce some of the observed structures. Unfortunately, Collie's work was more or less ignored and largely forgotten for nearly a half century, only to be reimagined and expanded in the middle of the century by A. J. Birch, another pioneer whose proposals met with considerable initial resistance. But unlike his predecessor, Birch ultimately prevailed by providing experimental results that supported a comprehensive theory of the biochemical origin of the group of compounds now universally known as "**polyketide**" natural products.

This structurally diverse family includes some of the most useful medicinal agents now known to us, with many members possessing powerful antibacterial, antifungal, anticancer, immunosuppressant, and even cholesterol-lowering biological properties. As we see in Fig. 5.1, such structures range from the relatively simple to the exceedingly complex and may include large macrocyclic lactone rings (macrolides) such as **erythromycin,** polycyclic ethers such as **monensin A**, polycyclic structures which may be partly or mostly aromatic as in **tetracycline, griseofulvin,** or **daunorubicin**, or nonaromatic polycyclics such as **tacrolimus** and **lovastatin**. Some also contain noncyclic linear components that may be saturated, oxygenated, or unsaturated, as seen in different regions of **amphotericin B** which, like erythromycin, daunorubicin, and many other polyketides, also possesses an **aglycone core** which has been **glycosylated** with a carbohydrate component at a specific position. But in spite of this range of structures, many polyketide compounds share some common features that ultimately become more evident upon closer inspection; six-membered rings (either aromatic or nonaromatic) and multiple oxygens which tend to appear in a repeating 1,3-relationship to one another on both acyclic, cyclic, and aromatic structures. These features will turn out to be important clues associated with Collie's and Birch's original work on understanding how polyketide natural products are actually formed.

But before we delve into the complex organic chemistry associated with polyketide biosynthesis, we must first spend some time examining a closely related process—fatty acid biosynthesis—whose mechanistic details and dependence on enzyme catalysis were understood well after Collie's time but also ultimately provided the inspiration for most of Birch's subsequent insights and investigations. As we will see, fatty acid biosynthesis is revealed as the original template upon which nature built the polyketide pathway, with only a few minor modifications or variations ultimately responsible for the vast structural differences observed in the array of different products produced by each pathway.

FATTY ACIDS: MULTIPLES OF TWO CARBONS, SATURATED OR UNSATURATED

Fats and oils, like steroids, are among the group of natural products referred to as **lipids**, compounds with substantial hydrocarbon character and limited water solubility. Two structurally similar types are the **glycerophospholipids** which are important cell membrane constituents and the **triacylglycerols** (or triglycerides) which constitute most of the familiar fats and oils obtained from plant or animal sources. Both of these types of lipids are esters derived from **glycerol** (glycerine or

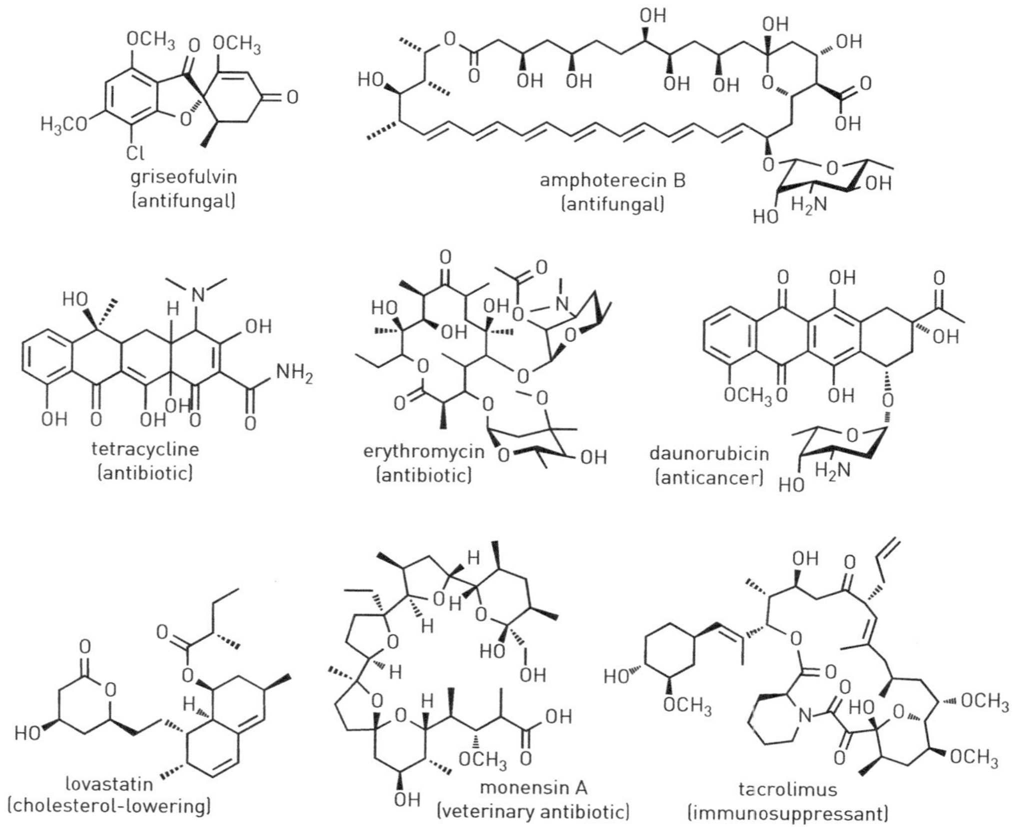

FIGURE 5.1
Some important examples of structurally diverse medicinal polyketide natural products.

1,2,3-propanetriol) which is linked to **fatty acids**—even-numbered, unbranched long-chain carboxylic acids containing between 12 and 20 carbon atoms, either with **saturated** chains (no double bonds) or **unsaturated** chains (one or more double bonds, almost always *Z* in configuration). A list of the most common fatty acids is given in Table 5.1.

Glycerophospholipids such as **phosphatidyl cholines** are based on a phosphate ester linkage between an amino alcohol such as choline, ethanolamine, or serine and **phosphatidic acid,** a monophosphate ester of glycerol which has been further esterified by two fatty acids, one of which is usually saturated and the other unsaturated. Triacylglycerols such as **tristearin** contain ester linkages between glycerol and three fatty acids, though the three need not be the same and can be a mixture of both saturated and unsaturated fatty acids (Fig. 5.2). Most fats or oils are not homogenous in composition but rather are complex mixtures of variously configured triacylglycerols. One important difference between animal and plant-derived fats is that animal fats are usually **saturated fats** which are often solids at room temperature, due in part to the ready aggregation or packing of their saturated chains. Conversely, plant-derived fats are usually oils at room temperature; these fats contain one or more nonconjugated double bonds in some or all of the chains; these are referred to as **monounsaturated** or **polyunsaturated** fats, depending on the number of double bonds present in the unsaturated chains. The nonlinear geometry of the *Z*-double bond in unsaturated fats tends to inhibit close aggregation or packing of the hydrocarbon tails and is the main reason for their lower melting points relative to saturated fats.

Most of us are familiar with the rather notorious ***trans*** **fats** which, like dietary saturated fats, have been shown to raise serum cholesterol levels (and the attendant

Table 5.1 Names and Structures of Some Common Saturated and Unsaturated Fatty Acids

Name	*C-atoms*	*π-bonds (Z)*	*Structure*
Lauric	12	0	$CH_3(CH_2)_{10}CO_2H$
Myristic	14	0	$CH_3(CH_2)_{12}CO_2H$
Palmitic	16	0	$CH_3(CH_2)_{14}CO_2H$
Stearic	18	0	$CH_3(CH_2)_{16}CO_2H$
Arachidic	20	0	$CH_3(CH_2)_{18}CO_2H$
Oleic	18	1: at C-9	$CH_3(CH_2)_7CH{=}CH(CH_2)_7CO_2H$
Linoleic	18	2: at C-9,12	$CH_3(CH_2)_4CH{=}CHCH_2CH{=}$ $CH(CH_2)_7CO_2H$
α-Linolenic	18	3: at C-9,12,15	$CH_3CH_2CH{=}CHCH_2CH{=}CHCH_2CH{=}$ $CH(CH_2)_7CO_2H$
γ-Linolenic	18	3: at C-6,9,12	$CH_3(CH_2)_4CH{=}CHCH_2CH{=}CHCH_2CH{=}$ $CH(CH_2)_4CO_2H$
Arachidonic	20	4: at C-5,8,11,14	$CH_3(CH_2)_4CH{=}CHCH_2CH{=}CHCH_2CH{=}$ $CHCH_2CH{=}CH(CH_2)_3CO_2H$

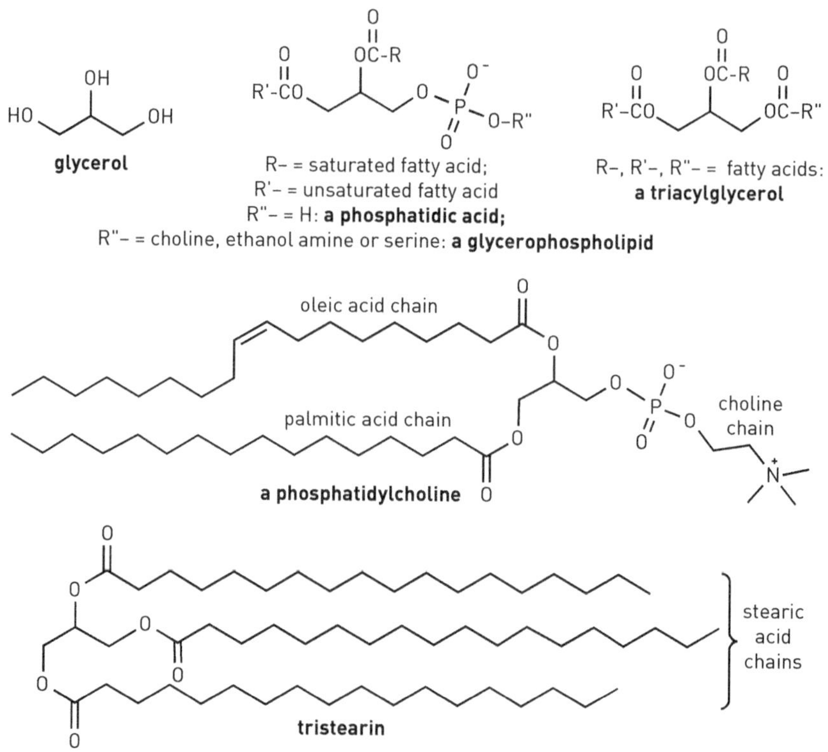

FIGURE 5.2
Structures of glycerol; a glycerophospholipid and a triacylglycerol.

risk of coronary artery disease). Most *trans* fats are actually food industry byproducts that arise from the catalytic hydrogenation of processed polyunsaturated fats ("partially hydrogenated vegetable oils"). For example, partial hydrogenation of soybean or peanut oils produces margarine or shortening with melting points similar to those of some saturated fats, not only because some chains are in fact saturated in the process, but also because *Z* to *E* isomerization of some π-bonds takes place during the high-temperature catalytic hydrogenation, typically producing fats with as much as 15% of the more thermodynamically favored transunsaturation (Fig. 5.3). And unlike *Z*-configured fats, an *E*-double bond in a fatty acid chain has a relatively linear conformation similar to saturated fat chains; this allows closer aggregation of the hydrocarbon tails and increased melting points. So why hydrogenate a vegetable oil and turn it into a (less healthy) synthetic fat with a higher melting point? As is usually the case, it's a combination of marketing and demand; some people prefer spreading a solid (like butter) on their bread rather than pouring an oil over it (even though doing so would probably be better for their health). Figure 5.3 illustrates the generation of a typical trans fat

Waxes are solid lipids produced by many plants, insects, and other animals and contain multiple components, many of which are high molecular weight esters derived from even-numbered fatty acids (16–36 carbons) and reduced "**fatty alcohols**"

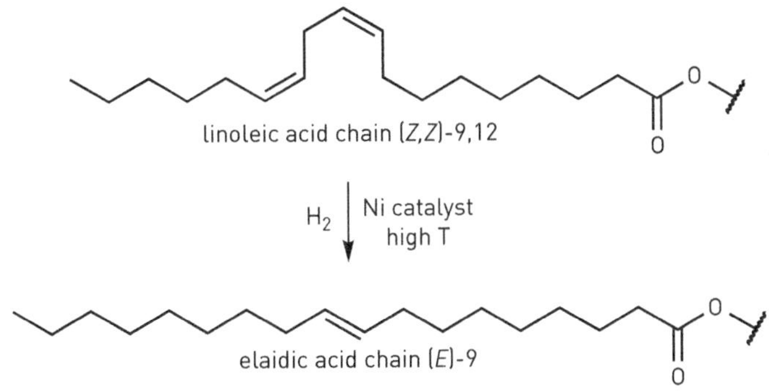

FIGURE 5.3
Formation of a trans fat during partial hydrogenation of a polyunsaturated fat.

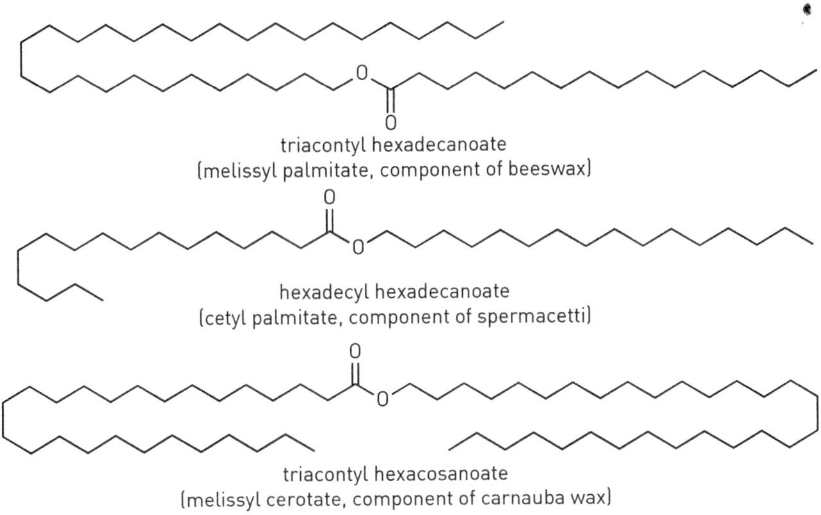

FIGURE 5.4
Structures of some common ester components of various natural waxes.

(16–36 carbons). Some common examples are **beeswax**, the main component of which is an ester derived from 1-triacontanol (melissyl alcohol, C_{30}) and hexadecanoic (palmitic, C_{16}) acid; **spermacetti,** derived from sperm whale oil and which consists primarily of the ester derived from 1-hexadecanol (cetyl alcohol) and palmitic acid; and **carnauba wax**, one of the hardest and most widely used waxes which among other components, contains the ester derived from melissyl alcohol and hexacosanoic (cerotic, C_{26}) acid (Fig. 5.4).

Phospholipids, triacylglycerols, and waxes all depend on **fatty acid biosynthesis** to provide the basic saturated hydrocarbon chains which can then be modified to yield the unsaturated, branched, or reduced components of these lipids. Much as the mechanism of nonphotosynthetic glucose synthesis (gluconeogenesis) was essentially the reverse of the catabolic degradation of glucose (glycolysis), fatty acid biosynthesis turns out to be essentially the reverse of the important catabolic process for degradation of fats known as **β-oxidation.** In that process, saturated fatty acids released by triacylglycerol hydrolysis are degraded two carbons at a time into individual acetyl-CoA units which can then enter the citric acid cycle for further catabolism to CO_2. Thus, one can think of triacylglycerols as energy storage units for on-demand release of fatty acids that can be degraded into acetyl-CoA for fuel. Unsaturated fatty acids are slightly more complicated to degrade, requiring additional enzymes at certain steps to isomerize double bonds into appropriate positions so that the normal β-oxidation process can continue. But we won't examine β-oxidation in detail; we should simply keep in mind when we consider the mechanistic details of saturated fatty acid biosynthesis, that we need only run that process in reverse to gain some understanding (at least mechanistically) of how fatty acids are ultimately metabolized in the β-oxidation process.

SATURATED FATTY ACID BIOSYNTHESIS: IT ALL STARTS WITH ACETYL-COA

Just as β-oxidation ends with acetyl-CoA, biosynthesis of saturated fatty acids *begins* with acetyl-CoA. And it is nature's use of this fundamental two-carbon building block that is ultimately responsible for the familiar even-numbered constitution of most fatty acids. We saw previously in Chapter 2 how acetyl-CoA reacts with N-carboxybiotin to undergo a carboxylation reaction via its enolate ion, leading to the important derivative, **malonyl-CoA** (Fig. 5.5). It is the combination of both acetyl-CoA and malonyl-CoA, acting in tandem that ultimately sets the process of fatty acid biosynthesis into motion.

As shown in Fig. 5.6, both acetyl-CoA and malonyl-CoA units undergo nucleophilic acyl substitution by thiol groups of an **acyl carrier protein** (ACP) and a second enzyme component, both of which are bound together within the **fatty acid synthase (FAS) complex,** a multifunctional enzyme system that contains within it all the catalytic components required to facilitate the entire process. The **loading** component of the process, mediated by a *malonyl/acetyl transferase* component or

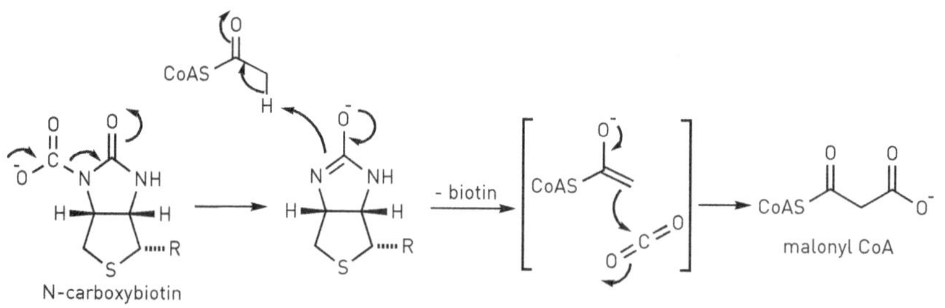

FIGURE 5.5
N-carboxybiotin and carboxylation of acetyl-CoA to give malonyl-CoA.

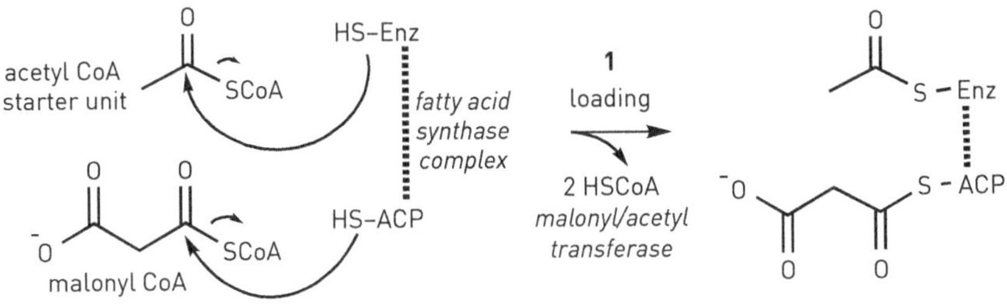

FIGURE 5.6
Loading of acetyl-CoA and malonyl-CoA units in step 1 of fatty acid biosynthesis.

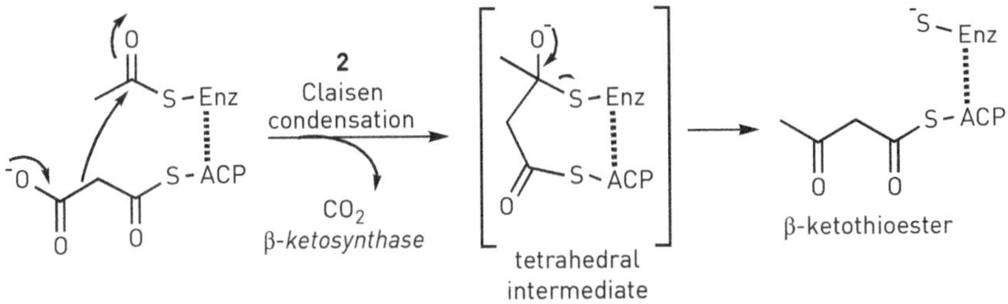

FIGURE 5.7
Claisen condensation in step 2 of fatty acid biosynthesis.

domain of the FAS enzyme, constitutes step 1 in the fatty acid biosynthesis process, producing two enzyme-bound thioesters, **acetylSEnz** and **malonyl-ACP.**

Once brought together within the complex, the malonyl-ACP unit undergoes a decarboxylation reaction to yield an ester enolate ion. This enolate acts as a nucleophile (Fig. 5.7) to attack the carbonyl group of the nearby **starter unit**, acetylSEnz, in what is formally a Claisen condensation in step 2. The domain of the FAS enzyme that mediates this essential C–C bond-forming step—*β-ketosynthase*—is aptly named, as it facilitates the formation of the usual product from a Claisen condensation, namely a β-ketoester (actually a **β-ketothioester** in this case).

Step 3 in the process is an NADPH reduction of the ketone function of the β-ketothioester (mediated by a *β-ketoreductase* domain) to yield a **β-hydroxythioester** intermediate. Note that delivery of hydride is stereospecific in this reaction, occurring at the *Si* face of the ketone function (Fig 5.8).

As might be expected, such β-hydroxycarbonyl compounds are susceptible to dehydration (Fig. 5.9), and this is the process seen in step 4, facilitated by a *dehydratase* domain, yielding the corresponding **α,β-unsaturated thioester** known as ***trans*-crotonyl-ACP**, a thioester of *trans*-crotonic acid (the common name for 2-butenoic acid).

In step 5, the *trans*-crotonyl-ACP from the previous step undergoes a reaction typical of α,β-unsaturated carbonyl systems; namely, an *enoyl reductase*-mediated **conjugate reduction**, as seen in Fig. 5.10. This involves conjugate nucleophilic addition of hydride from NADPH at the β-carbon of the thioester, affording a thioester

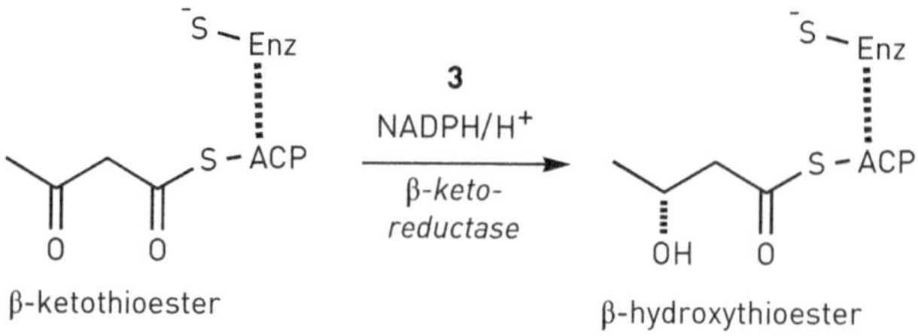

FIGURE 5.8
Ketone reduction by NADPH in step 3 of fatty acid biosynthesis.

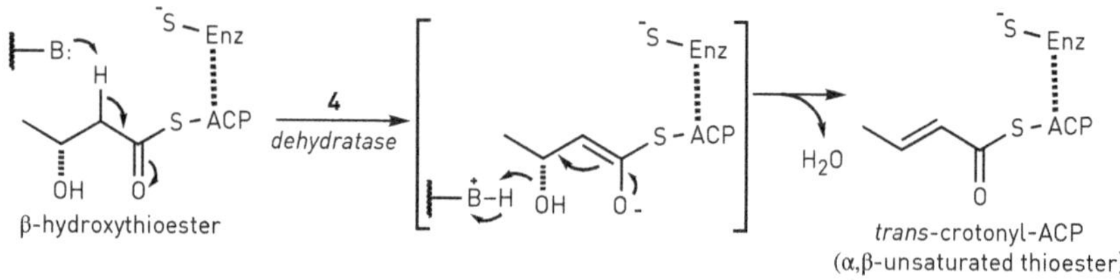

FIGURE 5.9
Dehydration to *trans*-crotonyl-ACP in step 4 of fatty acid biosynthesis.

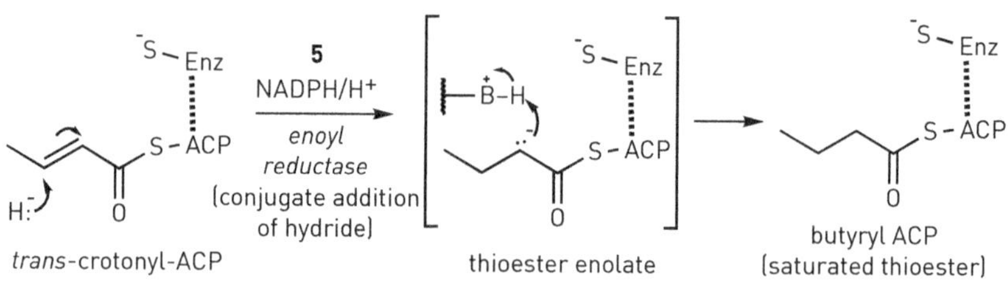

FIGURE 5.10
Conjugate reduction of *trans*-crotonyl-ACP to butyryl ACP in Step 5 of fatty acid biosynthesis.

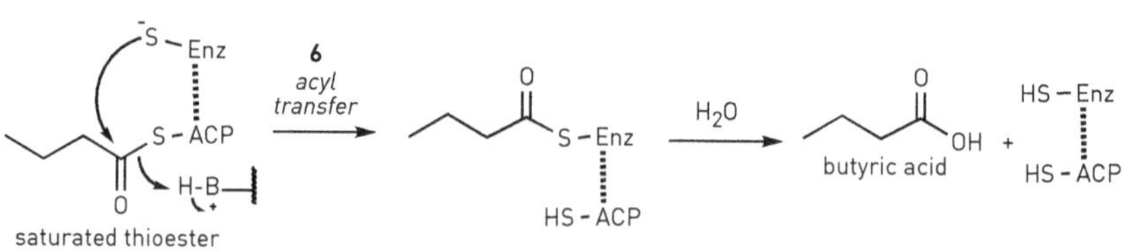

FIGURE 5.11
Acyl transfer in step 6 of fatty acid biosynthesis, followed by hydrolysis.

enolate intermediate which upon protonation yields the corresponding **saturated thioester** product, butyrylACP.

The sixth step of the sequence involves an intramolecular **acyl transfer** from the acyl ACP thiol function to the adjacent enzyme thiol function as shown in Fig. 5.11. What happens next depends on what outcome is required. For example, if the saturated acyl–SEnz bond were to undergo simple hydrolysis, the corresponding short-chain carboxylic acid, butyric acid, would be liberated.

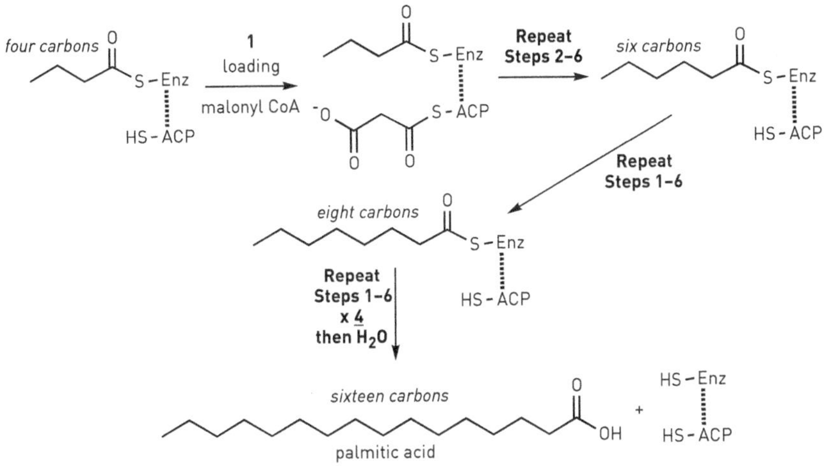

FIGURE 5.12
Sequence repetition for chain extension in biosynthesis of saturated fatty acids.

However, if loading of a new malonyl-CoA unit onto the fatty acid synthase complex should occur (i.e., a repeat of step 1) instead of hydrolysis, this would set the stage for subsequent repetition of steps 2 through 6, resulting in extension of the chain from four carbons to six carbons (Fig. 5.12). Another loading step followed by repetition of steps 2 through 6 would extend the chain further, to eight carbons. Finally, if an additional four cycles of malonyl-CoA loading followed by steps 2 through 6 culminated in a final hydrolysis step, the resulting carboxylic acid would contain a total of 16 carbons (palmitic acid). Thus, we see that by sequential loading of malonyl-CoA units and subsequent Claisen condensations, reductions, dehydrations, conjugate reductions, and acyl transfers, the chain grows by two carbons per cycle, always yielding a fatty acyl product with an even number of carbon atoms. While this may be a somewhat simplified representation of a complex biochemical system, the emphasis here on the involvement of a number of familiar organic reactions adequately represents the fundamental chemistry at the core of this important process.

Reversal of the steps outlined above constitutes the chemistry associated with the degradative β-oxidation of saturated fatty acids to acetyl-CoA. Figure 5.13 uses the degradation of palmitic acid (as palmitoylCoA) to illustrate the process, but note some important caveats: i) FAD → $FADH_2$ is used for the introduction of the enoyl thioester double bond and ii) NAD^+ is used for the alcohol oxidation step (FA synthesis carries out the corresponding reduction using $NADPH/H^+$). Otherwise, the organic chemistry of these two processes is nearly identical.

In some fatty acid synthase (FAS) systems, palmitic acid is the end product, though shorter chains may be obtained along the way by simple hydrolysis. Subsequent chain extension of palmitic acid can occur via its coenzyme A derivative (palmitoylCoA) as a starter unit along with malonyl-CoA units as extenders. In other systems, odd-numbered fatty acids, though rare, can be produced; one example is **margaric acid** (heptadecanoic acid), a trace component found in ruminant

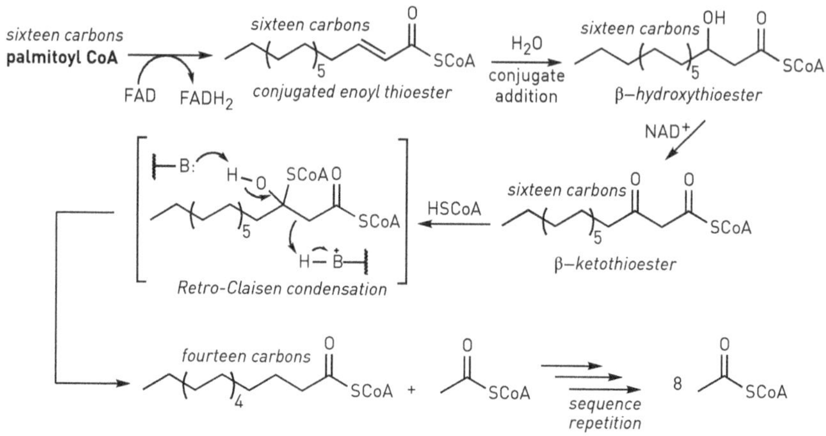

FIGURE 5.13
Reversal of steps in the β-oxidation pathway for degradation of palmitoylCoA to acetyl-CoA.

milk. The same sequence of reactions is involved as for palmitic acid biosynthesis, except that the starter unit is propionyl-CoA instead of acetyl-CoA. Subsequent cycles of malonyl-CoA chain extension via condensation, reduction, dehydration, and conjugate reduction will thus produce an odd-numbered fatty acid chain, regardless of the number of cycles. This is illustrated in Fig. 5.14 using bolded bonds for the three carbons derived from the propionyl-CoA starter unit.

BRANCHED FATTY ACIDS: DIFFERENT ROUTES AND DIFFERENT RESULTS

While most fatty acids are linear in constitution, some consist of chains containing one or more branches, usually methyl groups. A number of different routes to such structures are illustrated in Fig. 5.15. In route 1, a branched starter unit such as **isobutrylCoA**, **2-methylbutyrylCoA** or **isovalerylCoA** is loaded and extended in the usual fashion by sequential additions of malonyl-CoA units, leading to fatty acids with methyl branching either one or two carbons distant from the C-terminus of the chain. In route 2, acetyl-CoA is the starter unit, while **methylmalonyl-CoA** serves as the chain extender, leading to a hydrocarbon backbone with repeating methyl branches on alternate carbons. Route 3 involves the use of an existing unsaturated fatty acid and introduces a methyl branch via SAM alkylation of the unsaturated fatty acid double bond, followed by a 1,2-H shift and NAPDH reduction of the resulting carbocation.

Ring-containing fatty acids are rare but are also known, and Fig. 5.16 illustrates two examples. **Hydnocarpic acid**, an antimicrobial component of chaulmoogra oil which was used as an early and effective treatment for Hansen's diesease (leprosy), utilizes a cyclic starter unit, **cyclopent-2-enyl carboxylCoA** derived from the non-peptidic amino acid precursor cyclopentenyl glycine, followed by chain extension with five malonyl-CoA units and hydrolysis. Other terminal cycloalkyl fatty acids

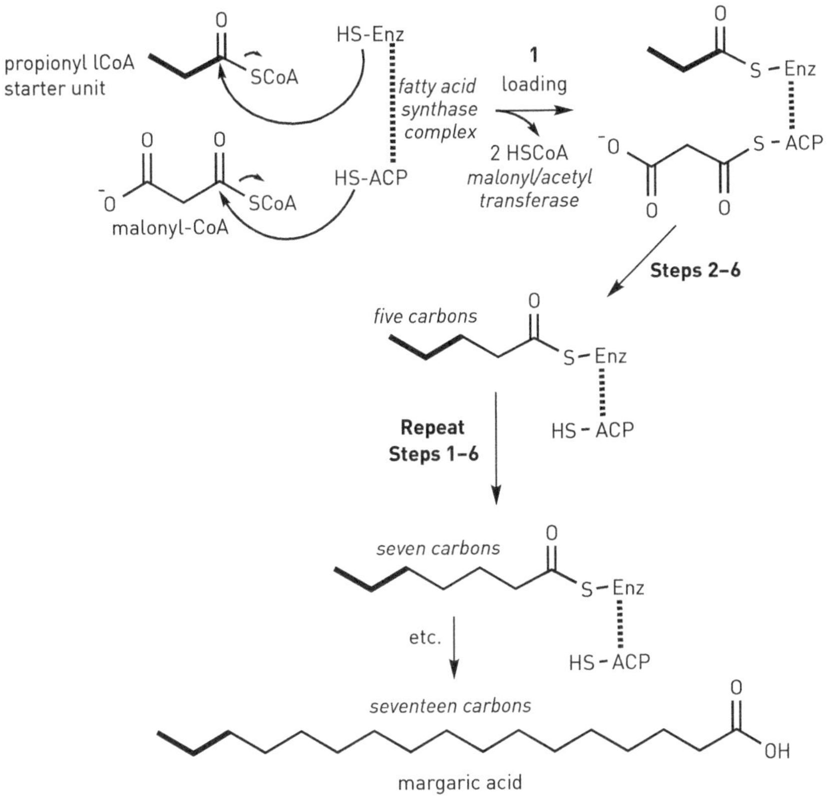

FIGURE 5.14
Propionyl-CoA starter unit and sequence repetition for biosynthesis of margaric acid.

are also known (cyclohexyl and cycloheptyl, for example) and arise in a similar fashion from chain extension of the corresponding cycloalkyl carboxylCoA starter units. Deuterium labeling studies show that alkylation of the double bond of **oleic acid** by SAM leads to the indicated carbocation which is quenched by methyl group deprotonation and subsequent cyclopropane ring closure rather than a 1,2-H shift and reduction as was previously outlined (Fig. 5.15), yielding **dehydrosterculic acid**. Also worth noting are the rare and highly strained **ladderane lipids** which include structures such as **pentacycloanammoxic acid**; all possess three or five fused cyclobutane rings at the C-terminus of a linear carbon chain. Produced by certain deep-sea ammonium-oxidizing (anammox) bacteria important to the oceanic nitrogen cycle, ladderane lipids are of unknown biosynthetic origin and may prove to be unique biomarkers for the anaerobic ammonium oxidation process.

MONO- AND POLYUNSATURATED FATTY ACIDS: PUTTING IN THE "ESSENTIAL" DOUBLE BONDS

Most unsaturated fatty acids are derived from saturated fatty acids by the introduction of one or more double bonds in a dehydrogenation process usually mediated by

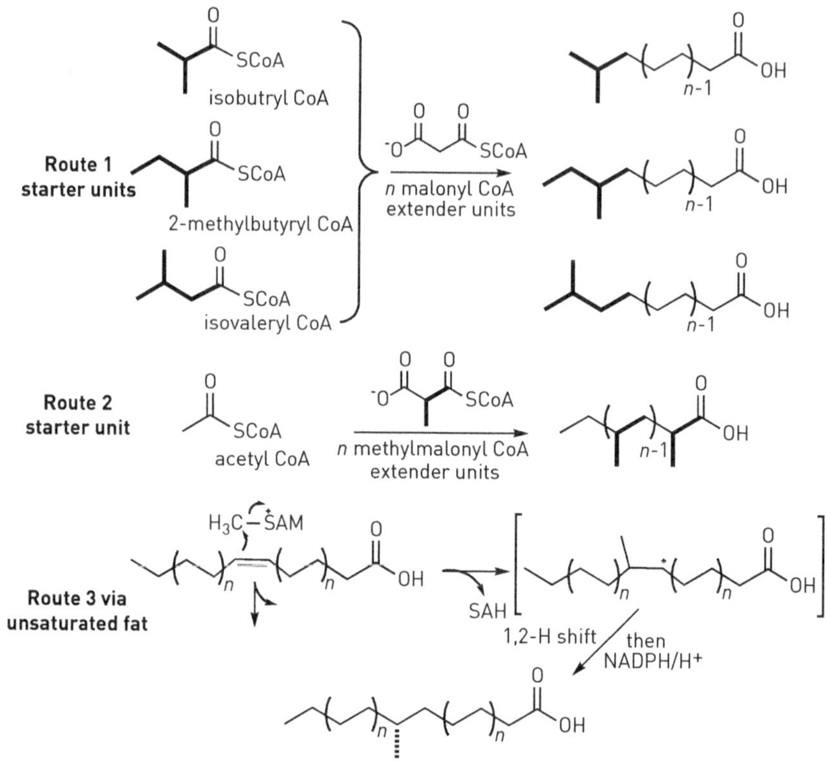

FIGURE 5.15

Different starter or extender units or SAM methylation for branched fatty acid biosynthesis.

so-called **desaturase** enzymes which also control the position at which the double bond will be introduced. As we will see, there are different mechanisms for the introduction of double bonds in fatty acid chains and these can vary depending on whether the producing organism is animal, plant, or bacteria. We will also see that animals and plants differ in their ability to introduce double bonds at specific positions in a fatty acid chain. But before we go there, let's take a brief look at the two most important methods for desaturation of fatty acid chains.

AEROBIC VERSUS ANAEROBIC ROUTES TO DESATURATION

The basic route to the introduction of unsaturation in a fatty acid via aerobic oxidation is illustrated for the conversion of **stearic acid** to **oleic acid** in Fig. 5.17. As the name suggests, the aerobic mechanism requires O_2 and is believed to involve diiron-oxygen complexes, though the proposed structures are speculative. Note the *syn* stereochemistry of this **oxidative** dehydrogenation process; the enzyme catalyzing the process is referred to as a Δ^9 **desaturase,** with Δ^9 referring to desaturation occurring at the 9-position relative to the carboxyl carbon. In plants, the fatty acid is ACP-bound for the desaturation process, while in animals and fungi it is CoA-bound.

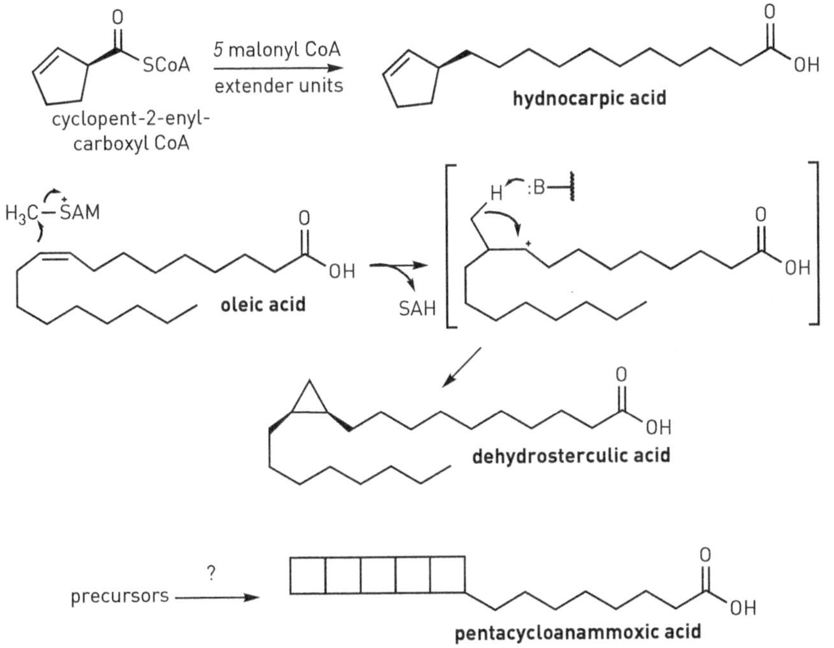

FIGURE 5.16
Ring-containing fatty acids: hydnocarpic acid, dehydrosterculic acid, and pentacycloanammoxic acid.

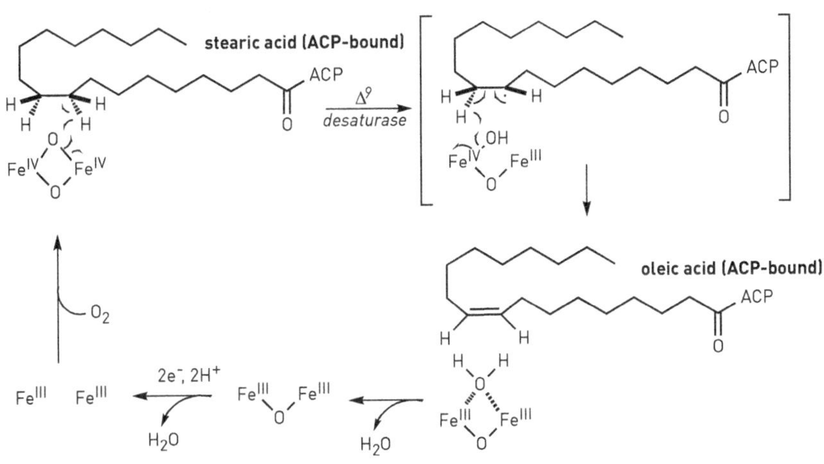

FIGURE 5.17
Possible Fe-mediated mechanism for aerobic fatty acid desaturation in plants.

The production of unsaturated fatty acids by anaerobic bacteria requires a **non-oxidative** mechanism; quite sensibly, such organisms utilize an intermediate already available during normal chain extension in the usual fatty acid biosynthetic process. As shown in Fig. 5.18, the process for the production of oleic acid involves interruption of the usual fatty acid chain extension process for 12 carbons at the α,β-unsaturated thioester stage. Instead of the extension process continuing with the usual

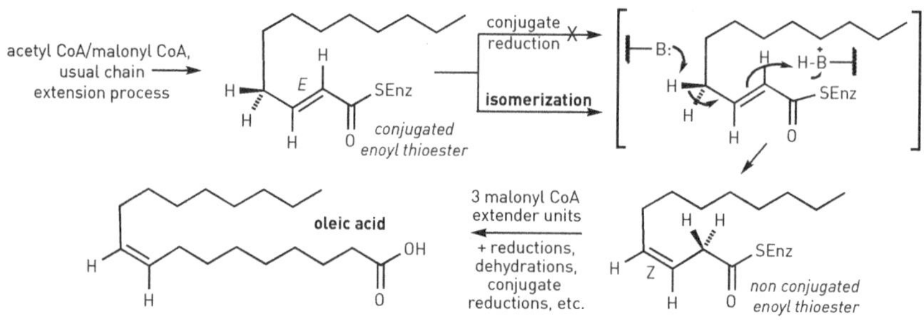

FIGURE 5.18

Nonoxidative desaturation mechanism for production of oleic acid by anaerobic bacteria.

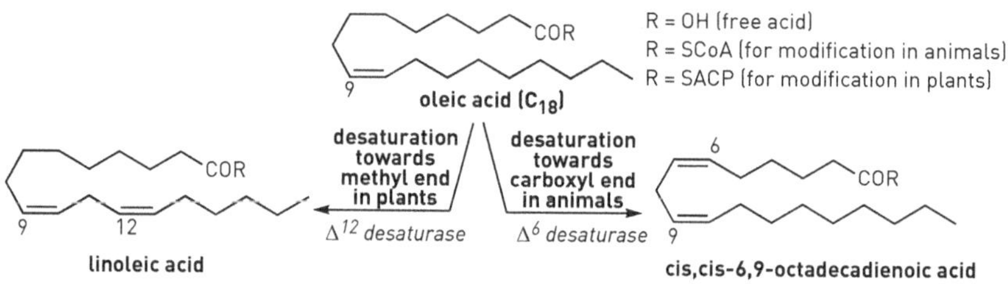

FIGURE 5.19

Divergence in plant vs. animal FA desaturation pathways and biosynthesis of linoleic acid.

conjugate reduction of the enoyl thioester intermediate, a double-bond isomerization takes place, converting the conjugated enoyl thioester into a nonconjugated enoylthioester with Z geometry at the double bond. At this point, normal fatty acid chain extension then resumes for three additional cycles of malonyl-CoA addition, with the attendant reductions, dehydrations, and conjugate reductions terminating at 18 carbons to ultimately produce oleic acid.

The anaerobic process can produce only monounsaturated chains, but the aerobic unsaturation process can proceed further from mono- to di-, tri- or polyunsaturated chains, usually with a nonconjugated relationship between the double bonds ("methylene-separated"). While both plants and animals can produce multiply unsaturated fatty acid (FA) chains, there is a distinct difference in positional selectivity for further double bond introduction. In plants, further oxidative desaturation of FAs (as their –S-ACP thioesters) usually occurs from an existing double bond toward the methyl end of the fatty acid chain, while in animals, further desaturation of FAs (as their –S-CoA thioesters) occurs from an existing double bond toward the carboxyl end of the fatty acid chain (Fig. 5.19). As it turns out, two fatty acids important to human health, **linoleic acid** and **α-linolenic acid,** are both derived from oleic acid; linoleic acid (shown in Fig. 5.19) is obtained by introduction of one additional double bond into oleic acid at the 12-position (i.e., toward the methyl end of the chain).

Further oxidative desaturation of linoleic acid at the 15-position (again, toward the methyl end of the chain) produces α-linolenic acid, the world's most prevalent

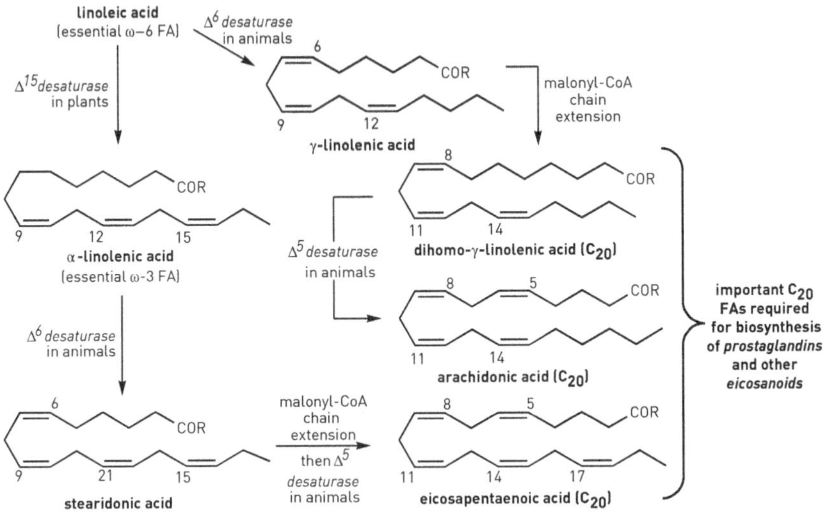

FIGURE 5.20
Biosynthetic modifications of linoleic, α-linolenic, and γ-linolenic acids.

fatty acid (Fig. 5.20). Since animals can only introduce additional double bonds toward the carboxyl end of the chain relative to an existing double bond, this means that humans cannot biosynthesize either of these two fatty acids (animals lack the required Δ^{12} and Δ^{15} desaturase enzymes). Thus, linoleic and α-linolenic acids are referred to as **essential** fatty acids, since they are essential to human health but can only be obtained in the diet from plant sources. In nutritional literature, linoleic acid is often referred to as an **omega-6 fatty acid** (or ω-6), using a Greek alphabet nomenclature system. Here, the omega-carbon of a fatty acid is identified as the one furthest from the carboxyl end of the chain; so in the case of linoleic acid, a double bond is located starting at the sixth carbon relative to the omega-carbon (methyl terminus) of the chain. Similarly, α-linolenic acid would be classified as an **omega-3 fatty acid** using this nomenclature system. Figure 5.20 also illustrates the divergence of further desaturation of fatty acids in plants versus animals and the sequences involved in producing some other important polyunsaturated fats. Once linoleic and α-linolenic are obtained in the diet, animals can further desaturate their chains toward the carboxyl end. Thus, linoleic acid (all cis-9,12-octadecadienoic acid) is converted by desaturation at the 6-position to give **γ-linolenic acid,** which can be used as a starter unit (as its SCoA derivative) for one additional cycle of fatty acid chain extension. The resulting C_{20} product, **dihomo-γ-linolenic acid**, can also undergo a subsequent desaturation at what is now the 5-position of the extended C_{20} chain (again toward the carboxyl end), leading to **arachidonic acid**. A similar sequence of chain extension and Δ^5 desaturation of α-linolenic acid yields **eicosapentaenoic acid**. All three of these C_{20} FAs are important as precursors for the production of **prostaglandins** and other structurally related C_{20} signaling molecules known as **eicosanoids** which are important compounds in several highly complex systems of the body, exerting control over inflammatory and immune responses among others. We will shortly take a look at some of the biosynthetic chemistry involved in their production, but first we conclude our examination of unsaturated fatty acids by briefly considering examples of some unusual acetylenic and heterocyclic structures that represent further modified fatty acids.

FURTHER DESATURATION OF FATTY ACIDS: TRIPLE BONDS AND RINGS

Further desaturation of fatty acids at existing double bond positions can lead to the introduction of one or more triple bonds. Such **acetylenic fatty acids**, though less common and far less stable than their olefinic precursors, are formed in an oxidative desaturation process using O_2 and a nonheme diiron complex analogous to that used in the introduction of double bonds and mediated by *acetylenase* enzymes, some of which have both *desaturase* and *acetylenase* functions. A typical system is shown in Fig. 5.21 in which the *cis*-12 double bond of linoleic acid is further desaturated to give the corresponding acetylenic derivative, **crepenynic acid,** which makes up about 70% of the content of *crepis alpina* seed oil. **Matricaria acid**, an antifeedant for both the pink bollworm (a serious cotton pest) and the tobacco hornworm, is derived from typical modifications of crepenynic acid as shown. Further desaturations and an allylic isomerization of the *cis* double bond from the 9- to the 10-position are followed by an allylic oxidation that introduces a ketone function at the 9-position.

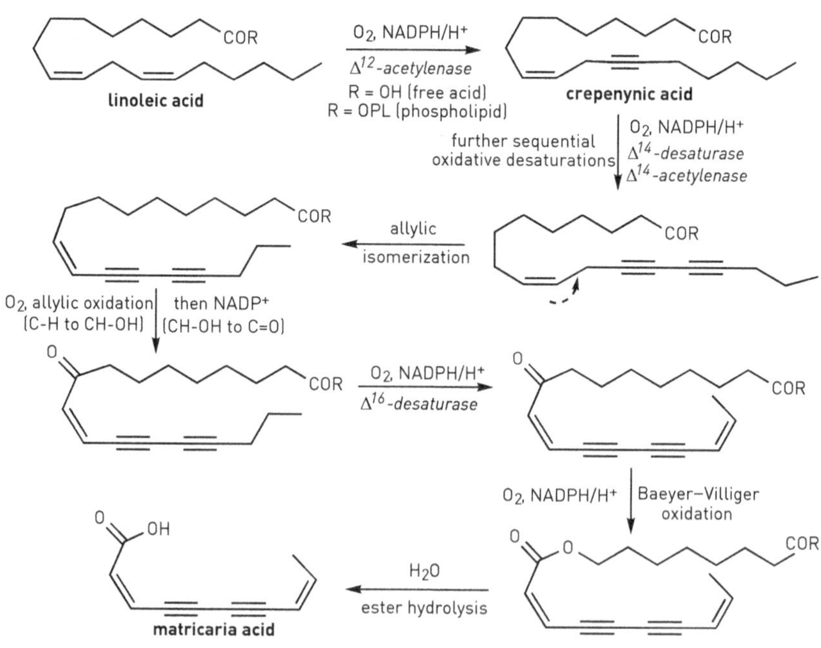

FIGURE 5.21

Biosyntheses of acetylenic FAs: crepenynic acid and its conversion to matricaria acid.

Subsequent desaturation to afford a second *cis*-double bond at the 16-position and a Baeyer–Villiger oxidative ketone-to-ester conversion terminates in ester hydrolysis, affording the polyunsaturated acetylenic FA natural product. The methyl ester of matricaria acid has also been found to be a powerful skin-bleaching agent, making it a potentially useful compound for the treatment of skin hyperpigmentation.

Other examples of polyacetylenic fatty acids include **13,14-dihydrooropheic acid**, isolated from the stem bark of the Indonesian plant *Mitrephora celebica* and which has shown remarkable activity against methicillin-resistant strains of *Staphylococcus aureus* (MRSA). Similar polyacetylenic FA derivatives (Fig. 5.22) are also thought to be involved in the biosynthesis of certain natural products containing O- and S-heterocyclic rings. **Wyerone**, a potent antifungal acetylenic keto ester isolated from faba bean shoots contains a furan ring which is believed to arise from H_2O addition across a conjugated diyne precursor as shown in Fig. 5.22. Finally, a phototoxic acetylenic dithiophene ester was isolated from *Dyssodia tageoides* along with what is believed to be its precursor, **3,5,7,9,11-dodecapentayn-1-yl-stearoate**; the dithiophene ring system presumably forms via sequential additions of H_2S across the conjugated polyacetylenic ester chain. While other furan and thiophene-containing FA derivatives are presumed to be derived in a similar fashion, a number are also known to be formed by alternate mechanisms.

There are many additional types of fatty acids, including those containing halogens, aldehyde, and ketone groups, hydroxyl and alkoxyl groups, amides, dicarboxylic acids, and others too numerous to treat here. But before we move on to the core of the polyketide pathway to natural products, we conclude our examination of modified fatty acids by returning to arachidonic acid and related C_{20} FAs and the

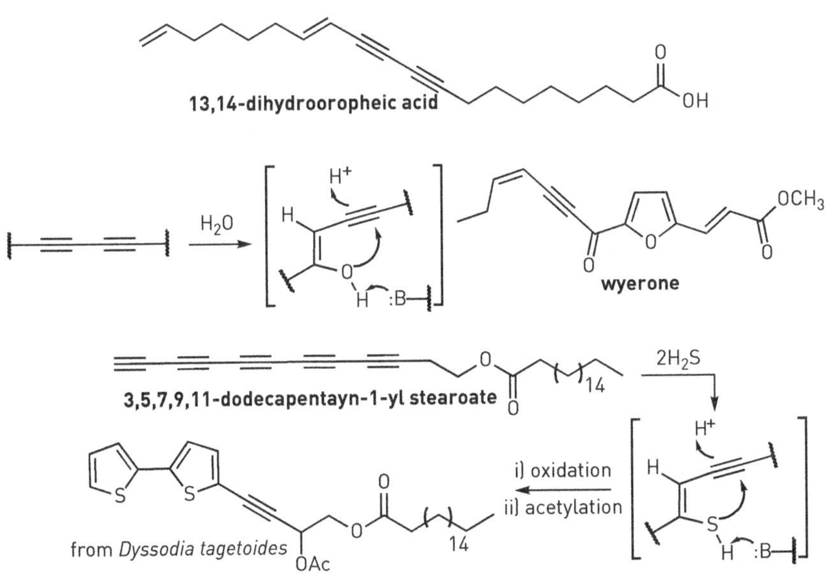

FIGURE 5.22
Dihydroorpheic acid and polyacetylenic precursors to furan and thiophene natural products.

role they play in the formation of the important group of signaling molecules known as the **eicosanoids**.

PROSTAGLANDINS, THROMBOXANES, AND LEUKOTRIENES: THE POWER OF OXYGENATED FAS

As seen in the previous section, oxidative transformations of various types can transform fatty acids into some unusually bioactive substances that bear only a passing resemblance to their relatively simple precursors. A most striking example of this is seen in the so-called **eicosanoids**, a group of powerful bioactive molecules derived from the essential C_{20} fatty acids **arachidonic acid, dihomo-γ-linolenic acid**, and **eicosapentaenoic acid** that we saw earlier in Fig. 5.20. Of these three, arachidonic is the most important player, as it serves as the essential precursor to most of the **prostanoid** family of eicosanoids which includes the **prostaglandins, prostacyclins**, and **thromboxanes**, examples of which are illustrated in Fig. 5.23 along with the standard abbreviations used for the constitutions of the various C-5 ring systems found in different prostaglandins (PGs) and thromboxanes (TXAs).

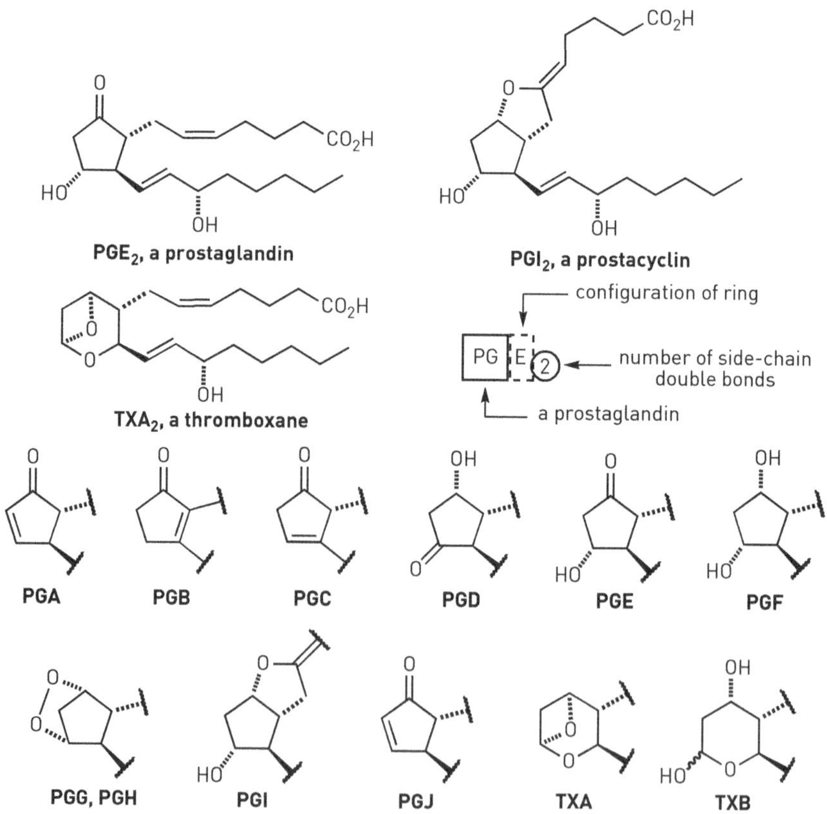

FIGURE 5.23
Examples of a prostaglandin, a prostacyclin, a thromboxane, and prostanoid ring abbreviations.

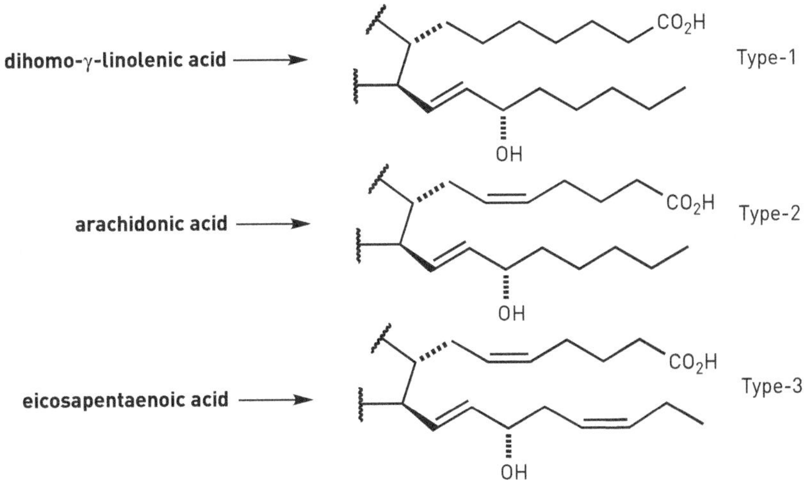

FIGURE 5.24
Origins of Type-1, Type-2 and Type-3 prostanoids from the three different C_{20} EFAs.

The examples shown in Fig. 5.23 are all members of the so-called "**Type-2**" prostanoids which are derived from arachidonic acid and are the most important of the three types. **Type-1** and **Type-3** prostanoids are derived from the two remaining C_{20} EFAs, as shown in Fig. 5.24. Type-1 prostanoids have only one side chain double bond, while the others have two and three side chain double bonds, respectively. Thus, the numeral 1, 2, or 3 following the abbreviation for a given PG or TX compound not only indicates the number of double bonds in the side chains but is also indicative of a given prostanoids EFA precursor.

The prostaglandins were originally isolated from the fluid of sheep prostate glands, hence the name "prostaglandin." These and other prostanoids have subsequently been found in most human tissues, though only in trace amounts. Prostaglandins exert their effects on many different systems in the body, including the constriction or dilation of smooth muscle, the induction of uterine contraction during labor, and the regulation of inflammatory response, among many others. While extremely potent compounds which exert their effects at very low concentrations, they are also relatively short-lived and so mainly act locally at the site of production.

The biosynthesis of the Type-2 prostaglandin **PGH_2** is outlined in Fig. 5.25. Following hydrolytic release of arachidonic acid from a phospholipid via *phospholipase A_2*, the process begins with an oxidative H-atom abstraction mediated by a *PGH_2 synthase* complex consisting of both a *cyclooxygenase (COX)* and a *peroxidase* component. The former may be one of two types, either COX-1 or COX-2, and it is their inhibition by aspirin or other COX-1/COX-2 inhibitors such as nonsteroidal anti-inflammatory agents (NSAIDs like ibuprofen, naproxen, or specific COX-2 inhibitors like Celebrex) that act to reduce inflammation through the inhibition of this biosynthetic pathway. Addition of molecular oxygen to the allylic radical and subsequent cyclization (hence the term *cyclooxygenase*) initiated by addition of a

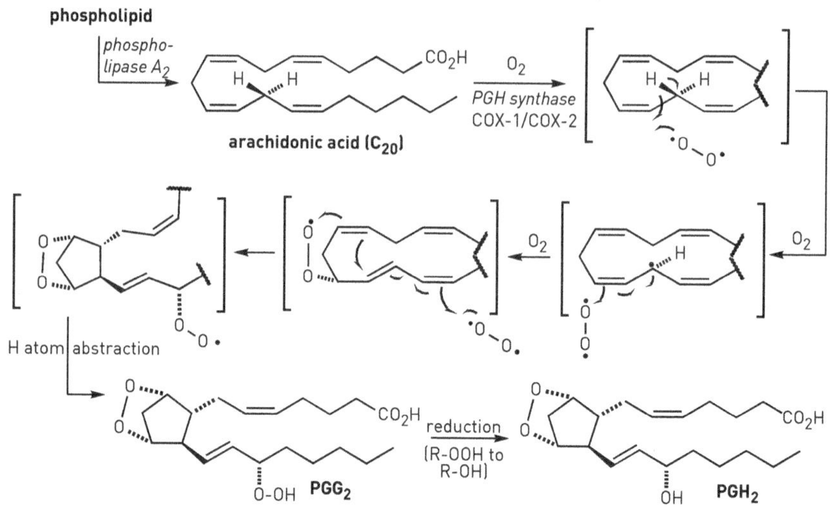

FIGURE 5.25
Oxidative transformations of arachidonic acid leading to a key prostanoid, PGH_2.

second O_2 molecule is followed by H-atom abstraction to produce the cyclic peroxide derivative **PGG$_2$** which undergoes a subsequent hydroperoxide-to-alcohol reduction to give **PGH$_2$**. Though unstable, PGH_2 is of great significance as it is a key intermediate which goes on to form several other important PG or TX derivatives by subsequent transformations.

Mechanisms for the conversion of **PGH$_2$** into **PGF$_2$** or **PGE$_2$** by nucleophilic ring opening of the cyclic peroxide component are shown in Fig. 5.26. Nucleophilic attack by hydride from NADPH and resultant cleavage of the O–O bond is accompanied by protonation to give the *cis*-1,3-cyclopentanediol component of PGF_2 (**PGF$_2$α** in prostanoid nomenclature, where α indicates the natural *S* configuration at the chirality center bearing the side chain hydroxyl group). Naturally occurring PGF_2 is also known in medicine as **dinoprost**; its function in causing uterine contractions makes it useful for the induction of labor at term. It can also act as a bronchodilator and an abortifacient.

For the formation of PGE_2, one proposed mechanism involves nucleophilic attack on the PGH_2 cyclic peroxide linkage by the sulfhydryl group of the important tripeptide **glutathione** (abbreviated GSH, as its anion GS^-), leading to an intermediate that undergoes oxidative elimination and regeneration of GS^- to afford the corresponding β-hydroxycyclopentanone ring system. PGE_2, known in medicine as the drug **dinoprostone**, also exhibits important effects in the induction of fever and, like PGF_2, can be used to induce labor at term or as an abortifacient. PGE_2 is also a vasodilator, causing the relaxation of smooth muscle. A number of synthetic analogues of both PGF_2 and PGE_2 have emerged over the years such as **carboprost** and **gemeprost** which have proven to be useful primarily in obstetric medicine, either for labor induction or as abortifacients.

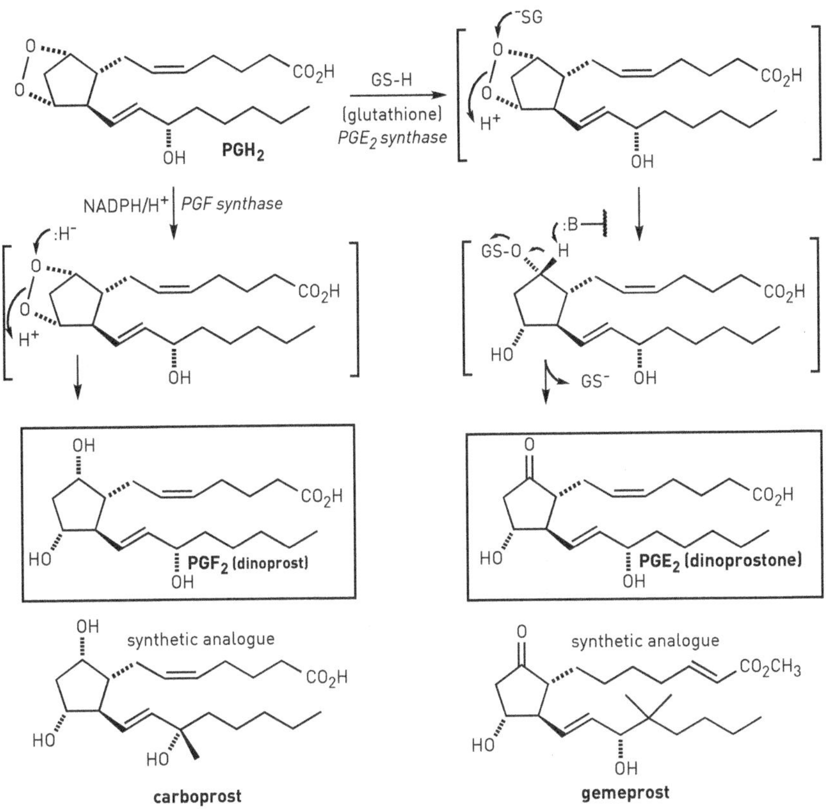

FIGURE 5.26
PGF_2 or PGE_2 via ring opening of PGH_2 plus synthetic analogues carboprost and gemeprost.

For the conversion of PGH_2 into **prostacyclin (PGI_2)**, a diradical mechanism via homolytic cleavage of the cyclic peroxide O–O bond, followed by cyclization and H-atom abstraction is usually invoked (Fig. 5.27). Prostacyclin is also known as the drug **epoprostenol** in medicine, being used as a very potent inhibitor of platelet aggregation; it was also the first prostaglandin the formation of which aspirin and other NSAIDs were shown to be able to limit through COX-1/COX-2 inhibition. Epoprostenol has also been found useful for the treatment of pulmonary hypertension, being administered either subcutaneously or by inhalation. As with PGE_2 and PGF_2, synthetic analogues of PGI_2 such as **iloprost** and **cicaprost** are finding increasing use for treatment of pulmonary hypertension and also certain autoimmune diseases such as scleroderma. Such analogues help to address the short half-life of the natural drug by removal of its labile enol ether function which, like an acetal, is subject to hydrolysis which deactivates PGI_2.

Finally, we consider the conversion of PGH_2 into the potent vasoconstrictor and platelet aggregation promoter, **thromboxane A_2 (TXA_2)**. Here, the mechanism is again formulated as a cleavage of the O–O bond of the cyclic peroxide ring accompanied by adjacent C–C bond cleavage to yield an alkoxyradical and an allylic

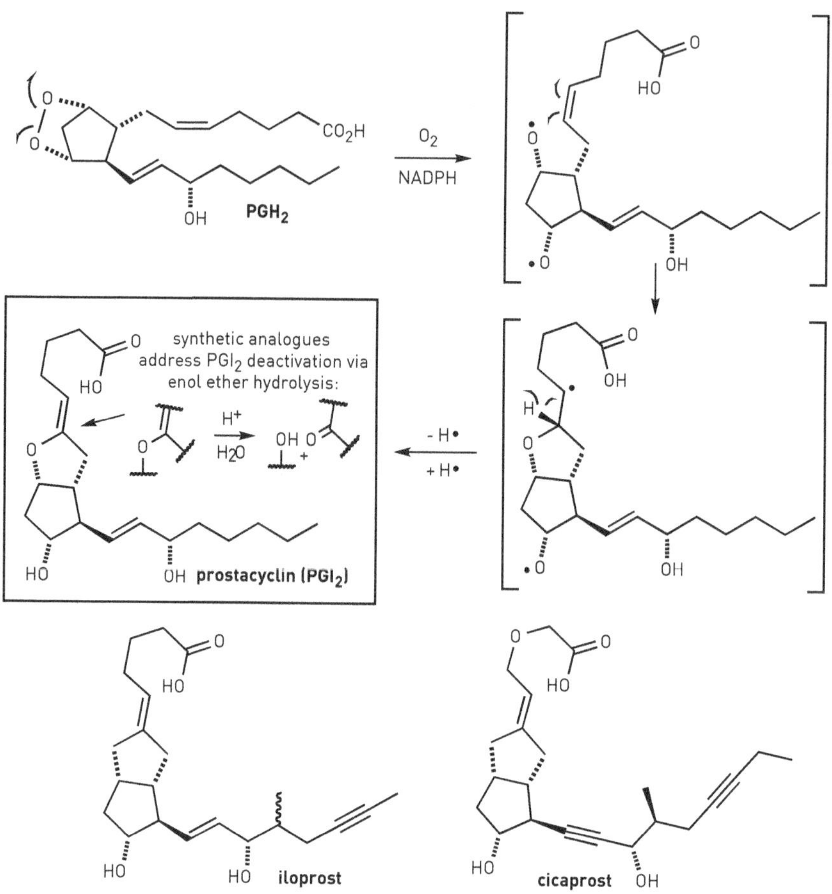

FIGURE 5.27

PGH_2 to PGI_2 conversion; two synthetic PGI_2 analogues, iloprost and cicaprost.

radical (Fig. 5.28). Single electron transfer from the allylic radical leads to a dipolar species composed of an allylic cation and an alkoxy anion, the latter of which acts as a nucleophile to attack the proximate aldehyde carbonyl. The resulting alkoxy anion then traps the allylic carbocation, leading to formation of the bicyclic acetal structure of **TXA_2**. This strained acetal has a very short half-life and is rapidly hydrolyzed to a bioinactive hemiacetal derivative, **TXB_2**. Because of its rapid hydrolysis, TXA_2 functions primarily in tissues surrounding its site of production. It is thought to be particularly important during tissue injury and inflammation, though its potent modes of action also implicate it in various blood-clotting (thrombosis) and angina-related events. Like other prostanoids derived from PGH_2, production of TXA_2 is also suppressed by aspirin or other COX inhibitors.

We conclude this section with a brief look at the formation of another group of arachidonic acid metabolites known as the **leukotrienes,** a group of compounds intimately linked to the effects of inflammatory response. The name is related to leukocytes (white blood cells) from which these compounds are produced; the name is also indicative of the characteristic presence of three conjugated double

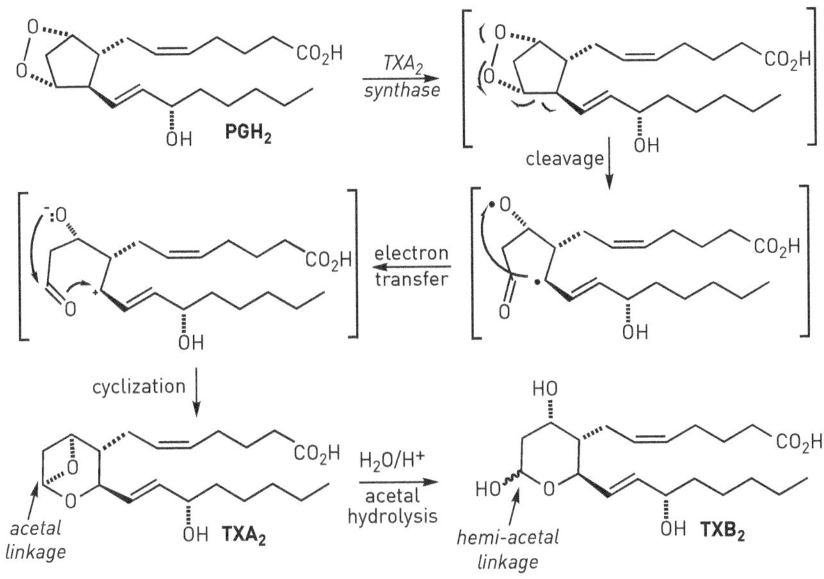

FIGURE 5.28
Conversion of PGH_2 to TXA_2 and its subsequent hydrolysis to inactive TXB_2.

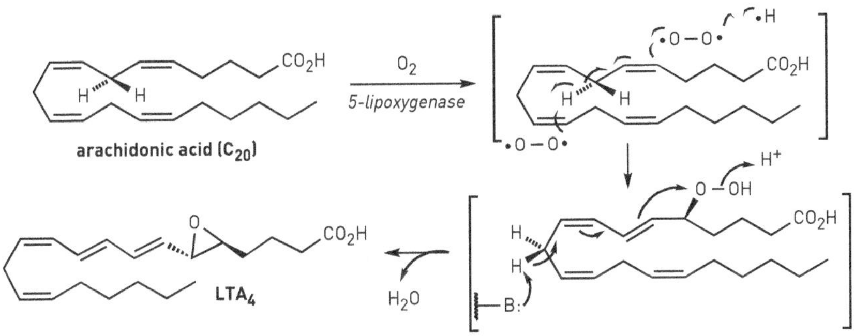

FIGURE 5.29
5-Lipoxygenase-mediated biosynthesis of leukotriene A_4 (LTA_4) from arachidonic acid.

bonds within the structures. Unlike the biosynthesis of prostanoids, which utilizes cyclooxygenase enzymes, initial leukotriene biosynthesis employs a **5-lipoxygenase** enzyme, the action of which initially leads to formation of **leukotriene A_4 (LTA_4)** as shown in Fig. 5.29 (as in the prostanoids, the numeric subscript indicates the total number of double bonds present within a given LT structure).

LTA_4 is subsequently converted to two other important leukotrienes, as illustrated in Fig. 5.30. The action of the enzyme *LTA_4 hydrolase* leads to remote nucleophilic attack by water at the furthest end of the conjugated triene system of LTA_4, resulting in conjugate addition and epoxide ring opening to produce **LTB_4**, a potent leukotriene associated with the immune response. LTB_4 acts in part by helping to direct white blood cells to affected tissues; however, elevated levels of LTB_4 have been linked to a variety of inflammatory and allergic conditions, especially asthma, ulcerative colitis, psoriasis, and rheumatoid arthritis among others,

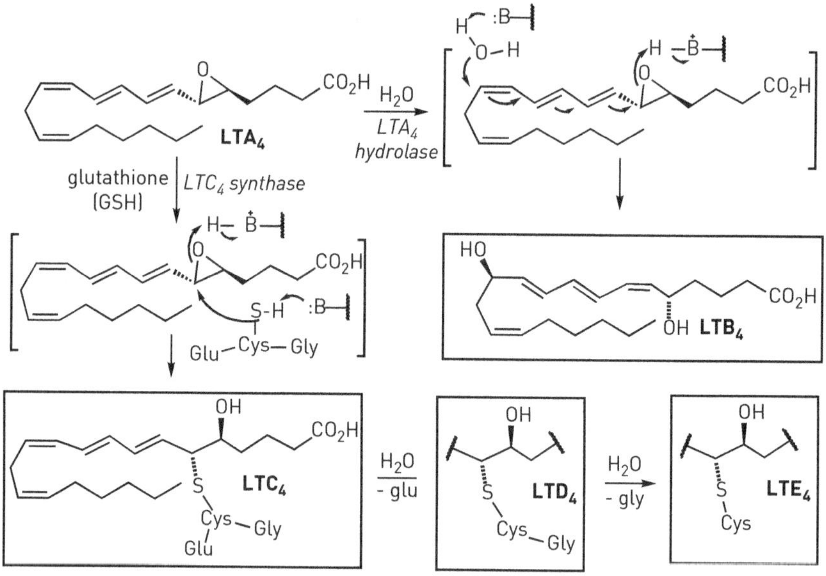

FIGURE 5.30
Conversion of LTA_4 to LTB_4 or LTC_4 and sequential LTC_4 hydrolysis to produce LTD_4 and LTE_4.

making the LTB_4 receptor an ongoing target for intensive drug research. Direct nucleophilic attack on the epoxide ring of LTA_4 by the cysteine thiol function of the tripeptide **glutathione** (Glu-Cys-Gly) leads to formation of **LTC_4** which, along with its metabolites **LTD_4** and **LTE_4**, make up the slow-reacting substance of anaphylaxis (SRS-A) that initiates this potentially life-threatening condition of extensive allergic inflammatory reaction. The SRS-A combination is estimated to be 5000 times more potent than histamine for triggering inflammatory response and though slower in onset, has a longer duration of action. LTD_4 and LTE_4 are produced from LTC_4 by sequential degradation of the glutathione component, first through hydrolytic loss of Glu to give LTD_4, then loss of Gly from LTD_4 to yield LTE_4, as shown.

We've presented here a bare minimum of examples of biosynthesis and subsequent modification of fatty acids in this section to at least have enough background to move to the main event in this chapter, namely the closely related chemistry associated with the biosynthesis of the diverse family of bioactive natural products derived from the so-called **polyketide pathway**.

POLYKETIDE BIOSYNTHESIS: MORE STARTER UNITS AND EXTENDER UNITS, BUT WITH A TWIST

As mentioned at the beginning of the chapter, **polyketide** research dates all the way back to the late 1800s with the work of J. N. Collie and others. In the following years, many new compounds that fit the structural peculiarities associated with this class of natural products were isolated and characterized from fungi and bacteria, mainly by organic chemists using standard organic laboratory techniques. But it wasn't until the early 1950s that A. J. Birch (also an organic chemist) began a serious examination of the possible biosynthetic mechanisms associated with production of these compounds, inspired in part by the known chemistry of fatty acid biosynthesis. Birch's initial proposal became known as the **acetate hypothesis**: "Polyketides are formed by the head-to-tail linkage of acetate units, followed by cyclization by an aldol reaction or by acylation to phenols" (Birch & Donovan, 1953). By using radiolabeling techniques in their early work, Birch and others were able to show that polyketide compounds derived from the same linkage of acetate units that was already known to be employed in primary metabolism pathways in the organisms producing the compounds. As we will see shortly, Birch's insight derived in part from his knowledge of the kinds of organic reactions a hypothetical polyketide compound might be expected to undergo and the resulting structural features that would result from those reactions. In other words, working backward from a natural product structure, he could visualize how standard organic transformations operating on a "polyketide" precursor could give rise to that product and that the origin of the "ketide" units in the polyketide would almost certainly be individual acetate units which had been linked together in some fashion.

The gist of this concept is illustrated in Birch's proposed biosynthesis of **orsellinic acid**, an aromatic compound isolated from the fungus *Penicillium griseofulvum* (Fig. 5.31). As can be seen, if the phenolic "enol" hydroxyl groups of orsellinic acid are envisioned in their keto forms, the remaining ring double bond could arise by a dehydration of a tertiary alcohol precursor which itself could be produced by an intramolecular aldol addition between carbonyl units of a folded "polyketide" precursor, which in turn could then be envisioned as arising from the linkage of four individual acetic acid units by some undetermined biosynthetic mechanism.

How could one support such a hypothesis experimentally? In Birch's early work, he made use of acetic acid (as acetate) which had been radiolabeled at the carboxyl

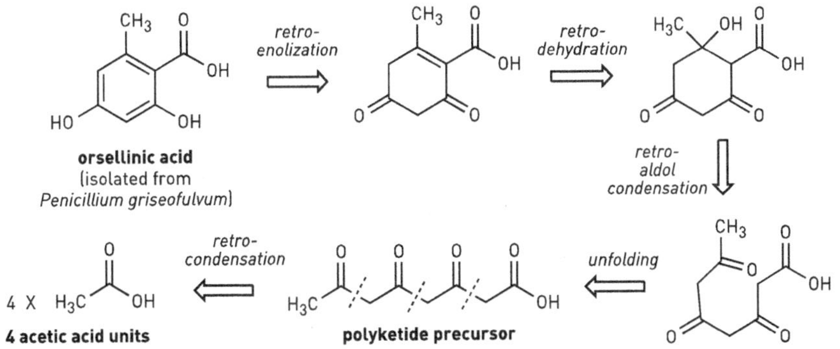

FIGURE 5.31
Retrosynthetic analysis of orsellinic acid leading to a "polyketide" derived from four acetic acids.

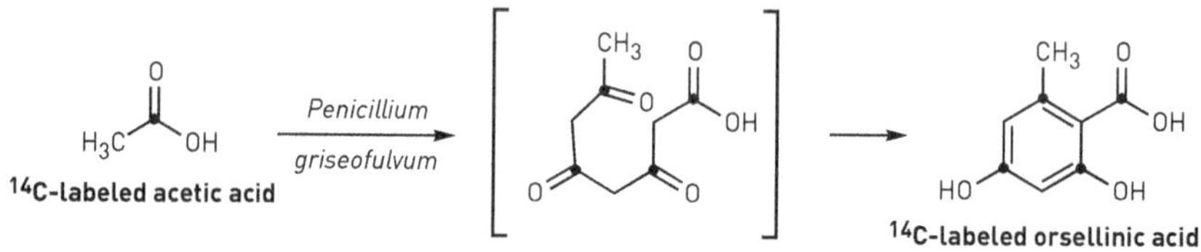

FIGURE 5.32
Location of ^{14}C atoms in orsellinic acid derived from ^{14}C-labeled acetic acid.

carbon (i.e., ^{14}C-labeled) and which was then provided to the organism producing the orsellinic acid. Such isotope feeding studies were then followed by isolation of the natural product which was then subjected to a series of systematic chemical degradation reactions. Careful measurement of the radioactivity content of the degradation products from different reactions allowed the unambiguous determination of the specific positions in orsellinic acid where ^{14}C atoms had been incorporated by the organism into the structure. As Fig. 5.32 illustrates, the pattern of ^{14}C substitution anticipated by the acetate hypothesis was confirmed in these experiments. In later work, patterns of incorporation of ^{13}C-labeled acetate into many other structures has been more conveniently analyzed through the use of ^{13}C NMR techniques; such isotope incorporation studies continue to be an essential part of ongoing studies of the chemistry of biosynthetic and metabolic pathways, as we will see shortly.

Of course, the above analysis does not address the more important question of *how* individual acetate units might become linked together to form the sorts of polyketide intermediates that could undergo subsequent condensations, aromatizations, and other reactions leading to various natural product structures. As it turns out, a likely answer to this question could be found by envisioning some simple modifications of the chemistry associated with the biosynthesis of fatty acids, a process which was already well understood by the time Birch began his initial investigations. As Fig. 5.33 illustrates, individual acetate units could be readily assembled into even-numbered polyketide chains of varying length via a modified fatty acid pathway in which the usual β-ketothioester intermediate arising from Claisen condensation of acetyl-CoA and malonyl-CoA units (enzyme-bound) is prevented from undergoing the usual sequence of ketone reduction, dehydration, and conjugate reduction that normally precedes acyl transfer and chain extension. In this way, malonyl-CoA extender units could be added sequentially to a growing **polyketide chain** rather than to a growing saturated fatty acid chain. While the scheme presented in Fig. 5.33 is general and rather simplified, it illustrates the basic chemistry associated with certain kinds of **polyketide synthase (PKS)** systems which, as it turns out, closely resemble fatty acid synthase (FAS) systems in both composition and chemistry.

We will find this modified fatty acid pathway operating mainly in bacteria, fungi, and plants rather than in animals (unlike the terpenoid pathway which is more widely distributed). Later on we'll sort out the different types of PKS systems that are found in various organisms and the different kinds of natural products that they can produce.

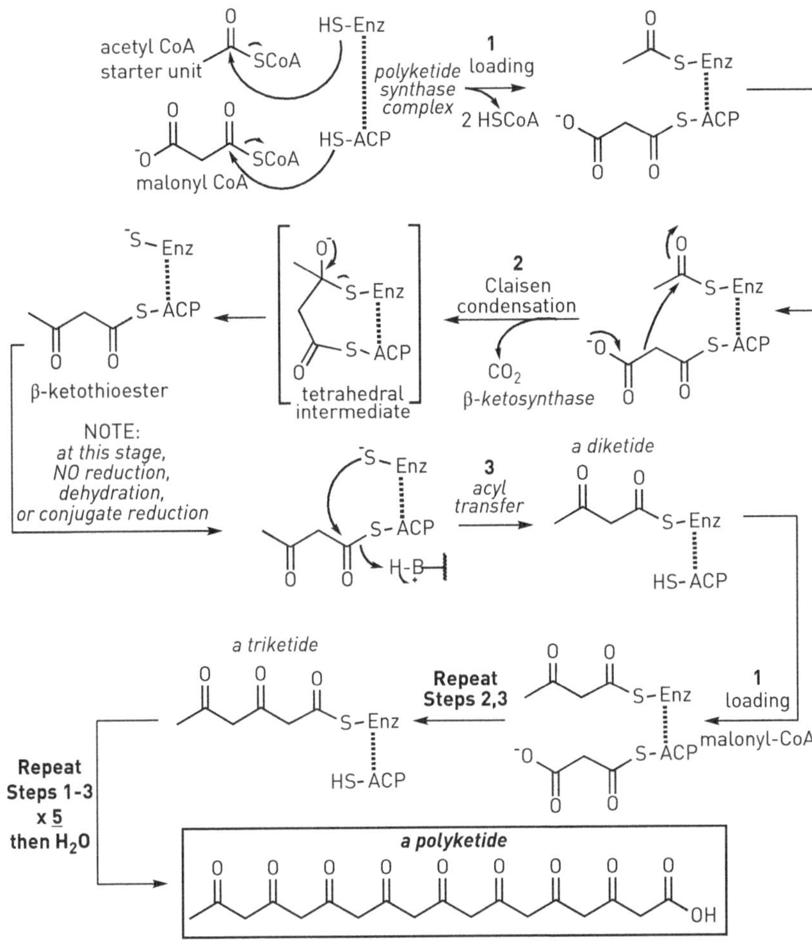

FIGURE 5.33

Simplified, generalized diagram of chain extension sequence in polyketide biosynthesis.

But for now, we focus initially on some simple polyketide structures assembled according to the simplified scheme above to learn how different kinds of reactions or even different chain folding patterns can lead to natural products of very different structure, even though they may arise from the same initial polyketide precursor.

AROMATIC POLYKETIDE NATURAL PRODUCTS: PHENOLS AND RELATED STRUCTURES

The polyketide pathway is one of nature's most important sources of aromatic compounds, especially phenols of widely varying structure; the versatility of polyketide cyclizations and aromatizations stems in part from the multifunctional nature of a given polyketide chain operating in conjunction with enzymes which may enforce different patterns of chain folding as well as positional selectivity for deprotonations and cyclizations. Thus, a single precursor polyketide chain may yield different

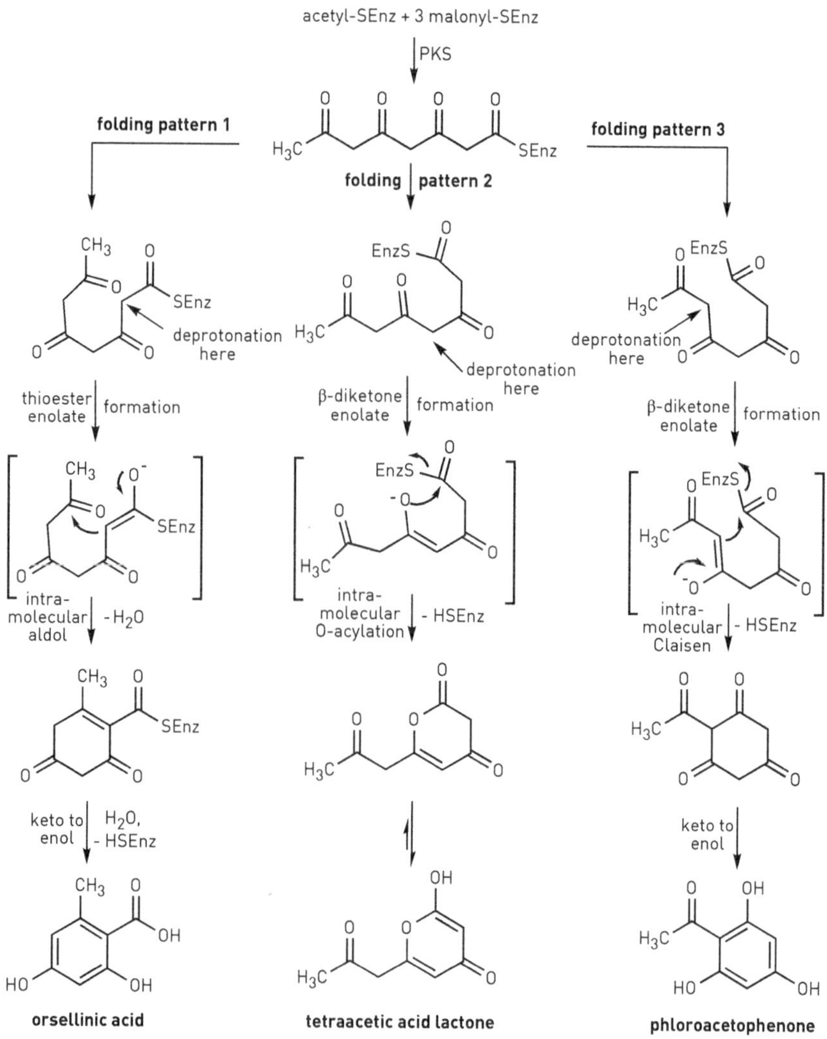

FIGURE 5.34
Different folding patterns and reaction modes for a tetraketide leading to different products.

products from different organisms, or a mixture of different products from the same organism via a thioester enolate or any one of several different β-diketone enolates that may be available via enzyme-specific deprotonation reactions. Different products will require different chain folding patterns which may then lead to different intramolecular condensations or other reactions.

As shown in Fig. 5.34, for **orsellinic acid**, folding pattern 1 can lead to formation of a thioester enolate ion which can cyclize via intramolecular aldol condensation followed by thioester hydrolysis and keto-to-enol tautomerization to produce the phenolic product. The same tetraketide precursor can follow folding pattern 2, undergoing intramolecular cyclization via O-acylation of the indicated β-diketone enolate ion, which leads directly to **tetraacetic acid lactone** (an α-pyrone) which has been isolated from a number of different sources such as *Penicillium* *stipitatum* along with small amounts of orsellinic acid and its decarboxylation product, **orcinol** (5-methylbenzene-1,3-diol). Folding pattern 3 and the indicated β-diketone enolate ion allows an intramolecular Claisen condensation, affording the cytotoxic fungicide **phloroacetophenone** after keto-to-enol tautomerization and aromatization. While the actual intermediates and precise sequence of events involved are undoubtedly more complex, a glance at this somewhat simplified analysis allows us to quickly see the predictive power of the acetate hypothesis and the synthetic power inherent in the polyketide pathway to natural products.

A somewhat more complex example of alternative polyketide chain folding patterns leading to different aromatic products is illustrated in Fig. 5.35. Both **alternariol**, a toxic metabolite of *Alternaria* fungi that is an important contaminant of fruits and grains, and **rubrofusarin,** an antimycobacterial compound with antiestrogenic properties from *Fusarium graminearum*, are derived from a common heptaketide intermediate that undergoes different enzyme-specific modes of cyclization and aromatization (with an additional postcyclization SAM methylation for rubrofusarin).

It is worth noting that the acetate hypothesis has often been quite useful in resolving conflicts over structural assignments when two alternative structures for a natural product are possible. Prior to the advent of highly sophisticated instrumental analysis techniques, choosing between or among similar proposed structures for a newly isolated natural product could be quite problematic (and occasionally still is). Using alternariol as an example, consider the two proposed structures, I and II, as shown in Fig. 5.36. These compounds are constitutional isomers of one another and either might be reasonable to assign to alternariol; both would have very similar IR, ^{1}H NMR and ^{13}C NMR spectra and would be difficult to distinguish from one another on that basis. Of course, one way to decide the matter would be to independently synthesize the proposed structures and then compare a sample of each with a sample of the naturally occurring compound for melting point, NMR spectra, and so on, (this was a common technique in earlier natural products chemistry and is still used for structure confirmation in some cases). However, this would involve a great deal of time and effort in the laboratory; fortunately, the decision is made much easier by application of the acetate hypothesis. One can readily derive a single folded polyketide chain precursor for structure I, while structure II cannot be derived without resorting to the use of two separate chains which would have to be linked together in a very specific and far more complicated sequence. While not an absolute proof, the acetate hypothesis, along with a bit of work with pencil and paper, quickly leads to a prediction of structure I as being far more likely than structure II for this compound.

ISOTOPIC LABELING STUDIES: BIOSYNTHETIC INSIGHTS VIA ^{13}C NMR

As mentioned earlier, one way to demonstrate how linear polyketide precursors are cyclized to yield aromatic products is through the use of ^{13}C NMR. One might expect

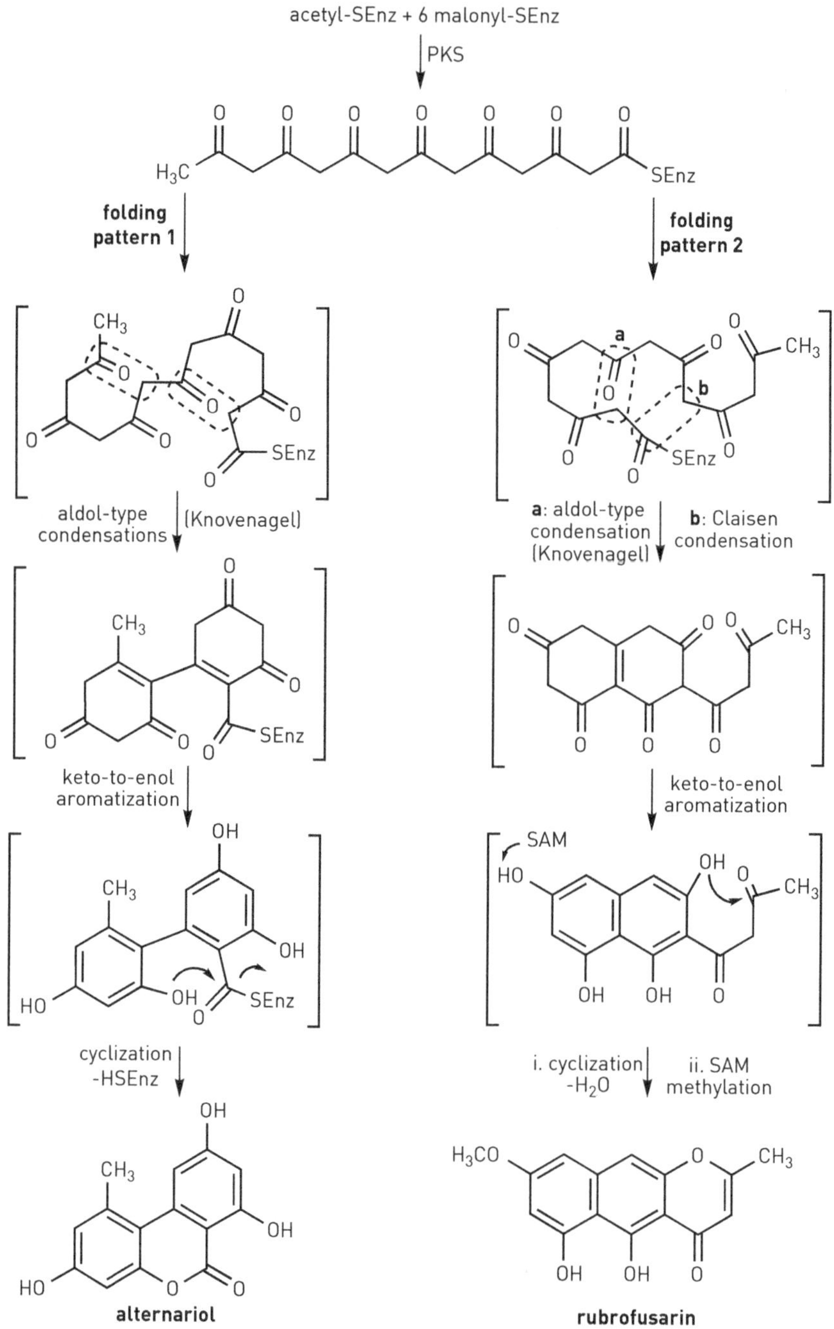

FIGURE 5.35
Different folding patterns and reaction modes leading to alternariol vs. rubrofusarin.

that adjacent ^{13}C atoms would be able "talk to each other" via spin-spin coupling, just as adjacent H atoms are able to do in ^{1}H NMR, but recall that the natural abundance of ^{13}C is only about 1.1%, so for ^{13}C NMR, ^{13}C–^{13}C coupling is insignificant since the probability of two ^{13}C atoms being adjacent to one another is vanishingly

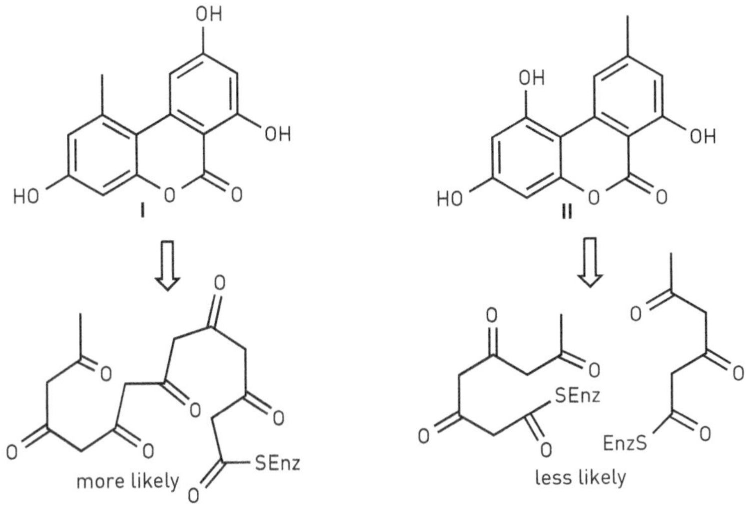

FIGURE 5.36
Two proposed structures for alternariol (I vs. II) clarified by the acetate hypothesis.

small. Also recall that ^{13}C–H coupling (which *is* significant) is usually removed by irradiation with a separate *rf* source, thereby eliminating any signal splitting due to such coupling, leaving a spectrum with single peaks for each nonequivalent carbon atom within a given structure. It is also useful to recall that signals from ^{13}C atoms which are not bonded to a H atom usually have a smaller intensity when compared to signals from ^{13}C atoms which *are* bonded to H atoms, often making these signals relatively easy to distinguish from one another (this is also one reason why the area under peaks in ^{13}C NMR spectra are not "integrated" as they are in ^{1}H NMR). With all the above in mind, consider Fig. 5.37, which provides a simple example of the use ^{13}C NMR in natural product studies.

In the simulated spectrum of natural **alternariol** (spectrum A) with C atoms numbered as shown, we can see a total of 14 signals, one for each C atom in the structure (since no two C atoms are equivalent in this compound). We can now compare this spectrum with simulated spectrum B, which is of alternariol isolated from a producing organism which had been fed **1-^{13}C-labeled acetate**. As expected, the signal for each of the odd-numbered carbon atoms in the structure has been enhanced in spectrum B. While not a dramatic result, it clearly illustrates how ^{13}C incorporation serves to confirm a polyketide pathway origin for the product.

Further use of ^{13}C NMR is illustrated in Fig. 5.38, again with two simulated spectra (A and B) of alternariol. Spectrum A is that of alternariol derived from a producing organism which has been fed **2-^{13}C-labeled acetate**. As expected, we now see enhanced signals for each of the even-number carbons in the structure. The more significant spectrum, however, is spectrum B which is the result of a feeding study employing acetate which has been doubly labeled (**1,2-^{13}C-labeled acetate**). This spectrum requires some additional explanation. First of all, when 1,2-^{13}C-labeled

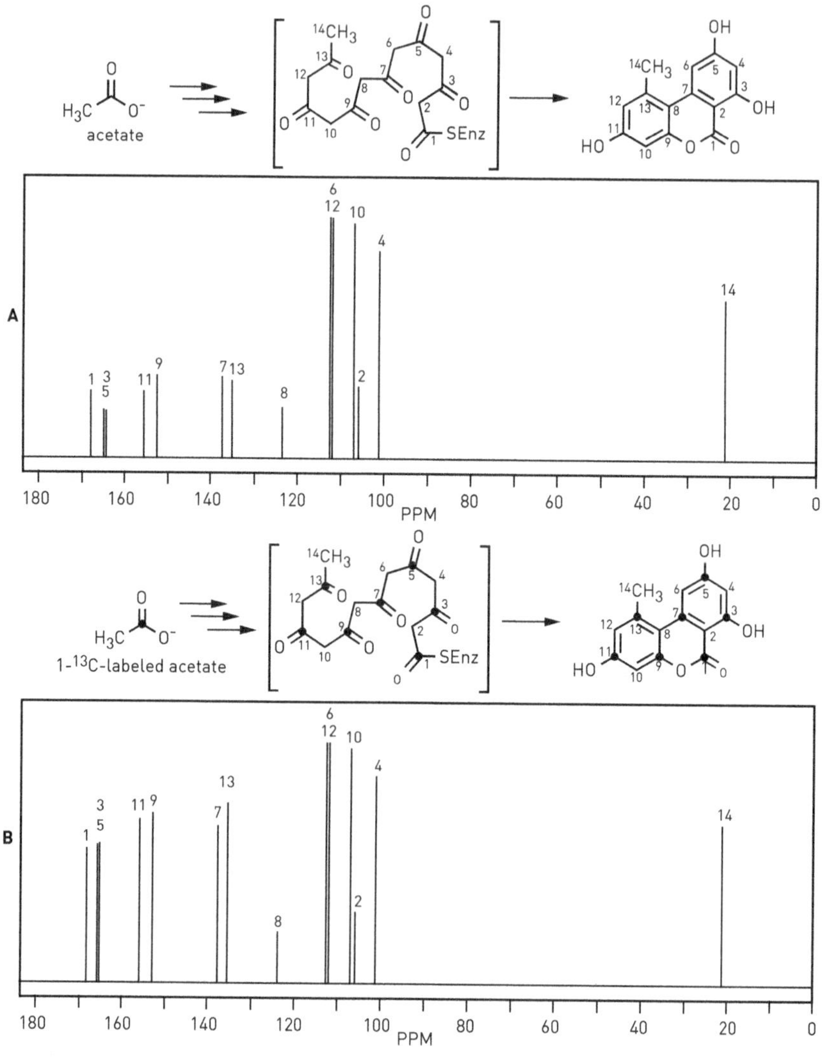

FIGURE 5.37
Simulated ^{13}C NMR of natural alternariol (A) and alternariol via 1-^{13}C-enriched acetate (B).

acetate is incorporated into a structure (depicted with bold bonds), we would not expect to see selected signal enhancements. But unlike the ^{13}C NMR spectrum of natural alternariol, in which no ^{13}C–^{13}C coupling would be expected, we should now see such coupling, since some of the compound will contain the doubly labeled acetate with adjacent ^{13}C atoms. Keeping in mind that the degree of incorporation of isotopically labeled precursors by organisms in such studies is usually very low compared to incorporation of normal starting materials, we will *not* expect to see ^{13}C–^{13}C coupling between *adjacent* doubly-labeled acetate units, since the probability of two doubly labeled units being incorporated next to one another within the same structure is vanishingly small. Thus, we should only see ^{13}C–^{13}C splitting between carbon atoms *within* a given doubly labeled acetate unit. It turns out that

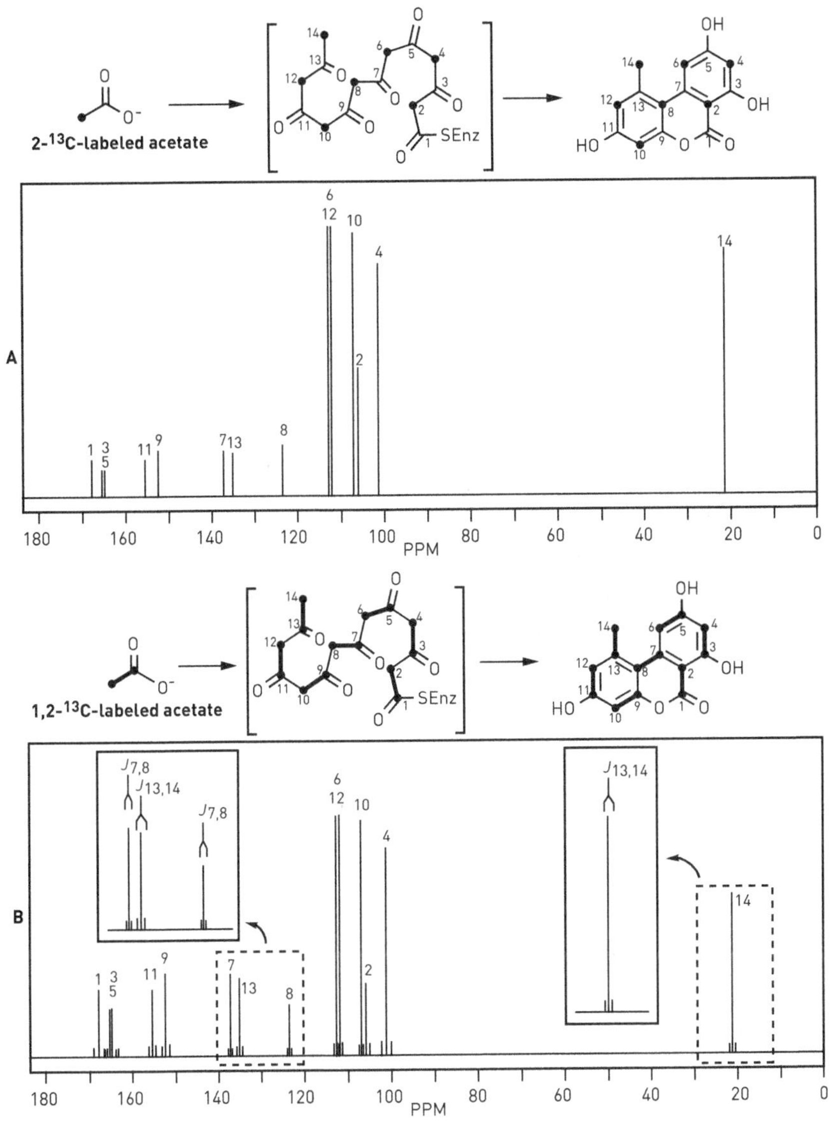

FIGURE 5.38
Simulated ^{13}C NMR of alternariol from 2-^{13}C-enriched acetate (A) and selected coupling constants from ^{13}C NMR of alternariol derived from 1,2-^{13}C doubly labeled acetate (B).

coupling constants for ^{13}C–^{13}C splitting are on the order of 50–80 Hz, so they are more than large enough to distinguish in such spectra and are sufficiently variable and environment-dependent that we should be able to match up a signal in one region of the spectrum to another to which it is coupled elsewhere in the spectrum, based on the unique line spacings involved. Spectrum B illustrates this for two selected cases: i) the coupling between the aromatic ring carbon at position 13 and the methyl group carbon at position 14 and ii) the coupling between the aromatic ring carbons at positions 7 and 8 which link the two aromatic rings together. For i), most of resonance at ~21 ppm from the methyl group comes from the natural abundance ^{13}C, but some must also come from doubly labeled acetate incorporated into the structure. This portion of the signal must in turn be split by the adjacent doubly labeled acetate carbon at position 13, and therefore should appear as a doublet with line spacing characteristic of the coupling constant $J_{13,14}$. If we look further down field for the resonance from the aromatic ring carbon at position 14, we again see the satellite peaks on either side of the main peak with the characteristic line spacing again corresponding to $J_{13,14}$ due to the expected splitting of this signal by the methyl carbon. A similar analysis applies to case ii) for the observed splitting for the signals associated with the aromatic ring carbons at positions 7 and 8.

We can see how this kind of doubly labeled acetate experiment could help sort out different possible folding patterns for certain polyketide cyclizations, as illustrated in Fig. 5.39. The polyketide precursor we saw earlier for biosynthesis of rubrofusarin should actually be able to adopt either of two possible folding patterns that would lead to the same product. Does the mediating enzyme for rubrofusarin production force the heptaketide precursor into folding pattern (1) or folding pattern (2), or is the product produced by a random mixture of the two possible folding patterns? This might seem like a very difficult question to answer, but it turns out that the ^{13}C NMR of the experimentally observed product from a 1,2-^{13}C-labeled

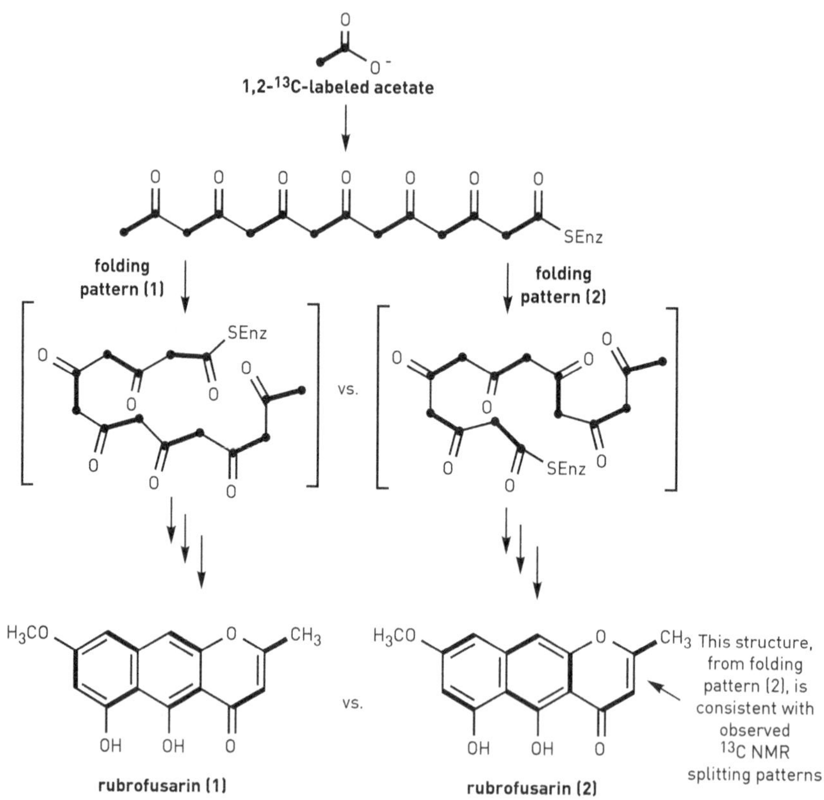

FIGURE 5.39

Different possible folding patterns for rubrofusarin; folding pattern (2) implied by ^{13}C NMR.

acetate feeding study shows isotope incorporation and coupling patterns consistent only with folding pattern (2) in this case. As might be expected, this has been just a small taste of how isotope incorporation experiments for NMR studies involving not only ^{13}C but also isotopes of H, O, N, and other atoms can be used to probe some of the most intimate details associated with biosynthetic mechanisms and pathways. We'll see more later on.

FURTHER MODIFICATION OF POLYKETIDES: ALKYLATIONS, OXIDATIONS, REDUCTIONS, AND DECARBOXYLATIONS

Relatively few aromatic polyketide-derived natural products arise from a process that terminates after simple polyketide cyclization and aromatization. In most cases, a variety of additional structural modifications take place either before or after cyclization/aromatization. For instance, we saw previously that following polyketide folding, cyclization, and aromatization, rubrofusarin biosynthesis required a final step involving S-adenosylmethione (SAM) methylation of a phenolic hydroxyl group (Fig. 5.35). Methylations of both O- and C-atoms are quite common components of various biosynthetic schemes, as in the final O-methylation of phloroacetophenone to give the antifungal **xanthoxylin** or the regiospecific C-methylation of the same tetraketide precursor followed by cyclization and aromatization to give **methylphloroacetophenone** (Fig. 5.40), a compound which we will see later in the biosynthesis of the antibiotic, **usnic acid**.

Another important and commonly observed alkylating agent is DMAPP which figured prominently in the terpenoid pathway and, like SAM, may serve as an alkylating agent for carbon or oxygen, though C-alkylations tend to predominate. An example of sequential polyketide alkylations is shown in Fig. 5.41 in which the triphenolic ketone derived from a pentaketide precursor may be directly alkylated by SAM to give **eugenone**, a fragrant component of oil of cloves, or may undergo further cyclization followed by DMAPP ring alkylation to give **peucenin**, a major chromone component of sneezewood and a potent acetylcholinesterase inhibitor.

Additional modifications of peucenin serve to illustrate some examples of some other typical postcyclization transformations, especially formation of benzofuran-type heterocycles. The dimethylallyl moiety is frequently a target of oxidative epoxidation in biosynthesis, leading to side chain cyclizations, as illustrated in Fig. 5.42. Both **khellin**, a diuretic herbal extract used in ancient Egypt as a remedy for renal colic, and **visnagin**, a potent vasodilator used in the treatment of angina and asthma, are obtained from a widely distributed member of the carrot family, *Ammi visnaga*. Epoxidation of peucenin followed by intramolecular nucleophilic cyclization leads to the indicated tertiary alcohol which may then undergo oxidation of the C5 ring at the benzyllic position to give the corresponding diol. A 1,2-elimination driven by aromatization and loss of acetone gives the intermediate benzofuran derivative

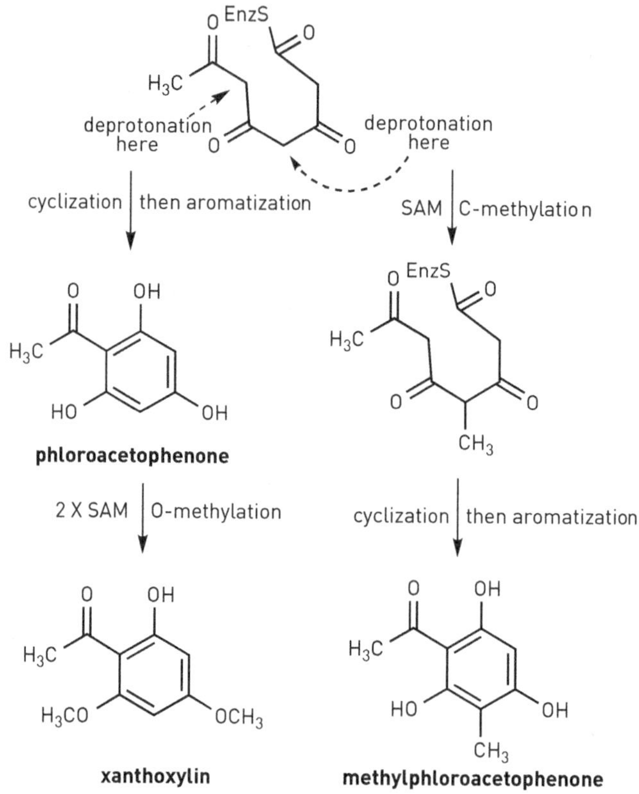

FIGURE 5.40
Examples of C- and O-methylations via SAM either before or after polyketide cyclization.

which may then undergo SAM methylation to yield visnagin or subsequent aromatic ring oxidation to a hydroquinone derivative which is then converted to khellin by SAM methylation of both phenolic hydroxyl groups. While use of khellin is quite limited due to a number of adverse side effects, the synthetic analogue **amikhelline** has seen use as a vasodilator and in the treatment of colic.

Another interesting and far more complex example of structural modification is seen in the biosynthesis of **tajixanthone,** a polyketide-derived antimicrobial compound of considerable interest (Fig. 5.43). Note the use of a ketone-to-alcohol reduction and dehydration sequence prior to polyketide cyclization—the usual pattern followed when aromatic structures appear in which oxygens have been removed (sequence A). After cyclization and aromatization (sequence B), introduction of oxygens via the usual C–H to C–OH oxidation (typically a postcyclization process) is followed by a hydroquinone-to-quinone dehydrogenation (sequence C) to yield the modified polyketide which then undergoes hydrolysis to the free carboxylic acid. A typical *ortho*-hydroxybenzoic acid-type decarboxylation followed by sequential DMAPP alkylations on both carbon and oxygen centers introduces additional side chains. A Baeyer–Villiger-type oxidation and a lactone carbonyl reduction are suggested for producing the intermediate hemiacetal (sequence E).

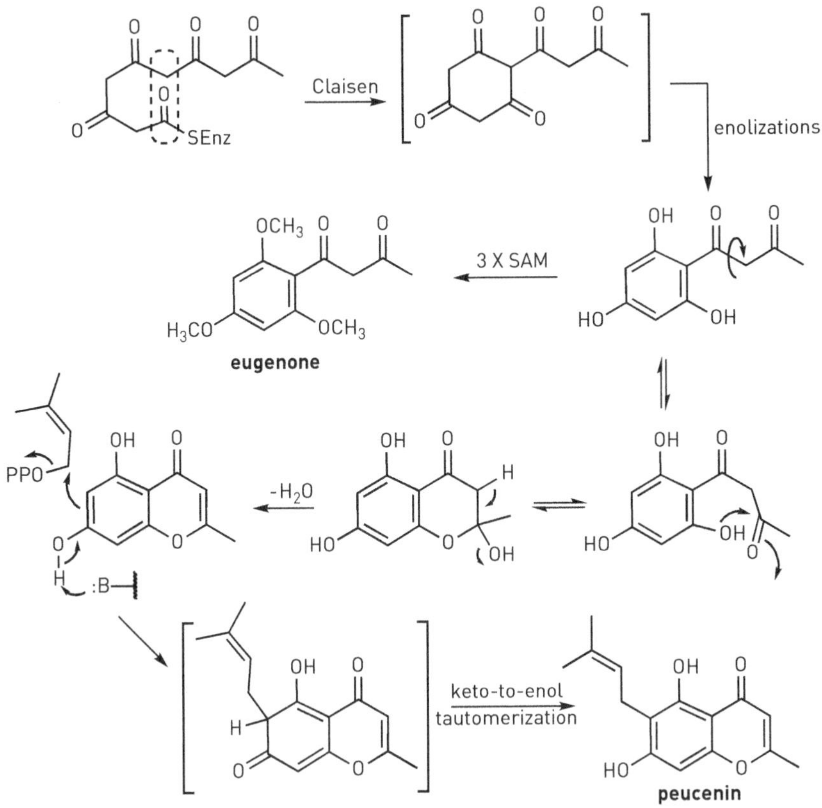

FIGURE 5.41
SAM O-methylation or DMAPP C-alkylation leading to eugenone or peucenin.

The open hydroxy aldehyde form must then undergo a conformational rotation to bring the aldehyde and the dimethylallyl functions into the proximity required for an "ene reaction" cyclization (somewhat like a Diels–Alder reaction) to produce the substituted benzopyran ring system; condensation between the now adjacent phenolic hydroxyls may be thought of as involving a conjugate-type addition of one hydroxyl group to the aryl ketone function of the adjacent ring followed by loss of water to produce the indicated xanthone core (sequence F). A final oxidative epoxidation completes the sequence of structural modifications to produce tajixanthone. This is a good example of a complex structure which at first glance might not appear to be of polyketide origin, but which contains several key structural features that arise via a series of structural modifications commonly seen in a variety of other natural product biosynthetic sequences.

OTHER OXIDATIVE MODIFICATIONS OF AROMATIC RINGS: EXPANSION OR CLEAVAGE PROCESSES

In the biosynthesis of tajixanthone, we saw an example of a Bayer–Villiger-type oxidation process leading to formation of the xanthone core of the structure. Oxidative

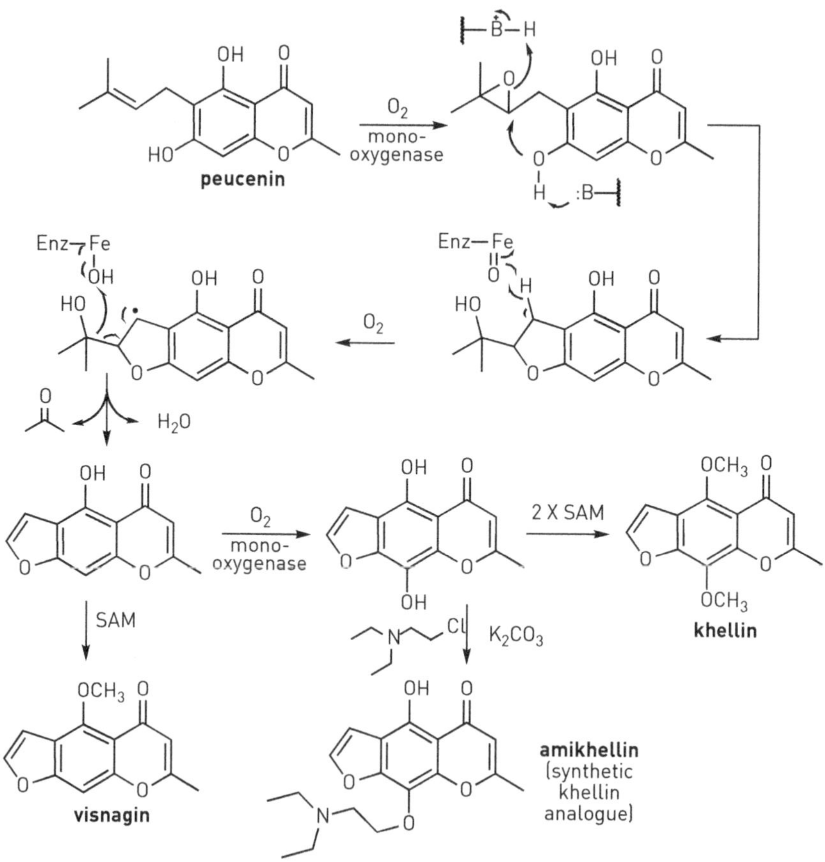

FIGURE 5.42
Modifications of peucenin to produce visnagin, khellin, and synthetic amikhelline.

ring expansions or cleavage reactions mediated by mono- or dioxygenase enzymes represent pathways for rather dramatic structural modification of other aromatic polyketide structures. For example, the tropolone core of the fungal metabolite **stipitatic acid** is derived from the simple aromatic polyketide, **3-methylorsellinic acid,** which undergoes an oxidative ring expansion via a pinacol–pinacolone-type rearrangement as shown in Fig. 5.44. Feeding studies show that orsellinic acid itself is not converted to stipitatic acid, indicating that SAM methylation of the polyketide precursor must precede cyclization and aromatization. The anhydride, **stipitatonic acid**, has been shown to be an intermediate.

A simple example of oxidative aromatic ring cleavage is found in the biosynthesis of the antifungal metabolite **xanthofusin**, as shown in Fig. 5.45. The addition of two methyl groups derived from SAM to a triketide precursor is followed by the usual cyclization and aromatization to give 2,4-dimethyl-1,3-benzenediol as an intermediate. Further ring hydroxylation is followed by a dioxygenase-mediated ring cleavage to a ketoacid which can undergo conformational rotation to bring an enol hydroxyl group into proximity with the carboxyl function. Cyclization with loss of

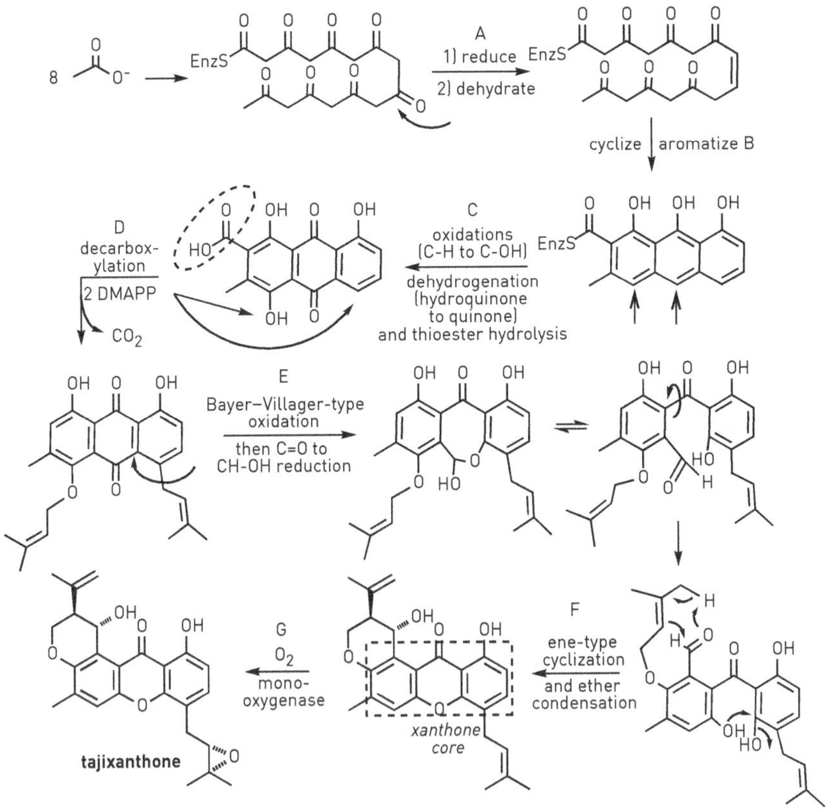

FIGURE 5.43
Octaketide cyclization and subsequent modifications to produce tajixanthone.

water leads to the keto-lactone product as shown. Once again, a quick glance at the structure of xanthofusin does not immediately suggest a polyketide origin, reminding us that relatively simple modifications following polyketide cyclization and aromatization can sometimes lead to dramatic structural changes.

OXIDATIVE COUPLING OF PHENOLS: FORMATION OF ARYL–ARYL BONDS

Formation of C–C bonds between aromatic rings has been a long-standing challenge in organic chemistry due to both the lack of available nucleophilic aryl systems and to resistance to many types of nucleophilic substitution reactions (S_N1, S_N2, etc.) for aryl rings bearing leaving groups typical for such reactions. Nature overcomes these limitations in many situations by linking aryl rings to one another through the coupling of aryl radicals which are readily formed via the oxidation of phenols. Such **phenolic coupling** reactions are quite common in biosynthesis and may occur in either an intermolecular or an intramolecular fashion. An example of the former type is shown in Fig. 5.46 for the oxidative dimerization of **methylphloroacetophenone**

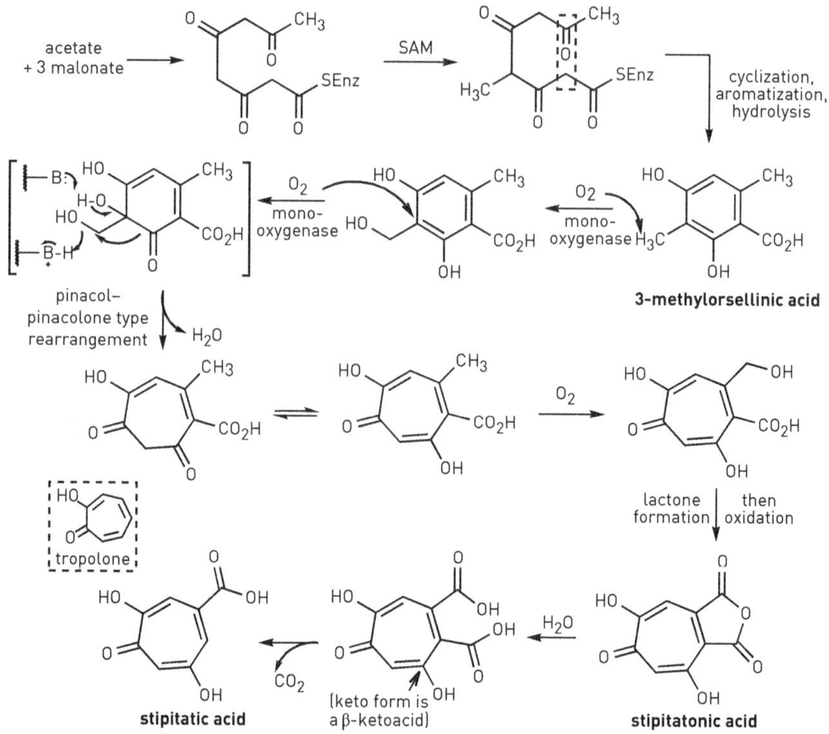

FIGURE 5.44
Oxidative ring expansion via pinacol–pinacolone rearrangement leading to stipitatic acid.

to produce **usnic acid**, a yellow, bitter-tasting compound produced by lichen which has potent antibiotic activity against gram-positive bacteria and also some antiviral, anti-inflammatory, and analgesic activity.

A typical O_2/CYP H atom abstraction is shown for generation of the required phenoxy radicals with their expected resonance contributors. The key C–C bond-forming step in the scheme is coupling of an *ortho* C-radical resonance form with a *para* C-radical resonance form, followed by keto–enol tautomerization (note that only one of the two rings can aromatize). The subsequent generation of the benzofuran ring system is represented as occurring via a cyclic hemiacetal formation followed by simple dehydration, though other mechanisms are plausible. This sequence produces the α,β-unsaturated cyclohexadienone ring system of usnic acid, which can exist as a mixture of keto–enol tautomers of what is formally the 6,6-disubstituted-4-cyclohexene-1,3-dione ring component of the structure.

An important example of *intramolecular* oxidative phenolic coupling is seen in the biosynthesis of **griseofulvin**, a potent antifungal agent that is widely used for treatment of ringworm, jock itch, and athlete's foot as well as fungal infections of the scalp, fingernails, and toenails (Fig. 5.47).

Note the unusual chlorination step (mechanism undetermined) that immediately follows the initial polyketide cyclization-aromatization steps and the sequential SAM

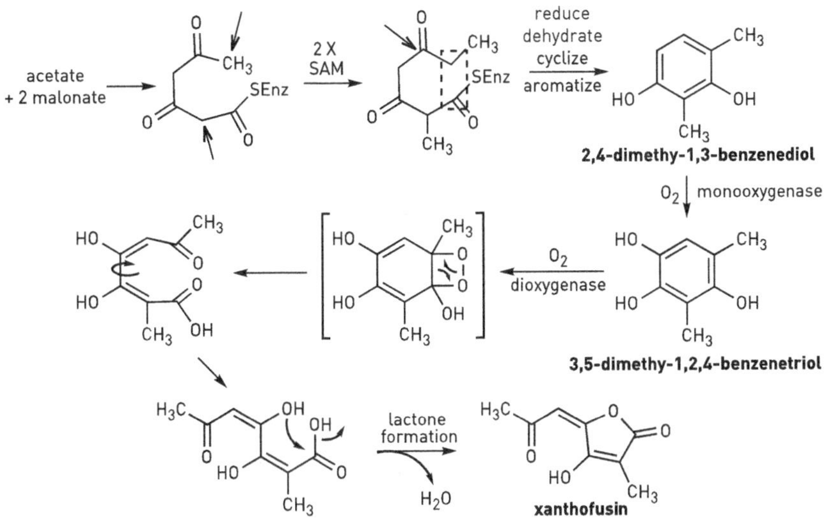

FIGURE 5.45
Oxidative aromatic ring cleavage and lactone cyclization in biosynthesis of xanthofusin.

methylations. The key C–O bond-forming step arises from phenol oxidations followed by an intramolecular trapping of a *para* C-radical resonance form by a phenoxy O-radical. NADPH conjugate addition of hydride to the resulting 2,5-cyclohexadienone ring system affords the natural product as a single diastereomer (Fig. 5.47).

Anthraquinone natural products are widely distributed examples of acetate-derived polyketides which can undergo a variety of postcyclization modifications, including various oxidative couplings. Biosynthesis of **emodin anthrone,** which illustrates a common route to precursors of natural anthraquinones, involves a poyketide cyclization and aromatization sequence with specifically timed steps for decarboxylation and dehydration which lead to a precursor anthrone structure (Fig. 5.48).

The various fates of emodin anthrone are illustrated in Fig. 5.49. As shown, direct oxidation of the anthrone methylene group (path A) leads directly to **emodin**, a natural anthraquinone derived from rhubarb root extracts and other plant sources which is sometimes used as an alternative medicine treatment for inflammation, constipation, or other conditions. An alternative oxidation of emodin anthrone (path B) leads to H atom abstraction and coupling of two resulting radicals to produce the dimeric form, **emodin dianthrone**. Further oxidative dehydrogenation of the dianthrone can then lead to formation of **emodin dehydrodianthrone**. With this nearly planar structure, adjacent phenols found in the upper and lower tricyclic ring systems are held in close proximity, favoring sequential intramolecular oxidative phenolic coupling reactions. The first coupling leads to C–C bond formation and production of **protohypericin**; this is then followed by a second oxidative phenolic coupling to produce **hypericin**, the main component and one of the active principals in extracts of St. John's wort (*Hypericum perforatum*), a formulation which has successfully competed in recent years with prescription medications as an effective

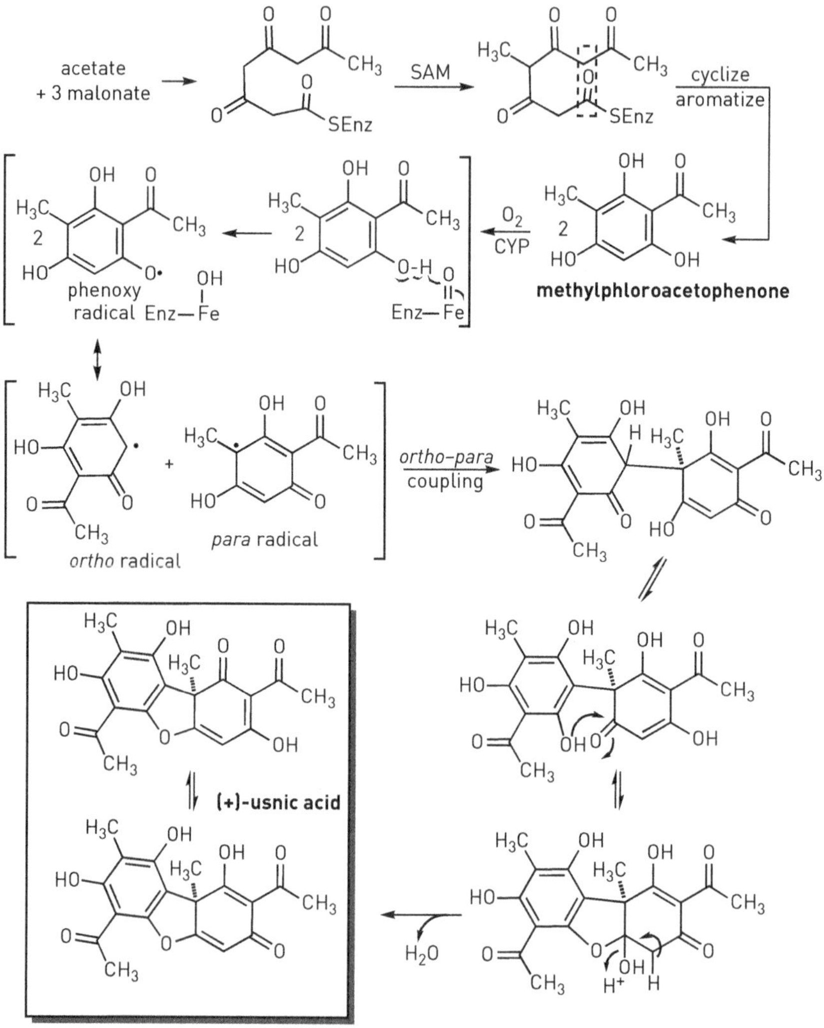

FIGURE 5.46
Intermolecular oxidative phenolic coupling of methylphloroacetophenone to produce usnic acid.

antidepressant. Various studies have reported other effects for hypericin including antineoplastic, antitumor, and antiviral activities (including against HIV and hepatitis C viruses), making hypericin a target of ongoing pharmacological research.

The forgoing introduction to phenolic oxidative coupling processes, while of obvious importance to the biosynthetic chemistry of the acetate pathway, will have even greater significance later on. As we move ahead in the next two chapters to examine the details of the shikimate pathway and the biosynthesis of alkaloid natural products, we'll see that oxidative phenolic coupling and some closely related oxidative processes play a central role in the formation of key C–C bonds in both of these important systems. Until then, we have some final crucial elements of the polyketide pathway to examine. In doing so, we will move beyond the forgoing families of natural products derived mainly from acetate and malonate precursors, eventually

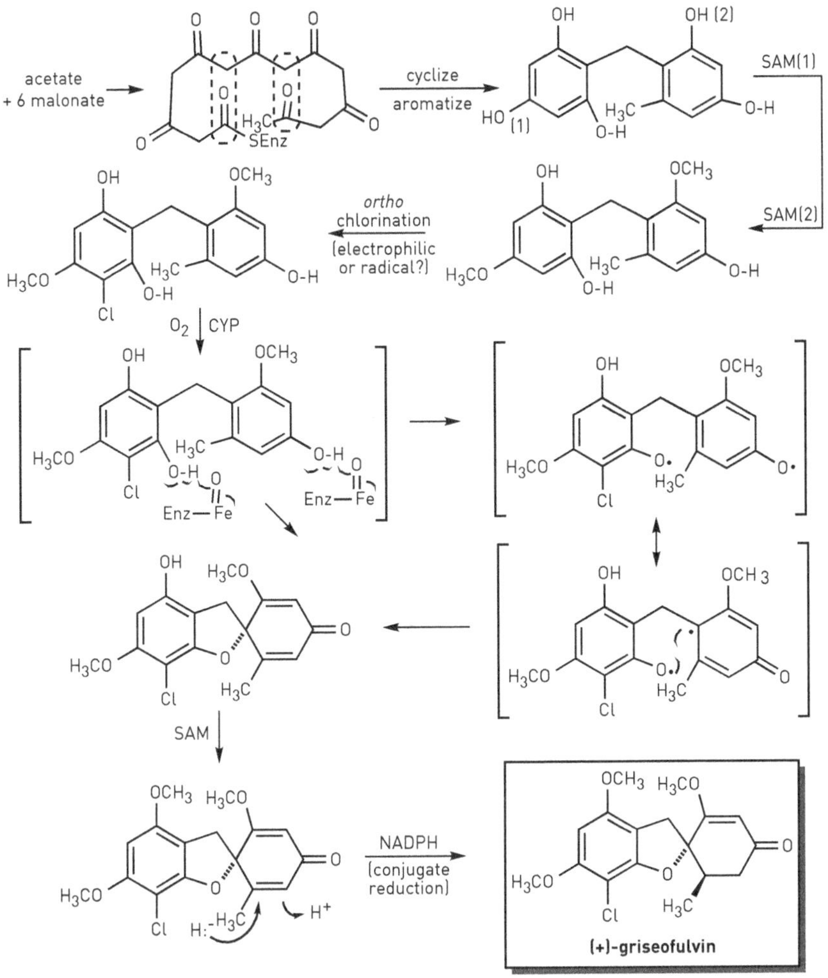

FIGURE 5.47
Intramolecular oxidative phenolic coupling in biosynthesis of the potent fungicide griseofulvin.

reaching out beyond aromatic structures to embrace branched polyols, large ring lactones, and polyether structures of special significance. To produce such a vast array of additional compounds requires further structural modifications of emerging polyketide chains as well as the use of different starter and/or extender units. To begin, we'll take a look at some simple examples of the latter modification and then move on to some more complex systems.

THE USE OF OTHER STARTER GROUPS: FROM CANCER DRUGS AND ANTIBIOTICS TO POISON IVY

One of the most important of all cancer chemotherapy drugs currently in worldwide clinical use is **doxorubicin**, a highly modified aromatic polyketide glycoside produced by fermentation of a mutant strain of *Streptomyces peucetius*, a soil microbe

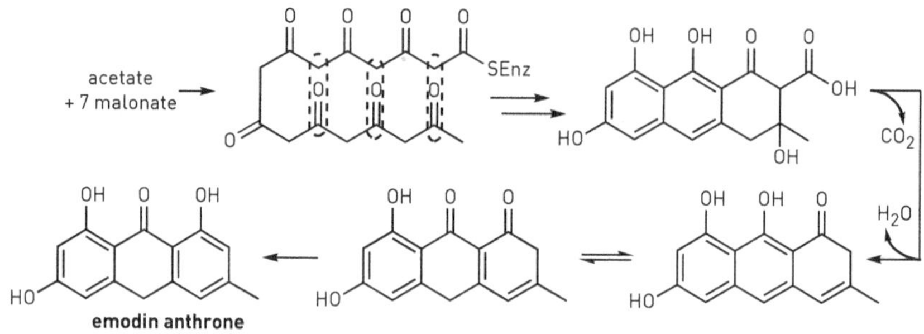

FIGURE 5.48
Polyketide cyclization and modifications in biosynthesis of emodin anthrone.

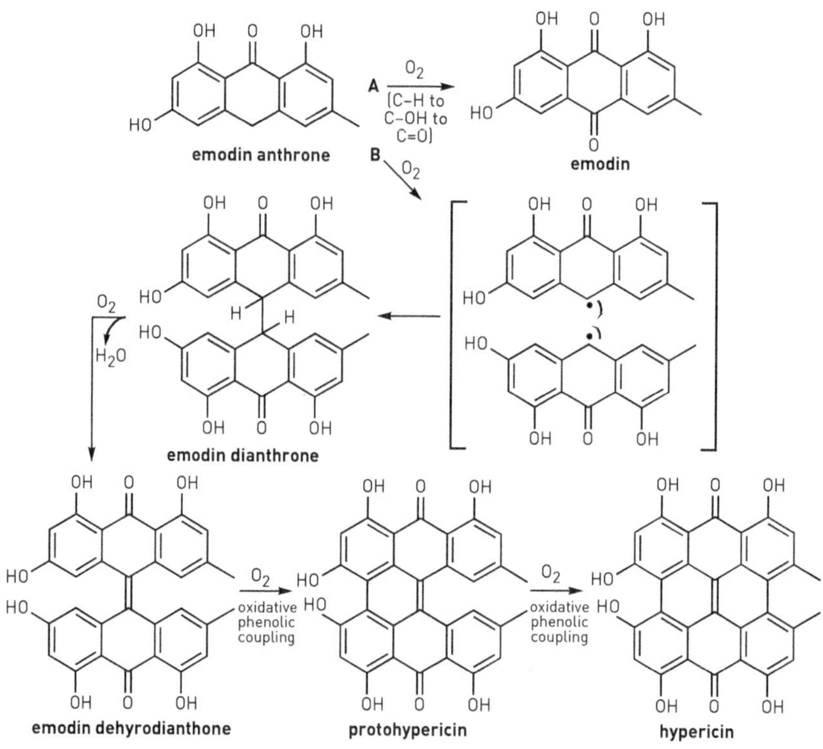

FIGURE 5.49
Oxidative transformations of emodin anthrone in biosynthesis of emodin and hypericin.

originally found in Italy. An abbreviated biosynthetic scheme for doxorubicin, which begins with a **propionate starter unit** instead of an acetate starter unit, is shown in Fig. 5.50. Note the initial formation of an anthrone ring which is then oxidized to an anthraquinone structure prior to aldol-initiated assembly of the cyclohexane (D) ring, which is the focus of most of the remaining modifications.

Oxidation of the C-ring and reduction of D-ring ketone is followed by glycosylation of the D-ring benzylic alcohol with the aminosugar **L-daunosamine** (via TDP-L-daunosamine) which is added prior to subsequent decarboxylation, further

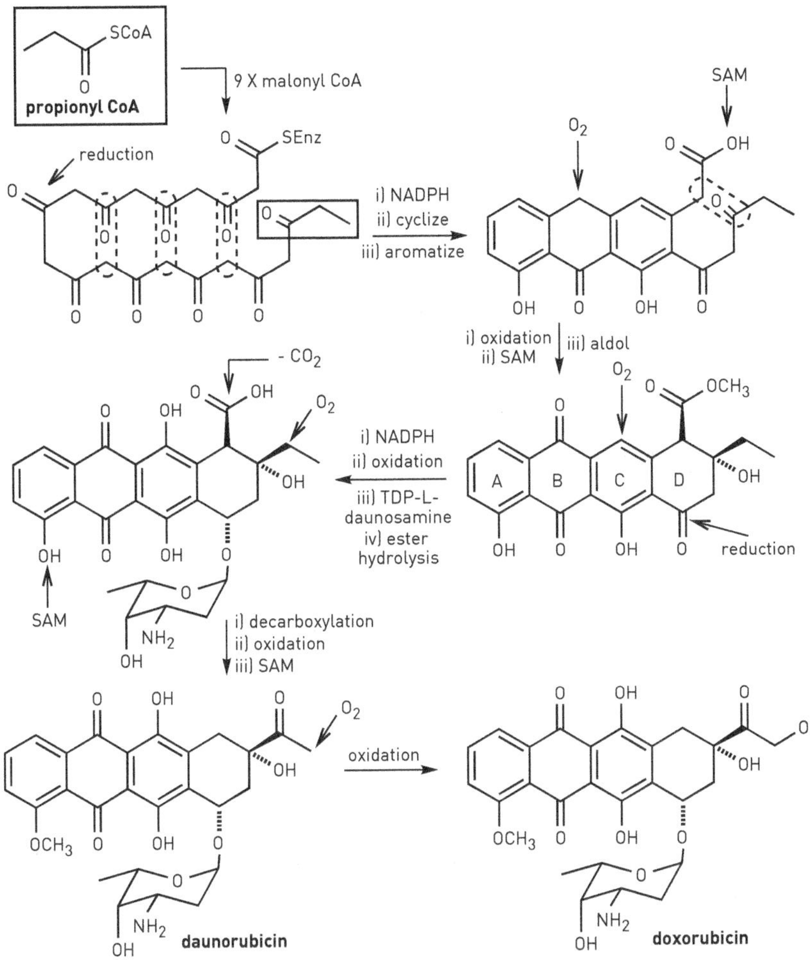

FIGURE 5.50
Abbreviated biosynthetic scheme for daunorubicin and its conversion to doxorubicin.

oxidation of the D-ring ethyl group (from the propionate starter unit) and SAM methylation of the A-ring hydroxyl to give **daunorubicin**, the original anthracycline drug produced by a nonmutant *S. peucetius* strain. A final oxidation of the D-ring methyl ketone function via the mutant strain produces the more powerful **doxorubicin**. Both drugs are members of a family of related polyketides collectively known as **anthracycline antibiotics** and represent some of the most effective anticancer agents known, in part because of their activity against so many different types of cancer. All work by binding directly to DNA base pairs via intercalation and also by inhibiting DNA repair. Doxorubicin is commonly used to treat many different types of cancers, including those of the bladder, stomach, lung, breast, and ovaries; it is also effective for treatment of both Hodgkin's and non-Hodgkin's lymphomas and certain forms of leukemia. Aside from the usual toxic side effects of cancer chemotherapy agents, most anthracyclines, including doxorubicin, exhibit a severe form of cardiotoxicity (causing heart damage) which limits the total amount of the drug

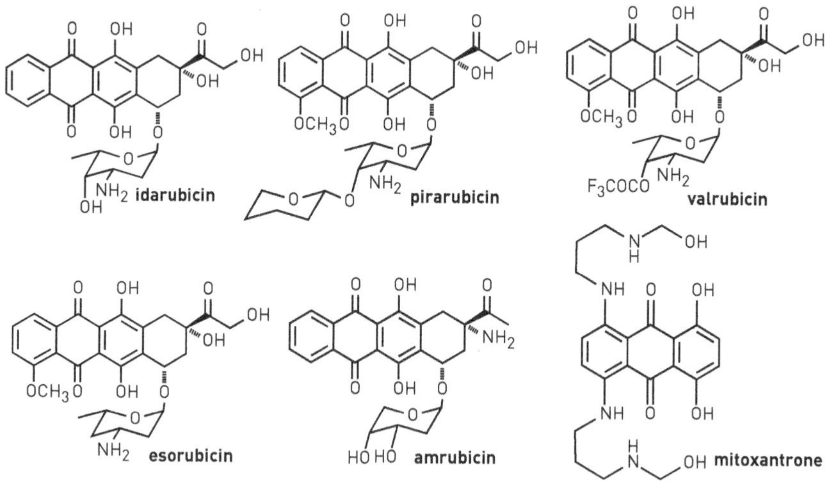

FIGURE 5.51
A group of synthetic and semisynthetic anthracycline analogues currently in clinical use.

which can be administered over a lifetime. Reduction of this cardiotoxicity has been the main focus of many years of research effort in the development of synthetic or semisynthetic analogues, several of which are shown in Fig. 5.51. All of these analogues are currently in clinical use.

Another group of biologically important polyketides are the **tetracycline antibiotics**. While somewhat similar in structure to the anthracyclines, both the starter unit and subsequent cyclization/aromatization patterns are different, with many additional transformations involved in producing the core structures (Fig. 5.52). The **malonamyl-CoA** starter unit, derived from malonyl-CoA, glutamine, and ATP, condenses with nine units of malonyl-CoA in the usual fashion to produce the polyketide precursor which, as in the anthracyclines, undergoes a single NAPDH reduction prior to cyclization and aromatization, though in this case of the tetracyclines, the folding pattern is different, leading to D-ring aromatization as shown.

The resulting aromatic tetracycle, **pretetramide**, then undergoes B-ring SAM methylation and a sequence of C-ring oxidations, the first to introduce a hydroxyl group and the second to dehydrogenate the resulting hydroquinone to the quinone oxidation state. The resulting **4-keto-6-methylpretetramide** then undergoes conjugate addition of water at the indicated position, followed by a PLP-mediated transamination of one of the C-ring ketone functions to introduce the amino group of **anhydrotetracycline**. This is followed by sequential SAM methylations of the amino group and oxidation at the methyl-bearing B-ring carbon to produce the corresponding tertiary alcohol of **5a,11a-dehydrotetracycline**. This key intermediate then either directly undergoes NADPH conjugate reduction at the indicated position to produce **tetracycline** (via path A), or undergoes C-ring oxidation to an alcohol prior to the aforementioned NADPH conjugate reduction (via path B) which then produces **oxytetratcycline**. A third member of the group, **chlorotetracycline**,

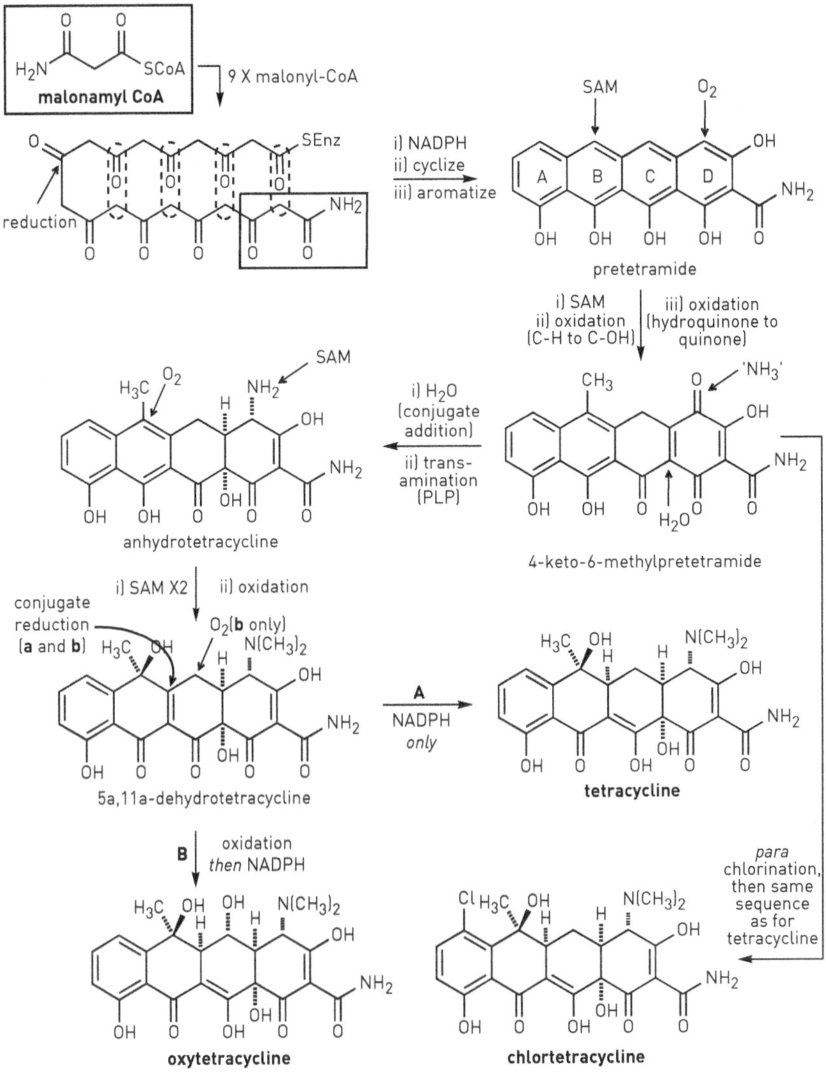

FIGURE 5.52
Abbreviated biosynthetic scheme for tetracycline and other tetracycline antibiotics.

is derived from 4-keto-6-methylpretetramide via A-ring chlorination *para* to the phenolic hydroxyl group followed by the same sequence of transformations involved in production of tetracycline; tetracycline itself is actually produced via a mutant bacterial strain that blocks the chlorination step leading to chlortetracycline, the first of these compounds to be isolated in the mid 1940s.

The tetracyclines are important broad-spectrum, orally active antibiotics which have been widely used in both human and animal medicine for many years. As with all such drugs, their widespread use has led to the development of tetracycline-resistant bacterial strains, creating a need for derivatives and analogues capable of overcoming resistance mechanisms. **Doxycycline** and **minocycline** (Fig. 5.53) are semisynthetic examples of so-called second-generation tetracyclines that, like

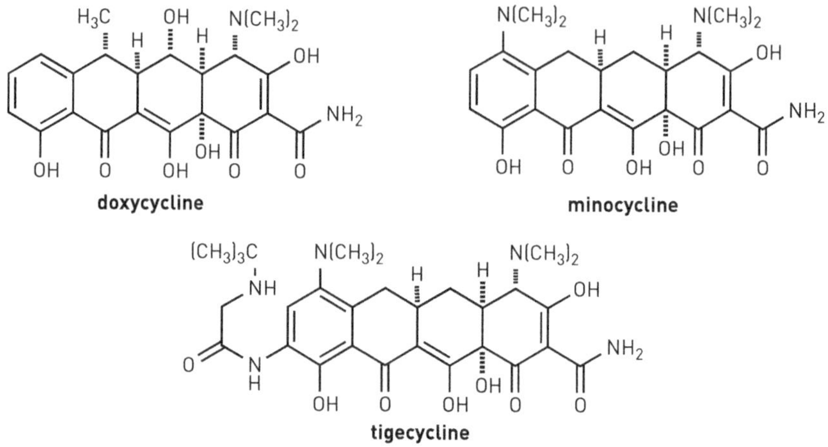

FIGURE 5.53
Second- and third-generation tetracyclines for antibiotic-resistant bacteria strains.

the natural tetracyclines, are orally active and have proven to be clinically valuable for treating a wide variety of antibiotic-resistant infections. More powerful third-generation derivatives like **tigecycline** have come into use only since the mid-2000s; tigecycline is especially valuable in its activity against methicillin-resistant *Staphylococcus aureus* (MRSA) strains, but must be given intravenously.

Long chain carboxylic acids are also sometimes employed as starter units for the production of aromatic polyketide products. One example of this is seen in the biosynthesis of the **urushiols,** a group of very potent allergen oils produced by a number of different plants including poison oak, poison ivy, and poison sumac (Fig. 5.54). Each member of this group of polyphenols consists of a basic catechol ring with a C_{15} or C_{17} side chain which may be either saturated or mono-, di- or triunsaturated, depending on the carboxylic acid starter unit. Mixture compositions vary with different plant sources, but structures with longer unsaturated chains tend to produce more significant allergic reactions in most individuals. The biosynthesis of the urushiol shown below begins with a monounsaturated C_{16} starter unit, **palmitoleic acid** (enzyme-bound) which is condensed with three units of malonyl-CoA in the usual fashion to produce the indicated triketide derivative. Carbonyl reduction followed by cyclization, dehydration, and aromatization produces the aromatic polyketide, **anacardic acid**. Decarboxylation of the *ortho*-hydroxybenzoic acid ring followed by *ortho*-oxidation to the corresponding catechol gives the unsaturated urushiol shown.

Use of a less familiar carboxylic acid starter unit, ***para*-hydroxycinnamic acid**, provides a preview of one of the many basic components available from the shikimate pathway (the subject of our next chapter) and how compounds derived from different pathways may intersect in some organisms to produce hybrid natural products of mixed biosynthetic origin. As shown in Fig. 5.55, enzyme-bound *para*-hydroxycinnamic acid may condense with three malonyl-CoA units as

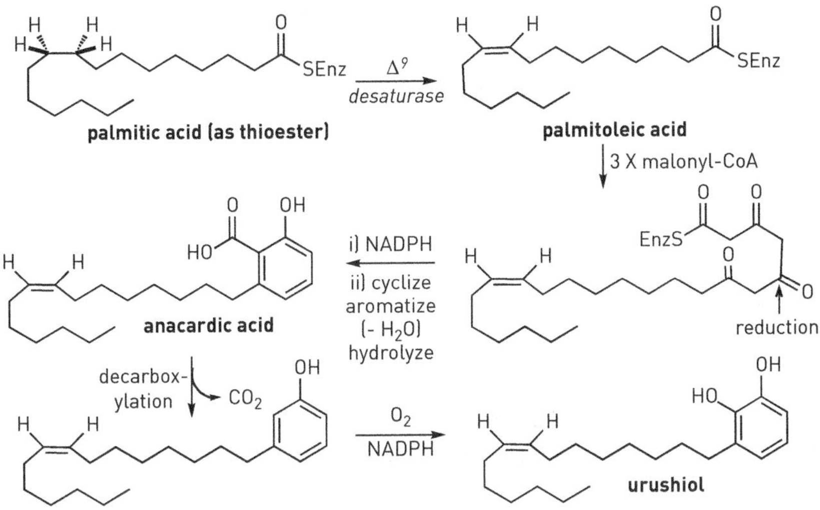

FIGURE 5.54
Biosynthesis of urushiol, the active principal in poison ivy.

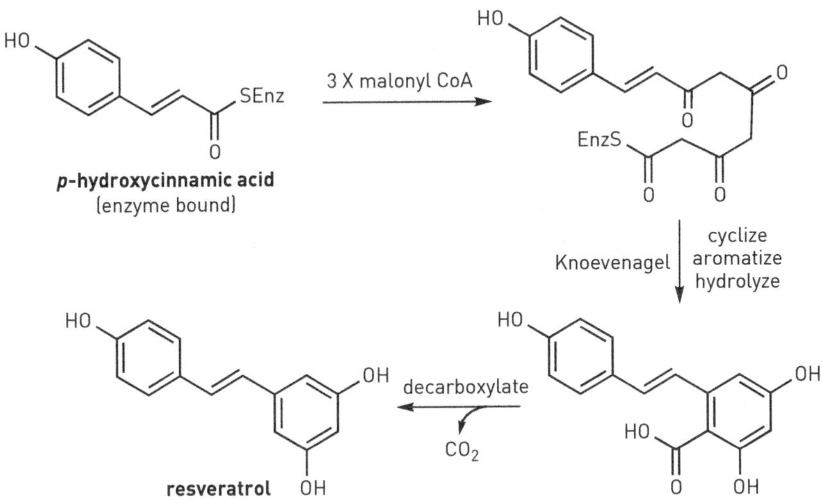

FIGURE 5.55
Biosynthesis of resveratrol, a mixed pathway polyphenolic natural product.

usual to produce the unique triketide structure shown. The usual Knoevenagel-type condensation and aromatization is again followed by decarboxylation of the ortho-hydroxybenzoic acid intermediate to produce the controversial polyphenolic metabolite, **resveratrol**.

Resveratrol is found in the skins of red grapes and in other fruits and has often been singled out as a component of red wine that may be partially responsible for the so-called "French paradox," the well-known observation of relatively low mortality from coronary heart disease among the French in spite of the fairly high levels of saturated fats in their diet. Resveratrol is just one of many red wine polyphenolic compounds which studies have shown to have antioxidant, anti-inflammatory, and other potentially beneficial effects, though the amounts required for such effects usually far exceed what would be obtainable from wine consumption. Even more controversial are claims made by dietary supplement suppliers that resveratrol consumption may lead to increased life span due to its antioxidant properties. In preliminary studies, resveratrol feeding has been found to extend the lifespans of certain worms, fruit flies, and fish, but there is only limited evidence at present to suggest that such effects will be seen in higher animals.

MORE ON POLYKETIDE SYNTHASE (PKS) SYSTEMS: INCREASING PRODUCT DIVERSITY

We noted earlier in the chapter the close relationship between fatty acid synthase (FAS) enzyme systems and polyketide synthase (PKS) systems. Certainly, both enzyme families catalyze Claisen condensations between acetyl-CoA and malonyl-CoA units to yield enzyme-bound β-ketoacyl thioesters. The simplified diagram shown earlier (Fig. 5.33) presented a PKS as essentially an FAS system lacking the usual components required for ketone reduction (β-*ketoreductase*), alcohol dehydration (*dehydratase*), and α,β-unsaturated thioester conjugate reduction (*enoyl reductase*), and this has been the basic pattern of polyketide biosynthesis up to this point. But there is much more to the story than starter units, extender units, and polyketide folding patterns leading to phenolic natural products. To better understand some of the more complex aspects of the biosynthesis of many nonaromatic polyketide natural products, we have to look a bit more closely at certain features of different types of PKS systems and how they work. The discussion that follows is limited and simplified, as it is intended mainly to provide a modest familiarity with some of the concepts, terminology, and abbreviations associated with the enzymology of PKS systems. While a great deal of important detail is omitted, it is hoped that what is presented will help to deepen the reader's understanding of the organic chemistry involved in the transformations catalyzed by these systems.

Most of the aromatic polyketide chemistry we have seen so far has been characteristic of bacteria employing a **type II PKS** system (as illustrated previously in Fig. 5.33) or by plants employing a **type III PKS** system. In these systems, one or more **modules** (essentially organized collections of enzyme active sites required for catalyzing the various reactions involved in the polyketide assembly or modification process) are employed in an **iterative** assembly mechanism; that is to say, the active sites of a given module are reused repeatedly to produce a growing polyketide chain. Recall that for the aromatic natural products we have seen, the PKS system loads a starter unit (usually acetyl-CoA), then loads an extender unit (always malonyl-CoA) adjacent to the starter unit, catalyzes a Claisen condensation between the two to produce a β-ketothioester which is then transferred back to the adjacent starter unit position to make way for the loading of a second malonyl-CoA extender unit, after which the entire process is repeated over and over to produce a growing polyketide chain. This kind of "back-and-forth" assembly mechanism is also employed by fungi using so-called **iterative type I PKS** systems. Table 5.2 summarizes some of the features of the different PKS systems currently known.

Another group of PKS enzymes used by bacteria, called **modular type I PKS** systems, employ a different kind of mechanism that is more like an "assembly line" process for generating a growing polyketide chain rather than the "back-and-forth" iterative mechanism. Here, the enzyme has multiple modules, each of which carries out one or more specific functions or modifications, after which the substrate molecule is passed on to the next module and so on (more on this later). The products produced by both iterative and modular type I PKS systems are often (but not always) nonaromatic compounds that may be linear, cyclic, polycyclic, or some combination of these. But it is important to keep in mind that all types of PKS systems can produce a variety of different product structures, so it is not really possible to classify polyketide natural product structures strictly according to the type of PKS systems involved in their production. Fortunately, much of the chemistry involved in all types is very similar, relying primarily on Claisen condensations of repeated acetate or propionate units for the basic assembly process, often accompanied by familiar FAS-type modifications.

Recall that in the formation of most aromatic polyketides, we only occasionally saw additional modifications to the growing polyketide chain (such as a reduction or a methylation, etc.) prior to cyclization and aromatization; most of the modifications were postcyclization/aromatization processes. With type I PKS systems, it is more common to see growing polyketide chains undergoing modifications more

Table 5.2 Types and Features of Different Polyketide Synthase (PKS) Systems

PKS Group	*Structure*	*Assembly Mechanism*	*Used by*	*Products*
Iterative type I	Single protein, single module	*Iterative*; repeated re-use of module active sites by growing chain	Fungi	Linear, cyclic, others
Modular type I	Single protein, multiple modules	*Linear*; single use of module active sites, growing chain passed on to next module	Bacteria	Linear, cyclic macrolides
Type II	Multiprotein, single module per protein	*Iterative*; single or repeated use of module active sites by growing chain	Bacteria	Aromatics
Type III	Single protein, multiple modules	*Iterative*; repeated re-use of module active sites by growing chain	Bacteria plants, fungi	Aromatics, others

akin to those we saw earlier in FAS systems, such as reductions, dehydrations, or conjugate reductions as the chain grows. We say that these PKS systems typically employ one or more modules with multiple **domains** of catalytic activity. If modules are described as organized groups of different domains (enzyme active sites), it is helpful to think of a domain as simply a region or part of a given enzyme structure that functions more-or-less independently from the rest of the enzyme, performing a specific catalytic function (often requiring various coenzymes) to facilitate a specific process such as loading, condensation, reduction, dehydration, and so on. A list of some familiar catalytic domains of significance in various polyketide synthase systems, along with commonly used abbreviations, is given in Table 5.3.

A convenient way to think about iterative PKS systems is to divide them into fully reducing, partially reducing, and nonreducing groups. A fully reducing PKS system, one in which KR, DH, and ER domains operate will closely resemble FAS systems, though different starter or extender units may be employed. A partially reducing PKS would produce products with only partially reduced structural segments (alcohols or alkenes) while a nonreducing PKS would lead to "true" polyketides, that is, multiple ketone functions as seen in our earlier examination of precursors to aromatic polyketides. Modular PKS systems, as we will see, can produce products with a mixture of structural segments that may be fully reduced, partially reduced, or unreduced.

To illustrate an excellent example of a fungal *iterative* type I PKS system in operation (another "back-and-forth" construction system), we will look at biosynthesis of the cholesterol-lowering drug, **lovastatin** (trade name *Mevacor*), a nonaromatic polyketide natural product obtained from oyster mushrooms. Lovastatin and similar statin drugs have a powerful inhibitory effect on **HMG-CoA reductase,** the enzyme

Table 5.3 Some Catalytic Domains and their Functions in PKS Systems

PKS Domain	*Abbrev.*	*Synthetic Function of Active Site*
Acyltransferase	AT	Loads starter or extender units
Acyl carrier protein	ACP	Anchors growing polyketide chain as thiol ester
β-Ketoacyl synthase	KS	Promotes Claisen condensation between starter/extender units
β-Ketoacyl reductase	KR	Reduces β-ketothioester to β-hydroxythioester (via NADPH)
Dehydratase	DH	Dehydrates β-hydroxythioester to α,β-unsaturated enoyl thioester
Enoyl reductase	ER	Promotes conjugate reduction of enoyl thioester to saturated thioester (via NADPH)
Methyltransferase	MT	Transfers methyl groups to a growing polyketide chain (via SAM)
Thioesterase	TE	Releases final polyketide product from enzyme

which mediates NADPH reduction of hydroxymethylglutarylCoA, a key intermediate in the production of mevalonic acid (MVA) that, you will recall, is required for terpenoid (and therefore cholesterol) biosynthesis (see Fig. 4.5). All of the statin drugs shown in Fig. 5.56 are obtained by fermentation processes with the exception of atorvastatin (trade name Lipitor) which is a synthetic drug retaining some of the key structural features of the natural statin products.

Lovastatin biosynthesis is mediated by **lovastatin nonaketide synthase (LovB)**, a multifunctional type I PKS enzyme that contains active KS, AT, DH, MT, KR, and ACP domains but an inactive ER domain. Because of this, a separate ER enzyme known as **LovC** is required whenever an α,β-unsaturated thioester reduction step (via NADPH) is required during the biosynthesis. The process begins with formation of a simple diketide (Fig. 5.57) via loading of acetyl-CoA and malonyl-CoA units onto LovB as usual, but in this case, the KS (Claisen condensation) operation is followed by ketone reduction (KR) and dehydration (DH) to give an α,β-unsaturated diketide intermediate. This same sequence of steps (which bears obvious similarity to steps in fatty acid biosynthesis) is repeated with a second malonate unit to give the doubly unsaturated triketide intermediate as shown. After addition of a third malonate unit, the KR and DH domains operate as usual, but in this step the LovC enzyme is employed for NADPH reduction of the α,β–double bond of the triene tetraketide, leaving a conjugated diene component intact (a segment of the molecule that will be of paramount importance later in the sequence). This is followed by action of LovB's MT (methyl transferase) domain which installs a methyl group (via SAM alkylation of the β-dicarbonyl enolate ion) at the α-carbon, producing the final form of the tetraketide intermediate. Repetition of the previous malonyl-CoA, KS,

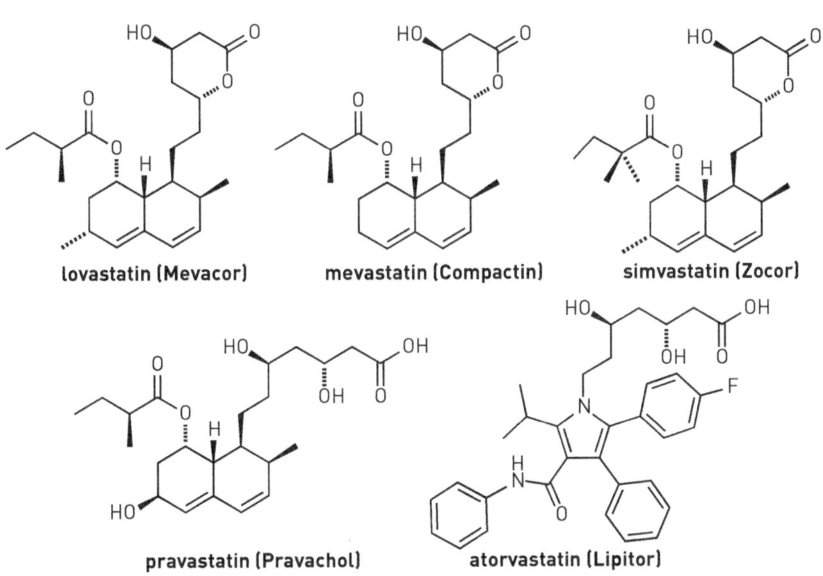

FIGURE 5.56
Structures of some important statin drugs (HMG-CoA reductase inhibitors).

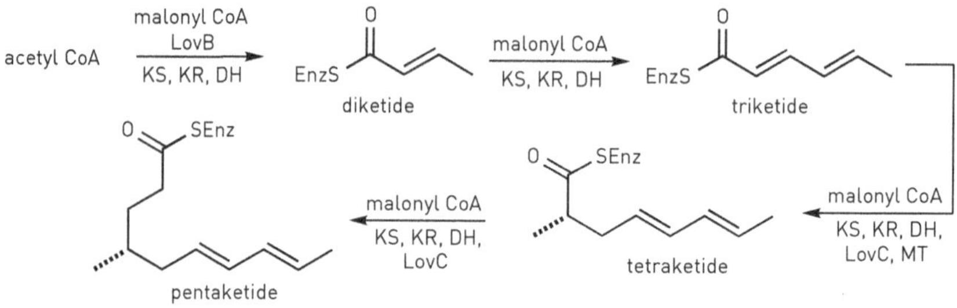

FIGURE 5.57
Early steps leading to production of the pentaketide intermediate in lovastatin biosynthesis.

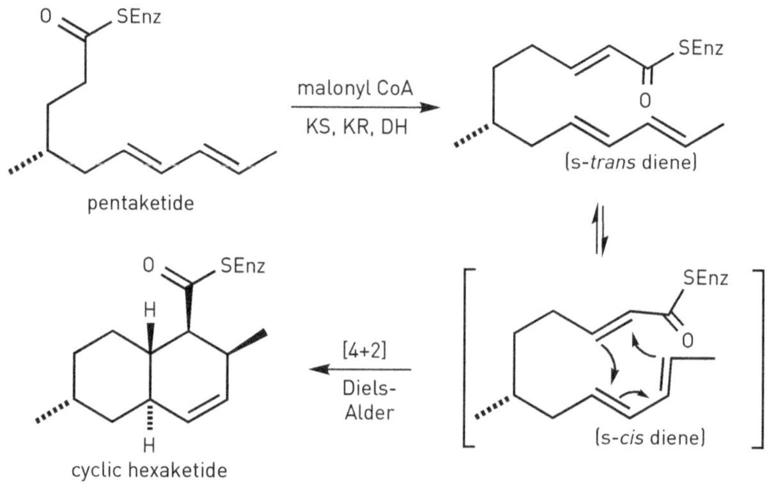

FIGURE 5.58
Intramolecular Diels–Alder leading to a *trans*-decalin-based lovatstatin hexaketide intermediate.

KR, DH, and LovC conjugate reduction steps then produces a pentaketide intermediate with the indicated structure.

Continuing on, we see one of the key steps (Fig. 5.58) in the overall biosynthetic sequence. Extension of the pentaketide chain introduces a sixth malonate unit, accompanied by ketone reduction and dehydration to the α,β-unsaturated thioester intermediate. At this point, a folding mode which brings the s-*cis* conformer of the conjugated diene portion of the hexaketide into close proximity with conjugated enoylthioester double bond leads to an **intramolecular Diels–Alder** reaction, producing the decalin ring system, a structural feature that will be carried all the way through to the end of the biosynthetic scheme. While somewhat surprising, this is just one of a number of examples of the use of biological Diels–Alder processes in the construction of various natural products. Note that the sequence pairs an electron-rich diene component (alkyl substitution at each end) with an electron-poor dienophile component (thioester electron-withdrawing group) as typically seen in more familiar laboratory Diels–Alder reactions.

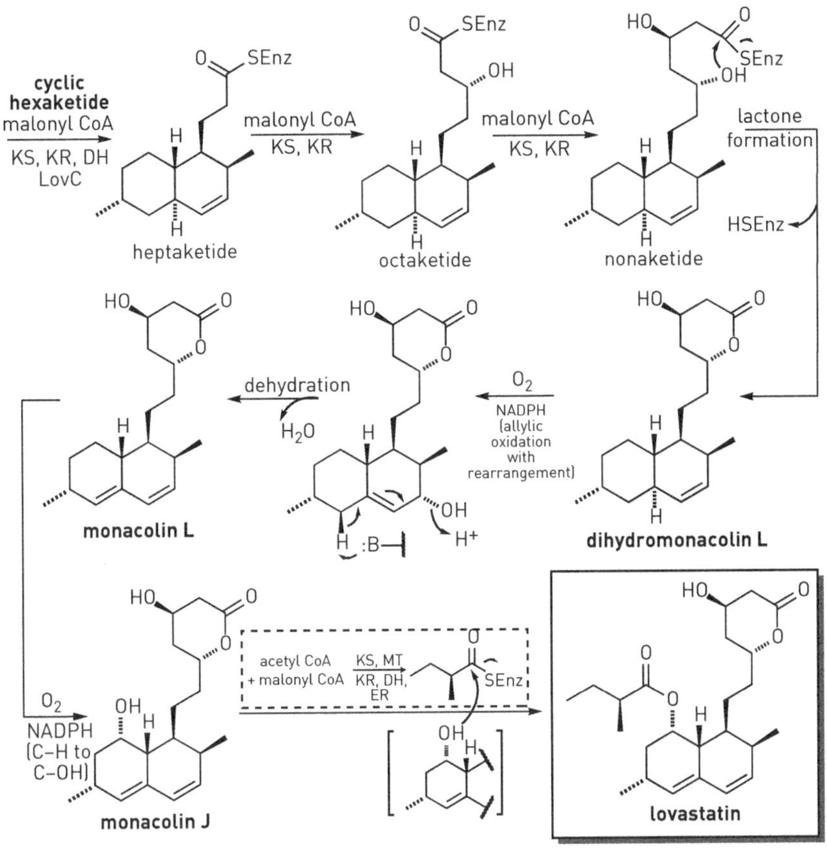

FIGURE 5.59
From hepta- and octa- to nonaketide and postcyclization modifications to give lovastatin.

As the process continues, we see a third LovC-mediated enoyl reduction (ER) step occurring at the heptaketide stage, resulting in a total of three fully reduced segments within the growing polyketide chain (Fig. 5.59). In moving through the remaining two stages of malonyl-CoA addition, we see that only KS and KR domains are employed, leading to sequential ketone-to-alcohol reductions and thus a final nonaketide that retains two secondary alcohol functions in the usual 1,3-relationship. It is at this stage that the nonaketide is released from the LovB enzyme via intramolecular nucleophilic acyl substitution with expulsion of LovB-SH to form the lactone **dihydromonacolin L.**

At this juncture, a C–H to C–OH oxidation takes place with allylic rearrangement to produce an intermediate allylic alcohol which then undergoes a 1,4-type dehydration process to introduce the conjugated diene component of the advanced intermediate **monacolin L.** A final C–H to C–OH oxidation produces **monacolin J** by introduction of a hydroxyl group adjacent to the decalin ring junction. The biosynthesis is completed by esterification of this final hydroxyl group by a methyl-substituted diketide intermediate which is produced in a separate biosynthetic operation, leading directly to **lovastatin** as shown.

MODULAR TYPE I PKS COMPLEXES AND MACROLIDE ANTIBIOTICS: ERYTHROMYCIN BIOSYNTHESIS

Perhaps the most interesting type of polyketide assembly mechanism is found in the modular type I PKS systems. These "assembly line" multienzyme complexes are involved in the production of a class of antibiotics known as the **macrolides** which are essentially large-ring (macrocyclic) polyfunctional lactone structures, the most important of which is **erythromycin.** Among the antibiotics most useful for those who have allergic reactions to penicillin, erythromycin is the most widely employed antibiotic of its type, with a complex mechanism of action that ultimately interferes with the ability of bacteria to produce certain proteins essential to their growth and reproduction. The PKS enzyme system responsible for the biosynthesis of **6-deoxyerythronolide B**, the key polyketide precurusor to erythromycin, is **6-deoxyerythronolide B synthase** or **DEBS**, a diagram of which is shown in Fig. 5.60. The enzyme consists of three linked polypeptide subunits, DEBS-1, DEBS-2, and DEBS-3, each of which is, in turn, composed of two segments or **modules** with specific catalytic domains. DEBS-1 also contains a loading module with AT and ACP domains for introduction of a propionyl-CoA starter unit, while DEBS-3 contains an "end" module with a TE domain that terminates the biosynthesis by facilitating the liberation of the final polyketide from the enzyme via lactone formation.

Each module in a DEBS subunit is capable of adding only a single *S*-methylmalonyl-CoA extender unit to the growing polyketide chain and therefore must pass the chain along to the next module, then to the next subunit, and so on until reaching the end module. Because of this modular "assembly line" mechanism, it is possible to "read" the enzyme's modular organization and predict the length of the final polyketide chain that will be produced prior to lactone cyclization. Thus in the case of DEBS, since there are six assembly modules present and each can catalyze only one Claisen condensation (via its KS domain) for chain growth, the final product prior to cyclization must be a heptaketide (module 1 produces a diketide, module 2 a triketide and so on in going from left to right). Furthermore, by "reading" the domain organization of each module, one can readily identify the chain modifications that will occur going forward as the assembly progresses. A diagram of the DEBS-1 subunit with its loading and extension modules and the reactions they carry out is shown in Fig. 5.61, along with a depiction of the actual nature of the long, flexible acyl-S-ACP linkage (denoted by a wavy bond in the ACP domain) used by these PKS systems (the same linkage is used in FAS systems).

As can be seen in Fig. 5.62, the long, flexible acyl-S-ACP linkages are required not only to allow the loaded methylmalonyl-CoA to reach across the system to engage the adjacent thioester units for Claisen condensations, but also to allow the growing chain to reach the KR domains in modules 1 and 2 for catalysis of the ketone reductions (via NADPH). Note the inversion of configuration at the methyl-bearing carbon of the diketide (which is subsequently epimerized back to the *S* configuration

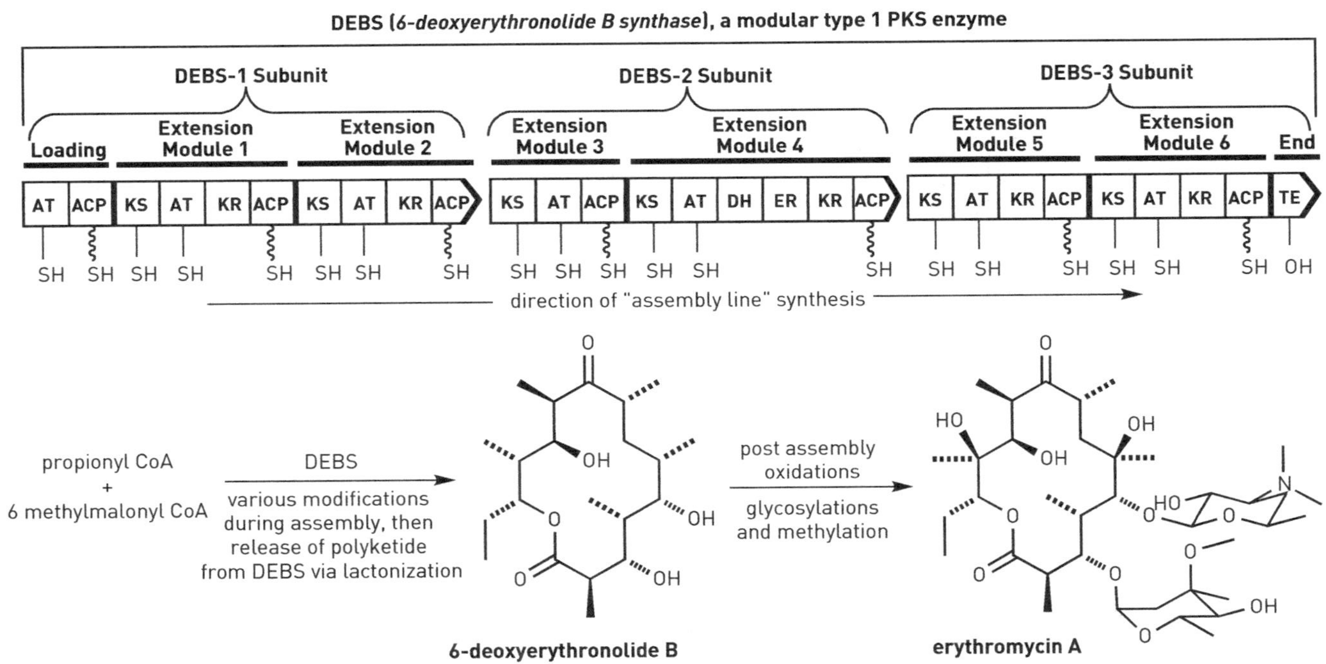

FIGURE 5.60
Overview of organizational structure of DEBS enzyme for biosynthesis of 6-deoxyerythronolide B.

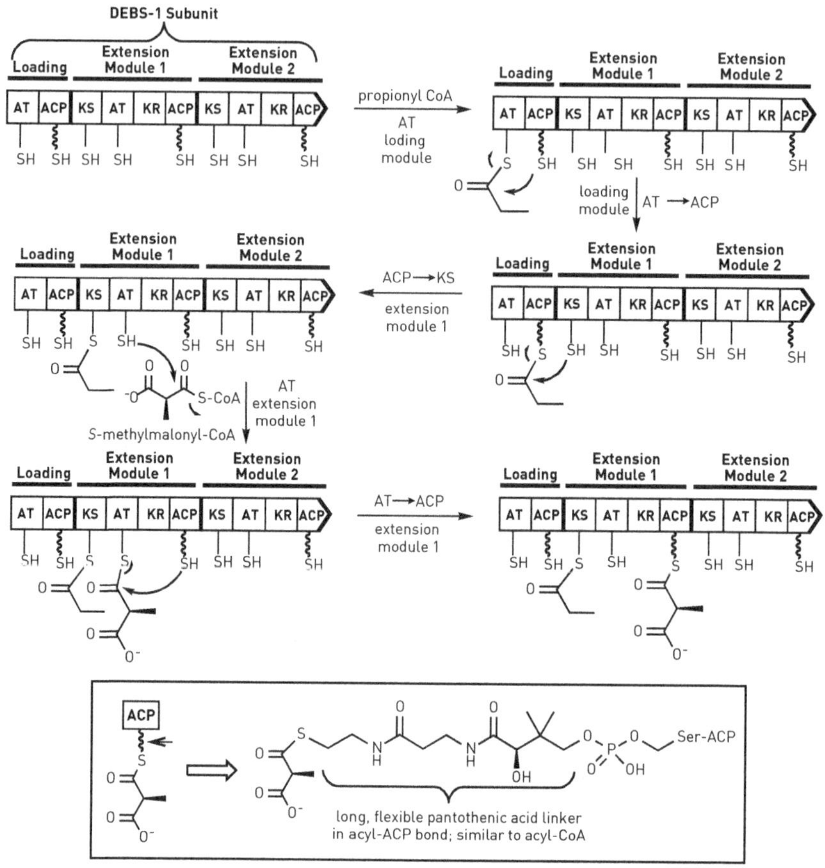

FIGURE 5.61
Initial loading in extension module 1 of DEBS-1 subunit and nature of acylACP thioester linkage.

prior to module 2 chain extension) and the stereospecific ketone reductions leading to the triketide intermediate. Note also that inversion at the methyl-bearing carbon does not occur during the second Claisen condensation.

At this point in the synthesis, the reduced triketide must now be passed forward from module 2 of DEBS-1 to module 3 of the DEBS-2 subunit, as shown in Fig. 5.63. Note that module 3 is relatively short as it lacks a KR domain. This means that only chain extension via the Claisen condensation can occur in this module, leaving the resulting tetraketide intermediate with its β-ketone function intact. Now the chain is ready to be passed further along the "assembly line" to extension module 4.

Moving forward, we now notice (Fig. 5.64) that extension module 4 not only has a KR domain as previously seen in modules 1 and 2, but also has active DH (dehydratase) and ER (enoyl reductase) domains. This is the same familiar set of catalytic domains found in the fatty acid synthase (FAS) systems we examined earlier. And once again, the importance of the long, flexible ACP thioester linkage is evident, as the pentaketide from the Claisen condensation must also be able to reach the KR, DH, and ER catalytic domains for the required ketone reduction, alcohol

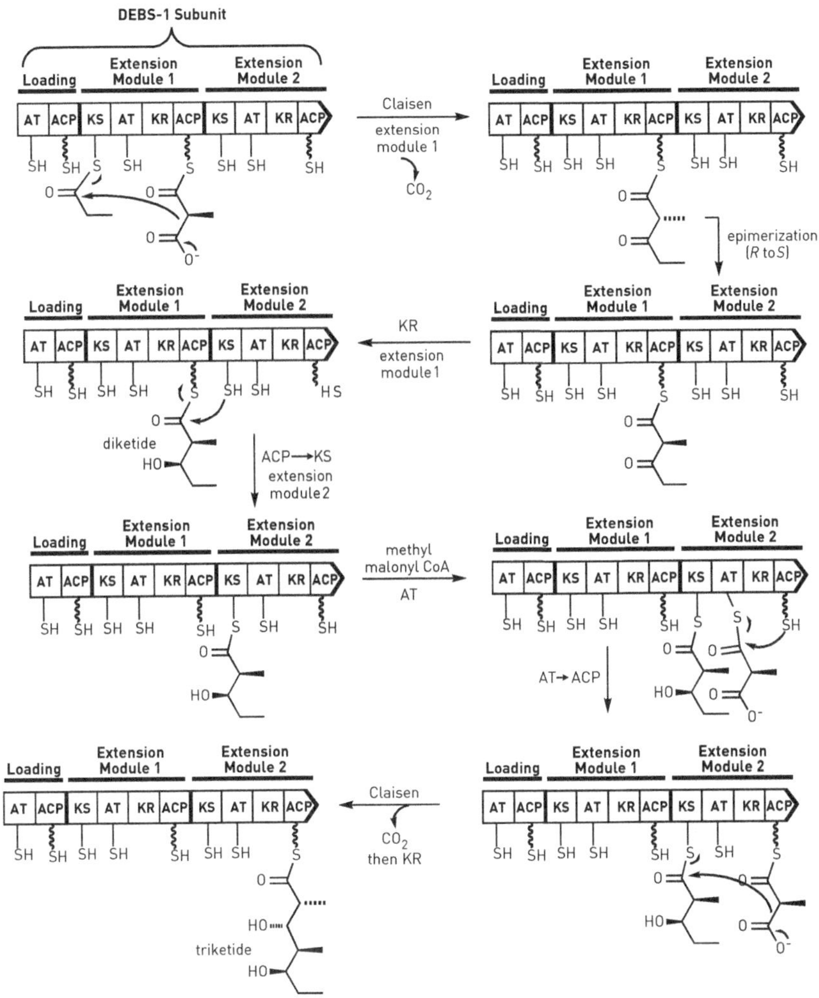

FIGURE 5.62
Diketide and triketide transformations for extension modules 1 and 2 of the DEBS-1 subunit.

dehydration, and conjugate reduction steps. This sequence of events leads to a segment of the growing chain that is now fully reduced (no oxygen at the β-carbon).

The growing chain is ready to be passed along to extension module 5 of the DEBS-3 subunit. If we read the domain organizations correctly, we should be able to determine that the transformations occurring in modules 5 and 6 will be more or less identical to those we saw previously in modules 1 and 2 (except for methyl group stereochemistry) as the process proceeds from a penta- to a hexa- to a final heptaketide of readily predictable structure as shown in Fig. 5.65. After the final ketone reduction step, this heptaketide is finally transferred by nucleophilic acyl substitution to the final TE (thioesterase) domain of the terminal module. Note that this final linkage to the heptaketide chain is not a thioester but rather a simple ester linkage formed via the side chain primary alcohol function of a serine residue contained within the final TE domain.

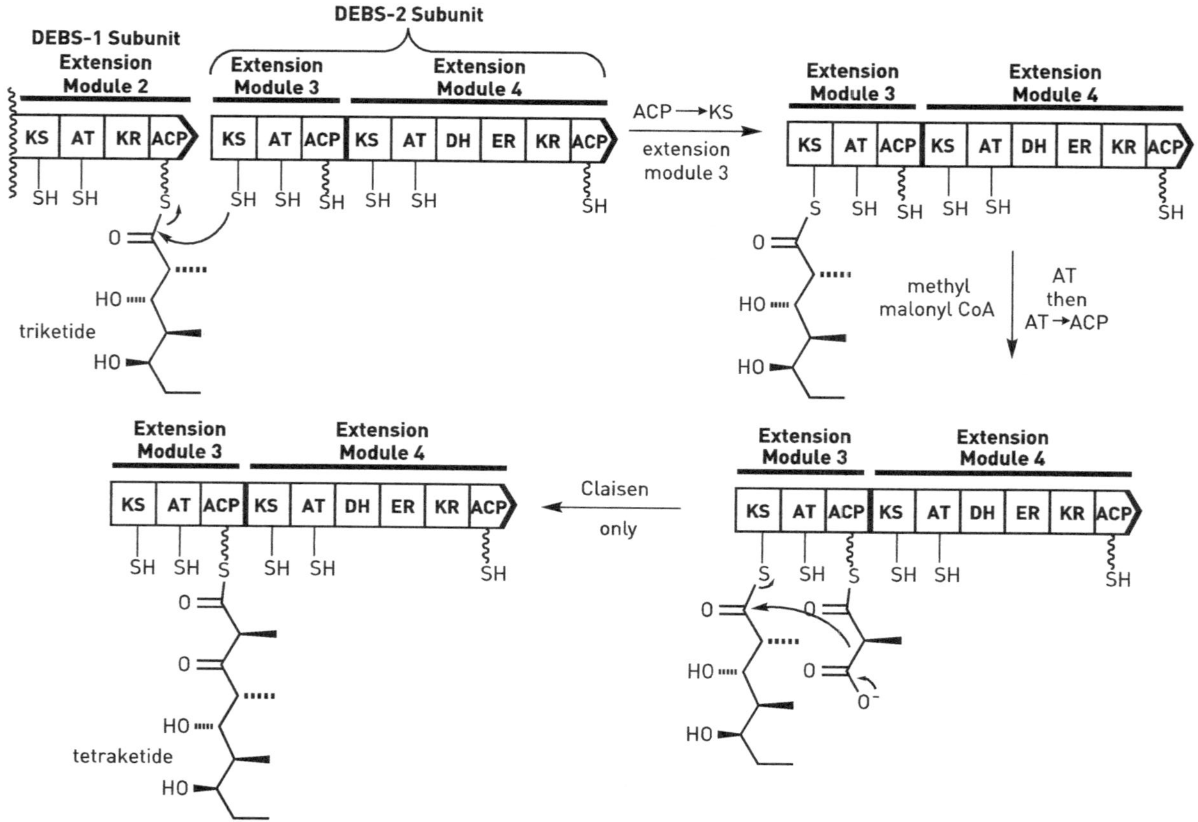

FIGURE 5.63
Transfer from DEBS-1 to DEBS-2 and module 3 triketide-to-tetraketide chain extension.

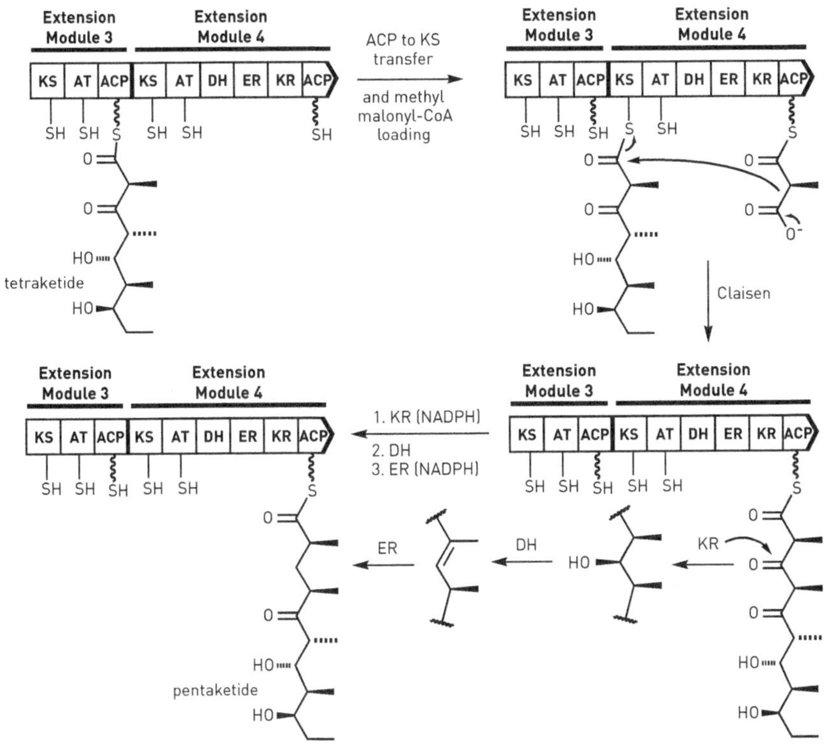

FIGURE 5.64
Module 4 chain extension with additional modifications via active KR, DH, and ER domains.

While this sequence has finally brought us to the end of the modular system, the enzyme still has one final and very critical operation to perform, and so the stage has been set for the final operation of the PKS assembly line, namely, the release of the polyketide chain from the enzyme structure itself.

As shown in Fig. 5.66, deprotonation of the indicated alcohol function leads to an alkoxide which can act as a nucleophile for intramolecular attack on the terminal ester carbonyl, resulting in lactone formation and production of the free macrolide, **6-deoxyerythronolide B.** The remaining oxidation, glycosylation, and methylation steps leading to **erythromycins A, C,** and **D** are all postcyclization processes catalyzed by individual enzymes. Thus, 6-deoxyerythronolide B is the aglycone precursor to all other members of the family. The unusual sugars employed for the glycosylation steps (both as TDP-sugars derived from glucose) are **L-mycarose** and **D-desoamine**; the final SAM step leading from erythromycin C to erythromycin A converts L-mycarose to its 3-*O*-methyl derivative **L-cladinose**.

GENETIC MANIPULATION OF MODULAR PKS SYSTEMS: RATIONAL DRUG MODIFICATION

It has long been a goal of drug discovery research to find ways to control or tailor the machinery of organisms that produce bioactive molecules in order to generate

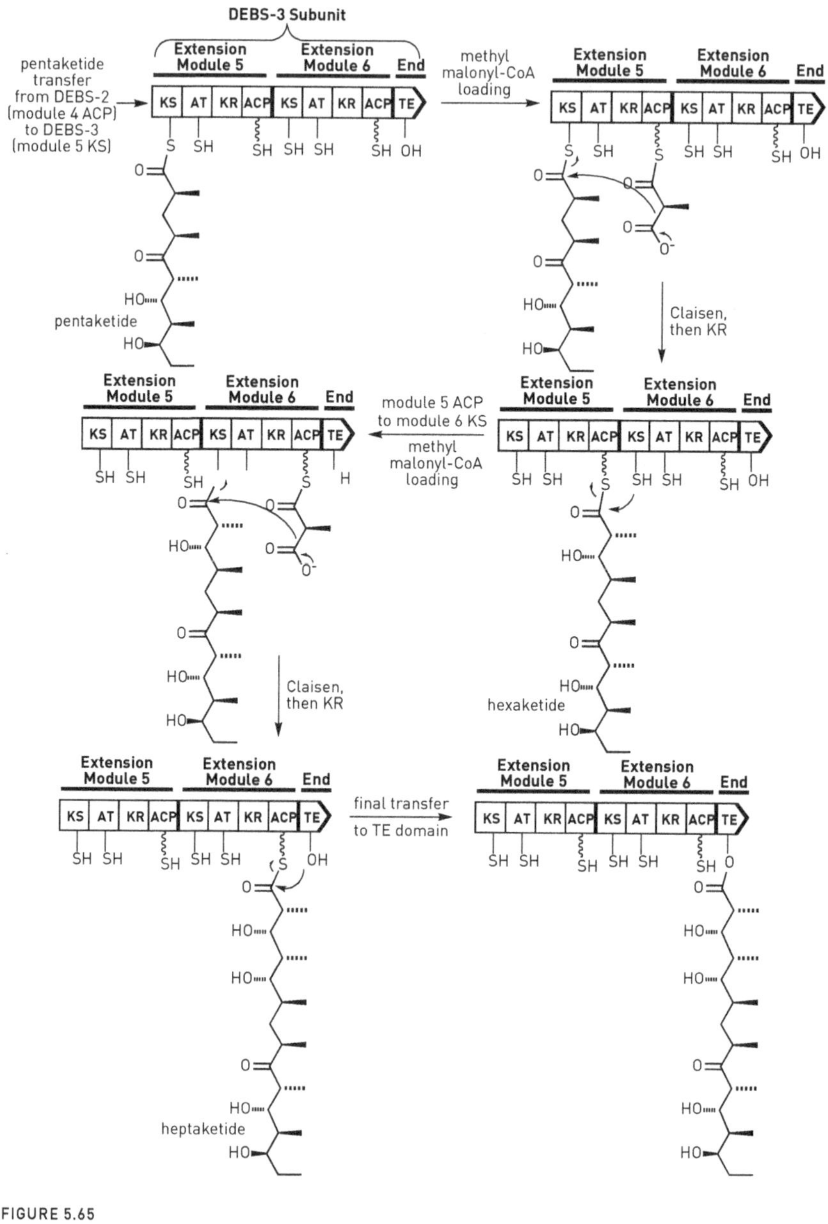

FIGURE 5.65
Final sequence of penta-to-heptaketide DEBS-3 chain extensions and transfer to TE domain.

analogues of known natural products or to produce, through biosynthesis, entirely new and novel compounds that may then be screened for potential therapeutic bioactivity. We briefly saw a relatively simple but successful example of this concept earlier when we noted that fermentation-based production of the highly active anthracycline anticancer compound **doxorubicin** was accomplished through use of an intentionally mutated strain of the organism that produced the original but less active anthracycline **daunorubicin**. In this context, the discovery of the modularity of certain PKS enzyme systems together with modern genetic engineering tools for

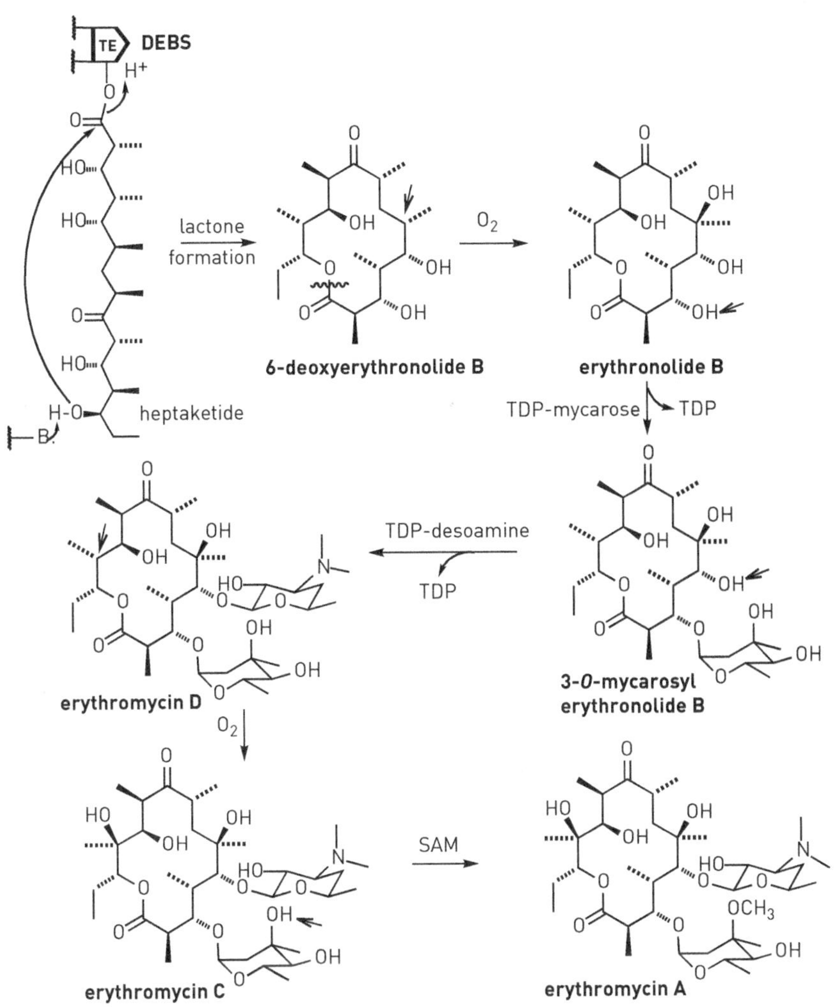

FIGURE 5.66
Erythromycins via release from DEBS, then oxidations, glycosylations, and a methylation.

manipulation of gene sequences involved in production of PKS enzymes has led to something of a revolution in the field of rational drug modification. This field is both large and quite complex, so a detailed examination is clearly beyond the scope of this chapter; nevertheless, a limited discussion of this crucial topic with some representative examples is both important and instructive.

Modular PKS systems are particularly amenable to genetic manipulation through a number of different approaches. Gene sequences responsible for generating specific catalytic domains can be modified or eliminated, thus "knocking out" certain domain activities within a given PKS module. For example, if the ER domain in module 4 of DEBS-2 is deactivated in this way, all remaining steps in the biosynthesis still occur as usual, leading to a 6-deoxyerythronolide analogue with unsaturation at the position indicated in Fig. 5.67. Similar macrolide structural modifications have been achieved

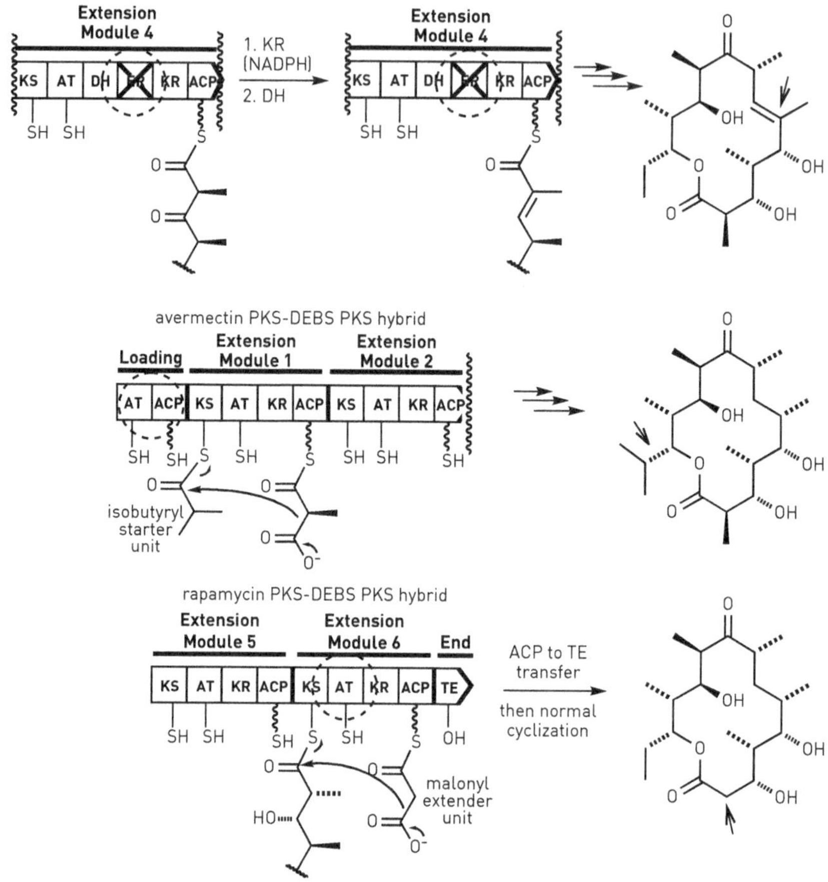

FIGURE 5.67
6-Deoxyerythronolide analogues via engineered DEBS modifications.

through use of so-called **hybrid modular PKS systems** which can be engineered by splicing portions of different natural PKS gene clusters together, leading to production of hybrid PKS enzymes with novel domain organizations. For instance, replacing the loading module (AT, ACP) of DEBS with the loading module from another natural PKS enzyme (one which produces *avermectin*) leads to a hybrid PKS enzyme system that, like the *avermectin* PKS, is selective for an isobutyrylCoA starter unit rather than the usual propionyl-CoA starter unit. This loading domain substitution ultimately leads to production of a 6-deoxyerythronolide analogue that incorporates the isobutyryl group at the expected position, as shown. A similar modification involving substitution of the DEBS-3 module 6 AT domain with the AT domain from module 2 of a PKS enzyme responsible for *rapamycin* synthesis leads to a hybrid enzyme with an AT6 domain which, like the *rapamycin* AT2 domain, is specific for incorporation of a malonyl-CoA unit rather than the usual *S*-methylmalonyl-CoA unit. This domain substitution leads to production of a 6-deoxyerythronolide analogue lacking a methyl group at the indicated position. It is worth noting that when this compound was isolated and fed to an organism capable of completing the

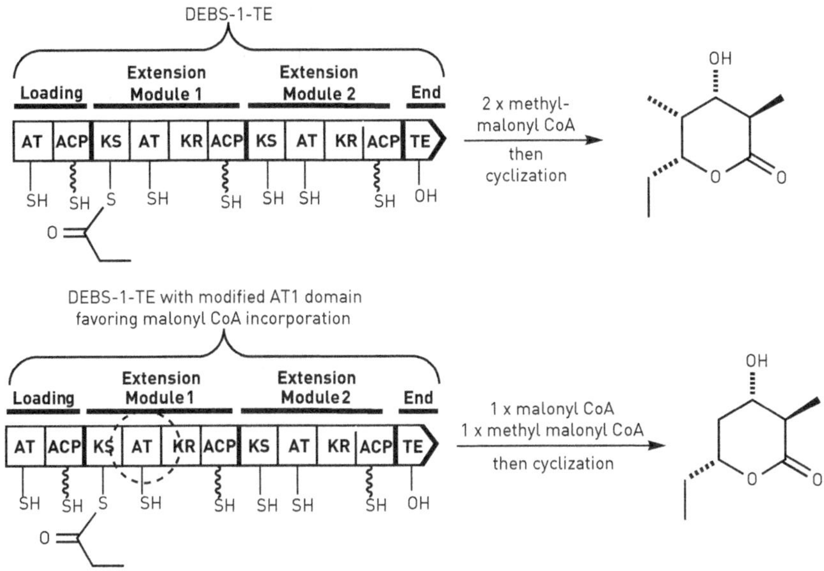

FIGURE 5.68

A truncated DEBS-1-TE and an AT1-modifed version leading to small-ring lactones.

required oxidation and glycosylation steps, the resulting erythromycin analogue retained significant antibacterial activity. Further advances in cloning and characterization of PKS-containing gene clusters for most of the major polyketide product groups has led to many additional examples of such hybrid construction.

Further interesting work has been carried out using truncated versions of the DEBS system in which a TE domain is incorporated onto the end of the DEBS-1 component which therefore terminates polyketide formation after only two cycles of chain extension. The resulting cyclized triketide lactone (Fig. 5.68) represents a potentially useful chiral building block for laboratory organic synthesis, as it is produced as a single diastereomer carrying four contiguous chirality centers. Similarly, genetic modifications have been employed to replace the AT1 domain with another which favors loading malonyl-CoA over methylmalonyl-CoA as an extender unit, ultimately leading to formation of the indicated chiral lactone lacking a methyl group at C-4. These examples serve to illustrate that while still in its infancy, the field of genetic engineering for the production of enzyme systems designed to produce specifically tailored polyfunctional organic compounds will surely play a key role in the future of industrial and pharmaceutical organic synthesis.

SOME FINAL PKS PRODUCTS OF MEDICINAL IMPORTANCE

With our additional insight into how different PKS stems operate, we can now take a brief look at some final polyketide natural products of relatively complex structure and be able to infer many of the details associated with their assembly without the need for extensive explanation or analysis. We begin with the assembly of two polyene antifungal agents of special significance, **amphotericin B** and **nystatin A**. As shown in Fig. 5.69, the two are structurally very similar, differing only in their oxygenation patterns as referred to in the positions indicated at R_1–R_4. Both utilize a combination of acetyl-CoA, malonyl-CoA and methylmalonyl-CoA units assembled in the indicated sequence with each of KR, DH, and ER domains operating alone or in combination in identical ways at various locations throughout the structures. The only exception to this is variation in oxygenation patterns at positions R_1–R_4 which are determined by operation of different domain combinations associated with each of the respective synthase enzymes, **amphotericin PKS** or **nystatin PKS**. Once the polyketides are assembled, release from the PKS enzyme is again achieved through lactone formation which is accompanied by intramolecular hemiacetal formation at the position indicated. The postcyclization transformations involve sequential oxidations at the indicated methyl group for both, with additional C–H to C–OH oxidation occurring on the ring at position R_2 for amphotericin B or at position R_4 for nystatin A. Both undergo a final glycosylation at the allylic alcohol function via a GDP-derivative of the aminosugar **mycosamine**.

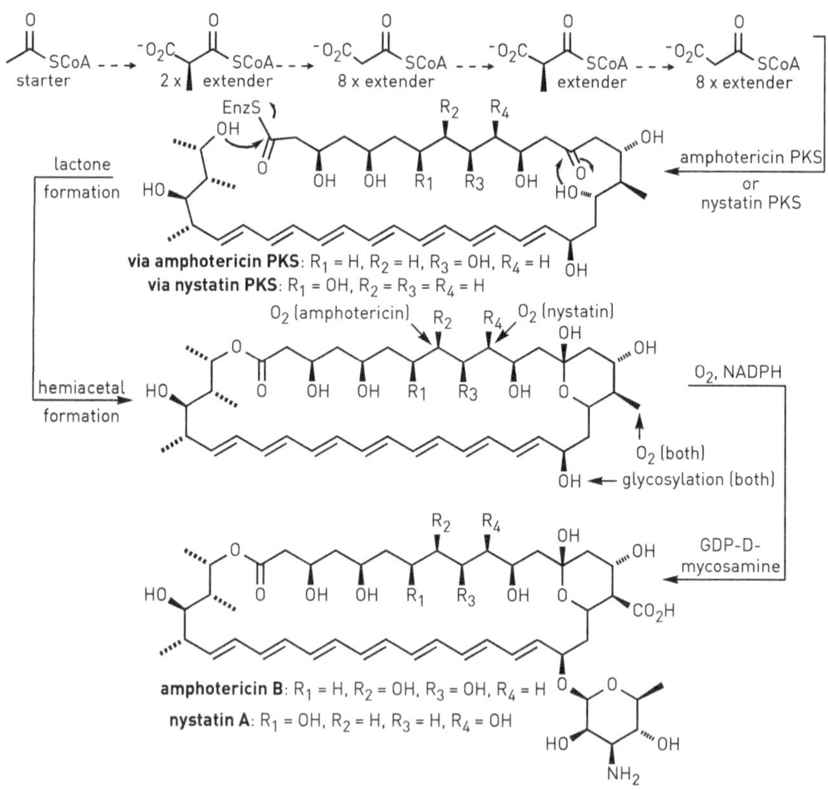

FIGURE 5.69
Biosynthesis of the antifungal polyene macrolides amphotericin B and nystatin A.

Originally extracted from *Streptomyces nodosus*, amphotericin B is so named because of amphoteric properties associated with the presence of both an acidic carboxyl function and a basic amino group (on the sugar moiety). The drug targets sterol-containing fungal cell membranes and is administered intravenously. While an important treatment for certain systemic and life-threatening fungal infections such as valley fever in the southwest United States, the potentially serious side effects of amphotericin B therapy make careful control of dosage essential. By comparison, the toxicity of nystatin A is much lower since it is mainly administered orally and its absorption by this route is minimal, making it especially useful for treatment of intestinal fungal infections. It is also a common component of topical creams for the treatment of fungal infections of the skin, usually in combination with a steroidal component used to reduce associated inflammation.

Monensin A, a polyether antibiotic isolated from *Streptomyces cinnamonensis,* is an **ionophore** that can form stable complexes with a number of different monovalent metal cations such as Li^+, Na^+, K^+, Rb^+, and Ag^+. The compound exhibits antibiotic, antimalarial, and other biological activities, and its valuable antibacterial properties have led to its wide use, mainly in the beef and cattle industries. Its biosynthesis is outlined in Fig. 5.70.

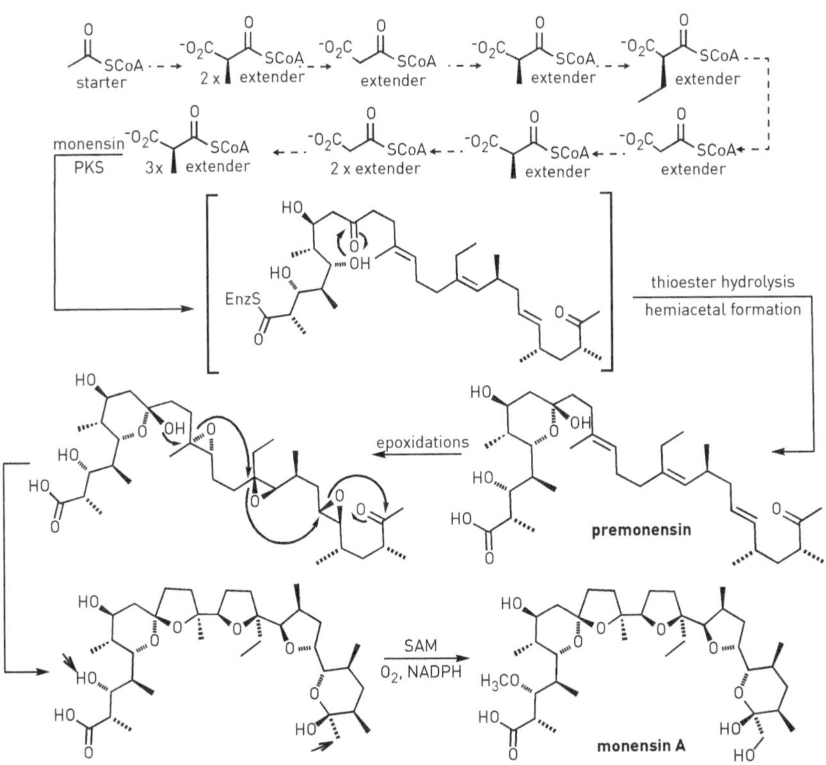

FIGURE 5.70
Biosynthesis of the polyether antibiotic monensin A.

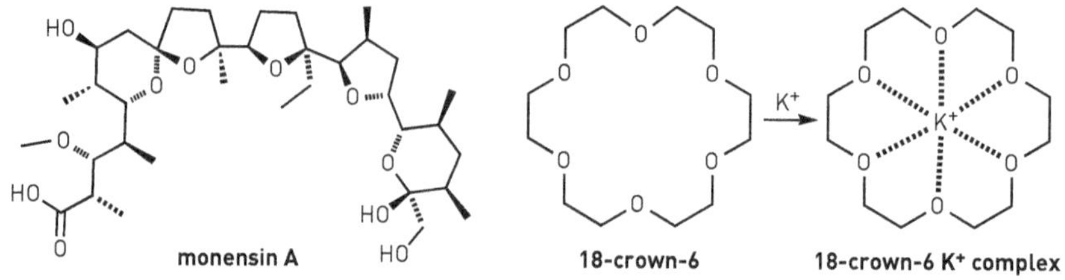

FIGURE 5.71
Comparison of monensin A with a crown ether potassium ion complex.

Ionophores are able to bind and transport metal ions across hydrophobic lipid membranes. In that context, note the structural similarity between monensin A and **18-crown-6**, a member of the "crown ether" group of synthetically prepared ionophore macrocyclic polyethers developed by Charles Pedersen of DuPont, for which he received a share of the Nobel Prize in chemistry late in his life. Binding occurs via interaction between the metal cation and ether oxygen lone pairs (Fig. 5.71).

LOOKING AHEAD

The foregoing examination of various classes of polyketide natural products represents a mere tip of the iceberg with regard to product diversity from this remarkably complex and ever-expanding area of biosynthetic chemistry. There is so much more to explore, particularly with respect to the many additional important compounds that may be produced when different biosynthetic pathways intersect. We saw a simple example of this earlier when we looked at the biosynthesis of poison ivy's active principal, urushiol (Fig. 5.54) which utilized an aromatic component, cinnamoylCoA, as a starter unit in the polyketide assembly process. At that point, we mentioned that this aromatic component was derived from a different pathway—the **shikimic acid pathway**. And it is no coincidence that this hybrid polyketide product was produced by a plant rather than by the bacteria or fungi which produce most of the polyketides we have seen. To better understand this connection, we move on in our next chapter to explore this unique biosynthetic pathway that operates only in plants and that represents the second major source of aromatic compounds produced by nature.

STUDY PROBLEMS

1. Propose a series of experiments to rule out biosynthesis of methylphloroacetophenone via a direct C-methylation of phloroacetophenone (see Fig. 5.40).
2. Suggest a biosynthetic sequence to produce 6-methylsalicylic acid.

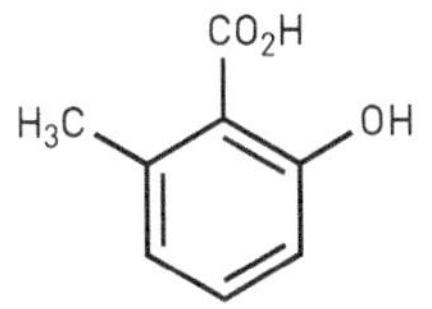

6-methylsalicylic acid

3. What are the R/S stereochemical assignments of the chirality centers in griseofulvin? The NADPH conjugate addition of hydride in the final step occurs at which face of the prochiral ring pi-bond?

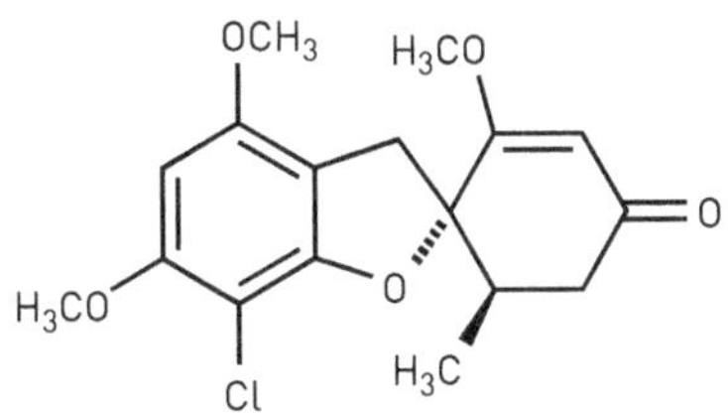

(+)-griseofulvin

4. Propose a triketide-based biosynthesis of 6-methoxymellein, an antifungal metabolite isolated from carrots.

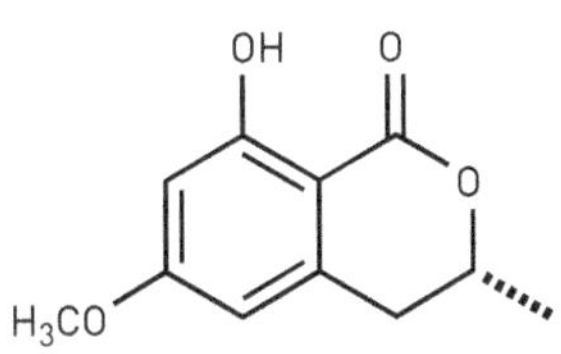

6-methoxymellein

5. Propose a mechanism for the laboratory oxidative conversion of 4-methylphenol to yield Pummerer's ketone.

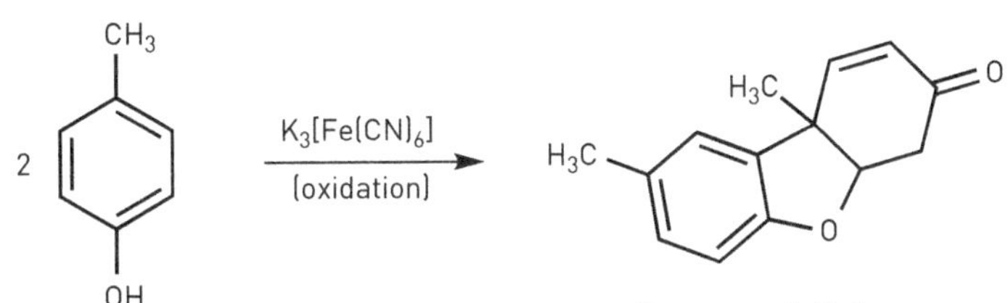

Pummerer's Ketone

6. Propose a reasonable mechanism for the oxidation step involved in conversion of anhydrotetracycline to 5a,11a-dehydrotetracycline.

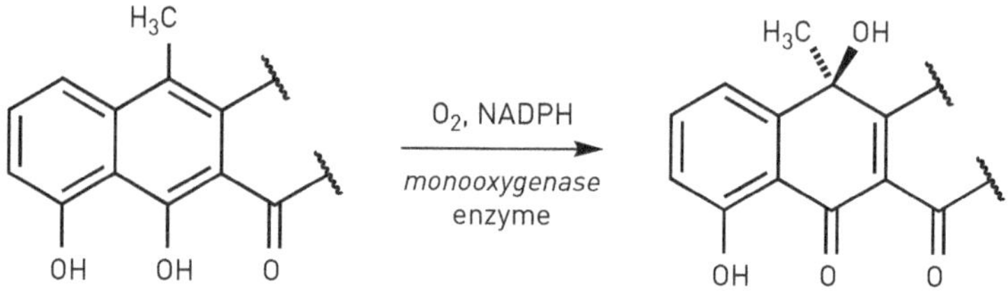

7. *Emodin* is a precursor to *skyrin*, a chiral natural product. Show how *skyrin* is formed from *emodin* and explain how this compound can be chiral even though it has no chiral carbons.

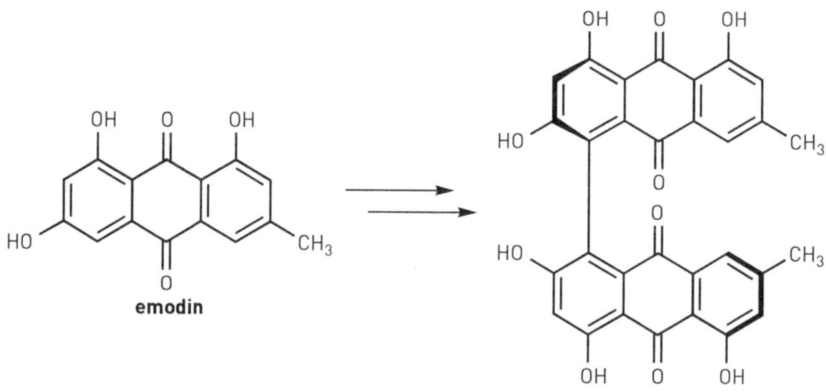

8. The acetate hypothesis has proven useful in resolving questions related to correct structural assignments for many natural products when alternate structural assignments were consistent with available spectroscopic data. Which structure (**A** or **B**) would you predict to be the correct structure of natural product **X**, assuming it is biosynthesized in a normal way via the polyketide pathway? (Note that a methylation and an oxidation would be required to produce **X** after polyketide cyclization.)

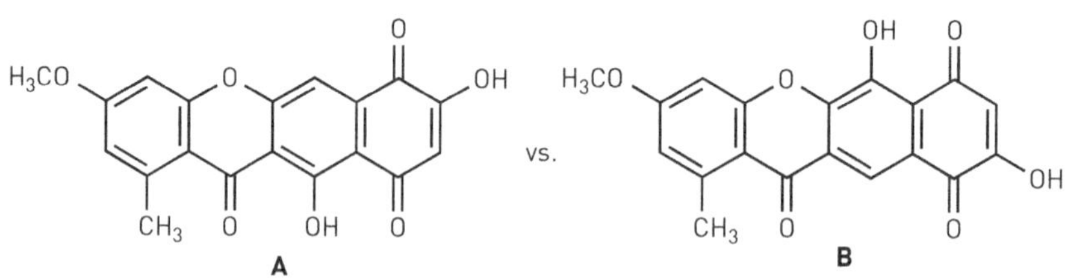

9. Compound **B** is formed from compound **A** which is initially produced via the polyketide pathway.

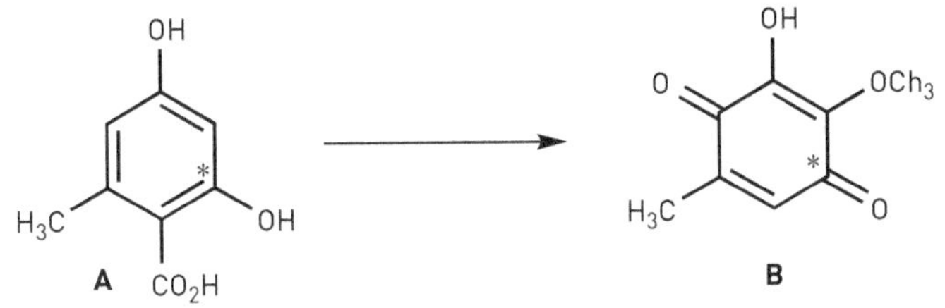

a) Outline the likely steps involved in the biosynthesis of **A**.

b) When the indicated carbon of **A** is labeled with $^{13}C(*)$ and fed to the organism producing the metabolite, only ^{13}C-labeled **B** is produced. Keeping in mind the potential symmetry of certain possible intermediates, what does this result say about the reactions involved and especially about the order of the steps in which **A** is probably transformed into **B**?

10. Provide the missing structures to complete the following reaction scheme (hint: the second step involves conjugate addition of water as a nucleophile).

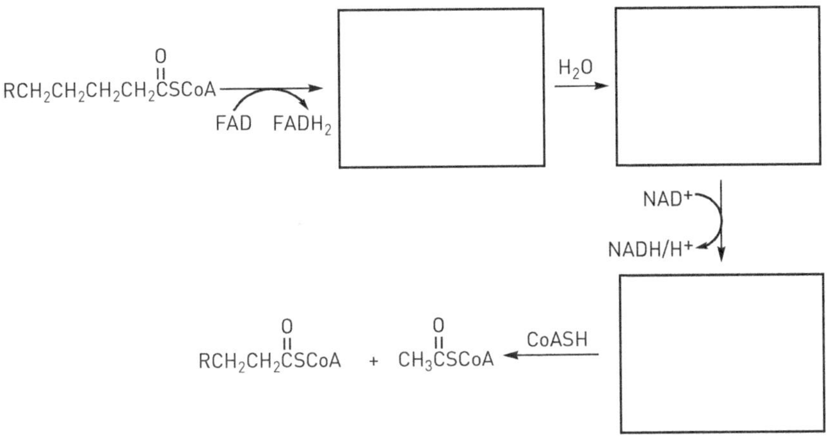

a) In the first step of this process, a dehydrogenation takes place which involves initial formation of a thioester enolate ion, followed by delivery to FAD of hydride as the pro-*R* β-hydrogen atom of the thioester enolate. The resulting FADH$^-$ anion then accepts a proton from a suitable donor to complete the process. Outline the correct structures and mechanisms involved in this overall transformation (use an abbreviated structural representation for FAD that shows the key structural elements involved in the dehydrogenation mechanism).

b) The last step is a retro-Claisen condensation. Show a mechanism including key intermediates.

11. The compound on the left reacts with geranyl diphosphate (GPP) to give the compound shown. Propose a mechanism for this biosynthesis of the phenolic component, then show its reaction with GPP and subsequent steps involved in converting it to the compound shown.

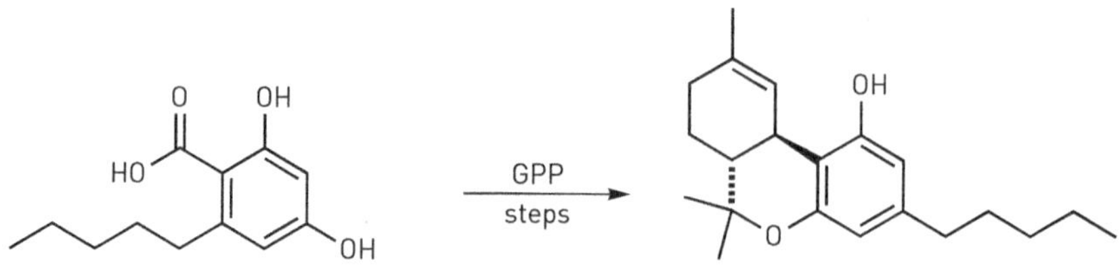

12. The polyketide metabolite **A** has recently been isolated from a lichen. Indicate a biosynthesis of **A** starting from acetyl-CoA. In your answer indicate the sequence of events and a mechanism for aromatic and lactone ring formation.

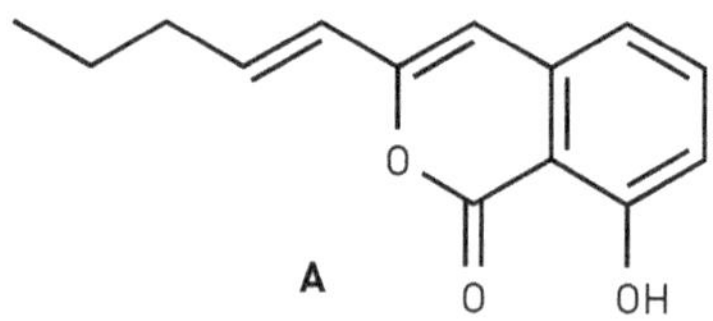

A

13. Suggest a reasonable polyketide-based biosynthesis of the macrocyclic lactone shown, produced by *Steptomyces venezuelae* (indicate starter unit, extension units and any transformations of the intermediate polyketide required prior to lactone cyclization).

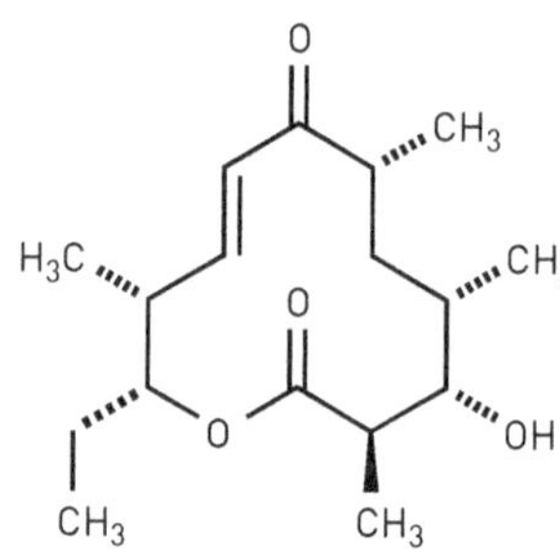

14. Prostacyclin (PGI_2) is a potent vasodilator and inhibitor of platelet aggregation produced by vascular endothelial cells. Decreased production of prostacyclin is associated with platelet aggregation and the formation of blood clots (thrombosis), a primary cause of heart attacks and stroke. Outline the biosynthesis of prostacyclin from PGH_2, and write a mechanism to show how it is transformed, in vivo, into the compound shown.

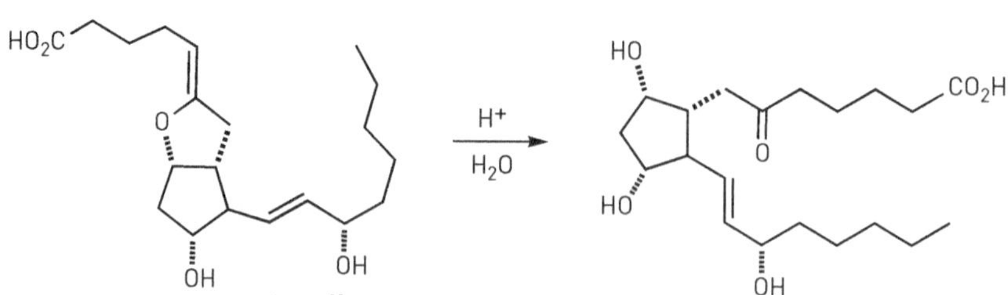

15. Mechanistically similar to the hydroperoxidation of arachidonic acid, linolenic acid forms a hydroperoxide in the presence of the enzyme *13-lipoxygenase*. This hydroperoxide can undergo several transformations, one of which involves a 1,2 C-to-O cationic migration somewhat similar to those seen in Baeyer–Villiger oxidations. Such rearrangements are the source of numerous C_6 aldehydes and alcohols found in nature. Outline a reasonable mechanism for how such a rearrangement of hydroperoxide **A** in aqueous solution would lead to the two aldehyde products B and C shown (the migrating carbon is starred in both **A** and **B**). Start it off by protonating the OH group of the hydroperoxide.

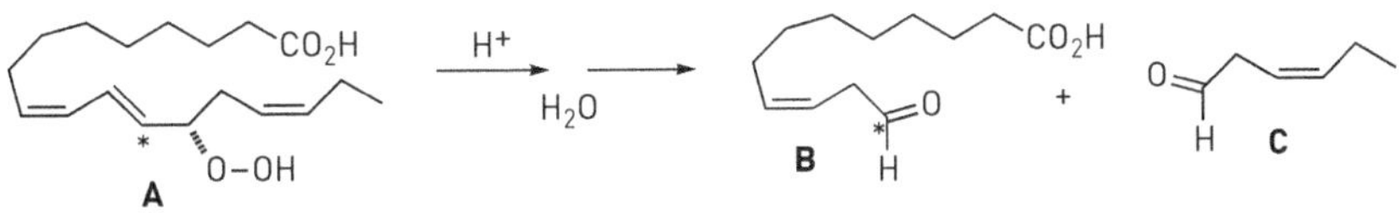

16. Use equations to show how a sequence of 1) oxidation, 2) dehydrogenation and 3) Baeyer–Villiger oxidation could transform **A** into **B**, a precursor to penicillinic acid. Show mechanistic details of the Baeyer–Villiger oxidation step.

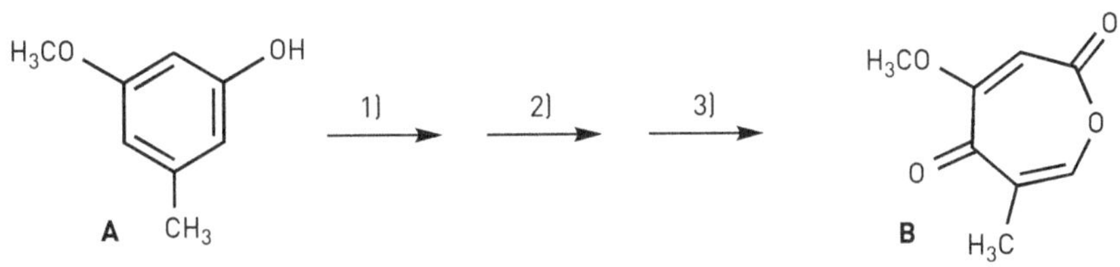

17. Assuming the producing organism was fed doubly labeled ^{13}C acetate, show the expected polyketide folding pattern and as well as the expected labeling pattern in the product shown below (for labeling, use bolded bonds as shown in Fig. 5.39).

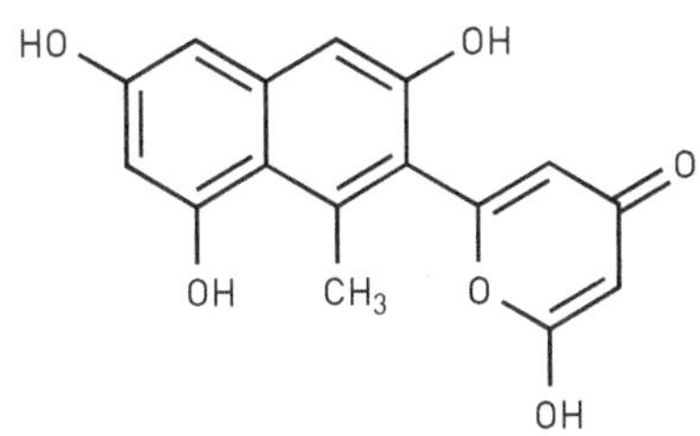

6 The Shikimate Pathway

Biosynthesis of Phenolic Products from Shikimic Acid

I think my strongest suit in science has been critical, logical analysis, leading to a single but decisive experiment, although a systematic program, pursuing the shikimate pathway, has probably contributed most to my scientific reputation
—Bernhard Davis (*Microbiologist, 1916–1994*)

Like other amino acids, the aromatic amino acids phenylalanine, tyrosine, and tryptophan are vitally important for protein synthesis in all organisms. However, while animals can synthesize tyrosine via oxidation of phenylalanine, they can synthesize neither phenylalanine itself nor tryptophan and so these essential amino acids must be obtained in the diet, usually from plant material. Though many other investigators made significant contributions in this area over the years, it was Bernhard Davis in the early 1950s whose use of mutant stains of *Escherichia coli* led to a full understanding of the so-called shikimic acid pathway that is used by plants and also by some microorganisms for the biosynthesis of these essential amino acids. The pathway is almost completely devoted to their synthesis for protein production in bacteria, while in plants the pathway extends their use to the construction of a wide array of secondary metabolites, many of which are valuable medicinal agents. These secondary metabolites range from simple and familiar compounds such as vanillin (vanilla flavor and fragrance) and eugenol (oil of clove, a useful dental anesthetic) to more complex structures such as pinoresinol, a common plant biochemical, and podophyllotoxin, a powerful cancer chemotherapy agent (Fig. 6.1).

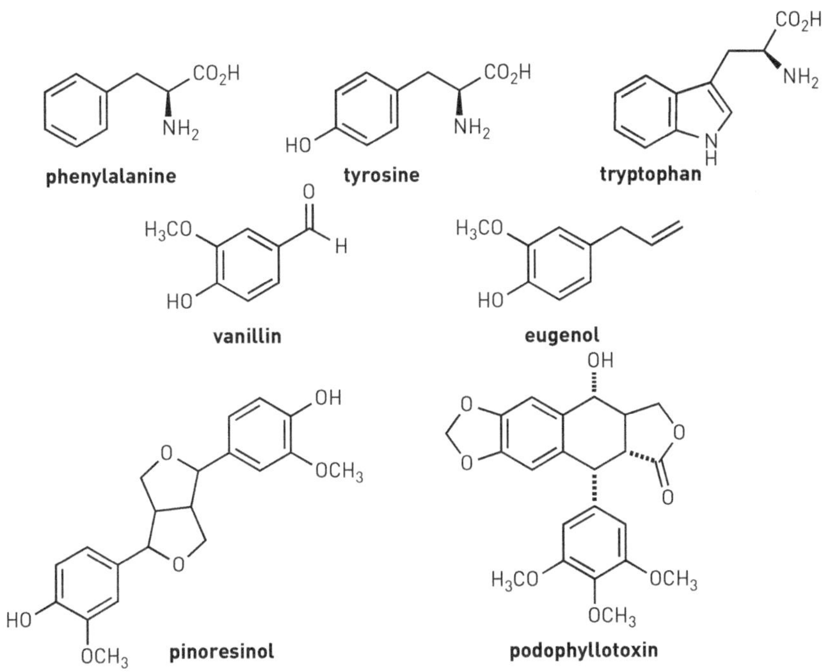

FIGURE 6.1
Aromatic amino acids and some secondary metabolites from the shikimic acid pathway.

WHAT IS SHIKIMIC ACID?

Earlier in Chapter 3, we encountered two important intermediates, **erythrose-4-phosphate** and **phosphoenolpyruvate (PEP)**, each of which was derived from a different pathway utilized in carbohydrate metabolism. Erythrose-4-P was an intermediate in one of the steps of the pentose phosphate pathway while hydrolysis of PEP to pyruvic acid was the final step in glycolysis (Fig. 6.2).

These two simple intermediates provide the seven carbon atoms required for construction of shikimic acid itself. The two are linked to one another via a sequence of enzyme-mediated aldol-type reactions, the first being a bimolecular reaction and the second an intramolecular variant that ultimately leads to a cyclic precursor of shikimic acid known as **3-dehydroquinic acid** as shown in Fig. 6.3.

Subsequent dehydration of **3-dehydroquinic acid** leads to **3-dehydroshikimic acid** which then leads directly to **shikimic acid** via NADPH reduction. It should also be noted that this intermediate may undergo several additional transformations (Fig. 6.4) such as dehydration, one of several routes to **protocatechuic acid**, a naturally occurring antioxidant and anti-inflammatory polyphenol, or dehydrogenation by the action of $NADP^+$ to give **3-didehydroshikimic acid**, an unstable tautomer rapidly converted to **gallic acid**, a major component (along with glucose) of **tannic acid** and other tannins derived from wood barks and other sources and used in the textile, wood, and food industries. Decarboxylation of gallic acid either

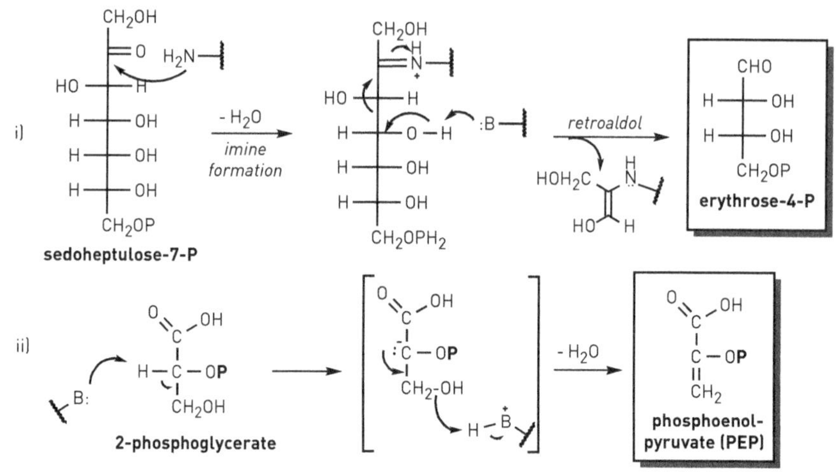

FIGURE 6.2

Production of: i) erythrose-4-P from the pentose phosphate pathway and ii) phosphoenolpyruvate (PEP) from the glycolytic pathway.

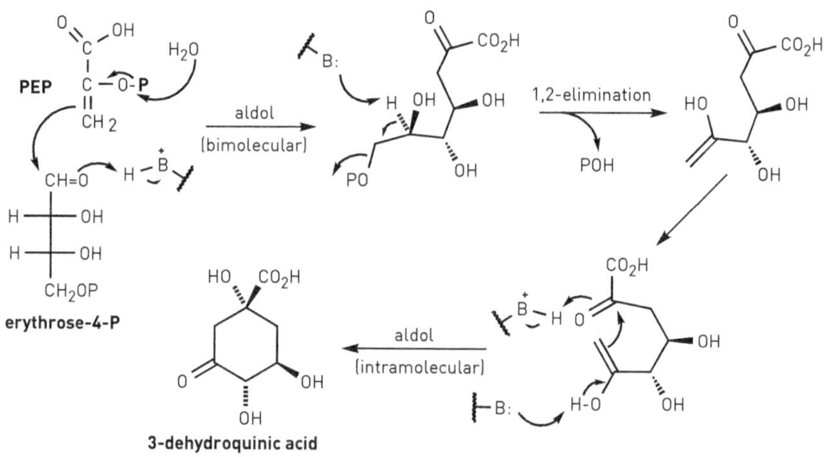

FIGURE 6.3

Sequential aldol additions in biosynthesis of 3-dehydroquinic acid.

enzymatically or by heating leads to **pyrogallol**, a phenolic reducing agent used in early photographic processes and still widely employed as a stoichiometric O_2 scavenger.

SHIKIMIC, CHORISMIC, AND PREPHENIC ACIDS AT THE HEART OF THE PATHWAY

Getting back to shikimic acid itself, we find that this key intermediate must undergo a number of additional transformations on its way to finally being converted into phenylalanine or tyrosine. Some have even suggested that the pathway might

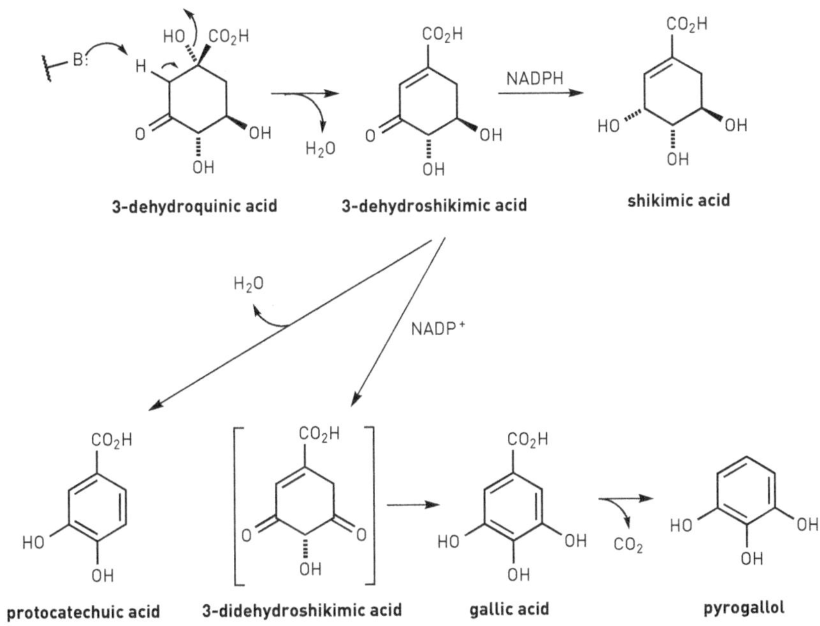

FIGURE 6.4
Formation of shikimic acid and other transformations of 3-dehydroshikimic acid.

be more suitably named after one of the subsequent intermediates, but names tend to stick over the years, and no matter how we look at it, the end products of the pathway ultimately lead back to shikimic acid as the pivotal source. The first of the subsequent intermediates, **chorismic acid**, is formed by initial ATP phosphorylation to give **shikimic acid-3-phosphate**. Subsequent addition of the 5-hydroxyl group across the π-bond of phosphoenolpyruvate (PEP) gives **5-enolpyruvylshikimic acid-3-phosphate**; sequential 1,2- and 1,4-eliminations of phosphate (POH) complete the conversion of shikimic acid to chorismic acid as shown in Fig. 6.5.

At this point, we stop to take a brief overview of the key reaction involved in the subsequent transformation of chorismic acid into prephenic acid, at which point we will be well on our way to the final targets, phenylalanine and tyrosine.

THE CLAISEN REARRANGEMENT: ALLYL VINYL ETHERS IN A CHAIR

One of the oldest and most important C–C bond forming reactions in organic chemistry is a process known as the **Claisen rearrangement**, originally reported by L. Claisen in 1912. As the first known example of the so-called **[3,3]-sigmatropic rearrangements**, the Claisen rearrangement is formally a **pericyclic reaction**, meaning that the transformation involves a cyclic transition state which facilitates the

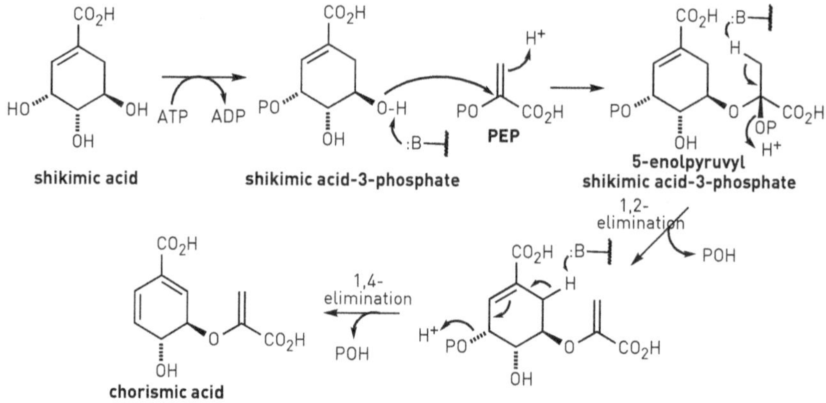

FIGURE 6.5

Conversion of shikimic acid to chorismic acid.

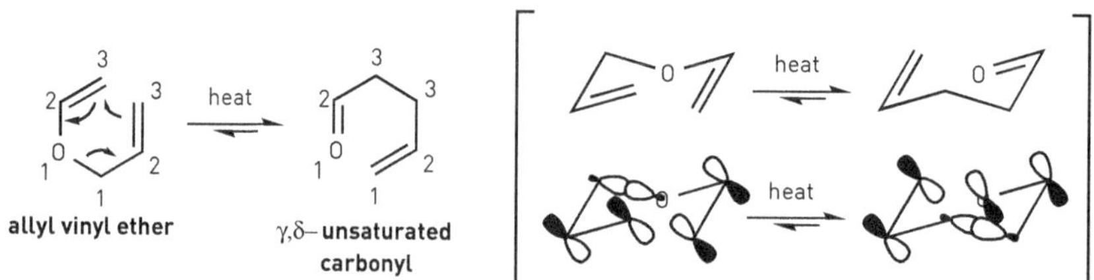

FIGURE 6.6

Some key features of the Claisen rearrangement of allyl vinyl ethers.

intramolecular rearrangement of bonding electrons among the atoms of the reactant. In the case of the Claisen rearrangement, the reactant is an **allyl vinyl ether** for which rearrangement of the electrons via the cyclic transition state leads to a γ,δ-unsaturated carbonyl derivative as shown in Fig. 6.6. Furthermore, it has been shown that the so-called **chair transition state** for the Claisen rearrangement is one which leads to the most rapid reaction via advantageous overlapping of orbitals in the reactant and product. Because of the thermodynamic stability of the C=O bond formed, the reaction is irreversible, unlike the related Cope rearrangement of 1,5-dienes. The [3,3]-sigmatropic rearrangement designation is related to the observed movement of a sigma bond between atoms designated 1 and 1 to the atoms numbered 3 and 3 in the diagram.

Like other sigmatropic rearrangements, there are no typical reactive intermediates involved in the Claisen rearrangement (i.e., no carbocations, no radicals, etc.). The reaction is a **concerted** one, meaning that all the bond-making and bond-breaking events occur simultaneously. This makes the Claisen rearrangement a **stereospecific** reaction: optically pure reactants will yield optically pure products when bonds to existing chirality centers are broken and new chirality centers are formed in the process.

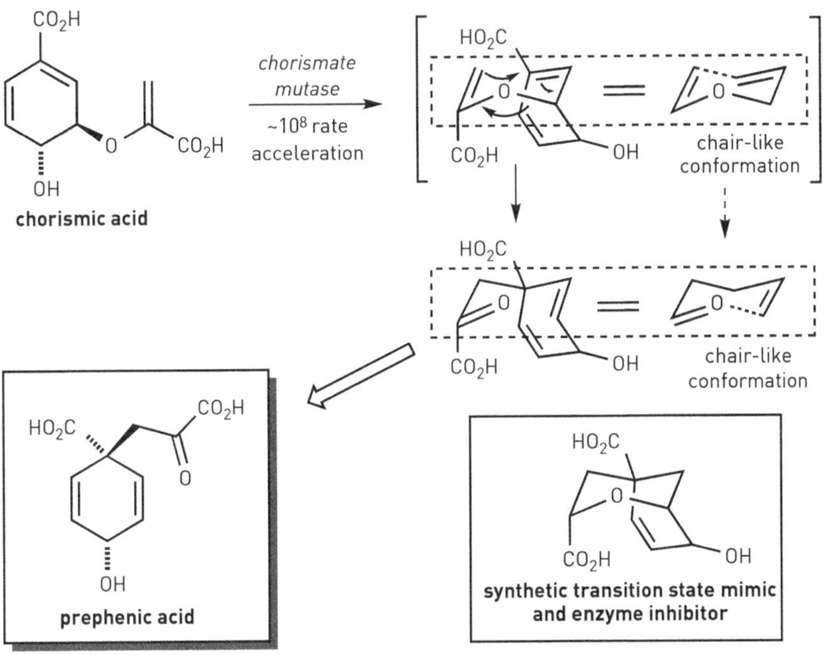

FIGURE 6.7
Chair-like transition state in the Claisen rearrangement of chorismic acid to prephenic acid and a transition state mimic and enzyme inhibitor.

CONVERSION OF CHORISMIC ACID TO PREPHENIC ACID

The *chorismate mutase*-catalyzed Claisen rearrangement of chorismic acid to **prephenic acid** (Fig. 6.7) is one of the few examples of an enzyme-catalyzed pericyclic reaction occurring in nature. The reaction shows an enormous rate acceleration—on the order of 10^8—relative to the uncatalyzed reaction, due in part to the ability of the enzyme to bind chorismic acid in the ideal chair-like conformation required by the Claisen rearrangement. Evidence to support this includes the significant rate retardation observed with the introduction of a synthetic enzyme inhibitor whose shape mimics the chair-like transition state believed to be involved. Again, note the stereospecificity of the reaction, with optically pure chorismic acid yielding optically pure prephenic acid, the key pathway precursor to both phenylalanine and tyrosine.

CONVERSION OF PREPHENIC ACID TO PHENYLALANINE OR TYROSINE

As shown in Fig. 6.8, prephenic acid serves as a branch point in the pathway, leading to either **phenylpyruvic acid** via a decarboxylative 1,4-dehydration or to **4-hydroxyphenylpyruvic acid** via a decarboxylative 1,4-dehydrogenation,

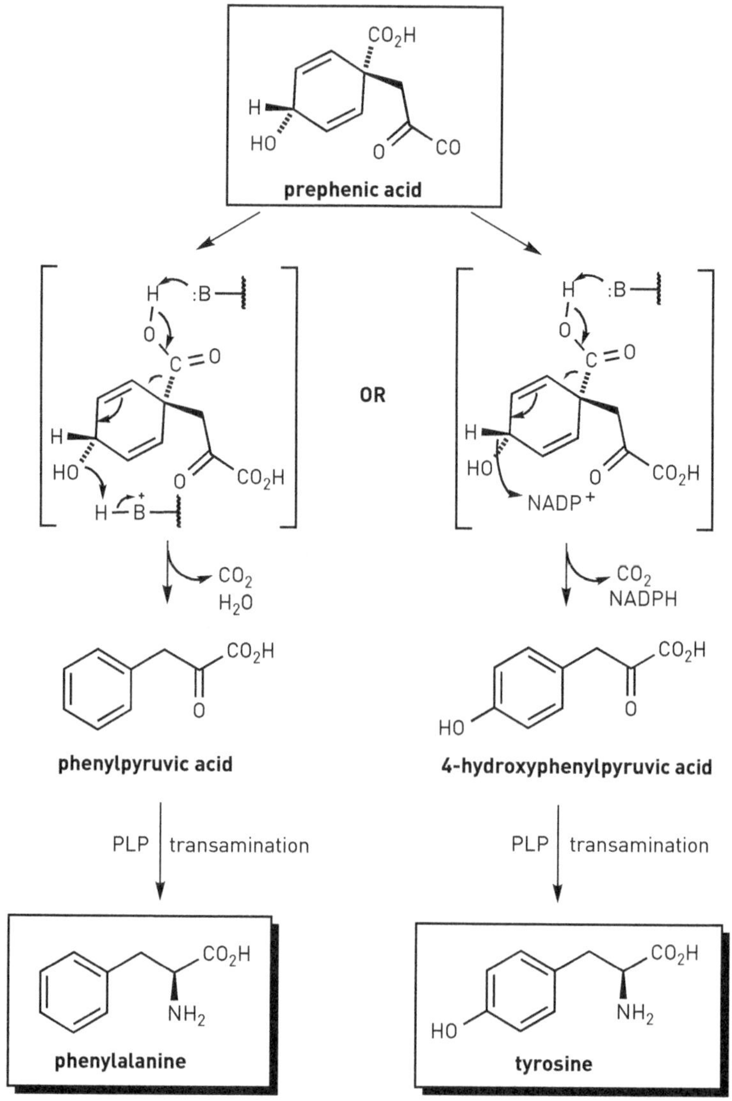

FIGURE 6.8
Steps in the conversion of prephenic acid to phenylalanine or tyrosine.

depending on which enzyme system is employed. Each of these products in turn may undergo PLP-mediated **transamination** in the usual fashion to convert the respective α-ketoacids to their corresponding L-α-amino acids, that is, **phenylalanine** or **tyrosine**. In some instances, transamination of prephenic acid (to give **arogenic acid**) has been found to precede the decarboxylative elimination steps, constituting yet another pathway to phenylalanine and tyrosine. Recall that animals lack the shikimic acid pathway but can synthesize tyrosine by oxidation of phenylalanine obtained in the diet.

MORE USES FOR CHORISMIC ACID

Before moving on to the many compounds derived from phenylalanine or tyrosine, we should briefly consider some additional transformations of chorismic acid, including its role in the biosynthesis of **tryptophan**, the third amino acid from this pathway. As shown in Fig. 6.9, chorismic acid may be diverted from the chorismic-to-prephenic acid Claisen rearrangement via several different routes (A–C among others), each of which ultimately leads to the formation of a biologically important benzoic acid derivative.

Path A (Fig. 6.9) involves a 1,4-addition–elimination sequence with water acting as both nucleophile and leaving group, leading to an intermediate that undergoes subsequent aromatization via a 1,2-elimination of pyruvic acid to afford **salicylic acid,** the well-known aspirin precursor and active agent in medicinal extracts of

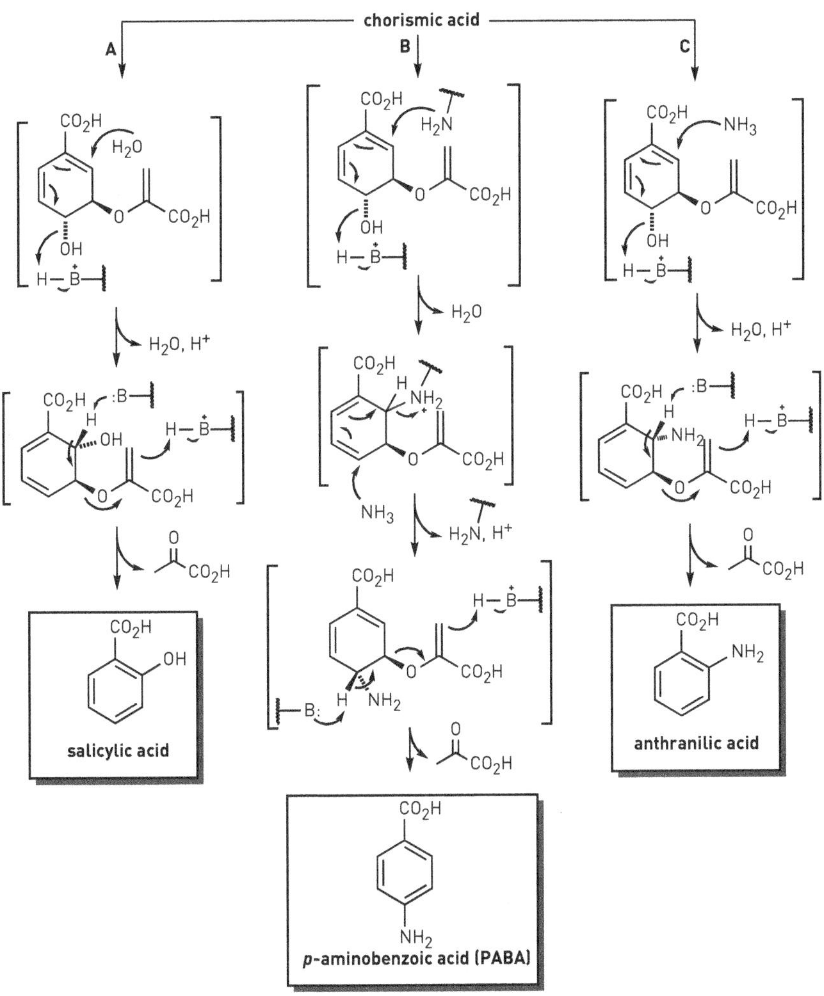

FIGURE 6.9
Other transformations of chorismic acid to salicylic acid, PABA, and anthranilic acid.

willow bark whose recorded use dates as far back as the 5th century B.C. Salicylic acid is also an important plant hormone with roles in photosynthesis, growth, and ion transport, among many others. Paths B and C each lead to simple aromatic amino acids: path B involves a 1,4-addition–elimination sequence with water as leaving group and an enzyme side chain amino group as nucleophile. The intermediate obtained then undergoes subsequent 1,4-addition–elimination with NH_3 (from glutamine) as nucleophile and the enzyme side chain amino function acting as leaving group. Subsequent 1,2-elimination and aromatization (again with loss of pyruvic acid) affords **para-aminobenzoic acid** (PABA), an important precursor to **folic acid** (vitamin B_9). Path C differs from path A only in the use of glutamine-derived NH_3 as nucleophile instead of water in the initial 1,4-addition–elimination sequence. 1,2-Elimination of pyruvic acid again leads to aromatization and formation of **anthranilic acid** which, as we will see, plays a key role in **tryptophan** biosynthesis in a series of reactions that initially converts anthranilic acid to the simple *N*-heterocycle **indole**, as shown in Fig. 6.10.

The sequence begins with a nucleophilic displacement of diphosphate from the anomeric carbon of **5-phosphoribose diphosphate** by the anthranilic acid amino group nitrogen to give an *N*-glycoside (i). Ring-opening protonation of the furan ring oxygen atom of (i) leads to an imine intermediate (ii) which may be envisioned

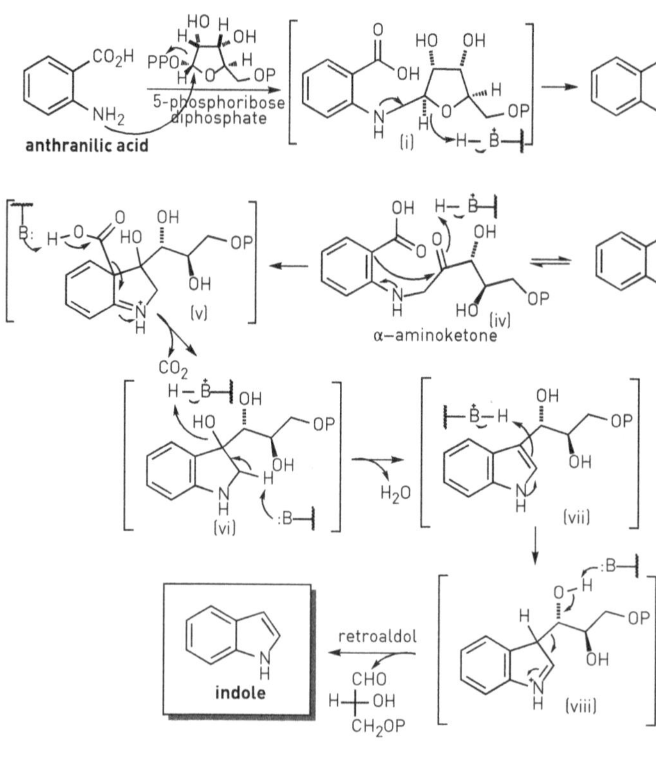

FIGURE 6.10
Conversion of anthranilic acid to indole.

as undergoing a series of tautomerizations, imine-to-enamine (iii), with the resulting enamine also serving as an enol due to the hydroxyl group present on the enamine double bond. This enol (iii) in turn can give rise to an enol-to-ketone tautomerization ultimately leading to an α-aminoketone intermediate (iv). To form the five-membered ring required for the indole nucleus, a subsequent decarboxylative ring-forming reaction may be viewed as involving an intramolecular enamine-type aldol reaction with nucleophilic attack by the C-6 ring "enamine" of (iv) at the adjacent ketone function to give cyclic intermediate (v). This is followed by decarboxylation of (v) to restore the aromaticity of the C-6 ring in intermediate (vi) and then by a simple dehydration to give the 3-substituted indole aromatic heterocycle (vii) as shown. Finally, protonation at the 3-position of (vii) gives an iminium ion (viii) which may be viewed as undergoing what is formally a retroaldol reaction to produce the parent **indole** structure with concurrent loss of the three carbon side chain as **3-phosphoglyceraldehyde**, a familiar intermediate previously seen in glucose metabolism (Fig. 3.25).

Next, we consider the final conversion of indole to tryptophan as shown in Fig. 6.11. This sequence begins with condensation of the amino acid serine with PLP (protonated form) to give the usual intermediate as shown. The attached PLP system acts essentially as an electron withdrawing function which facilitates a dehydration reaction (i.e., a 1,2-elimination) to give the unsaturated derivative which then undergoes an enamine-type nucleophilic attack by the indole C-5 ring to give the required C–C bond formation at position 3. Proton loss for restoration of the indole ring aromaticity is accompanied by protonation at the α-position of the carboxylic acid to give the familiar α-amino acid-PLP derivative. This imine linkage is essentially the same PLP derivative that would be observed in transamination of

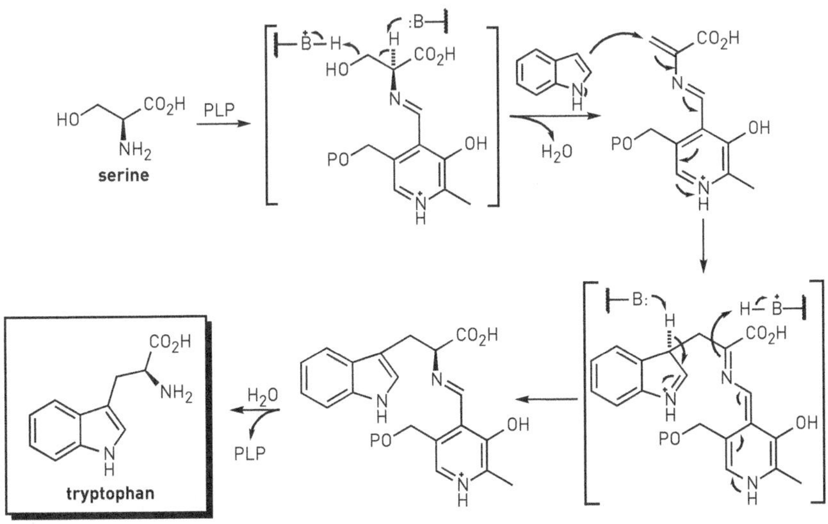

FIGURE 6.11
Final conversion of indole to tryptophan via serine and PLP.

tryptophan itself. Thus, subsequent hydrolysis of the imine linkage regenerates PLP and releases **L-tryptophan** as shown.

We will see a number of uses of this essential amino acid as a building block for alkaloid biosynthesis and for the production of neurotransmitters, neurohormones, and other important biochemicals when we consider alkaloid biosynthesis in the following chapter. But for now, we must return to consideration of the many fates that await the amino acids phenylalanine and tyrosine, as these are the predominant building blocks used for the construction of a remarkably diverse array of biologically and medicinally important secondary metabolite compounds that are produced by the shikimic acid pathway.

SHIKIMIC ACID PATHWAY PRODUCTS FROM PHENYLALANINE AND TYROSINE: AN OVERVIEW

A simplified overview of the pathway and how its intermediates are partitioned into various products is shown in Fig. 6.12. Aside from the alkaloid products that will be treated separately in the next chapter, we will see that both phenylalanine and tyrosine will undergo modifications to afford various **phenyl C_3 compounds** via reductions, deaminations, oxidations, and other transformations. Some phenyl C_3

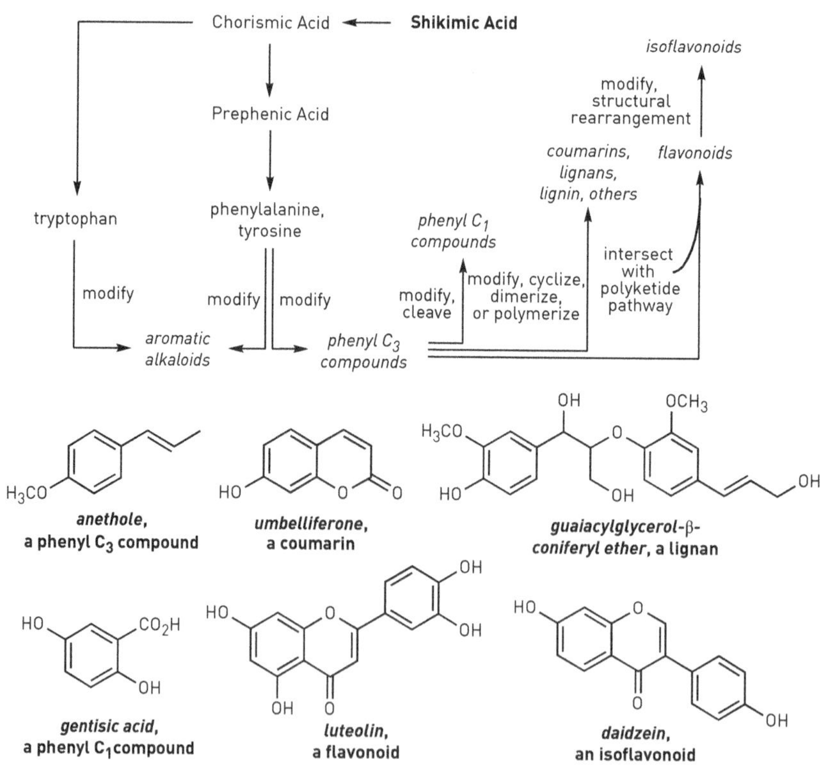

FIGURE 6.12
A brief overview of some product classes derived via the shikimic acid pathway.

compounds may then undergo further modifications to give cyclic products such as **coumarins**. Phenolic oxidative **dimerization** of certain phenyl C_3 compounds can lead to a wide variety of so-called **lignan compounds**, while their oxidative polymerization yields an abundant plant structural material known as **lignin**. Phenyl C_3 compounds may also undergo various modifications followed by oxidative cleavage to yield a variety of simpler **phenyl C_1 compounds**, while others may be employed as unique starter units for polyketide chain extensions, modifications, and cyclizations to yield compounds known as **flavonoids**, some of which may then be structurally rearranged to yield the so-called **isoflavonoids**. While other classes of compounds are known, we will focus primarily on those shown here since these include the majority of the most important products produced by this pathway.

PHENYLPROPANOIDS: A LARGE FAMILY OF PHENYL C_3 COMPOUNDS

As we will see, many of the products from the shikimic acid pathway are so-called **phenyl C_3 compounds** or **phenylpropanoids** which are derived from phenylalanine or tyrosine. The initial phenylpropanoid which serves as the building block for many others is ***trans*-cinnamic acid**, a product obtained via a 1,2-elimination (deamination) of the elements of NH_3 from L-phenylalanine as shown in Fig. 6.13. This *anti* elimination process is catalyzed by the enzyme **phenylalanine ammonia-lyase (PAL)**.

Figure 6.14 outlines the steps involved in biosynthesis of four other important shikimic acid pathway products, namely **4-coumaric acid, caffeic acid, ferrulic acid**, and **sinapic acid**. These compounds are sometimes referred to as cinnamic acid derivatives since they all share the same *trans*-1-phenylpropenoic acid structural core and are all derived from *trans*-cinnamic acid via a series of simple oxidations and SAM methylations (4-coumaric acid is also known as **4-hydroxycinnamic acid**). Note that while the main substrate for the PAL enzyme system is phenylalanine, it may also be employed for direct production of 4-coumaric acid via deamination of L-tyrosine, though the primary route to this product is via direct oxidation of *trans*-cinnamic acid.

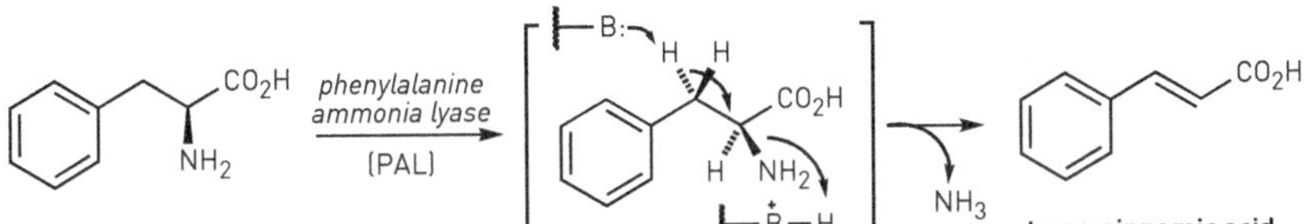

FIGURE 6.13
Biosynthesis of *trans*-cinnamic acid from phenylalanine.

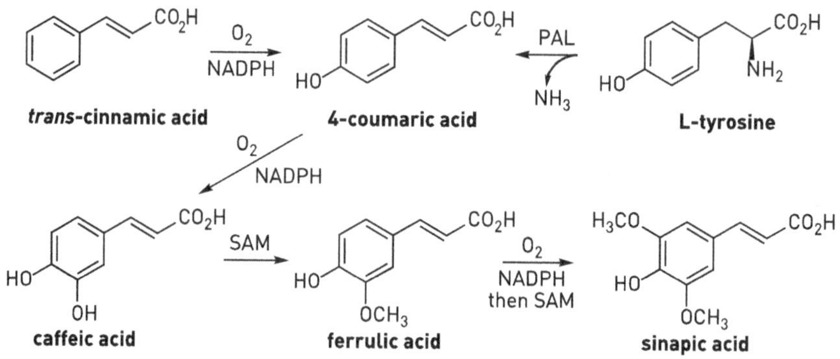

FIGURE 6.14
Biosynthesis of coumaric, caffeic, ferulic, and sinapic acids.

PHENYLPROPANOIDS: REDUCTION OF ACIDS TO PHENYL C_3 ALDEHYDES AND ALCOHOLS

In addition to simple esters such as ethyl cinnamate, plants produce a wide variety of other phenyl C_3 acid derivatives, many of which involve simple reduction of the carboxylic acid moiety. The transformations involved are illustrated in Fig. 6.15 for the conversion of *trans*-cinnamic acid into **cinnamaldehyde** and **cinnamyl alcohol**. The same sequence of reactions may be applied to the other phenyl C_3 acids to produce the corresponding aldehydes and alcohols as shown. Cinnamaldehyde is the organic compound responsible for the familiar flavor and fragrance of cinnamon which may be obtained from the bark of the cinnamon tree. It is also widely used as a fungicide in agriculture due to its relatively low toxicity.

Conversion of these acids to their aldehydes is achieved via NAPH reduction of the corresponding CoA thioesters to a thiohemiacetal which dissociates to give the aldehyde and HSCoA. Subsequent conversion of the aldehydes to the alcohols is likewise via the usual NADPH reductions. These alcohols are sometimes referred to as **monolignols**, since they serve as the monomers used in production of the plant biopolymer **lignin** which we will discuss shortly.

REDUCTION OF PHENYL C_3 ALCOHOLS TO PHENYLPROPENES

Phenylpropenes are widely distributed plant aromatics derived from phenyl C_3 alcohols (lignols) and are components of many different essential oils. All share the same basic phenylpropene (allyl benzene) core and differ mainly in the degree, position, and/or alkylation pattern of ring oxygens. Examples include **myristicin** from nutmeg oil, **estragole** found in basil and pepper, and its isomer **anethole**, isolated from anise and fennel and responsible for the familiar flavor and fragrance of licorice (Fig. 6.16).

Evidence suggests that in some instances, the NADPH reduction of lignols to phenylpropenes proceeds via initial conversion of the lignol to its acetate ester,

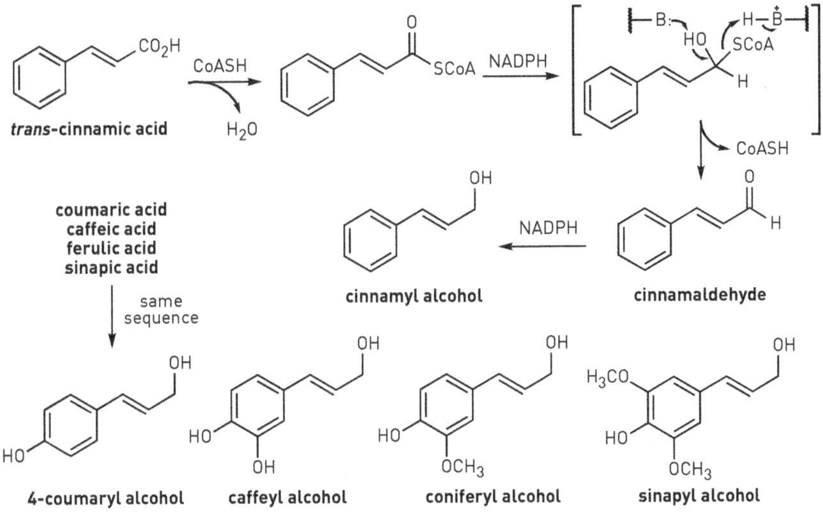

FIGURE 6.15
Reductions of cinnamic acids to their corresponding aldehydes and alcohols.

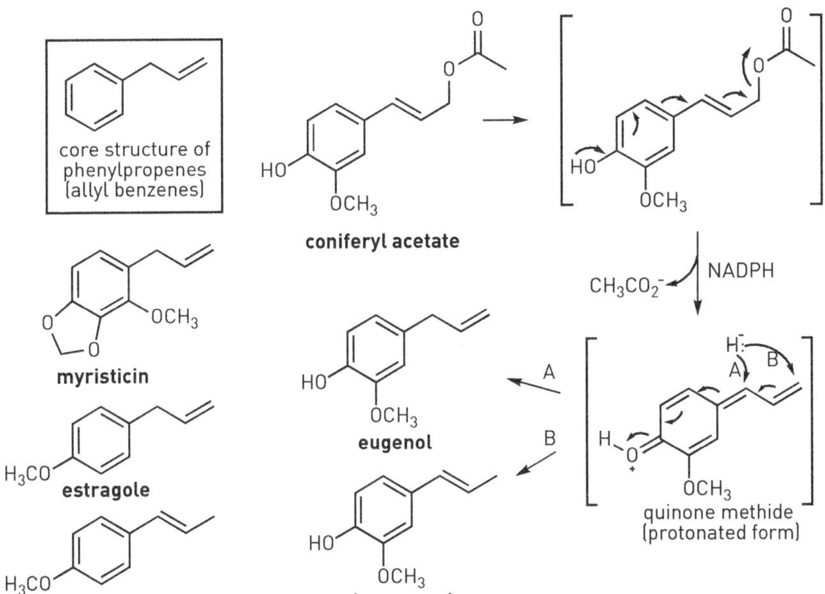

FIGURE 6.16
Some common phenylpropenes and conversion of coniferyl acetate to eugenol or isoeugenol.

followed by loss of acetate to give a **quinone methide** intermediate. Conjugate delivery of hydride from NADPH may occur via either path A or path B as shown in Fig. 6.16 for reduction of **coniferyl acetate**. Path A leads to the usual allyl benzene derivative, in this case **eugenol**, while path B affords the corresponding conjugated isomer **isoeugenol** which accompanies eugenol isolated from sources such as oil of clove or oil of nutmeg. Other phenylpropenes may arise by direct hydride displacement of lignol acetate or alcohol functions (in protonated form) rather than proceeding via the quinone methide mechanism.

LIGNANS AND LIGNIN: OXIDATIVE PHENOLIC COUPLING WITH A TWIST

We saw in the previous chapter an example of bimolecular oxidative phenolic coupling in the biosynthesis of **usnic acid** (Fig. 5.46). We now revisit this type of oxidative C–C bond-forming reaction within the context of the chemistry of the phenolic lignols we have just examined. Figure 6.17 shows how the initially formed phenoxy O-radical from any of the three indicated lignols leads to the usual *ortho* and *para* C-radical resonance forms, but in the case of the lignols, the propenyl side chain of the oxidized phenol permits further electron delocalization to yield an additional contributor referred to here as the *allyl* C-radical resonance form as shown.

These radicals may be coupled together in several different combinations to form C–C bonds. The resulting products are examples of **lignans**—essentially lignol dimers—that are widely distributed in nature and have many interesting properties. Several lignans have been found to act as **phytoestrogens** (compounds that can mimic the hormonal activity of estrogen) and are currently being studied as potential agents for cancer prevention, especially breast cancer.

CONIFERYL ALCOHOL OXIDATIVE COUPLING: *ALLYL* C-RADICAL + *ALLYL* C-RADICAL

The term "lignan" is sometimes specifically applied to products arising from lignol oxidative coupling at the central carbon of the phenyl C_3 side chain. When coupling

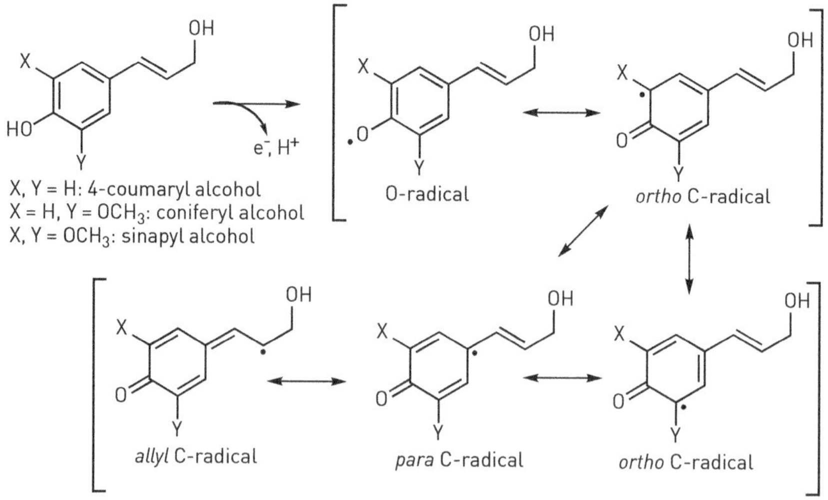

FIGURE 6.17
Resonance forms of lignol phenoxy radicals.

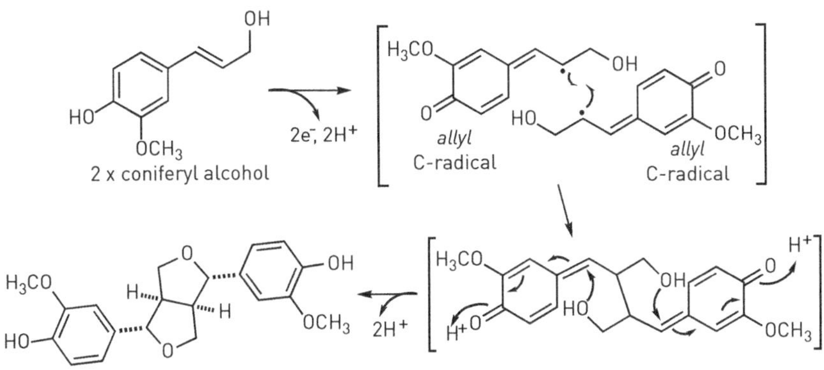

FIGURE 6.18
Biosynthesis of (+)-pinoresinol via allyl C-radical/allyl C-radical coupling.

follows some other combination of lignol radicals, the products may be referred to as **neolignans.** The distinction is not a terribly significant one, so we will stick with referring to all such products as lignans while pointing out examples of products for which the neolignan designation might be used.

One example of an important lignan arising via what we term *allyl* C-radical/ *allyl* C-radical coupling is the formation of **(+)-pinoresinol** from coniferyl alcohol as outlined in Fig. 6.18. In this scheme, C–C coupling of the radicals leads to two highly reactive intermediate quinone methide structures which then undergo intramolecular nucleophilic conjugate addition of the adjacent hydroxyl oxygens. This results in rearomatization of the phenolic rings accompanied by formation of the fused bicyclic tetrahydrofuran ring system shown. As we will see shortly, this lignan serves as the starting material for subsequent biosynthetic modifications leading to formation of the important cancer chemotherapy agent, **podophyllotoxin.**

CONIFERYL ALCOHOL OXIDATIVE COUPLING: *ORTHO* C-RADICAL + *ALLYL* C-RADICAL

Oxidative C–C dimerization of coniferyl alcohol via *ortho* C-radicals and *allyl* C-radicals leads to another quinone methide-type intermediate which undergoes subsequent aromatization via the intramolecular conjugate addition of an adjacent phenolic alcohol function to yield **(−)-dehydrodiconiferyl alcohol** as shown in Fig. 6.19. This 2-aryl dihydrobenzofuran compound and its many structurally related derivatives are promising lead structures for pharmaceutical research, having shown a broad spectrum of biological and pharmacological activities such as antioxidant, antitumor, antimicrobial, and antiHIV activities among others.

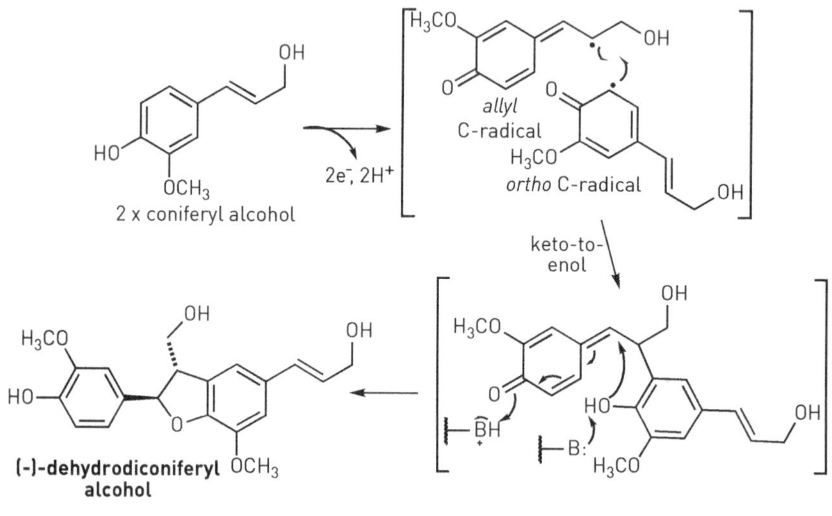

FIGURE 6.19
Biosynthesis of (−)-dehydrodiconiferyl alcohol via ortho C-radical/allyl C-radical coupling.

CONIFERYL ALCOHOL OXIDATIVE COUPLING: O-RADICAL + *ALLYL* C-RADICAL

As shown in Fig. 6.20, a different combination of coniferyl alcohol radicals can lead to formation of a C–O ether linkage rather than a C–C linkage, yielding yet another quinone methide intermediate that in this case is quenched and aromatized via conjugate addition of a water molecule to yield the lignan **guaiacylglycerol-β-coniferyl ether**. Unlike the other lignans, the stereochemistry at the chirality centers of this compound is usually revealed as a mixture of *erythro* and *threo* diastereomers and like dehydrodiconiferyl alcohol, this lignan and its derivatives continue to be evaluated as potential pharmaceutical lead compounds due a spectrum of biological activities.

LIGNIN: A PLANT POLYMER AND MAJOR SOURCE OF CARBON

As we have seen, radical combinations from oxidized lignols follow different paths of dimerization to yield some interesting organic compounds. Of course, other radical combinations from lignols are also possible and a variety of such linkages are involved in the formation of **lignin**, a plant biopolymer with structural segments closely related to the lignan structures we have already seen. A hypothetical structural segment of lignan is shown in Fig. 6.21 in which some of the individual linkages between lignol monomer units are readily identifiable. The detailed structure of lignin itself is difficult to specify, since it may vary from source to source and may also vary depending on techniques employed in its isolation or degradation.

Lignin is the second most abundant biopolymer after cellulose, making up about 30% of the dry mass of wood, the remainder being mostly cellulose along

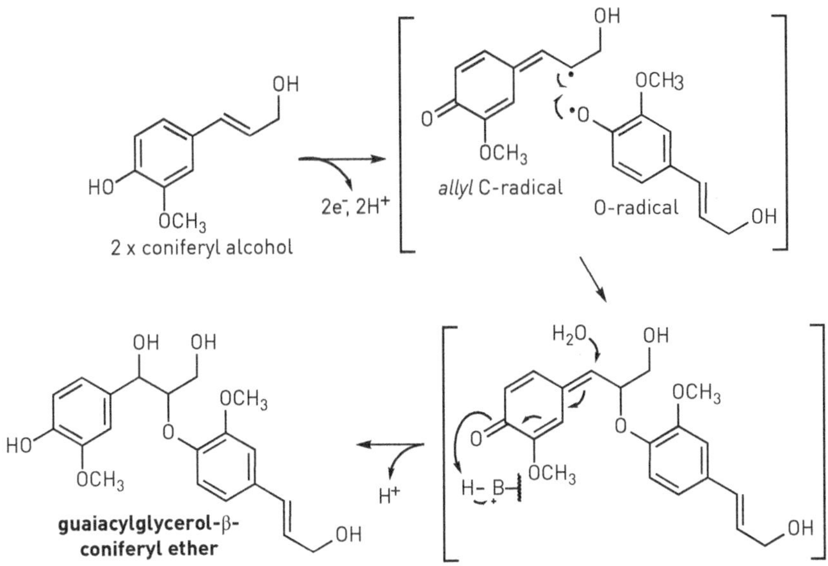

FIGURE 6.20
Biosynthesis of guaiacylglycerol-β-coniferyl ether via O-radical/allyl C-radical coupling.

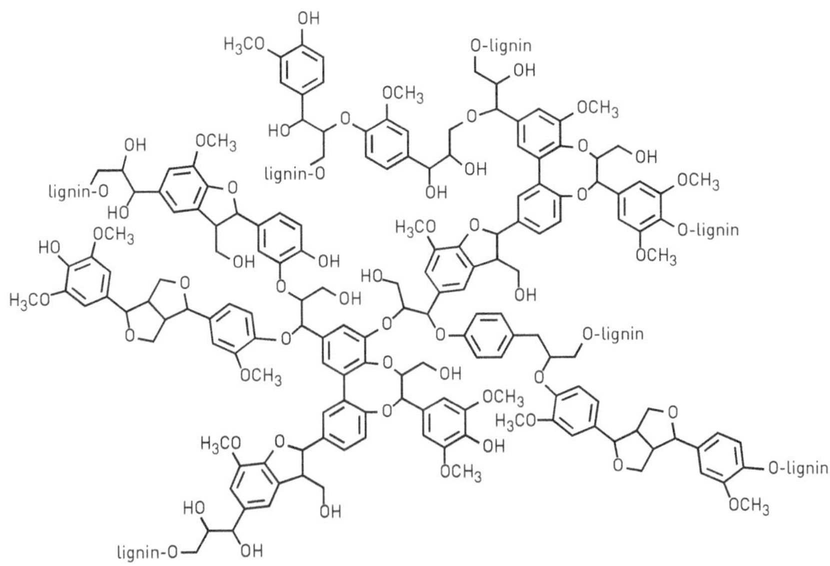

FIGURE 6.21
Hypothetical structural segment of the biopolymer lignin.

with smaller amounts of resins containing lower molecular weight organics such as terpenes. Unlike cellulose and other biopolymers, lignin has an irregular, nearly random structure. The three principal lignol monomers (**monolignols**), *p*-coumaryl, sinapyl, and coniferyl alcohols, are connected to one another via linkages that arise via oxidative coupling processes similar to those utilized in the formation of lignans. Unlike lignans however, whose structures are enzyme-specific and therefore enantiomerically pure, lignin stereocenters are racemic. So while the basic building blocks for lignans and lignin are the same, lignin itself is not built up from preformed lignans but rather from a stereorandom polymerization of monolignols. Given its structural elements, one would expect lignin to be a tough, durable polymer that would be difficult to break down. Certainly, wood owes much of its usefulness as a renewable source of building materials and fuels to its high lignin content.

PODOPHYLLOTOXIN BIOSYNTHESIS: ARYLTETRALIN LIGNANS FROM THE AMERICAN MAYAPPLE

We now turn to a brief examination of an important class of bioactive lignans known as the **aryltetralins.** The term is derived from *tetralin*, the common name for the familiar fused bicyclic arylcyclohexyl ring system which forms the core of these lignan structures. They are produced in significant quantities by plants of the genus *Podophyllum,* among which is the American mayapple, *Podophyllum peltatum* (also known as mandrake), which grows abundantly in eastern North America and produces an extremely important aryltetralin known as **podophyllotoxin.** Its principal biological activity involves the blocking of cell mitosis (cell division); it is widely used in topically applied creams for the treatment of certain viral warts.

The biosynthetic process for podophyllotoxin (Fig. 6.22) begins with sequential NADPH reductions of the fused tetrahydrofuran rings of the lignan **(+)-pinoresinol** whose structure and synthesis we described earlier. The resulting tandem reductions lead to the diol **(–)-secoisolariciresinol.** Selective $NADP^+$ oxidation of one of the alcohol functions leads to the corresponding aldehyde which forms a cyclic hemiacetal or **lactol.** Further $NADP^+$ oxidation of the lactol yields the corresponding lactone **(–)-matairesinol**, a key intermediate product which undergoes a series of aryl ring oxidations and SAM methylations proceeding through the intermediate compounds **thujaplicatin** (see structure for position numbering) and **4',5'-dimethylthujaplicatin** whose *ortho*-methoxyphenol moiety then undergoes standard oxidative conversion to a methylenedioxy unit, yielding **(–)-yatein.** The precise details associated with the subsequent oxidative cyclization of yatein are not fully known, but an intramolecular electrophilic aromatic substitution is most likely involved, leading to **(–)-deoxypodophyllotoxin** which is then finally hydroxylated at the 4-position of the cyclohexane ring to yield **(–)-podophyllotoxin** as shown.

Most significantly, podophyllotoxin is used as the starting material for production of the semisynthetic cancer chemotherapy drugs **etoposide** and **teniposide** (Fig. 6.23). While podophyllotoxin acts as a mitosis inhibitor, etoposide and teniposide have an entirely different mode of action, interfering with the action of the enzyme *topoisomerase II* which assists in the unwinding and reconnection of DNA strands.

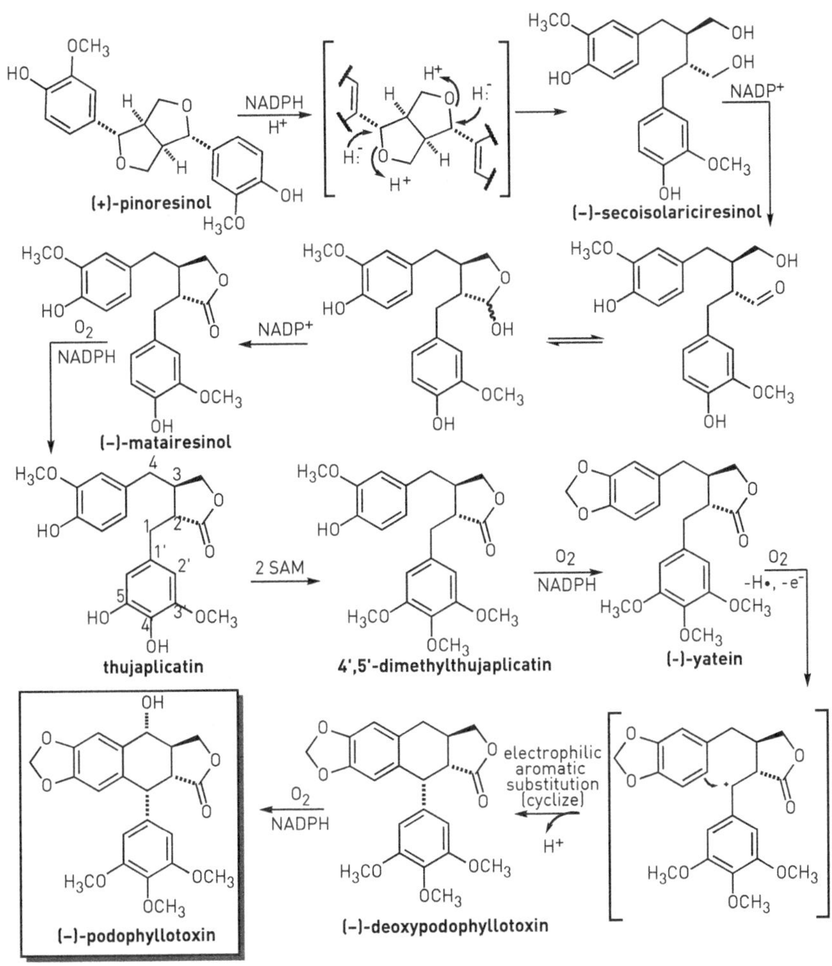

FIGURE 6.22
Biosynthesis of (–)-podophyllotoxin from (+)-pinoresinol.

These important chemotherapy agents have seen wide clinical use for the treatment of small and large cell lung, testicular, pancreatic, and stomach cancers, as well as in the treatment of myeloid leukemias, lymphomas, and Kaposi's sarcoma. The key step in the production of these drugs is conversion of podophyllotoxin to **4'-demethylepipodophyllotoxin**. This is achieved synthetically by demethylation of podophyllotoxin's 4'-methoxyl group and epimerization at the 4-hydroxyl group position via treatment with gaseous HBr followed by aqueous $BaCO_3$; the remaining steps involve protection of the phenolic hydroxyl group, formation of a specific glycoside linkage at the 4-hydroxyl group and final deprotection back to the free 4'-phenol. Ultimately, epimerization at the 4'-position has been shown to be responsible for the remarkable shift in mode of action from mitosis inhibition for podophyllotoxin to *topoisomerase II* inhibition for both etoposide and teniposide.

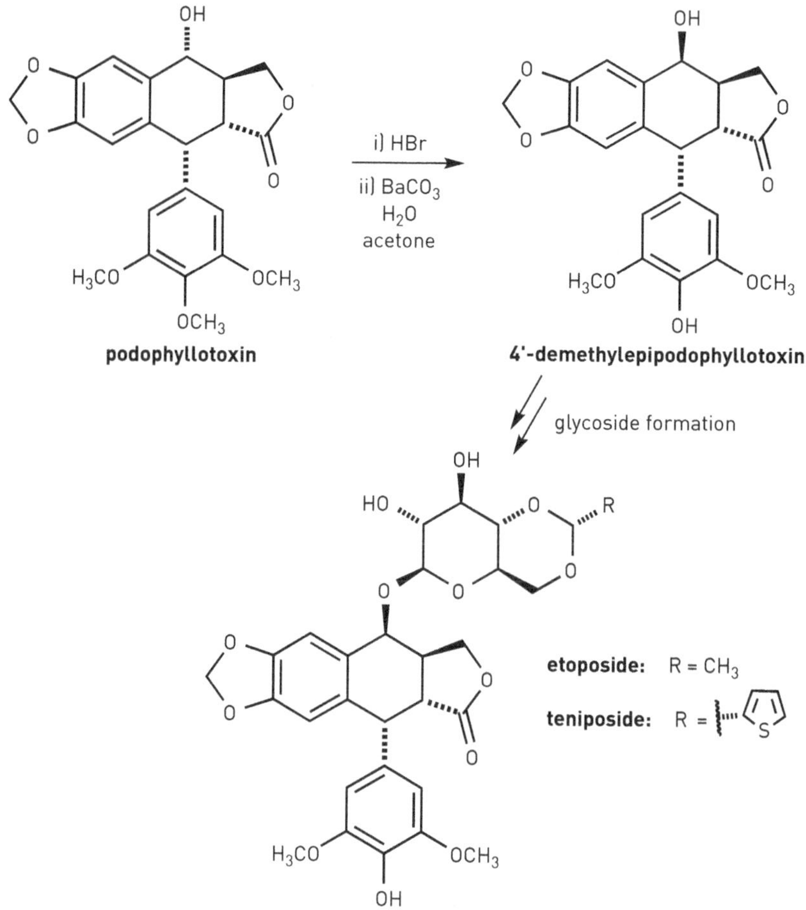

FIGURE 6.23

Conversion of podophyllotoxin to 4'-demethylepipodophyllotoxin, etoposide, and teniposide.

CLEAVAGE OF CINNAMIC ACIDS TO PHENYL C_1 COMPOUNDS: DIFFERENT ROUTES, SIMILAR OUTCOMES

In addition to the formation of phenyl propenes and other phenyl C_3 compounds, the cinnamic acid group also serves as a source of simple phenyl C_1 compounds that may be obtained by oxidative cleavage of propenoic acid double bond. One commonly cited cleavage sequence is illustrated in Fig. 6.24 for biosynthesis of both **vanillin**, the compound from vanilla bean extract responsible for vanilla flavor and fragrance, and **vanillic acid**.

This cleavage mode proceeds via the corresponding CoA thioester derivatives which undergo conjugate addition of water, $NADP^+$ oxidation to the β-ketothioester then, in a sequence analogous to the β-oxidation process involved in fatty acid metabolism, addition of CoASH to the ketone function to give a hemithioacetal which then undergoes a retro-Claisen condensation. The resulting phenyl C_1 thioesters

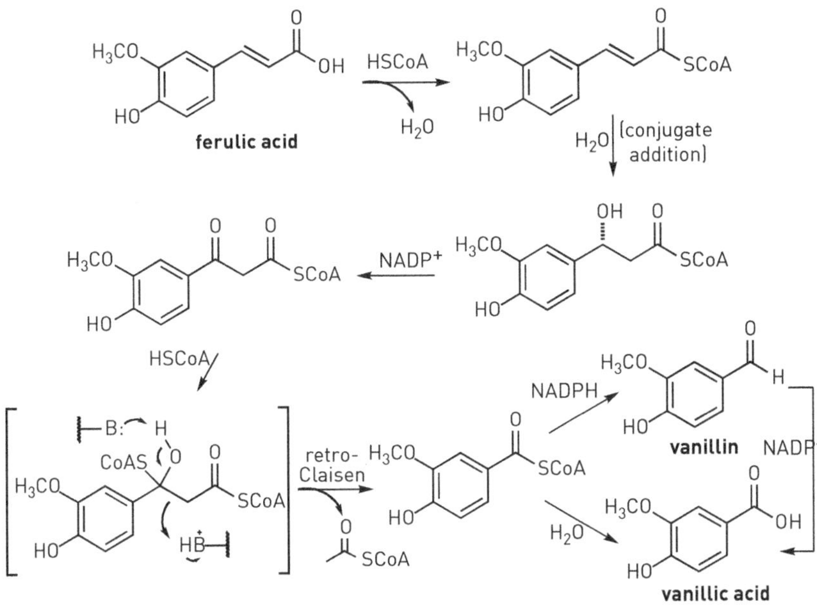

FIGURE 6.24
Proposed routes for cleavage of ferulic acid to vanillin and vanillic acid.

may then undergo NADPH reduction to the corresponding aldehydes or simple hydrolysis to yield the acids as shown in Fig. 6.24, though these acids may also be formed via oxidation of the aldehydes themselves.

A second, often cited, cleavage mechanism also involves conjugate addition of water to the cinnamic acids, but does not involve CoA thioester derivatives and requires a retroaldol rather than a retro-Claisen condensation. Two alternative routes utilizing this sequence are illustrated in Fig. 6.25 for the conversion of cinnamic acid to **salicylic acid** from which **aspirin** (acetyl salicylic acid) is commercially prepared by treatment with acetic anhydride. Biosynthetically, SAM methylation of salicylic acid leads to **methyl salicylate** (wintergreen oil) which serves as both a natural flavor and a fragrance additive for candies and gums and as well as a "deep heating" agent and analgesic used in various topically applied creams and liniments.

COUMARINS: SWEET-SMELLING BENZOPYRONES

Another group of compounds derived from cinnamic acids may be formed when the aryl ring position *ortho* to the propenoic acid side chain is oxidatively hydroxylated. As shown in Fig. 6.26, such *ortho*-hydroxylation of cinnamic acid leads to **2-coumaric acid** as the usual (*E*) isomer. However, conjugated π-bonds such as these are susceptible to reversible photochemical isomerization (by ambient sunlight in plants), and when such isomerization occurs for 2-coumaric acid, the *Z*-isomer may then undergo irreversible intramolecular condensation to yield the corresponding lactone known as **coumarin**, which is an example of a **benzo-2-pyrone** (the

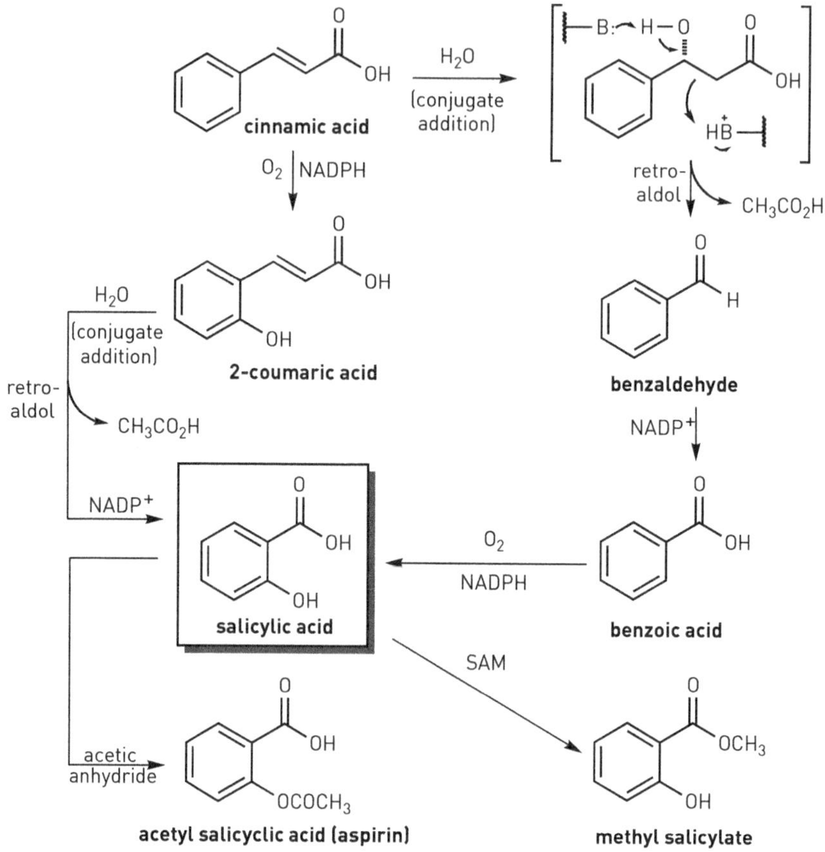

FIGURE 6.25
Conversions of cinnamic acid and 2-coumaric acid to salicylic acid and derivatives.

carbonyl carbon is at the 2-position relative to the ring oxygen). Derivatives with the same basic ring system are often referred to generically as substituted coumarins. Coumarin itself, produced by many different kinds of plants, is responsible for the familiar sweet smell of newly mown hay and is thought to be produced by plants as a defense chemical, acting as deterrent to grazing animals due to its bitter taste. A similar sequence for 4-coumaric acid leads to another coumarin known as **umbelliferone**, a powerful UV-absorbing compound found in plants such as coriander and carrots and which is used in some sunscreen preparations. It also serves as the biosynthetic precursor to a variety of substituted derivatives with interesting biological properties (see Problem 6.4).

When moldy grasses are acted upon by certain fungi, the presence of coumarin can lead to formation of **4-hydroxycoumarin** (Fig. 6.27) via conjugate addition of water to 2-coumarylCoA followed by NADP$^+$ oxidation and cyclization. A subsequent aldol condensation in the presence of formaldehyde can then produce an unsaturated cyclic β-ketoester which undergoes conjugate addition by the enolate of a second molecule of 4-hydroxycoumarin, producing a compound that is formally two 4-hydroxycoumarin units linked together by a bridging methylene unit. This

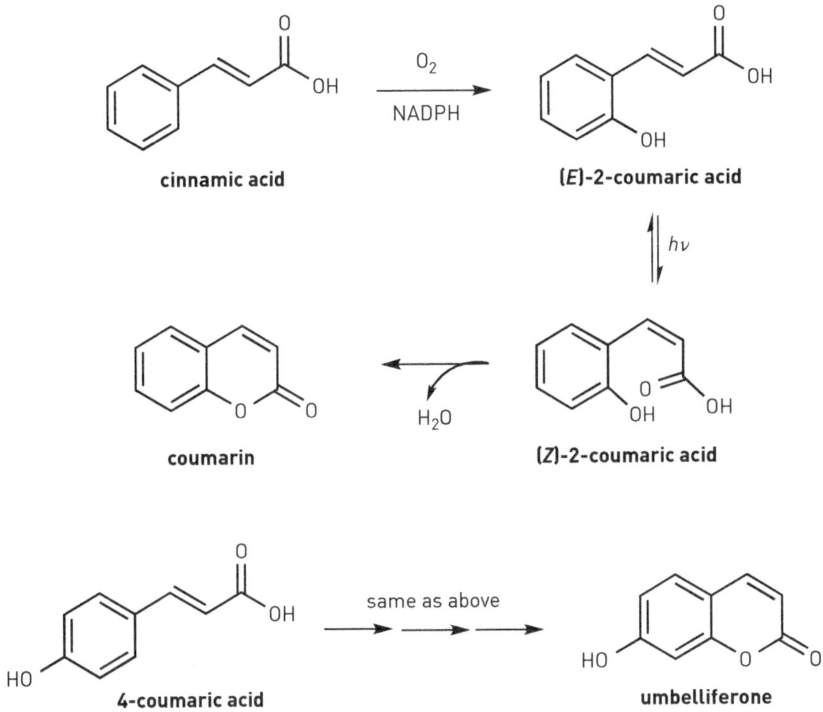

FIGURE 6.26
Formation of coumarins from cinnamic acid or 4-coumaric acid.

compound is known as **dicoumarol**, an important anticoagulant agent that acts by depleting levels of vitamin K (clotting factor) in the blood. The presence of dicoumarol in contaminated sweet clover or hay was found to be responsible for many historical cases of fatal internal bleeding in grazing cattle; this discovery led to extensive research into its mode of action and ultimately to its use as an anticoagulant for the treatment of deep vein thrombosis in humans, though it has since been largely displaced by the use of heparin in conjunction with various synthetic derivatives of 4-hydroxycoumarin such as **warfarin** (trade name **Coumadin**). It should be noted that coumarin itself has no anticoagulant activity.

MIXED PRODUCTS: COMBINING THE SHIKIMATE, POLYKETIDE, AND TERPENOID PATHWAYS

We saw in the previous chapter an example of the use of *para*-hydroxycinnamic acid (4-coumaric acid) as the starter unit for a polyketide chain extension, cyclization, aromatization, and decarboxylation sequence in the production of the mixed polyphenolic stilbene (1,2-diphenylethylene) derivative **resveratrol** (see Fig. 5.55). As it turns out, a remarkably diverse array of secondary metabolites make use of shikimic acid pathway products as starter units for familiar polyketide construction sequences and further structural elaboration via terpenoid building blocks. So we are now in

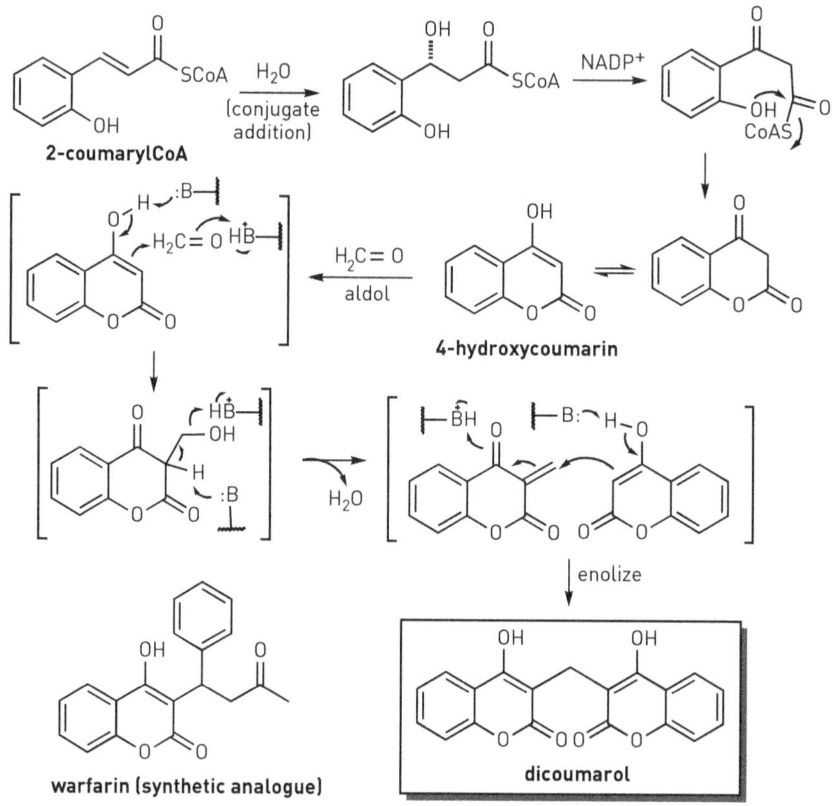

FIGURE 6.27

Formation of dicoumarol from 4-hydroxycoumarin; structure of warfarin.

a position to examine a number of such products in light of what we have already learned so far about the shikimate, polyketide, and terpenoid biosynthetic pathways.

KAVALACTONES: NATURAL SEDATIVES FROM THE SOUTH PACIFIC

Kava is the common name for the western Pacific pepper plant, *Piper methysticum*, the roots of which are extracted to produce an intoxicating or sedating beverage of the same name (kava or kava-kava) consumed by much of the native island populations in the South Pacific. The active ingredients in these extracts are members of a group of compounds known as **styrylpyrones** (styrene + pyrone), structurally similar to stilbenes but with a pyrone ring in place of one of the aryl rings. Several of these so-called **kavalactones** (or kavapyrones) are shown in Fig. 6.28. While many examples are known, the main structural types are represented by **yangonin, kavain** (which has a reduced pyrone ring), **dihydrokavain** (with reduced pyrone and styryl groups), and **methysticin** (with a methylenedioxy-substituted aromatic ring).

A typical biosynthetic scheme is shown in Fig. 6.29. 4-Coumaric acid undergoes two cycles of malonyl-CoA chain extension followed by intramolecular cyclization

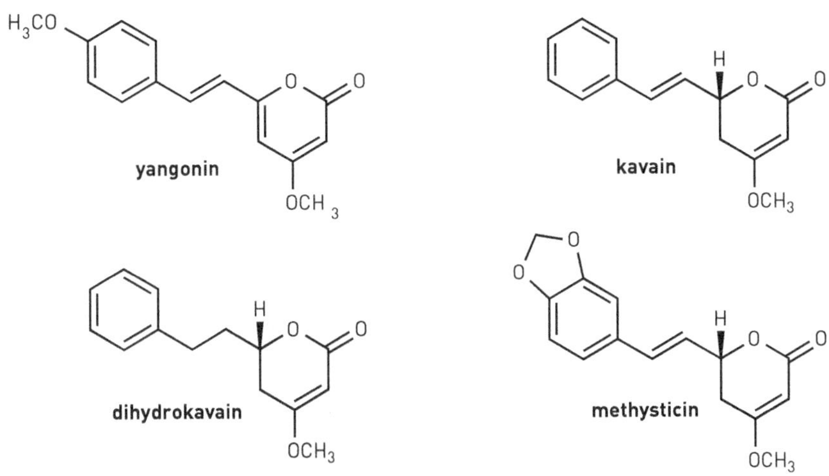

FIGURE 6.28
Representative kavalactone structures: yangonin, kavain, dihydrokavain, and methysticin.

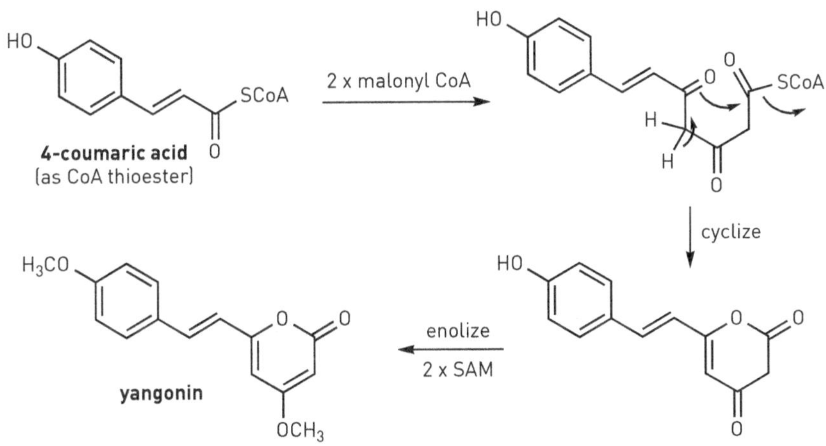

FIGURE 6.29
Biosynthesis of the kavalactone yangonin, a mixed pathway natural product.

to give the pyrone structure; this is followed by enolization of the pyrone ketone carbonyl and SAM methylation of the enol and phenolic hydroxyl groups to afford **yangonin.**

The effects of these secondary metabolites are similar to those associated with antianxiety benzodiazepines such as Xanax, Valium, or Librium. However, kavalactones are currently regulated or banned in many countries due to reports of significant liver damage in some users. Interestingly, recent data has suggested that dietary supplements, many of which are herbal extracts, now account for nearly 20% of all drug-related liver injuries seen in U.S. hospitals, raising new concerns about the lack of regulation of such compounds for use as dietary supplements or in alternative medicine. In some instances, "natural" products can lead to adverse interactions with prescription medications and should always be used with caution and under the supervision of a physician.

FLAVONOIDS: STRUCTURALLY DIVERSE PLANT POLYPHENOLICS

The flavonoids constitute a very large class of phenolic natural products derived mainly from condensation of 4-coumarylCoA and malonate units. Over 4500 examples have been identified up to the present time in a variety of forms representing some of the most abundant of all plant-derived polyphenolics. Many contribute to plant and flower coloration which can serve a role in the attraction of pollinating insects, while others may act as bitter-tasting antifeedants, antifungal, or antibiotic agents, or even as UV absorbers. Given their remarkable abundance, flavonoids likely have other properties or play other roles in plant biochemistry that are yet to be discovered which makes them obvious targets for investigation as potential medicinal agents. The diversity of principal flavonoid structural types and their biosynthetic relationships are illustrated in Fig. 6.30.

Many flavonoids have already been intensively studied in terms of their potential as antiviral, antibiotic, and anticancer agents. For instance, some studies suggest that the lower incidence of both breast and prostate cancers in vegetarians may be related

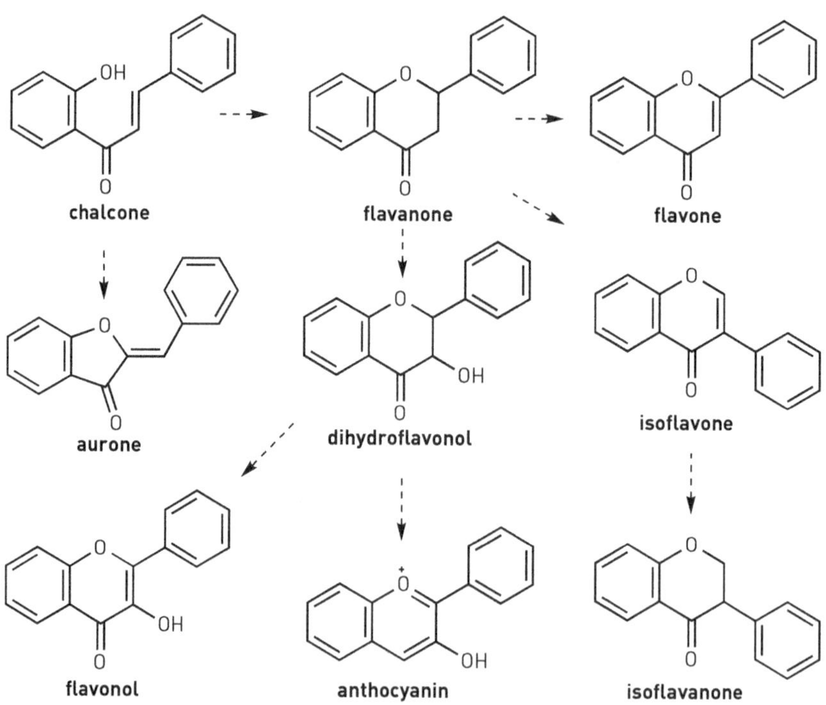

FIGURE 6.30
Principal structural types and relationships in the flavonoid series.

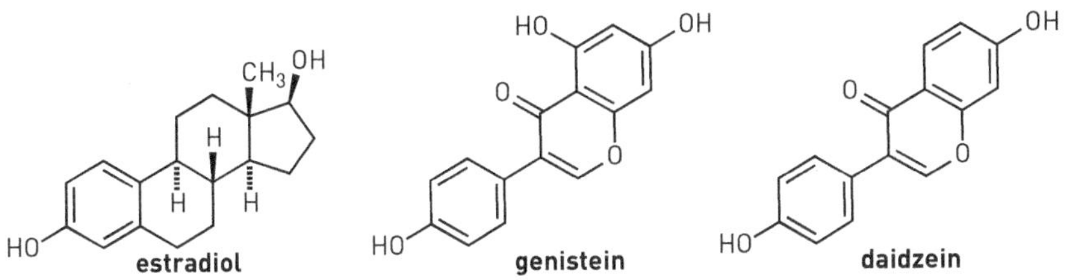

FIGURE 6.31
Structural similarities of estradiol with the phytoestrogens genistein and daidzein.

to consumption of dietary isoflavonoids such as **daidzein** and **genistein.** These compounds, abundant in sources such as soybean, tofu, and fava bean, are two examples of plant products known as **phytoestrogens**, so-called because of their structural resemblance to the estrogen hormone **estradiol** and their ability to bind to an estrogen receptor site (Fig. 6.31). Clearly flavonoids and isoflavonoids are of significant medicinal potential, so the following sections will explore their biosynthetic origins and some further transformations.

THE CHALCONE-TO-FLAVANONE-TO-FLAVONE SEQUENCE: FORMATION OF APIGENIN

Given the vast number of known flavonoids and their structural diversity, we can only briefly consider representative examples from specific classes to help illustrate some of the general biosynthetic details involved in their production. To begin, **apigenin** will serve as a particularly interesting example of a flavone compound under active investigation due to its potentially valuable biological activity. Its biosynthesis is outlined in Fig. 6.32. The sequence again begins with a polyketide chain extension of 4-coumarylCoA with three units of malonyl-CoA. Note that the resulting polyketide is the same one employed in the biosynthesis of **resveratrol** (see Fig. 5.55), but instead of a folding pattern leading to cyclization via a Knoevenagel-type condensation, the **chalcone synthase**-mediated folding pattern leads to cyclization via a Claisen-type condensation, affording **apigenin chalcone**.

The chalcone then undergoes an intramolecular conjugate addition to the α,β-unsaturated ketone function by the adjacent phenolic hydroxyl group, leading to a flavanone derivative known as **naringenin.** Conversion of this flavanone to the flavone apigenin is accomplished via an O_2/2-oxoglutarate-type C–H to C–OH oxidation. The intermediate β-hydroxyketone thus formed then undergoes a dehydration to afford the α,β-unsaturated derivative, **apigenin**, a flavonoid found in many fruits and vegetables including celery and parsley. Like many other flavanones and flavanols, apigenin has been shown to be a potent inhibitor of **CYP2C9**, a P450 enzyme involved in oxidative metabolism of various pharmaceutical and medicinal agents in the body. It is also one of only a few compounds known to activate proteins responsible for the transport of

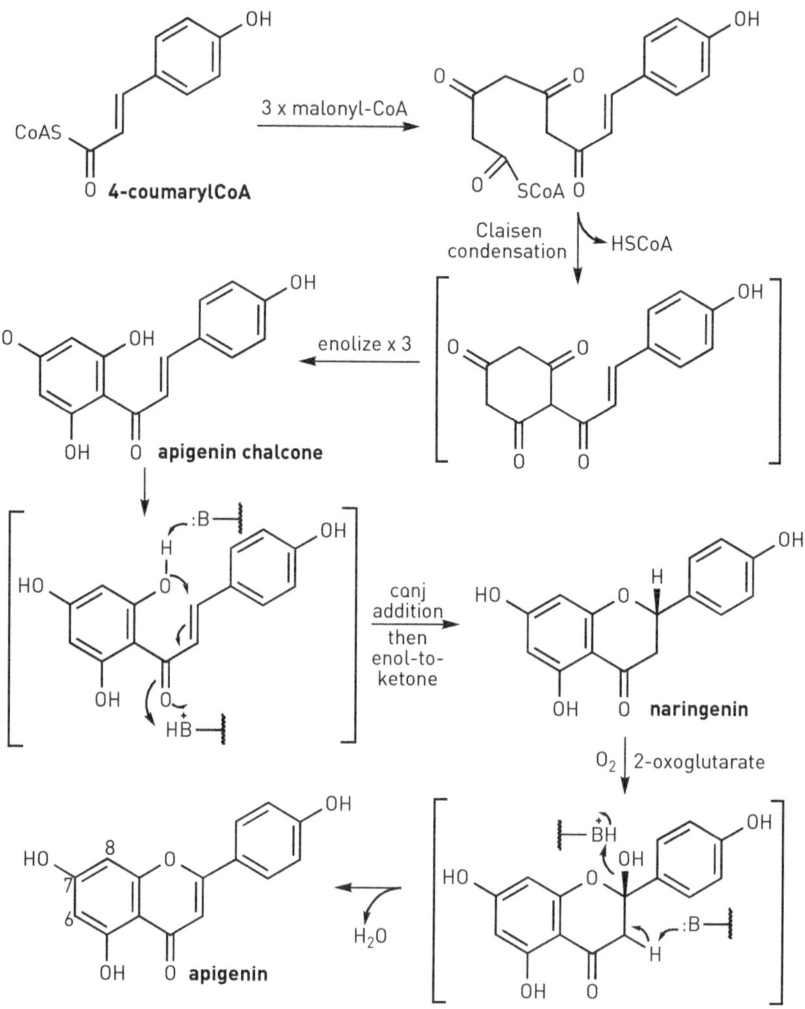

FIGURE 6.32
Formation of apigenin chalcone, naringenin, and apigenin.

monoamine neurotransmitters such as serotonin, dopamine, and norepinephrine. Structurally, apigenin also serves as the aglycone component of a number of known O-glycosidic derivatives such as **apigetrin** (apigenin 7-O-glucoside), and two unusual C-glucoside examples, **vitexin** (apigenin 8-C-glucoside) and **isovitexin** (apigenin 6-C-glucoside; see Fig. 6.32 for apigenin ring position numbering).

THE FLAVANONE-TO-DIHYDROFLAVONOL-TO-ANTHOCYANIN SEQUENCE: FORMATION OF PELARGONIDIN

Many flavonoids serve as color-producing components in various flowers and fruits. An example is the variety of shades of yellow or orange that can result

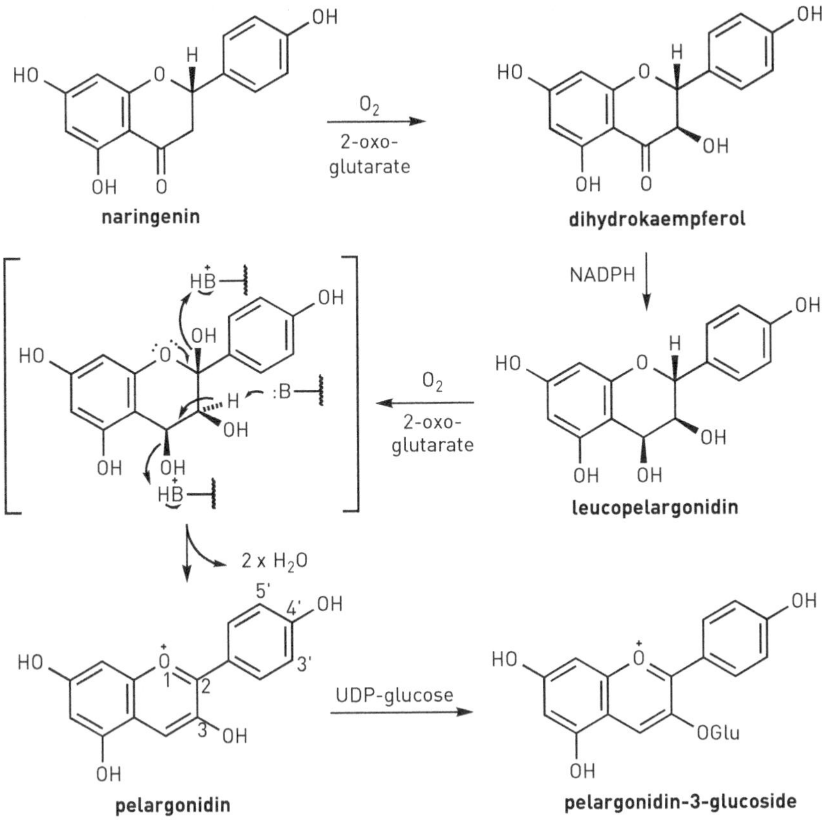

FIGURE 6.33
Formation of pelargonidin (an anthocyanidin) and its 3-glucoside (an anthocyanin).

from variations in substitution patterns present in a number of different flavones. Perhaps the most striking color-producing flavonoids are the **anthocyanidins**, an oxonium ion-containing group of flavonoids that come in virtually every color but green. The biosynthesis of **pelargonidin**, a salmon-pink anthocyanidin, is outlined in Fig. 6.33.

The sequence begins with a 2-oxoglutarate-dependent C–H to C–OH oxidation of the saturated ring of **naringenin** to give **dihydrokaempferol** whose ketone function undergoes NADPH reduction to afford **leucopelargonidin.** This compound also undergoes a 2-oxoglutarate-dependent C–H to C–OH oxidation which is followed by sequential dehydrations to yield the **anthocyanidin** derivative **pelargonidin**. Glucosylation at the 3-position hydroxyl group gives the corresponding glucoside. Such sugar-substituted anthocyanidins are known as **anthocyanins**. Both groups of compounds are largely responsible for many of the vibrant shades of red, violet, and blue found in flowers and fruits, with the anthocyanins being especially important in the coloration of autumn leaves. Interestingly, flower colors have been successfully manipulated commercially by altering some of the genes and enzymes involved in flavonoid biosynthesis.

THE FLAVANONE-TO-ISOFLAVANONE-TO-ISOFLAVONE SEQUENCE: FORMATION OF GENISTEIN

Isoflavonoids are essentially regular flavonoids which have undergone a structural rearrangement in which the 4-coumarylCoA-derived phenol ring of a flavanone has undergone an oxidatively driven 1,2-aryl migration to yield an isoflavanone. Subsequent dehydration of the isoflavanone yields the unsaturated isoflavone. A typical sequence, illustrated in Fig. 6.34, begins with **naringenin** (a flavanone) undergoing a typical C–H to C–OH oxidation via O_2/NADPH at the ketone α-position. This leads to a radical intermediate that, prior to capture of hydroxyl oxygen, undergoes a 1,2-aryl shift via a bridged radical intermediate. The resulting rearranged radical is then quenched as usual to give the hydroxylated product **2,4',5,7-tetrahydroxyisoflavanone**. A final dehydration step gives the isoflavone **genistein**. An analogous sequence starting with the flavanone **liquiritigenin** leads to formation of the isoflavone **daidzein**.

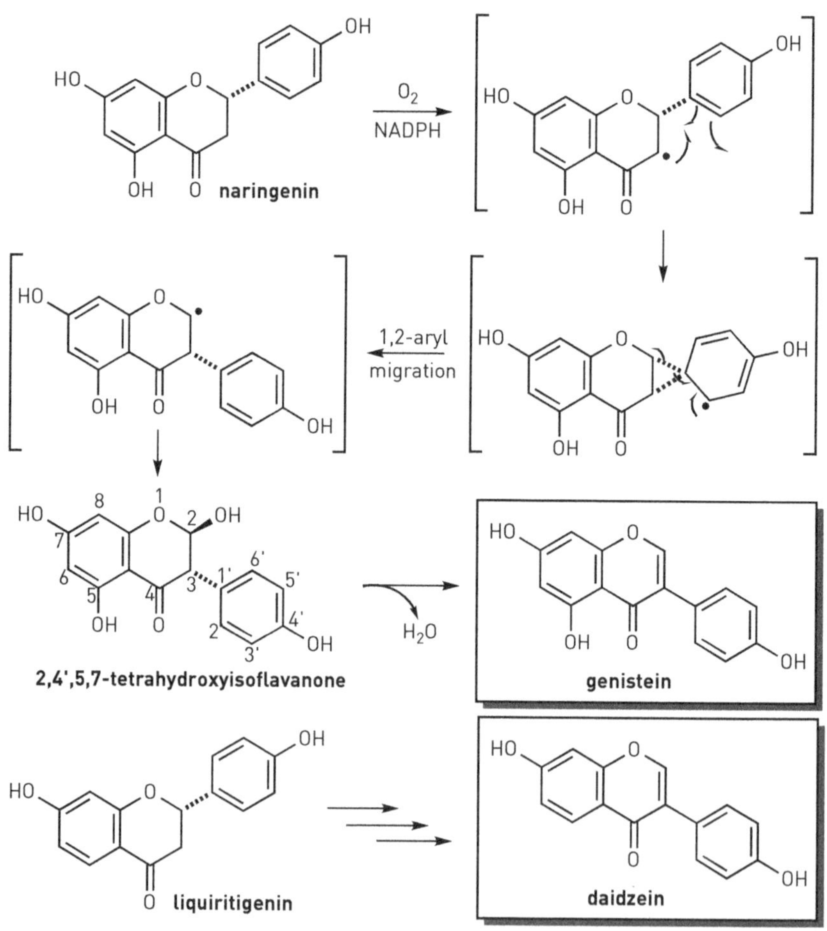

FIGURE 6.34

Isoflavonoid biosynthesis via 1,2-aryl migration; formation of genistein and daidzein.

ISOFLAVANOID STRUCTURAL MODIFICATIONS: PRODUCTION OF ANTIMICROBIAL PHYTOALEXINS

The isoflavonoids may be further metabolized to yield a remarkable array of interesting structures, many of which have significant biological activity. A series of some typical transformations is illustrated in Fig. 6.35, beginning with **2,4',7-trihydroxyisoflavanone**. While dehydration leads directly to **daidzein**, SAM methylation of the 4'-hydroxyl group followed by dehydration yields the isoflavone **formononetin**. Aryl ring hydroxylation of formononetin yields **2'-hydroxyformononetin** which then undergoes both conjugate reduction and ketone-to-alcohol reduction by NADPH to give a triol intermediate which cyclizes with loss of water to yield **medicarpin**, a member of a large class of such tetracyclic isoflavonoids known as **pterocarpans**. The pterocarpans are examples of **phytoalexins**, antimicrobial phenols, or polyphenols that are produced by plants at the site of an attack by pathogens. Their antifungal and antibiotic properties have made them obvious targets for extensive pharmaceutical research over the years.

The parallel sequence shown in Fig. 6.35 involves the oxidative hydroxylation and conjugate reduction of daidzein to yield **2'-hydroxydihydrodaidzein**. This compound forms a cyclic hemiacetal derivative which readily undergoes dehydration to yield **3,9-dihydroxypterocarpene**, which is effectively an unsaturated pterocarpan. Further oxidation of the cyclic ether linkage of the pterocarpene (mechanism as yet undetermined) yields **3,9-dihydroxycoumestan**, more commonly known as **coumestrol**, found mainly in red clover, alfalfa, soybean, and kudzu leaf. Its structural similarity to estradiol is obvious and like other phytoestrogens, it has a binding affinity for the estrogen receptor (higher than that of genistein). Its mode of action is related to the inhibition of certain enzymes involved in steroidal hormone biosynthesis, leading to modulation of their production. In model studies, coumestrol has shown some potential for the reduction of osteoporosis-related bone loss.

A recurring theme in isoflavonoid biosynthesis is the formation of products which have undergone alkylation with one or more prenyl groups from the terpenoid biosynthetic pathway (Chapter 4). These groups may then undergo further modifications via oxidations and/or cyclizations to yield a remarkable array of higher molecular weight derivatives, some of which show promise as therapeutic agents. A sampling of such compounds is shown in Fig. 6.36.

An interesting geranyl derivative of daidzein is **corylifol A**, a compound found in the seeds of *Cullen corylifolium*. **Glabroisoflavanone A**, a minor constituent found in extracts used in the production of licorice, is just one of many examples of prenyl-substituted isoflavanones in which a prenyl side chain has undergone subsequent cyclization to yield a pyran-type ring. **Eryvarin S**, a prenyl-substituted isoflavone and **erypoegin H**, a prenylated pterocarpene, have both exhibited activity against multiple methicillin-resistant *Staphylococcus aureus* (MRSA) strains; erypoegin H

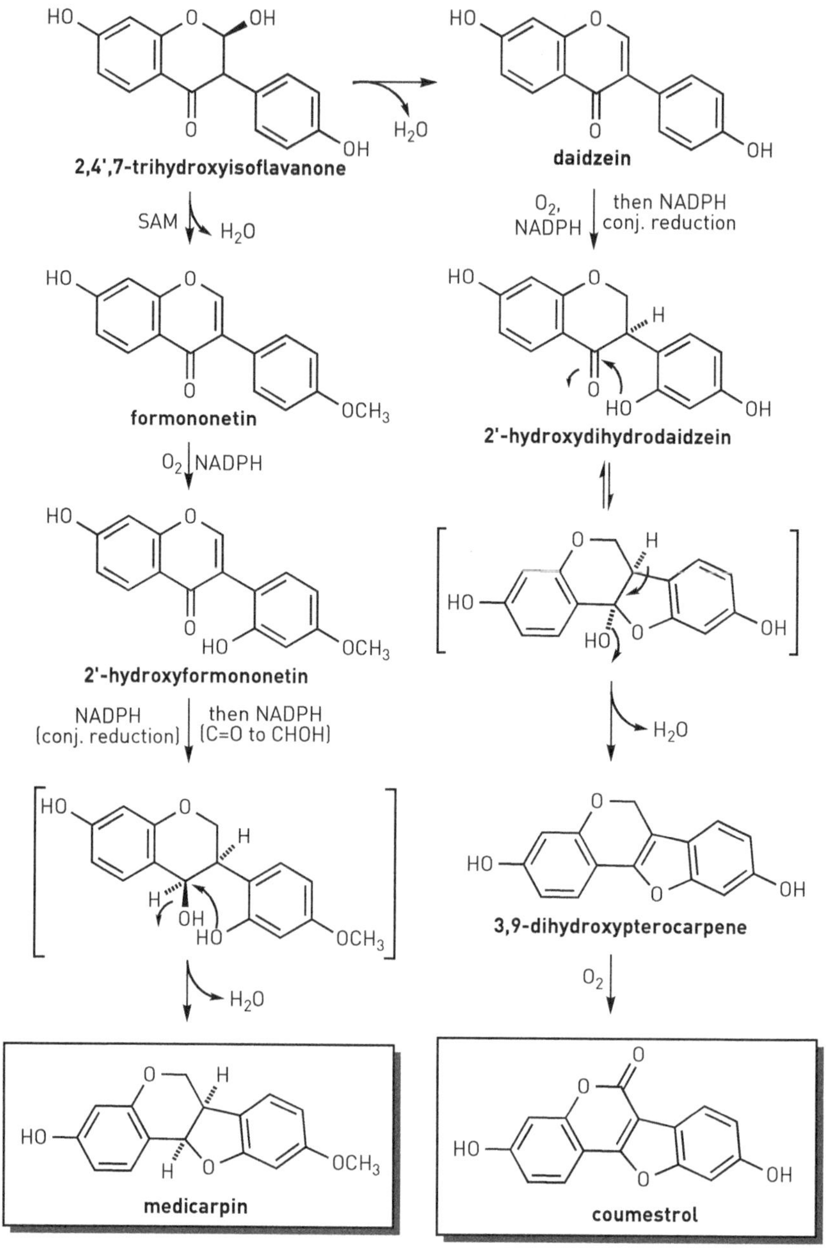

FIGURE 6.35
Transformation of formononetin and daidzein into medicarpin and coumestrol.

has also shown promising activity against vancomycin-resistant enterococci (VRE) strains. Many examples of prenylated pterocarpans have also been isolated; **rautandiol B** shows how such prenylation may be followed by oxidative cyclization leading to formation of furan-type ring systems. Like coumestrol, **hedysarimcoumestan D** is an example of a phenolic **coumestan**, but with additional prenyl substitution. Mild anthelmintic activity (activity against parasitic worms) has been demonstrated

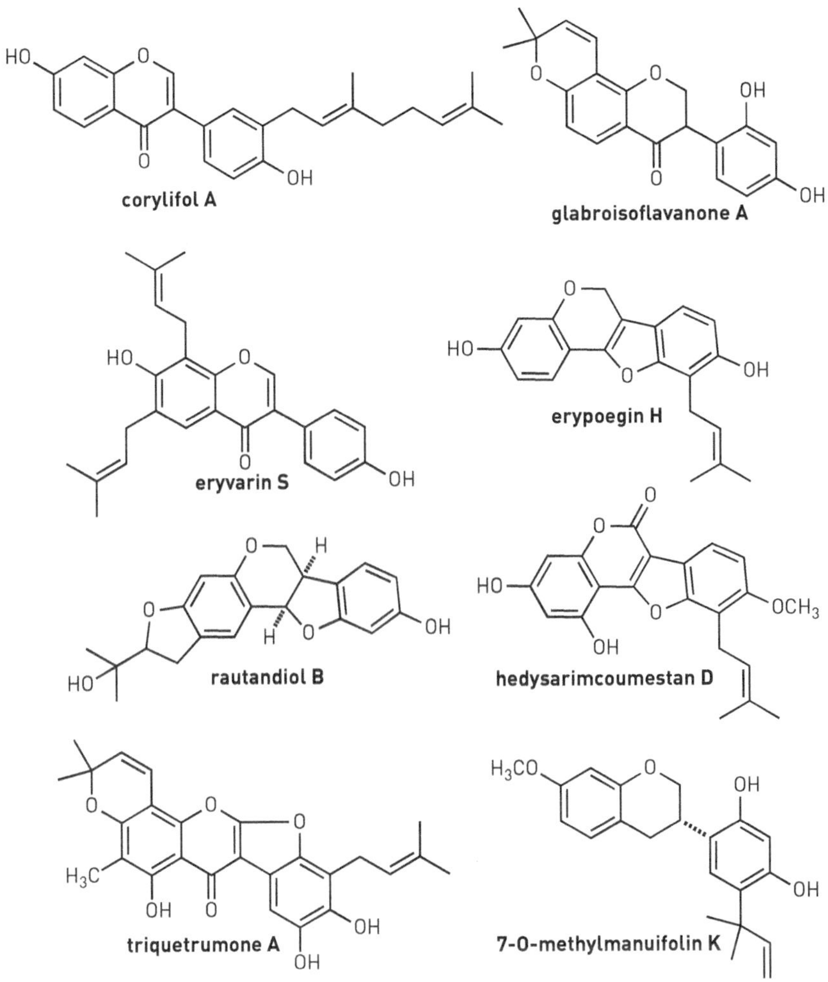

FIGURE 6.36
Some examples of structural variety in alkylated isoflavonoids.

by **triquetrumone A**, a member of the **coumaronochromone** group of isoflavonoids which shows how both cyclized and open prenyl chains may be incorporated within a single structure. Finally, the less common incorporation of a tertiary prenyl group is shown in the structure of **7-O-methylmanuifolin K**, an isoflavan that has been shown to be comparable to **amphotericin B** in its activity against *Naegleria fowleri*, an amoeba responsible for a rare but often fatal disease, primary amoebic meningoencephalitis.

The oxidative cyclizations employed in conjunction with prenyl side chain incorporation usually involve epoxidation followed by nucleophilic ring opening and subsequent 1,2-elimination of water. A reasonable mechanism summary leading to **dihydrobenzofuran**, **benzopyran**, and **benzofuran** ring systems is shown in Fig. 6.37. Note that the mechanism leading to the benzofuran nucleus also involves a benzylic oxidation followed by loss of acetone and water.

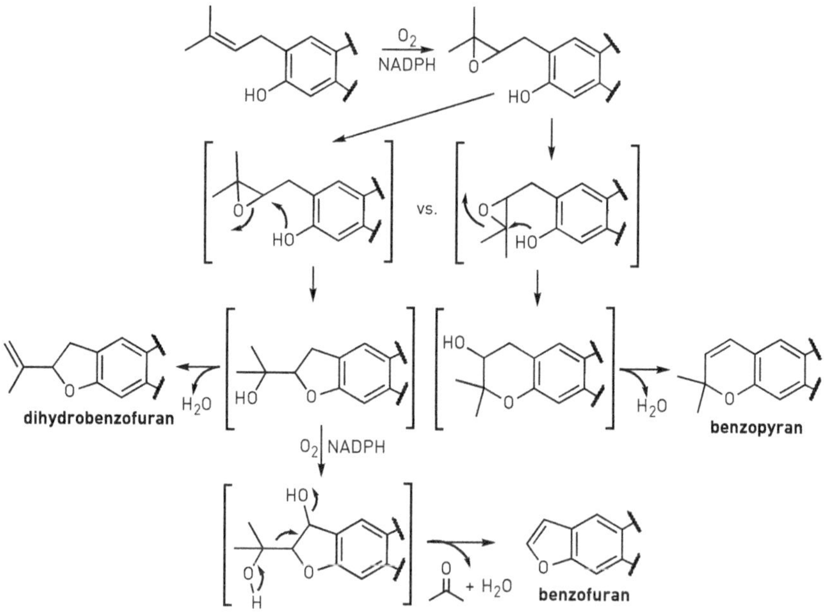

FIGURE 6.37
Prenyl side chain cyclization and modification mechanisms.

ROTENOIDS: FISH POISONS FROM ISOFLAVONES

The **rotenoids** are a class of plant compounds derived from isoflavones that play a role in plant defense against insect attack. Six plant-derived rotenoids are known to occur naturally in *Derris eliptica* (Southeast Asia) and either *Lonchocarpus utilis* or *L. urucu* (South America) and all are similar in chemical structure. The most common member of the family is **rotenone** which has been used for centuries by native populations as a potent fish poison. Its biosynthesis from **formononetin** is shown in Fig. 6.38.

After two sequential aryl ring hydroxylations and SAM methylations, an additional oxidation takes place at a methoxyl methyl group. One proposed sequence for subsequent cyclization of the resulting radical proceeds via its addition to the adjacent α,β-unsaturated enone and subsequent radical quenching by a H-atom donor to yield the ketopyran ring as shown (other cyclization mechanisms have also been proposed). The remaining steps involve ortho-alkylation of the phenolic ring by DMAPP to yield **rotenoic acid**. This is followed by a final oxidative cyclization of the prenyl side chain to give the 5-membered ring of **rotenone.** Another rotenoid known as **duegelin** differs from rotenone only in that the final prenyl side chain cyclization step leads to a 6-membered pyran rather than a 5-membered furan ring system. The cyclization mechanisms involved have not been fully determined but appear to be more complex than those involved in formation of similar ring systems we have seen in other isoflavonoid derivatives.

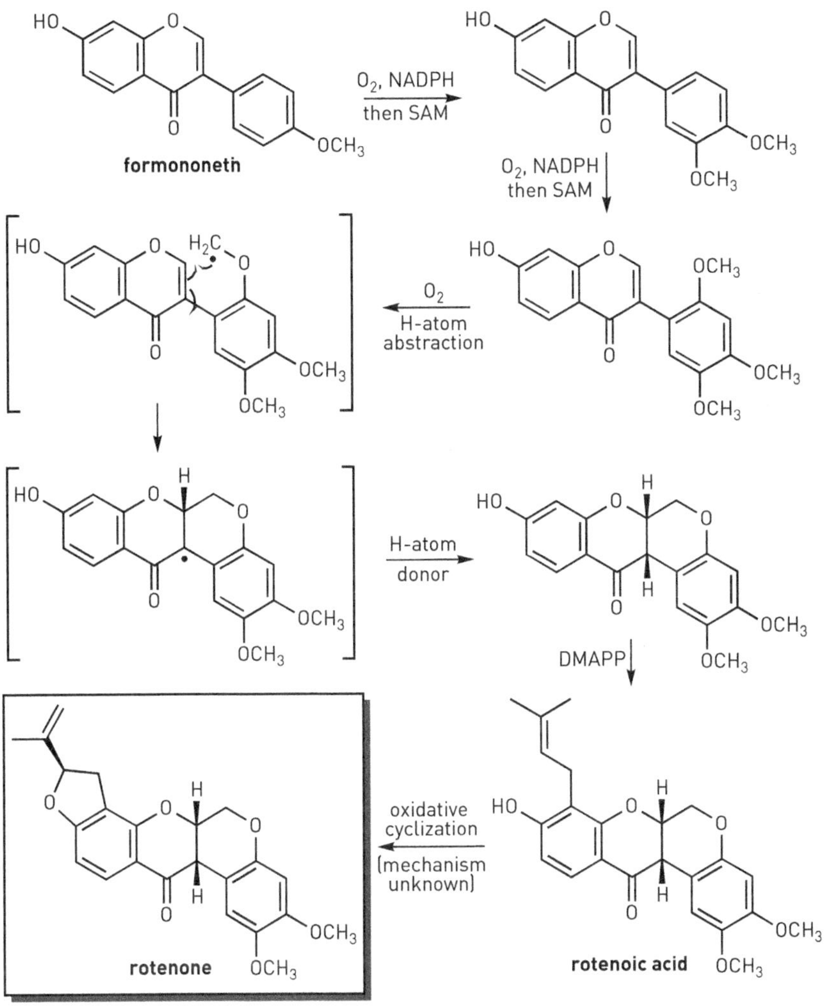

FIGURE 6.38
Biosynthesis of the fish poison rotenone from the isoflavone formononetin.

A dusty solid when isolated, rotenone is also useful as a biodegradable insecticide that is effective for control of a variety of beetles and other common garden pests. It is also used in veterinary medicine for treatment of mites, but because it is irritating to the skin, eyes, and upper respiratory tract, it has seen only limited use in human medicine as a topical treatment for head lice and other parasites. In animal studies, rotenone has been shown to interfere with both the electron transport chain and with tubulin self-assembly and can also promote Parkinson's-like symptoms.

LOOKING AHEAD

We have only just scratched the surface of the incredible diversity of organic compounds produced by the shikimic acid pathway. Now that we are armed with some of its basics and have a better understanding of how certain amino acids can be transformed into remarkably complex organic structures by relatively simple organic reactions, we can move on to Chapter 7 which deals with another group of compounds that utilize amino acids as building blocks and which are historically the most important of all secondary metabolites in terms of their impact on human health: the alkaloids.

STUDY PROBLEMS

1. Many 2-arylbenzofuran natural products such as the one shown here are potent anti-inflammatory and cytotoxic agents. Using cinnamoyl-CoA (from the shikimic acid pathway) as a starter unit, three malonyl-CoA units and any other building blocks required, propose a reasonable biosynthesis of the indicated compound.

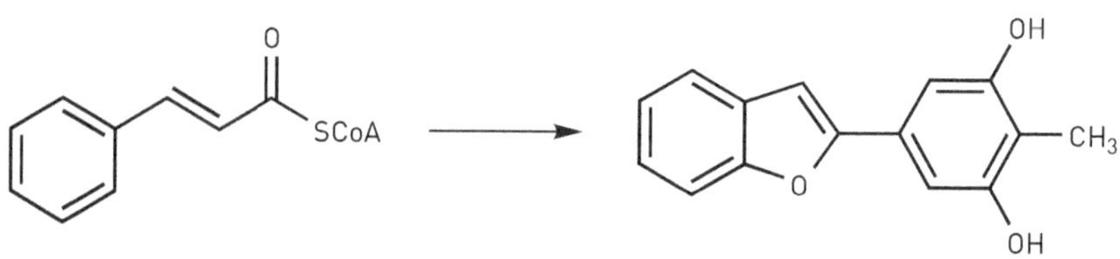

2. In the naturally occurring form of podophyllotoxin, the C_6–C_5 lactone ring junction is not the most thermodynamically stable form and is readily isomerized to its more stable form. Suggest a structure and a mechanism for the isomerization.

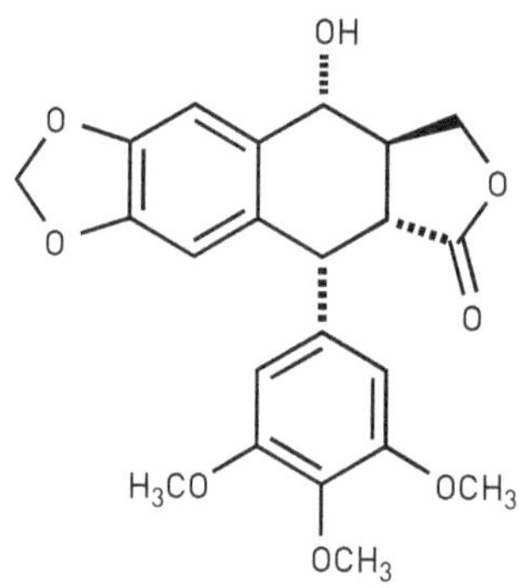

podophyllotoxin

3. The cyclization of yatein (Fig. 6.22) to deoxypodophyllotoxin involves a benzylic carbocation intermediate. A quinone methide-type carbocation intermediate has also been invoked for this process. Draw a likely structure for it. Would this structure be more stable than the simple benzylic carbocation shown in Fig. 6.22? Why or why not?

4. Bergaptol is an example of one of several furanocoumarins that occur naturally in grapefruits and are responsible for the so-called "grapefruit juice effect" that arises from their interference with certain CYP-type enzymes that assist in the metabolism of various drugs including statins (such as Liptor and Mevacor) and benzodiazepams (such as Valium and Xanax) among others. Outline a likely biosynthesis of bergaptol from umbelliferone, DMAPP and other co-factors.

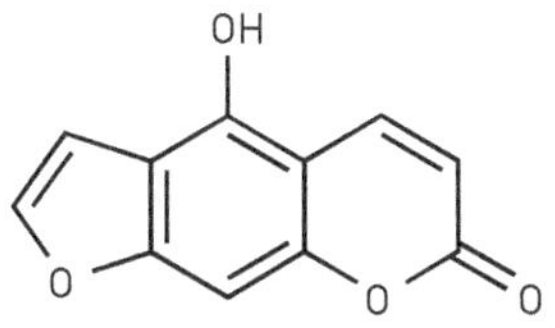

bergaptol

5. Warfarin (brand name Coumadin) is a coumarin derivative originally produced and used as a rat poison in the 1940s. It is a synthetic analogue of dicoumarol (Fig. 6.27) and derives its name from the original source of funding for its discovery (Wisconsin Alumni Research Foundation plus *-arin* from coumarin). Warfarin is widely used in human medicine as a highly effective anticoagulant for the prevention of thrombosis (blood clots). It can be synthesized from 2-hydroxyacetophenone, diethyl carbonate, benzaldehyde, and acetone as shown.

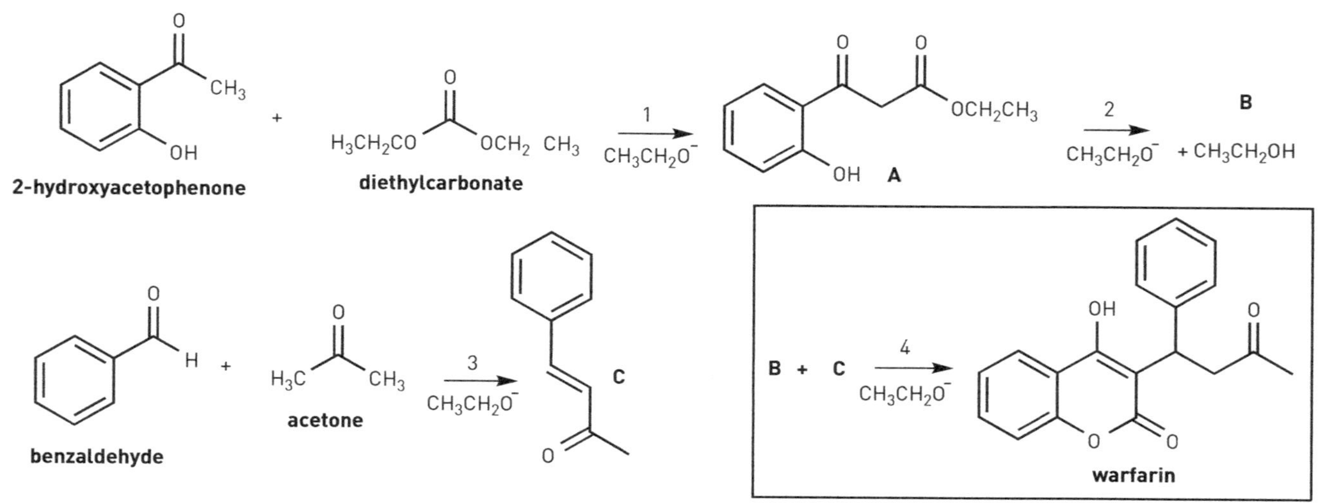
2-hydroxyacetophenone
diethylcarbonate
1
CH3CH2O-
A
2
B
+ CH3CH2OH
benzaldehyde
acetone
3
C
B + C
4
warfarin

Given that all the laboratory reactions can be carried out using a simple base such as ethoxide ion, answer each of the following:

a) Provide a mechanism for the condensation involved in Reaction 1 to produce **A**.

b) Determine the structure of **B** derived from **A** via Reaction 2.

c) Provide a mechanism for the condensation involved in Reaction 3 to produce **C**.

d) Provide a likely mechanism for reaction of **B** with **C** to produce warfarin.

6. **Gingerol** ($C_{27}H_{26}O_4$) is a spicy, pungent yellow oil isolated from fresh ginger.

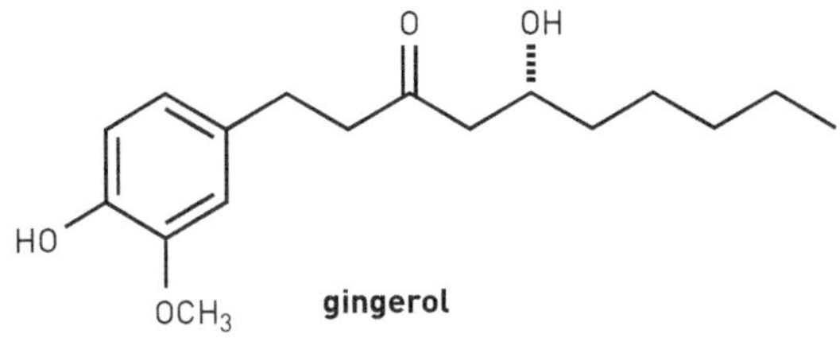

gingerol

a) Propose a reasonable sequence for biosynthesis of gingerol from ferulic acid CoA and 3-ketooctanoic acid CoA (among the required reactions are a Claisen condensation and NADPH reductions). Think carefully about sequence.

b) The cooking process can transform gingerol into **zingerone** ($C_{11}H_{14}O_3$). The reaction involved is a retroaldol addition. What is the likely structure of zingerone?

c) **Shogaol** ($C_{27}H_{24}O_3$) is structurally related to gingerol but much spicier. It is formed when ginger is heated and dried. What is a likely structure for shogaol?

7. Propose a reasonable sequence for biosynthesis of **liquiritigenin**. What is its configuration as shown?

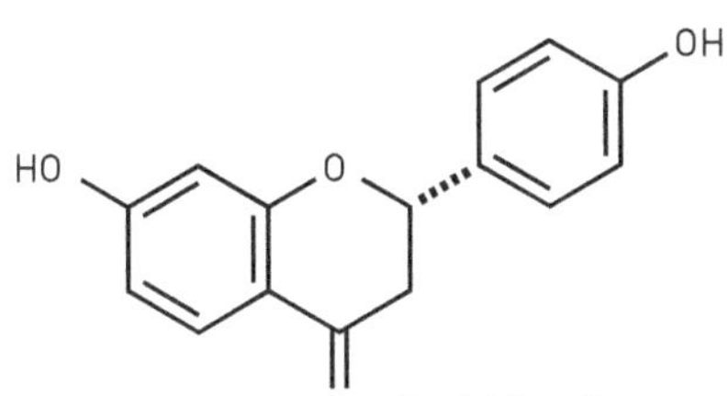

liquiritigenin

8. **Silibin A** is a so-called flavonolignan compound that has been shown to help protect liver cells from certain toxins and has also been shown to possess anticancer activity against breast, prostate, colon, and lung cancer cells. Its structure is a hybrid of the dihydroflavonol **taxifolin** and the lignol **coniferyl alcohol** which have been joined via an oxidative coupling process similar to that used in the formation of lignans. Propose a mechanism for biosynthesis of silibinin A.

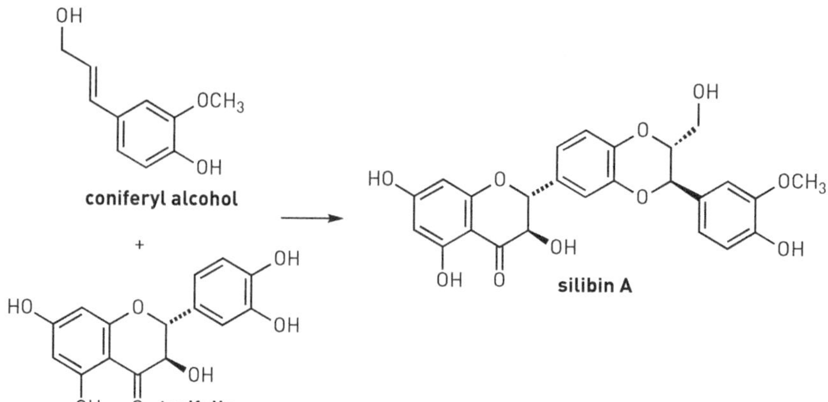

coniferyl alcohol

taxifolin

silibin A

9. Isolation of **rautandiol B** (Fig. 6.36) also yields a constitutional isomer, **rautandiol A**. Both are pterocarpans formed in a single step from the same immediate precursor and both have a phenolic hydroxyl group. The remaining hydroxyl group of rautandiol B is a tertiary alkanol, but for rautandiol A, it is a secondary cycloalkanol which can exist as two stereoisomers. Propose likely structures for the rautandiol A stereoisomers and a sequence for their biosynthesis from **vestitone**.

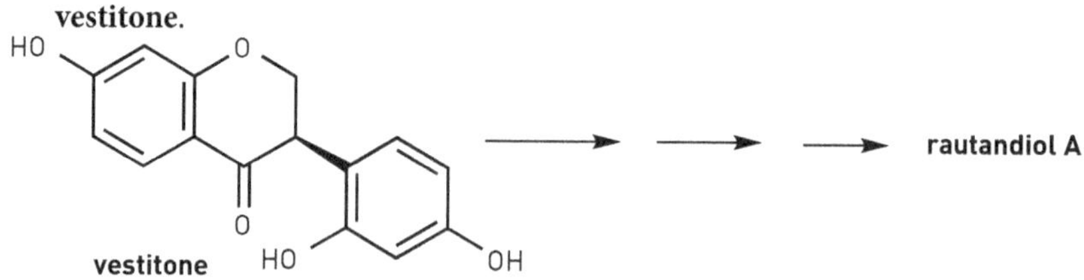

vestitone

10. **Cyanidin** can undergo an NADPH reduction to afford **epicatechin**.

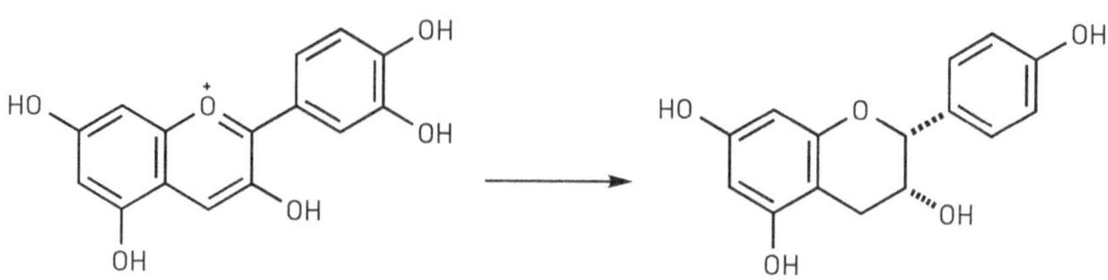

cyanidin

epicatechin

a) Show a reasonable mechanism for this hydride reduction.
b) Show a plausible mechanism for the reaction of cyanidin with epicatechin to give the dimeric structure shown.

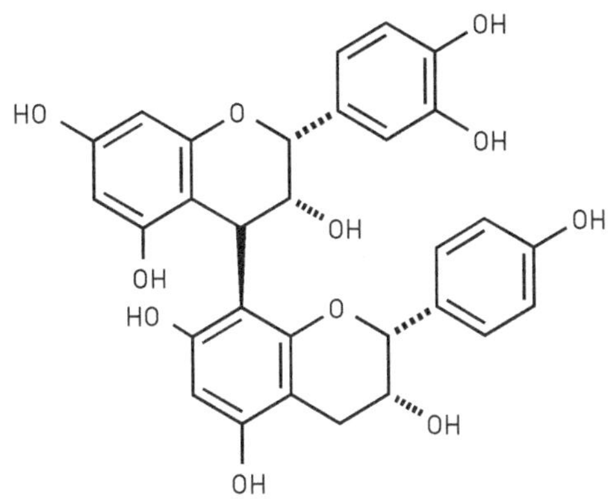

11. Eugenol and isoeugenol are involved in the biosynthesis of a number of interesting lignan compounds. One unusual example is **guaianin**, shown here. Show how the two isoeugenol units could be joined together via oxidative coupling, then propose a structure for compound **A** that is a likely transient intermediate in the biosynthesis and finally show how **A** can then be further transformed into guaianin. Note that the alcohol function of guaianin is derived from the H_2O incorporated into A.

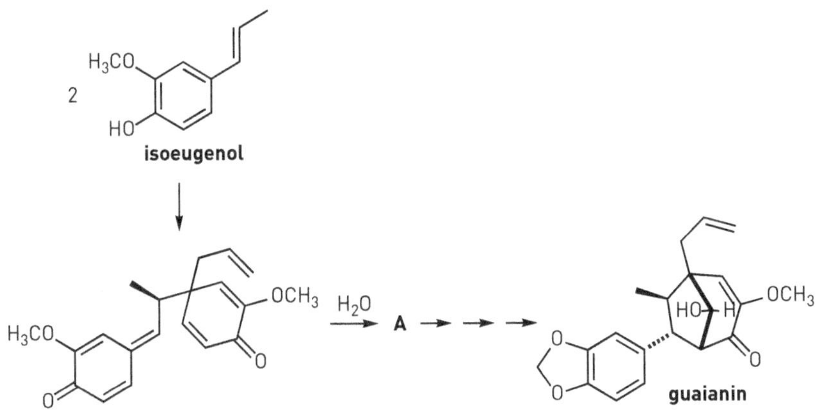

7 Biosynthesis of Alkaloids and Related Compounds

I could imagine his [Sherlock Holmes] giving a friend a little pinch of the latest vegetable alkaloid, not out of malevolence, you understand, but simply out of a spirit of inquiry in order to have an accurate idea of the effects. To do him justice, I think that he would take it himself with the same readiness. He appears to have a passion for definite and exact knowledge . . . of botany—well up in bella-donna, opium, and poisons generally; of chemistry—profound
—Sir Arthur Conan Doyle (*A Study in Scarlet, 1887*)

Though definitions may vary from source to source, the term **alkaloid** generally refers to members of a large set of naturally occurring, slightly basic (i.e., alkaline) nitrogen-containing organic compounds. Generally excluded from this group are amino acids, peptides, proteins, *N*-containing carbohydrates, and nitrogenous bases used in the construction of nucleotides. Though a small number are produced by animals or microorganisms, the vast majority of alkaloids are plant-produced compounds possessing a remarkably diverse range of structural features, from simple cycloaliphatic amines to highly complex polycyclic *N*-heterocycles. Some representative alkaloids are shown in Fig. 7.1.

Alkaloid-containing plants and their extracts have been used by humans for thousands of years, mainly on the basis of their stimulant, therapeutic, or poisonous properties. References to plants containing compounds such as **morphine** (from opium poppies), **strychnine** (from seeds of the *Strychnos nux-vomica* tree), **ephedrine** (from the plant *Ephedra chinensis*), and **coniine** (from the poison hemlock plant) may be found in some of our earliest known writings. Today, it has been estimated that the health care of over 5 billion people worldwide benefits from the use of

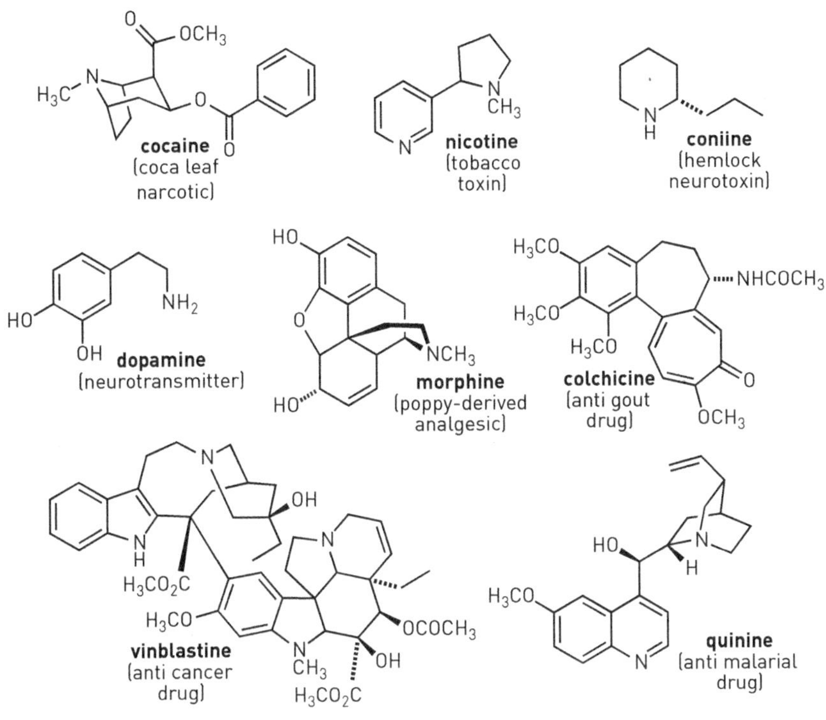

FIGURE 7.1
Some representative structures of alkaloids with significant biological activities.

plant-based medicinal agents, many of which are alkaloids. With that in mind, it is worth noting concerns that deforestation, environmental damage, large-scale development, and unregulated harvesting programs may ultimately lead to the extinction of hundreds of known medicinal plants and perhaps even more whose medicinal properties have yet to be discovered, thereby endangering the prospects for future discoveries of new curative agents for the benefit of all humankind.

As a scientific field, alkaloid chemistry itself dates back to the early 1800s with the first isolation of pure crystalline morphine from opium. This milestone achievement allowed the delivery of accurate, therapeutic doses of a drug that was immensely valuable for the relief of pain but which could also lead to fatal overdoses when administered from simple extracts of variable composition and strength. The subsequent rapid development of increasingly sophisticated techniques for the isolation and purification of the active components (often alkaloids) from many other medicinal plants essentially spawned the field of organic chemistry. By 1886, less than 70 years after the isolation of morphine, laboratory synthesis of the alkaloid coniine had been accomplished; further triumphs in the synthesis of increasingly complex natural products quickly followed, ultimately launching the fields of synthetic organic and medicinal chemistry. By the end of the first decade of our own century, well over 12,000 different alkaloid structures had been identified. Such a staggering number obviously requires us to limit ourselves here to an abbreviated introductory approach only; readers are therefore strongly encouraged to consult some of the outstanding general reference works listed at the end of the book for a more detailed and thorough treatment of this vast and challenging topic.

ALKALOID STRUCTURE: THE IMPORTANCE OF *N*-HETEROCYCLES

Alkaloids are generally derived from modification of various amino acids, though not always the more familiar amino acids used in peptide and protein formation. Among many others, shikimate-derived amino acids are important precursors in the construction of alkaloids, with terpenoid or polyketide components also incorporated in many instances. Thus, the alkaloids themselves are not necessarily thought of as being derived from a distinct biosynthetic pathway of their own, but rather from a set of bioorganic transformations operating on amino acid building blocks and other components provided by each of the three main biosynthetic pathways we have already encountered. Because alkaloid structures are quite variable and not readily classified, it is often convenient to group them according to the amino acid precursor from which they were originally derived. As noted earlier, most transformations of amino acid precursors ultimately lead to alkaloid structures containing one or more *N*-heterocyclic rings. With these points in mind, Table 7.1 summarizes the most important amino acids involved in alkaloid biosynthesis along with the most commonly encountered *N*-heterocyclic ring systems that each of them produces.

The above listing is by no means exhaustive, but it does present a reasonable overview of some of the main sources of alkaloid structure. Amination reactions of acetate- and terpene-derived precursors as well as purine-derived alkaloids and selected β-lactam antibiotics will also be treated briefly.

ALKALOIDS NOT DERIVED FROM AMINO ACIDS: AMINATION REACTIONS, POISONS, AND VENOMS

Before we delve into the chemistry of alkaloids that make extensive use of amino acids, we begin our examination with some examples of simple alkaloids derived from the acetate pathway, with nitrogen being provided by transamination. One such alkaloid is the neurotoxin **coniine**, which we encountered previously (Fig. 7.1). The biosynthesis of this structurally simple 2-alkylpiperidine from octanoic acid is outlined in Fig. 7.2. The process begins with oxidation of **octanoic acid** to give a ketone at position 5; this is followed by the usual sequence for thioesterification and reduction of carboxylic acid functions to give **5-oxo-octanal**. Alanine transamination of the aldehyde function then yields **5-oxo-octylamine** which undergoes intramolecular imine formation to give **γ-coniceine**. Finally, NADPH reduction of

Table 7.1 Heterocyclic Alkaloid Ring Systems from Amino Acid Precursors

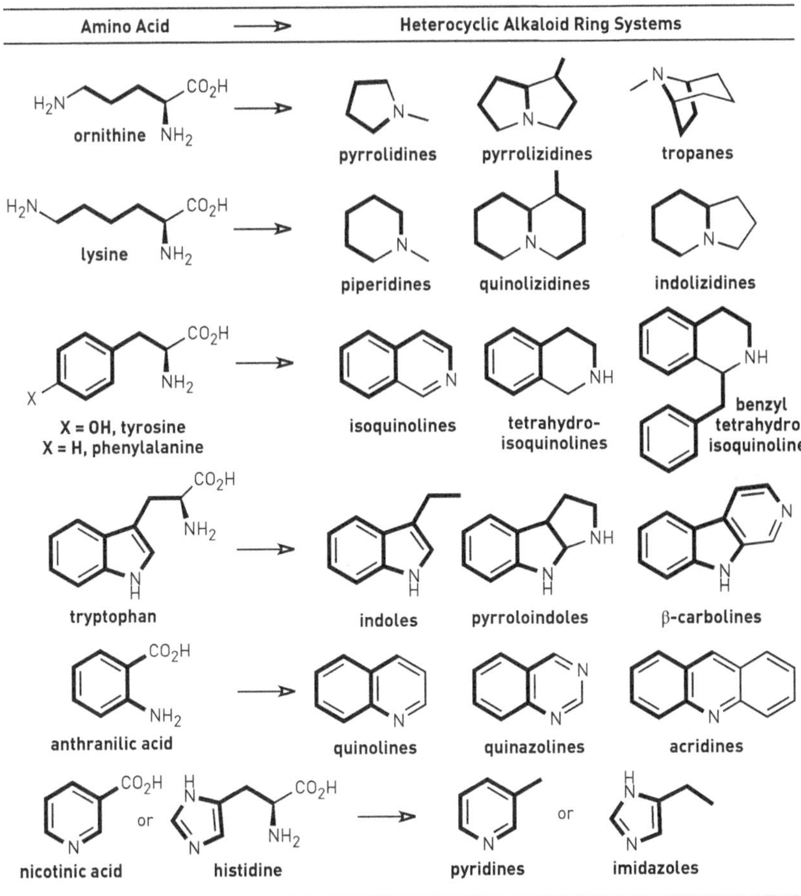

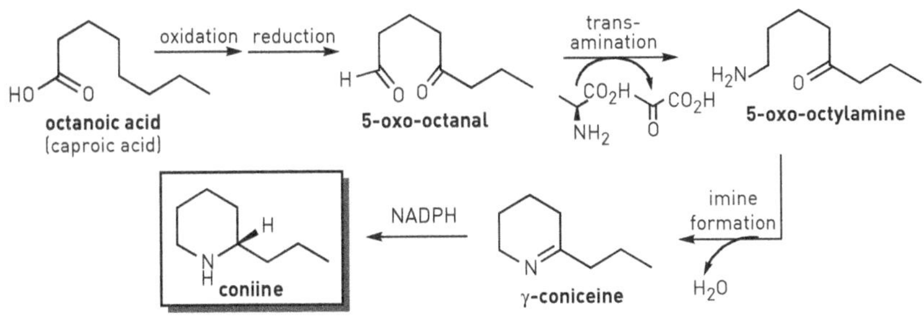

FIGURE 7.2
Biosynthesis of the neurotoxin alkaloid coniine.

the imine linkage affords coniine as shown. Note also that SAM methylation of coniine affords the related compound **N-methylconiine** (not shown) which is also often present, though in much smaller amounts.

Most notable as the poisonous hemlock alkaloid used in the death of Socrates in 399 B.C., coniine is extremely toxic to animals with a fatal dosage of less than 200 mg for humans. Notorious for causing death by slow suffocation, coniine consumption

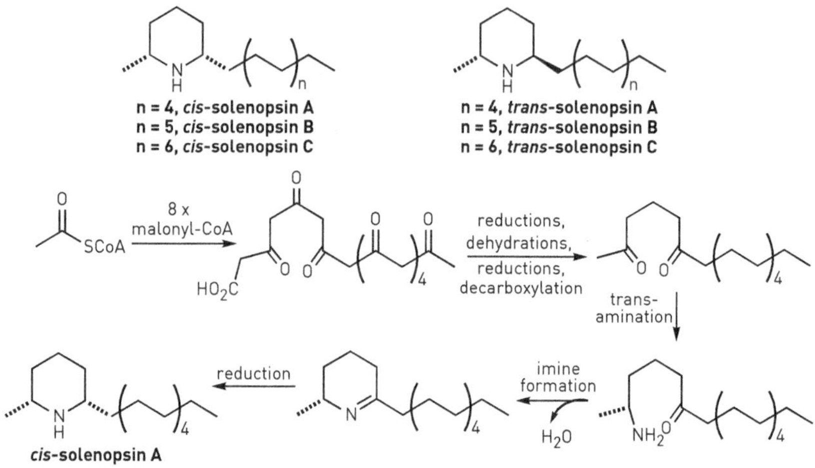

FIGURE 7.3
Proposed biosynthetic scheme for *cis*-solenopsin A, an alkaloid component of fire ant venom.

induces a gradual sequence of drowsiness, tingling, or burning sensations of the skin, lack of muscle control, nausea, excessive salivation, and rapid pulse eventually followed by fatal respiratory paralysis. Salts of coniine have reportedly shown some local analgesic action in the symptomatic relief of intense itching, but the toxicity of coniine generally makes its use for such purposes far too dangerous.

Polyketide-derived precursors can also play a role in alkaloid biosynthesis. An interesting example is a group of 2-methyl-6-alkyl-substituted piperidines known as the **solenopsins**. These alkaloids are unusual in that they are among the very few insect-produced alkaloids known to us. As the toxic defense components of fire ant venom, these compounds are ultimately responsible for the extremely painful welts produced by fire ant stings. While the details of their biosynthesis are not fully known, careful studies have confirmed their polyketide origin and that they are produced by the insects themselves rather than being present due to ingestion from a plant source. A reasonable biosynthetic sequence similar to that of coniine is outlined in Fig. 7.3.

In addition to painful welts, allergic reactions to the solenopsins in fire ant stings can lead to anaphylactic shock if not properly treated. Nevertheless, these compounds have been investigated extensively for their surprising ability to inhibit formation of new blood vessels (angiogenesis). The solenopsins have also been found to have insecticidal, antibacterial, and antifungal properties and also show promise as inhibitors of nitric oxide production, comparing favorably in potency with other widely used inhibitors.

AMINO ACIDS AND MANNICH REACTIONS: IMPORTANT KEYS TO ALKALOID BIOSYNTHESIS

A great deal of postassembly modification of alkaloids occurs via oxidative processes and these can vary a great deal depending on class and plant source. Nevertheless, there are some fundamental transformations employed in the initial assembly of alkaloid structures that are used over and over again in various biosynthetic contexts. Several of these reactions are outlined in general terms in Fig. 7.4. In order to utilize amino acids as building blocks for alkaloid biosynthesis, they are usually first converted either into **amines** via **PLP-mediated decarboxylation**, or into **α-ketoacids** via PLP-mediated **transamination**. Transamination is usually then followed by **TPP ylide-mediated decarboxylation** to produce the corresponding **aldehydes**. With primary amines and aldehydes thus readily available, a simple condensation reaction leads to the formation of **imines** which play a central role in the construction of many higher-molecular weight amines from simpler building blocks. Beyond this simple beginning, we then find one of the most important C–C bond forming reactions in alkaloid biosynthesis—the so-called **Mannich reaction**. The classic laboratory version of this reaction makes use of three components: 1) an amine or ammonia; 2) a nonenolizable aldehyde or ketone; and 3) an enolizable aldehyde or ketone. The first two components condense to form an imine, while the third component serves as a carbon nucleophile, usually as an enol or enolate ion, which attacks the electrophilic imine linkage, thereby forming a carbon–carbon bond and the resulting amine product as shown in Fig. 7.4.

In alkaloid biosynthesis, the first two components—an amine and an aldehyde derived via the decarboxylation and transamination chemistry shown in Fig. 7.4—condense to form the corresponding imine required for a biological Mannich reaction. The nucleophilic component (represented here by a generic nucleophile, $Nu:^-$) may be a β-dicarbonyl enol, a thioester enolate from acetyl CoA or similar sources, or the nucleophilic enol of a phenol ring. The combination of such components leads to a Mannich-derived alkaloid amine. As we will see, this combination of components can be used in both inter- and intramolecular versions of the biological

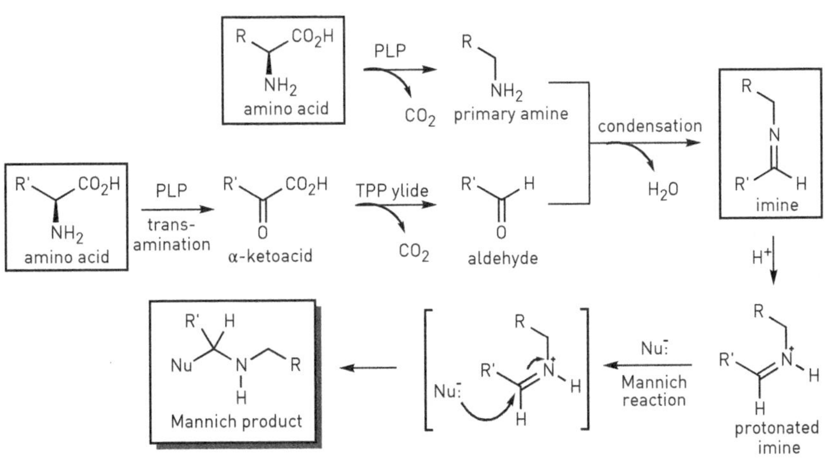

FIGURE 7.4
Mannich reaction via amino acid transformations, imine formation and nucleophilic addition.

Mannich process, resulting in formation of alkaloid products with a variety of unique structural features.

ALKALOIDS FROM ORNITHINE: TROPANES VIA THE MANNICH REACTION IN ACTION

In biochemistry, the **urea cycle** (sometimes also referred to as the **ornithine cycle**) is a metabolic pathway that allows for the excretion of excess nitrogen via production of urea. One of the reactions of this pathway is the *arginase*-mediated production of urea from the peptidic amino acid **L-arginine**. Extrusion of urea from its structure converts arginine into **L-ornithine**, a nonpeptidic amino acid which, in addition to other important roles, serves as a precursor to a number of important alkaloids, including those of the tropane family. Some of the reactions involved in biosynthesis of **tropinone** and **tropine** are outlined in Fig. 7.5.

Initial PLP-mediated decarboxylation of ornithine gives the 1,4-diamine **putrescine** whose foul (putrid) odor is associated with the smell of decaying flesh. SAM methylation leads to *N*-methylputrescine which is converted to the corresponding aminoaldehyde by a *diamine oxidase*-mediated oxidative process. Intramolecular imine cyclization leads to formation of an important 5-membered ring intermediate, ***N*-methyl-Δ^1-pyrrolinium cation**, which is the substrate for a subsequent Mannich reaction using malonyl-CoA as the carbon nucleophile. Addition of a second malonyl-CoA via Claisen condensation extends the side chain by two carbons to give a diketide-substituted *N*-methylpyrrolidine derivative. Oxidation at the 2-position of the pyrrolidine ring gives a hydroxyl-substituted derivative (shown here in a more useful conformation) which then undergoes dehydration to afford a substituted *N*-methylpyrrolinium ion. Hydrolysis of the thioester yields a β-ketoacid which serves as the nucleophile for a second intramolecular Mannich reaction, forming a bicyclic β-ketoacid which decarboxylates as expected to afford the key alkaloid ketone known as **tropinone**. A final NADPH reduction of the ketone function affords **tropine** as shown.

A number of synthetic alkaloid derivatives of tropine have been prepared and marketed, mainly as anticholinergic drugs for the treatment of Parkinson-like symptoms. Tropine also serves as the biosynthetic precursor to the natural tropane alkaloids **hyoscyamine** (atropine) and **hyoscine** (scopolamine) as outlined in Fig. 7.6. The hydroxyl group of tropine initially undergoes condensation with the CoA thioester of phenyllactic acid. The resulting ester, **littorine**, then undergoes an unusual oxidative rearrangement of the hydroxyl ester core to yield hyoscyamine. Subsequent sequential oxidations of the tropine ring of hyoscyamine yield the epoxide function in the structure of hyoscine. Note the change from an axial to an equatorial position for the N-methyl function in the hyoscyamine-to-hyoscine transformation.

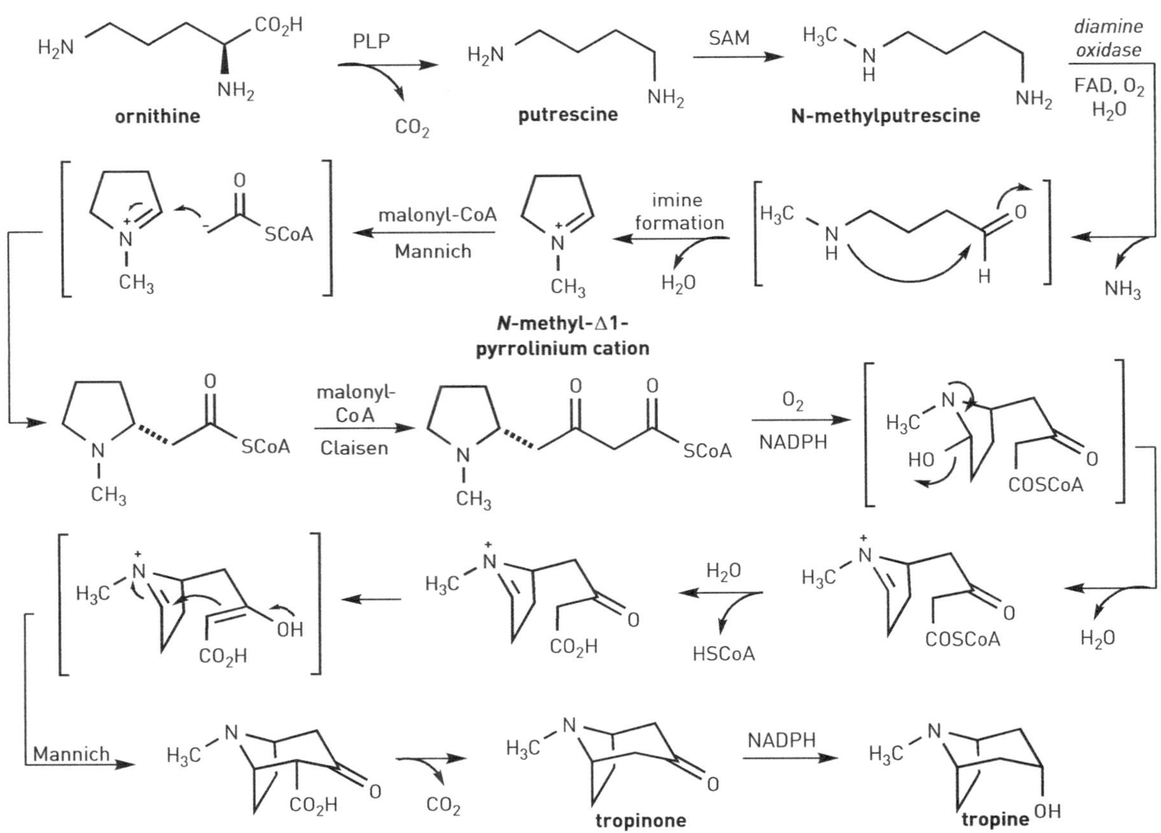

FIGURE 7.5

Conversion of ornithine to tropine via N-methyl-Δ^1-pyrrolinium cation and Mannich reactions.

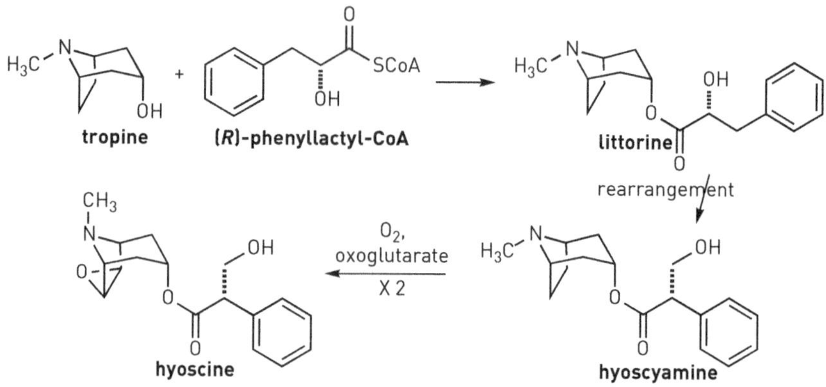

FIGURE 7.6
Conversion of tropine to hyoscyamine (atropine) and hyoscine (scopolamine).

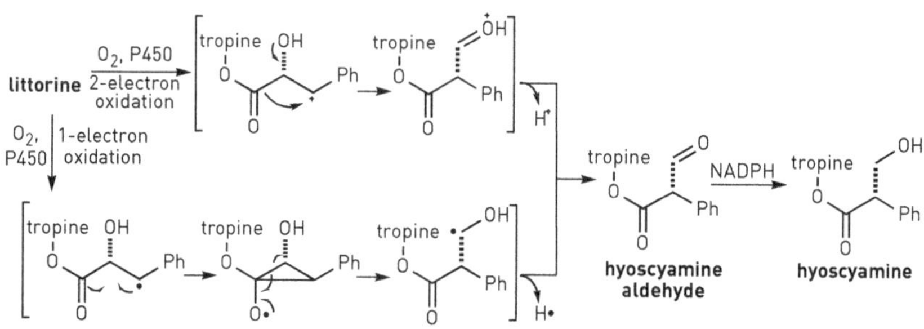

FIGURE 7.7
Proposed mechanisms for the oxidative rearrangement of littorine to hyoscyamine.

The conversion of **littorine** to **hyoscyamine**, as shown in Fig. 7.7, is an interesting rearrangement which has been studied extensively. The process has been shown to occur via a P450-type benzylic oxidation that initially leads to formation of **hyoscyamine aldehyde** as an intermediate. The aldehyde is subsequently reduced to the corresponding alcohol via the usual NADPH reduction process. Some evidence suggests that hydroxylation at the benzylic position of littorine may occur prior to the rearrangement, but the precise mechanistic details of the oxidative process remain to be determined. Given that examples of P450 oxidations involving two-electron as well as one-electron oxidative processes are now known, reasonable mechanisms for this rearrangement have been proposed that involve either a carbocation or a radical as a likely intermediate. The carbocation process would involve a 1,2-shift and proton loss similar to that seen in pinacol–pinacolone-type rearrangements, while the radical process would require formation of a transient cyclopropane intermediate which would fragment to a rearranged radical and loss of a H atom. Both processes would lead to the same aldehyde intermediate as shown in Fig. 7.7.

Hyoscyamine and hyoscine are sometimes referred to as the "nightshade alkaloids," owing to deadly nightshade (*Atropa belladonna*) as a primary plant source, though these compounds may also be found in mandrake, jimson weed, and henbane among others. They have a long history in folk medicine, beginning with the use of mandrake extracts for anesthesia by the early Greeks. More notorious were the potent hallucinatory effects which these alkaloids provided when used in the so-called "flying ointments" allegedly employed in witchcraft practices of the Middle Ages. Interestingly, the term **belladonna** (beautiful woman) was originally associated with the pupil dilating effects of **atropine** (the racemic form of hyoscyamine) and its use by women during the Renaissance to achieve a striking cosmetic look for their eyes. In modern ophthalmology, use of atropine for pupil dilation has diminished in favor of compounds whose effects are of shorter duration. Atropine is still widely used as a preanesthetic medication to reduce salivary secretions during surgeries, as an antidote for certain types of poisonings, and for the treatment of colon spasms. Hyoscine, also known as **scopolamine**, is sometimes used for the treatment of severe nausea and motion sickness, usually via a transdermal patch. From 1915–1960, a mixture of morphine and hyoscine was commonly used in hospital anesthesia to block the pains associated with childbirth. Popularly known as "twilight sleep," administration of this mixture gradually fell into disfavor as "natural childbirth" methods gained wider acceptance as a means of reducing the exposure of both mother and infant to the systemic effects of such anesthetic agents.

Certainly the most infamous of the tropane alkaloids is **cocaine**, an addictive CNS-stimulant narcotic derived from leaves of the South American coca plant. Its biosynthesis, outlined in Fig. 7.8, again involves the ornithine-derived *N*-methyl-Δ^1-pyrrolinium cation and its elaboration via sequential malonyl-CoA additions,

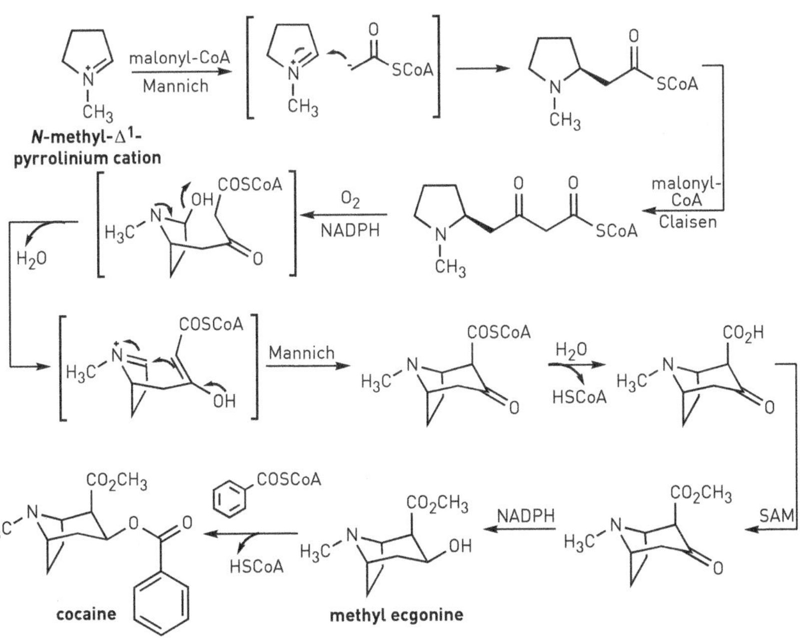

FIGURE 7.8
Conversion of *N*-methyl-Δ^1-pyrrolinium cation to the coca leaf alkaloid cocaine.

but yielding the diketide of opposite configuration at the pyrrolidine ring side chain relative to the diketide formed in tropine biosynthesis (Fig. 7.5). Though oxidation and dehydration to a substituted pyrrolinium ion proceeds as before, the subsequent Mannich cyclization proceeds via the CoA ester rather than via the free acid, and subsequent hydrolysis to the β-ketoacid is immediately followed by SAM methylation rather than decarboxylation to give a methyl ester. NADPH reduction of the ketone function yields the equatorial alcohol derivative known as **methyl ecgonine**. A final esterification of the alcohol with benzoylCoA gives the corresponding benzoate ester, **cocaine**.

Coca leaves have been chewed by South American natives for hundreds of years primarily for stimulation, relief of hunger, and for use in various religious ceremonies. Cocaine itself was not isolated until 1860 but in little more than twenty year's time it found clinical use in ophthalmic surgery as the first truly modern local anesthetic. The subsequently recognized toxic and addictive effects of cocaine and its derivatives ultimately led to strict limitation of its use, though it still finds application in ophthalmic procedures and in surgeries involving the ear, nose, and throat. As an alkaloid narcotic of abuse, cocaine is rivalled only by heroin, with as many as 14% of U.S. adults having reportedly used the drug at least once. As an amine salt, such as the hydrochloride, cocaine is fairly water soluble and it is in this form that it is "snorted" or dissolved in water and injected. So-called "crack" cocaine is actually the corresponding "free base" of the amine rather than a salt and it is in this form that cocaine is "smoked" by heating and inhaling its vapors. The effects of cocaine are many and varied, ranging from an increased sense of energy, general euphoria, and feelings ranging from superiority to hyperactivity, pupil dilation, rapid speech, and elevated pulse among others. Tolerance, intense cravings, and addiction usually follow extended use and withdrawal is usually multisymptomatic and exceedingly disagreeable.

PYRROLIZIDINE ALKALOIDS: POISON PLANTS AND INSECT DEFENSE

Alkaloids of the pyrrolizidine group are also derived from putrescine, but unlike the tropanes, are surprisingly common antiherbivore defensive compounds found in over 6000 plants worldwide, which are regarded as the most common poisonous plants known to affect wildlife, livestock, and humans. Of the approximately 660 known compounds in this group, some show carcinogenic or mutagenic properties, but roughly half of them are readily metabolized to compounds which are significantly toxic to the liver, representing a potentially important health problem for humans using certain medicinal herbs such as borage leaf, comfrey, or coltsfoot which are known to contain small quantities of pyrrolizidines. The biosynthesis of **retronecine**, the core compound from which many other pyrrolizidines are formed, is shown in Fig. 7.9.

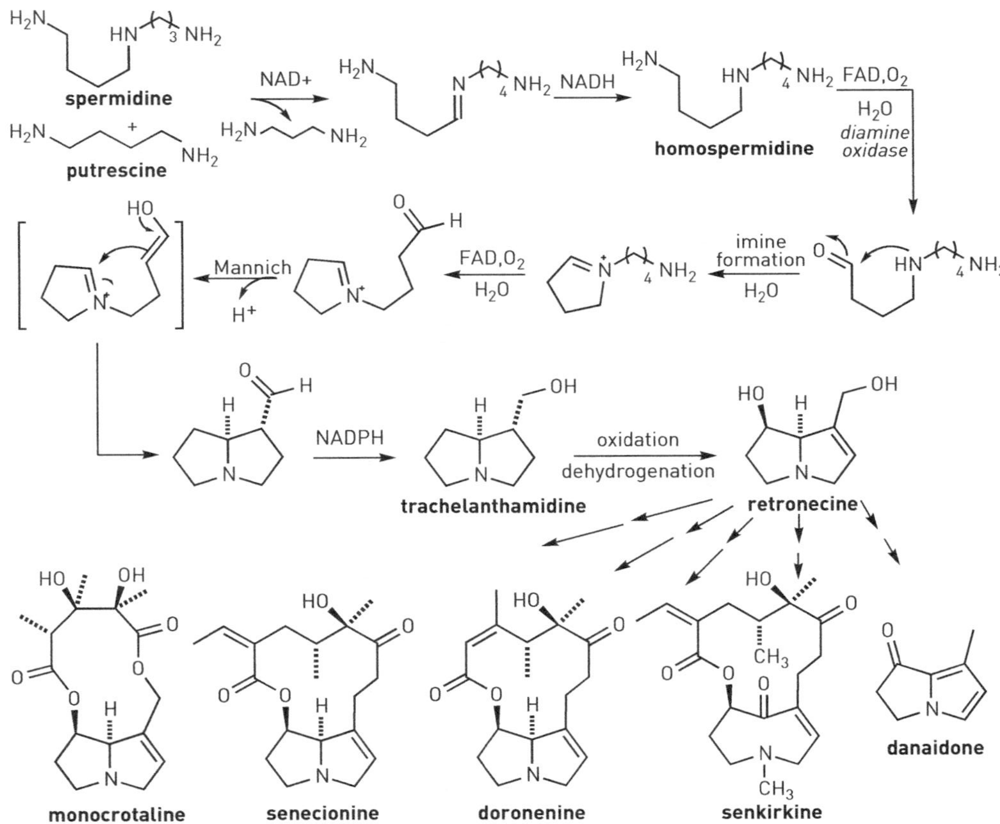

FIGURE 7.9
Biosynthesis of retronecine and some of its pyrrolizidine alkaloid derivatives.

Again, intramolecular imine formation and the Mannich reaction figure prominently here as ring-forming processes. Though other pyrrolizidine structures are known, many are diester derivatives of retronecine and related structures generally referred to as **necines**. The various carboxylic acids with which necines condense to form these esters are known as **necic acids.** Compounds such as **senecionine** and **doronenine** provide representative examples of these toxic diesters, while the carcinogenic **senkirkine** serves an example of a pyrrolizidine with cleavage of the usual bicyclic ring junction. Aside from their general toxicity, some pyrrolizidines are also known to play an ecological role as antifeedants in insects such as the larvae of the ornate moth which feeds on plants containing **monocrotaline**, secreting the stored compound as a toxic defense against predators. Other pyrrolizidines are consumed and converted biochemically by certain male butterflies into derivatives such as **danaidone** which acts as a female "flight arrestant" for mating purposes.

PIPERIDINE-TYPE ALKALOIDS DERIVED FROM LYSINE

We saw earlier how the piperidine alkaloid **coniine** was derived via an amination reaction of 5-oxo-octanoic acid. Most other piperidine alkaloids are actually derived from the amino acid **lysine** via a PLP-mediated decarboxylation, oxidation, and imine formation sequence involving a 6-membered rather than a 5-membered cyclic iminium ion intermediate, as illustrated in Fig. 7.10 for the biosynthesis of the piperidine alkaloids **pelletierine** and **pseudopelletierine**.

Decarboxylation of lysine affords the 1,5-diamine known as **cadaverine** whose foul smell, like that of putrescine, is associated with decaying flesh. A *diamine oxidase*-mediated oxidation of cadaverine affords the corresponding aminoaldehyde which then undergoes intramolecular imine formation and protonation to yield **Δ^1-piperidinium cation**, the key 6-membered ring intermediate. The remaining steps are analogous to those shown earlier (Fig. 7.5) for the formation of tropinone and its precursors. **Pelletierine** is a mild poison that is active against intestinal tapeworms; **pseudopelletierine**, a homologue of tropinone, is usually isolated along with pelletierine and although a mixture is often administered for tapeworm treatment, the antiparasitic activity resides only in pelletierine.

Some significant medicinal compounds contain piperidine-type alkaloids which have been used as building block components of larger structures. A good example of this is the incorporation of **L-pipecolic acid** in the structure of the polyketide macrolide **rapamycin** (sirolimus), a powerful immunosuppressant drug used in organ transplants and as a coating in coronary stent formulations. The biosynthesis of L-pipecolic acid from lysine, and its incorporation in the structure of rapamycin are shown in Fig. 7.11. Note the retention of the lysine carboxyl group throughout which distinguishes this route to piperidines from those in which decarboxylation of lysine precedes alkaloid formation.

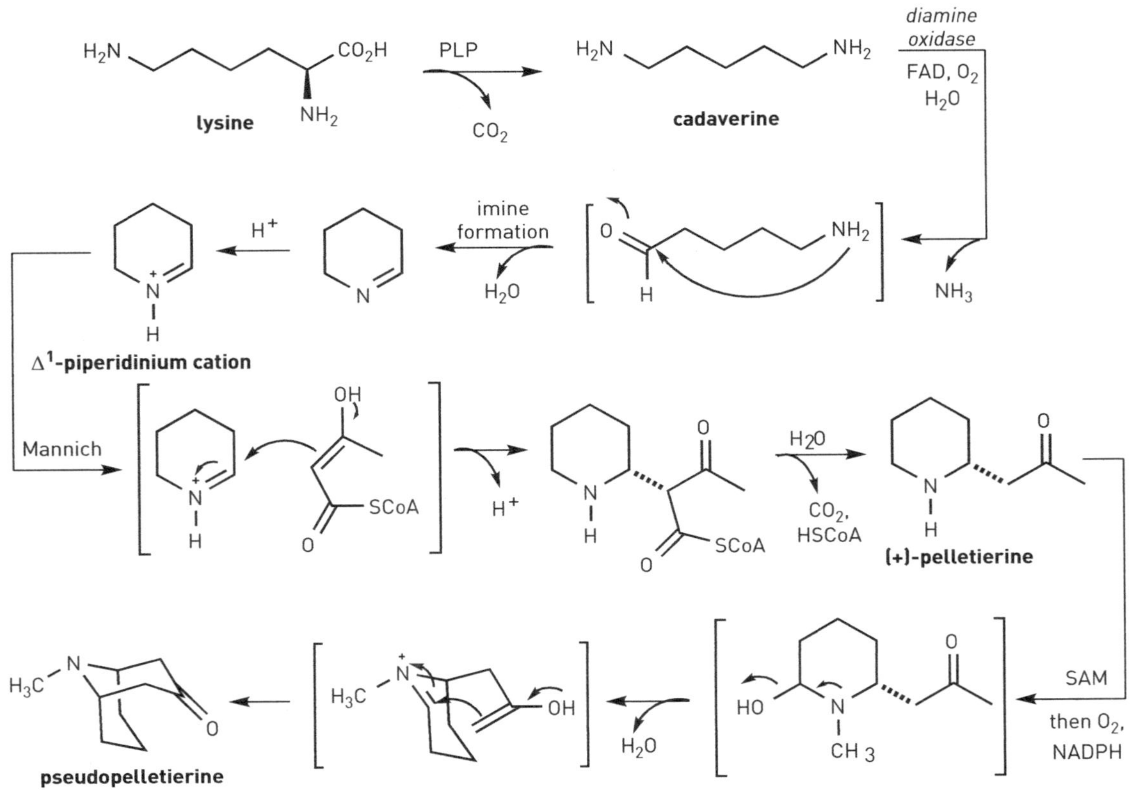

FIGURE 7.10
Conversion of lysine to cadaverine and biosynthesis of pelletierine and pseudopelletierine.

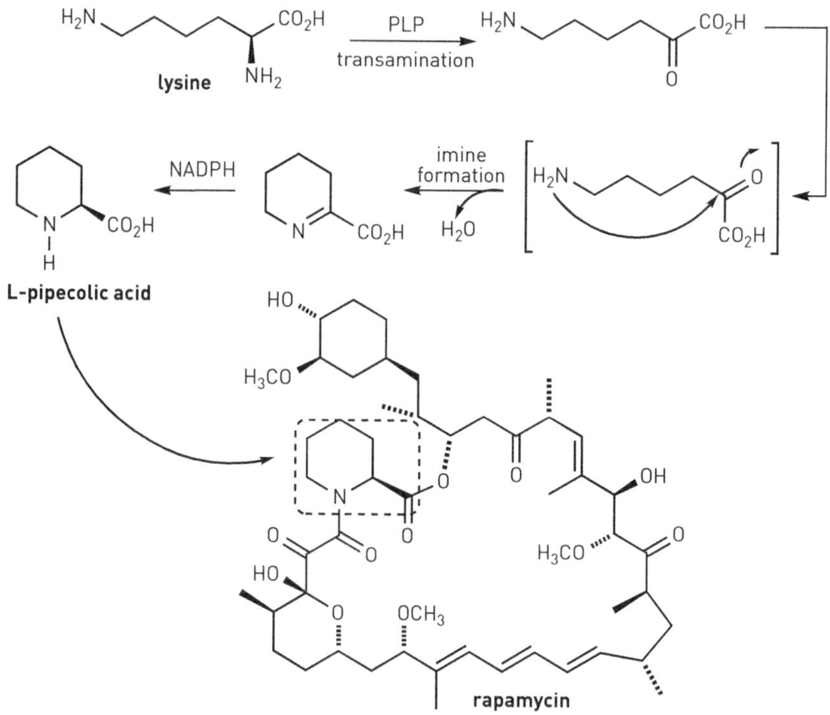

FIGURE 7.11
Origin of L-pipecolic acid in the macrolide immunosuppressant rapamycin.

As might be expected from its structural simplicity, L-pipecolic acid is not exclusively a plant secondary metabolite as it is produced in a variety of organisms via metabolism of lysine. However, it serves as the basic building block for certain **indolizidines**, a relatively small alkaloid group of pharmaceutical interest primarily because of the tri- and tetrahydroxy alkaloids **swainsonine** and **castanospermine**. Abbreviated biosynthetic routes are shown in Fig. 7.12.

Beginning with **L-pipecolylCoA**, one malonyl-CoA extension gives the diketide derivative which is reduced to a ketoaldehyde. Iminium ion formation and reduction give **1-indolizidinone** which is further reduced to diastereoisomeric alcohols. One diastereomer undergoes three sequential C–H oxidations to give the tetrahydroxy alkaloid **castanospermine**, while the other alcohol disastereomer is first oxidized to a diol and then to an iminium ion which upon reduction inverts the stereochemistry at the ring junction. A final C–H oxidation gives the trihydroxy alkaloid **swainsonine**.

The intoxicating and toxic gastrointestinal effects of "locoweed" on grazing animals are due in part to the swainsonine and castanospermine content of certain prairie plants. Both are known inhibitors of enzymes responsible for hydrolysis of glycoside linkages, especially *N*-glycoside linkages associated with glycoprotein formation, and their ability to interfere with formation of the viral glycoprotein coating of HIV has generated considerable interest. Swainsonine can also act as an appetite suppressant and has shown some promise both as an anticancer agent and in the

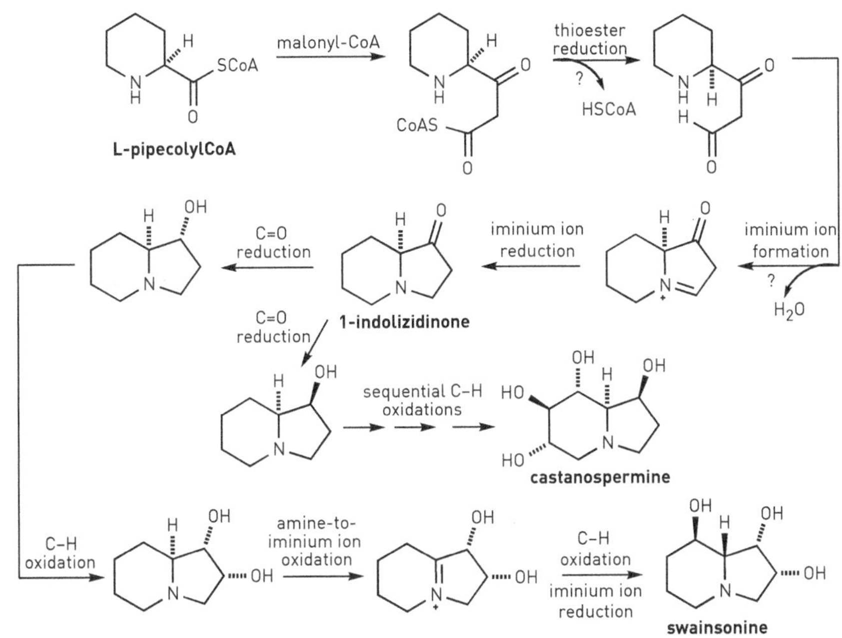

FIGURE 7.12
Proposed routes to the indolizidine alkaloids castanospermine and swainsonine.

promotion of postchemotherapy bone marrow replenishment. Both compounds have generated substantial activity among synthetic and medicinal chemists over the years as lead structures in a continuing quest for analogues with increased activity and diminished toxicity.

QUINOLIZIDINE ALKALOIDS: LIVESTOCK POISONS FROM CADAVERINE

The quinolizidine alkaloids, also known as the lupin alkaloids, have one or more nitrogen atoms shared by two rings and so are similar in structure to the indolizidines. Plant producing species are abundant and often act as poisons for grazing livestock, especially sheep. The presence of these alkaloids in lupin bean makes what might otherwise be a useful protein source problematical for human consumption. Unlike the indolizidines, studies have shown that the quinolizidines incorporate two or more cadaverine molecules as symmetrical precursors via use of **Δ^1-piperidinium cation** or its corresponding enamine. While many biosynthetic details remain to be worked out, some of the proposed transformations involved in production of **lupinine**, **lupanine**, and **sparteine** from various studies are shown in Fig. 7.13.

Cadaverine, which is first converted to its aminoaldehyde via oxidative deamination, condenses to form the cyclic imine. The protonated form, **Δ^1-piperidinium cation,** then participates in an initial Mannich reaction with its enamine tautomer; hydrolysis of the resulting iminium ion gives an amino aldehyde side chain on the newly formed piperidine ring. Oxidative deamination of the side chain to a dialdehyde is then

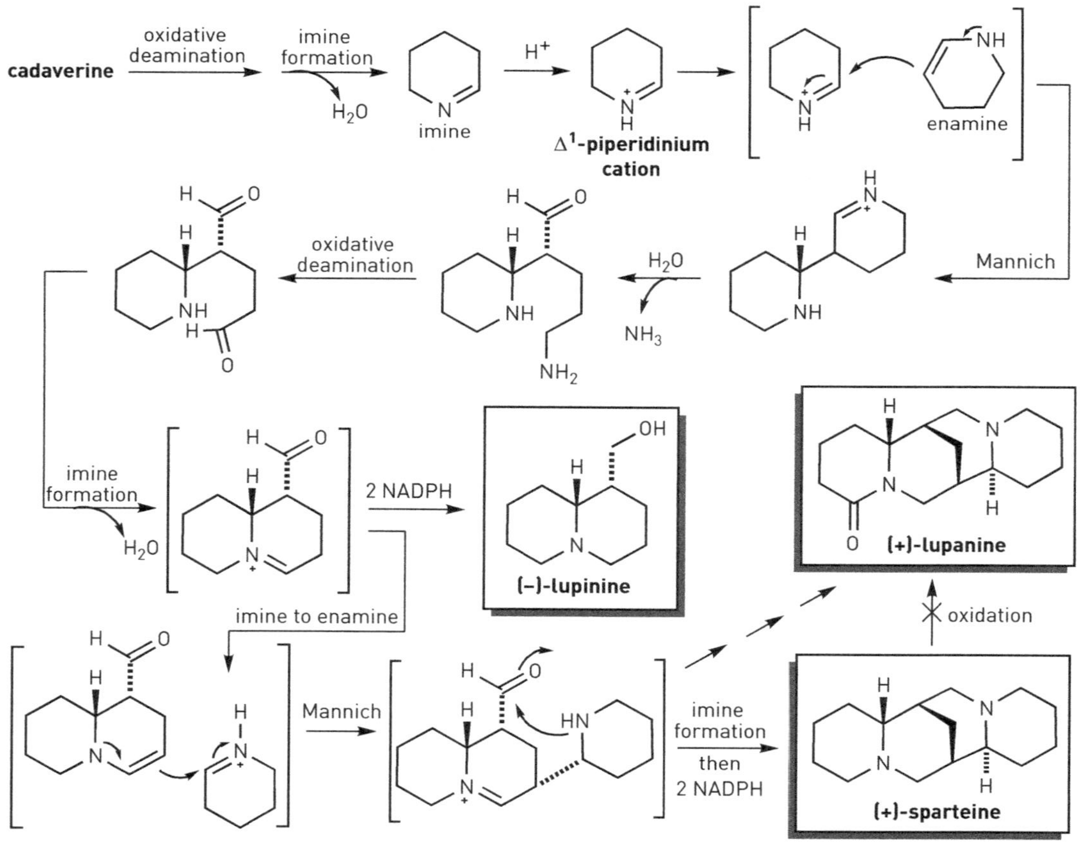

FIGURE 7.13

Proposed conversion of cadaverine to the quinolizidine alkaloids lupinine, lupanine, and sparteine.

followed by cyclization and imine formation. The resulting aldehyde-iminium ion is then either completely reduced by NADPH to produce **lupinine** or is isomerized to its corresponding enamine tautomer which may then serve as the nucleophile in a second Mannich reaction with Δ^1-piperidinium cation to form an amino-aldehyde-iminium ion that is further cyclized and reduced to give **sparteine**. The structurally related quinolizidine **lupanine** would initially appear to be derived from oxidation of sparteine, but apparently arises via oxidation of iminium ion precursors. Sparteine itself was formerly used as a diuretic and cathartic but sees little use for these purposes in recent times. All members of this class are bitter tasting alkaloids that play a role in the defense mechanisms of plants by acting as effective feeding deterrents for herbivores.

ALKALOIDS FROM PHENYLALANINE: FROM NEUROTRANSMITTERS TO DECONGESTANTS AND NARCOTICS

Phenylalanine and tyrosine are closely related to one another structurally, and in animals, tyrosine can be produced by oxidation of phenylalanine. On the other hand, plants produce these amino acids separately via a common intermediate in the shikimic acid pathway. In terms of plant alkaloid biosynthesis, those derived from tyrosine constitute a much larger group and are of special significance. Though fewer in number, plant alkaloids derived from phenylalanine are nevertheless quite abundant and are of particular interest from a pharmacological perspective.

The Chinese described *ma huang* and its medicinal uses as far back as 2700 B.C. First isolated in 1887, derived from *Ephedra* plant species with the active component **ephedrine**, this simple alkaloid came into widespread use only during the last century. Most producing plants yield a mixture of alkaloid diastereomers, with ephedrine itself usually the predominant component. Studies have demonstrated that phenylalanine provides the phenylC_1 component via benzoylCoA, with the remaining two carbons incorporated via TPP-mediated decarboxylation of pyruvic acid and nucleophilic acyl substitution to produce **1-phenyl-1,2-propanedione** as shown in Fig. 7.14. A transamination process then produces **cathinone** which undergoes carbonyl reduction to produce **norephedrine** and **norpseudoephedrine**, with the traditional "nor" prefix in alkaloid nomenclature indicating the absence of an *N*-methyl group relative to the parent alkaloid structures which, in this case, are those of **ephedrine** and **pseudoephedrine** formed as usual via subsequent SAM methylation. An additional SAM methylation produces the remaining pair of diastereomeric derivatives known as ***N*-methylephedrine** and ***N*-methylpseudoephedrine**. Note the similarity in structure to the synthetic alkaloids **amphetamine** and **methamphetamine**, the latter of which (generically known as "crystal meth") is a devastatingly addictive and destructive drug of abuse.

Ephedrine is quite useful for the relief of symptoms associated with asthma due to its long lasting effect as a bronchodilator. Its effect on mucous membranes also makes

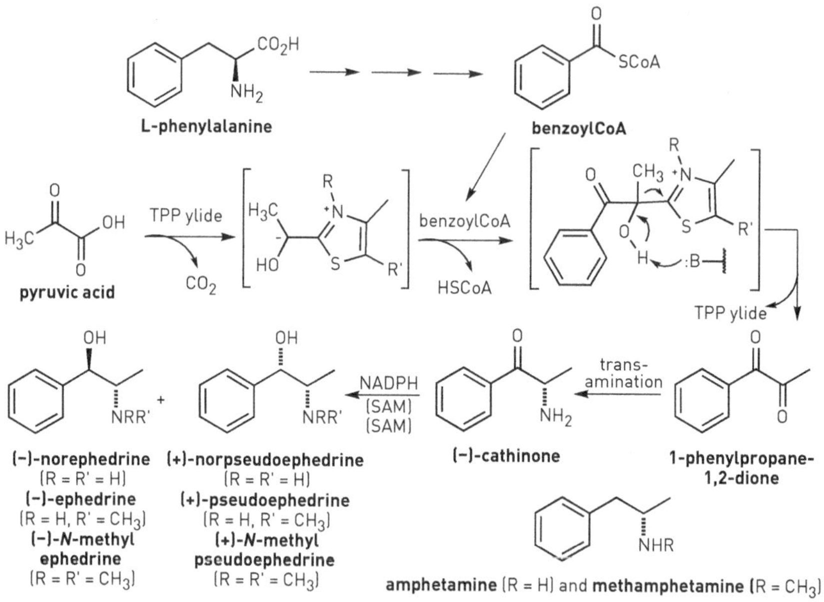

FIGURE 7.14
Biosynthesis of the ephedrine alkaloid group and structures of the synthetic amphetamines.

it useful as a nasal decongestant for the treatment of symptoms associated with hay fever and other allergies. Pseudoephedrine (trade name Sudafed) is likewise useful in this regard and is incorporated into many formulations for cough syrups and other cold remedies, though its availability is limited due to its ready conversion into **methamphetamine** via various reductive methods. A mixture of norephedrine and norpseudoephedrine, commonly known as **phenylpropanolamine (PPA)**, has similar properties and was once widely used in over-the-counter (OTC) cold medicine formulations in the United States but was banned in the mid-2000s due to an increased risk of stroke in younger women. PPA is still in use in some European countries and is also used in veterinary medicine for treatment of urinary incontinence in dogs.

ALKALOIDS FROM TYROSINE: THE PICTET–SPENGLER REACTION IN ALKALOID BIOSYNTHESIS

Tyrosine is the basic building block for an enormous number of different alkaloid structures. The simplest of such alkaloids are typified by compounds such as **tyramine, hordenine, dopamine**, and **adrenaline**, as shown in Fig. 7.15.

The biosynthesis of these **β-arylethylamines** is straightforward, beginning with PLP-mediated decarboxylation to the corresponding amines followed by alkylations or oxidations. Decarboxylation of tyrosine leads directly to **tyramine** which occurs widely in plants. Interestingly, while commonly found in the diet, tyramine-rich foods may cause dangerous hypertension crises in individuals taking antidepressant **monoamine oxidase inhibitors** (MAOIs), since MAOs are required to properly

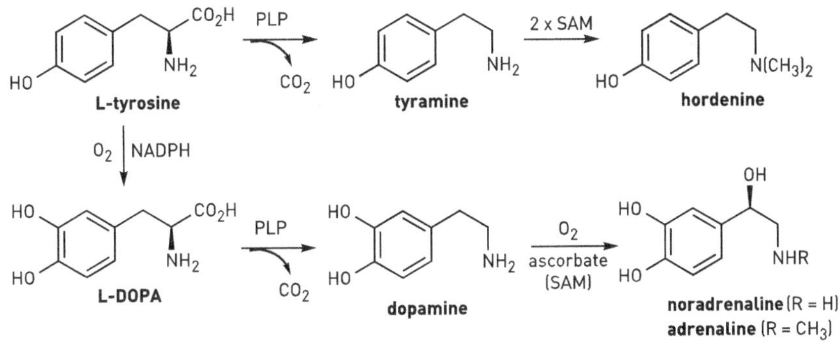

FIGURE 7.15
Conversion of tyrosine to tyramine, hordenine, L-DOPA, dopamine, noradrenaline, and adrenaline.

metabolize tyramine. In barley, double *N*-alkylation of tyramine via SAM affords **hordenine**, a common alkaloid that is widely sold as a nutritional supplement, usually marketed as a weight loss aid though there are no valid studies in humans to support such claims. In addition to being an essential drug for the treatment of Parkinson's disease, **L-DOPA**, derived from ring hydroxylation of tyrosine, is an important **catecholamine** neurotransmitter and precursor to the other important CNS catecholamines **dopamine**, **noradrenaline** (norepinephrine), and **adrenaline** (epinephrine) via subsequent decarboxylation (to give dopamine) followed by oxidation (to give noradrenaline) and then SAM *N*-alkylation (to give adrenaline).

The condensation of β-arylethylamines like tyramine or dopamine with various aldehydes or ketones leads to imine derivatives which may undergo a biological version of an old but very useful intramolecular cyclization known as the **Pictet–Spengler reaction**. The general process is outlined in Fig. 7.16 for the biosynthesis of two different tetrahydroisoquinoline alkaloids. **Lophocerine** (one of a group of dopamine-derived cactus alkaloids that includes the psychedelic phenethylamine **mescaline** as shown) is formed via initial SAM O-alkylation of dopamine followed by condensation with 3-methylbutanal. The resulting imine then undergoes the Pictet–Spengler cyclization which is essentially an intramolecular electrophilic aromatic substitution onto an electron-rich aryl ring by a protonated imine function serving as the electrophilic component. Similarly, reaction of dopamine with acetaldehyde (an ethanol metabolite) leads to a similar sequence of condensation and cyclization, affording **salsolinol**, traces of which may be detected in humans as the result of alcohol consumption.

(*S*)-RETICULINE: A VERSATILE PICTET–SPENGLER-DERIVED BENZYLTETRAHYDROISOQUINOLINE

Probably the most important alkaloid building blocks derived from the Pictet–Spengler process are the so-called **benzyltetrahydroisoquinolines**. In particular,

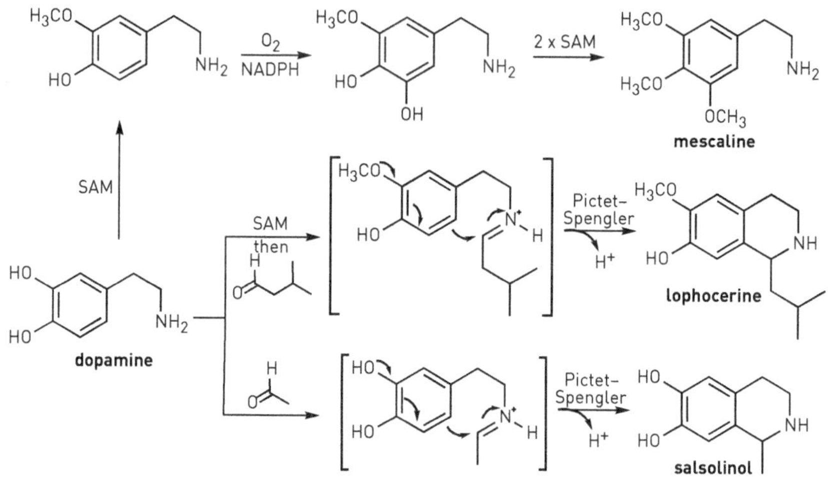

FIGURE 7.16
Dopamine-derived mescaline biosynthesis and Pictet–Spengler formation of lophocerine and salsolinol.

(*S*)-reticuline is a remarkably versatile member of this family that serves as a precursor to a remarkably diverse array of polycyclic alkaloid structures, as shown in Fig. 7.17.

Before briefly examining just a few of the important products derived from this essential compound, we take a look at how it is assembled from tyrosine as outlined in Fig. 7.18.

The process begins with conversion of tyrosine to **dopamine** as before. For the Pictet–Spengler reaction, a carbonyl compound is required for condensation with dopamine and in this case, tyrosine is again the source providing ***p*-hydroxyphenylethanal** via transamination to the corresponding α-ketoacid followed by TPP-mediated decarboxylation. After condensation, the resulting imine undergoes the usual Pictet–Spengler cyclization to yield the initial benzyltetrahydroisoquinoline product **(*S*)-norcoclaurine.** Sequential SAM *N*- and *O*-methylations followed by *ortho*-hydroxylation of the phenolic ring component affords **(*S*)-3'-hydroxy-*N*-methylcoclaurine**. A final SAM *O*-methylation leads directly to **(*S*)-reticuline** as shown.

OXIDATIVE COUPLING IN ALKALOID BIOSYNTHESIS: BIOSYNTHESIS OF CORYTUBERINE AND MORPHINE

An important route to a number of different alkaloids derived from benzyltetrahydroisoquinolines involves a process we have encountered before, namely **phenolic oxidative coupling.** A simple example of this route to polycyclic alkaloids is the direct intramolecular oxidative coupling of (*S*)-reticuline to produce **(*S*)-corytuberine**, a member of the so-called **aporphine** family of alkaloids, as shown in Fig. 7.19.

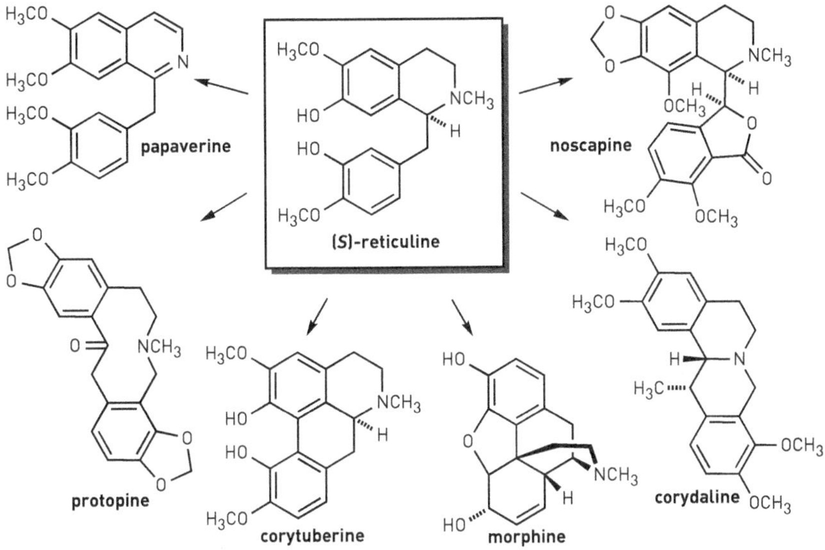

FIGURE 7.17
Structural diversity and complexity of polycyclic alkaloids derived from (*S*)-reticuline.

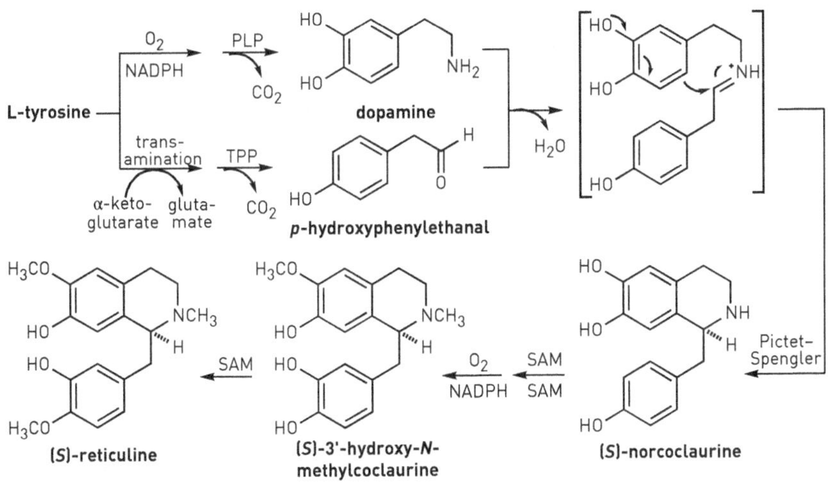

FIGURE 7.18
Conversion of tyrosine to the benzyltetrahydroisoquinoline alkaloid (*S*)-reticuline.

Assuming the indicated conformation for (*S*)-reticuline, oxidative formation of the phenoxy radical pair leads to an obvious opportunity for *ortho–ortho* resonance contributors to participate in intramolecular C–C bond formation. The resulting diketone rapidly tautomerizes to the corresponding diphenolic (*S*)-corytuberine product.

Other variations on this theme are well known in alkaloid biosynthesis, but certainly one of the most important is the intramolecular oxidative phenolic coupling process involved in the biosynthesis of opium-poppy-derived **morphine**, an extremely

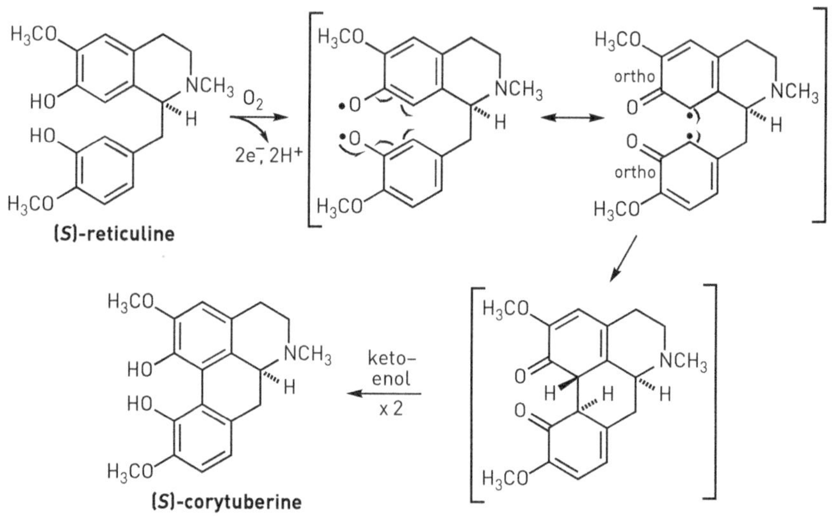

FIGURE 7.19
Intramolecular ortho–ortho oxidative coupling of (*S*)-reticuline to give (*S*)-corytuberine.

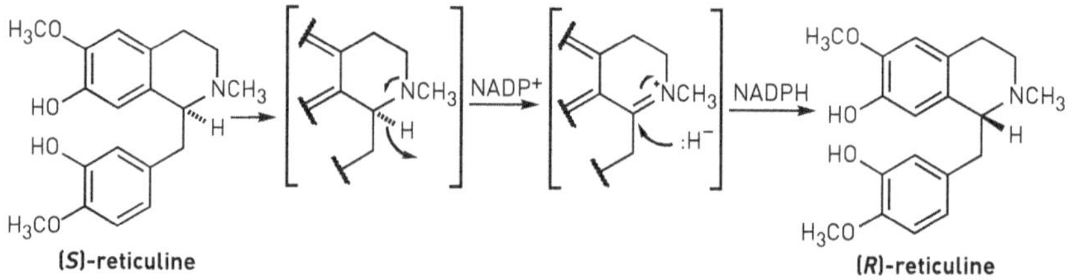

FIGURE 7.20
NADP+ oxidation of (*S*)-reticuline to an iminium ion and its NADPH reduction to (*R*)-reticuline.

potent analgesic alkaloid which, in spite of its highly psychoactive and dangerously addictive properties, will be found on any list of pharmacological compounds considered essential for basic health care delivery. As shown in Fig. 7.20, the biosynthetic route to morphine begins with a two-step inversion of configuration of (*S*)-to-**(*R*)-reticuline** via NADP$^+$ oxidation to an iminium ion which is then reduced by NADPH addition of hydride to the iminium ion *Si* face. While involved in biosynthesis of other alkaloids, (*R*)-reticuline finds its most important role within the context of morphine production.

The phenolic oxidative coupling step in morphine biosynthesis is illustrated in Fig. 7.21. It begins with (*R*)-reticuline in a conformation that brings the two phenolic rings together in an *ortho–para* orientation with respect to one another. For convenience, one methoxyl group is bolded to allow the reader to track how the mediating enzyme folds the usual representation of (*R*)-reticuline into the required conformation prior to oxidative coupling. As in the corytuberine biosynthesis, the product resulting from the intramolecular coupling is initially a diketone, but in this case, the 2,5-cyclohexadienone form of the lower ring cannot tautomerize from ketone to enol form since the 4-position is quaternary. Enolization of the upper ring

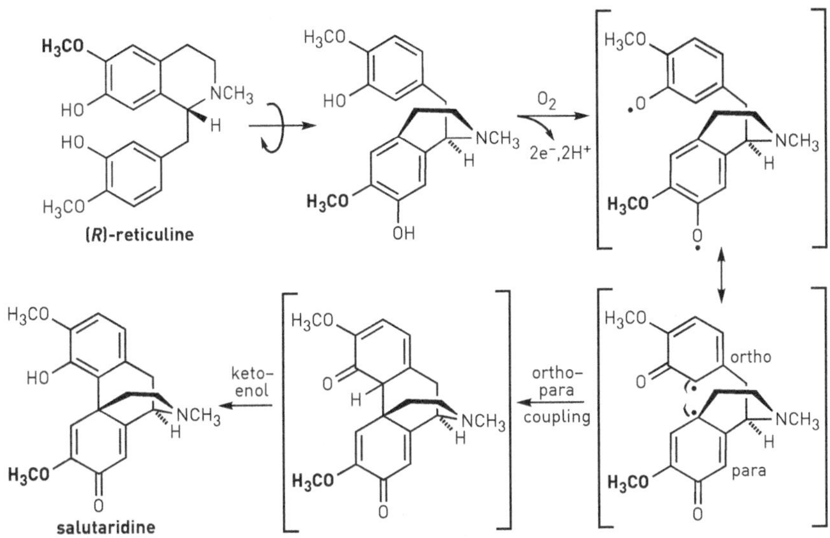

FIGURE 7.21
Intramolecular ortho–para oxidative phenolic coupling in formation of salutaridine from (*R*)-reticuline.

leads to the usual phenolic structure, affording the intermediate product **salutaridine** as shown.

To complete the transformation to morphine, the 2,5-cyclohexadienone ketone function of salutaridine is first reduced to the corresponding alcohol via NADPH, as shown in Fig. 7.22, affording **salutaridinol.** The alcohol function is then trans-esterified to an acetate derivative via reaction with acetyl-CoA. This intermediate acetate ester, **7-*O*-acetylsalutaridinol**, undergoes a spontaneous nucleophilic attack by the upper ring phenolic hydroxyl group, with the allylic acetate function serving as the leaving group in what is essentially an S_N2'-type nucleophilic substitution. The resulting product, **thebaine**, is a minor constituent of opium extract that is nevertheless significant since it serves as a useful precursor to a large number of important semisynthetic opiate derivatives such as **oxycodone** and **naloxone** (about which more will be said shortly). An O_2/2-oxoglutarate oxidation of the enol ether methyl group leads to demethylation and formation of an enol which tautomerizes to the corresponding ketone function of **neopinone**, a nonconjugated β,γ-unsaturated cyclohexenone which then undergoes spontaneous isomerization to give the conjugated α,β-unsaturated ketone function of **codeinone**. NADPH reduction of the codeinone ketone function affords the useful and familiar antitussive agent **codeine**, which upon further O_2/2-oxoglutarate oxidative demethylation of the upper ring methoxyl group finally affords **morphine** as shown.

Among the oldest of plant-derived alkaloids harvested for use by humans and the first plant alkaloid isolated in pure form (in 1804), morphine is found only in opium poppy (*Papaver somniferum*) and a few related species in which it is known to act (along with other opioid compounds) as a chemical defense against

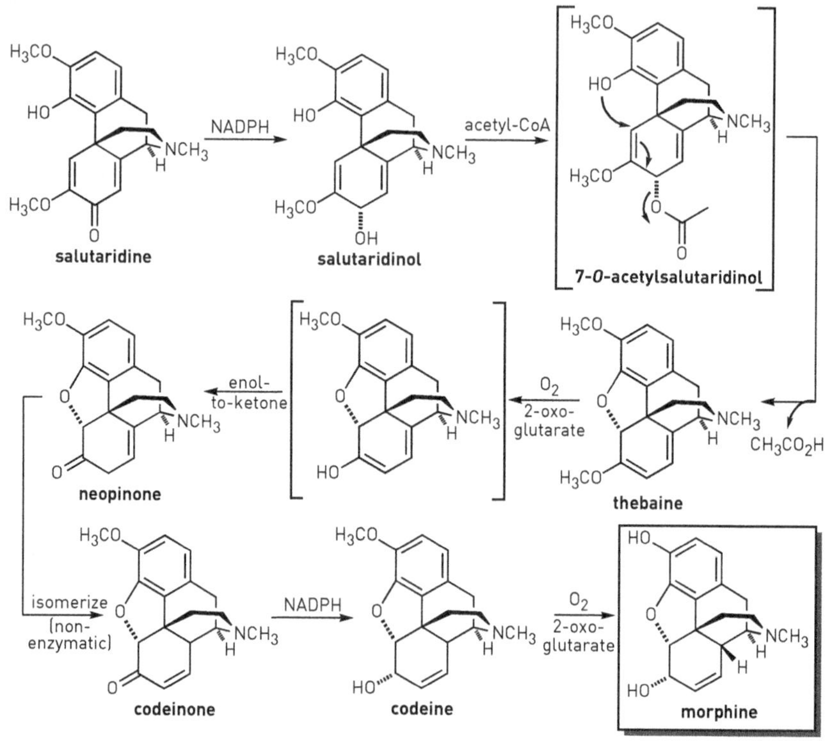

FIGURE 7.22
Conversion of salutaridine to thebaine and the principal route from thebaine to the opium poppy alkaloids codeinone, codeine, and morphine.

herbivores. Among its many physiological properties, morphine readily induces sleep in humans and its name was originally derived from Morpheus, the god of dreams in ancient Greek mythology. While both morphine and codeine are widely used in medicine, morphine's unrivaled ability to relieve severe and otherwise unbearable pain makes it one of the most indispensable drugs for post-surgical applications and for alleviation of the intense pain and respiratory anxiety often associated with certain late-stage cancers. This action is achieved by its ability to strongly bind to the so-called **opioid receptors** present in humans and other animals. While codeine is considered moderately addictive, morphine is a highly addictive substance associated with strong physical and psychological dependence and rapid development of increasing levels of tolerance with extended use. Interestingly, both morphine and codeine continue to be obtained primarily from plant extracts since commercially viable synthetic routes have yet to be developed in spite of many years of effort.

THE MORPHINE RULE

In addition to morphine, there are many structurally related opioid analgesics, both naturally occurring and synthetic. As it turns out, there are some basic structural features present in morphine that allow it to bind strongly to the opioid receptors and which partially account for its unusually potent analgesic properties. These fundamental structural

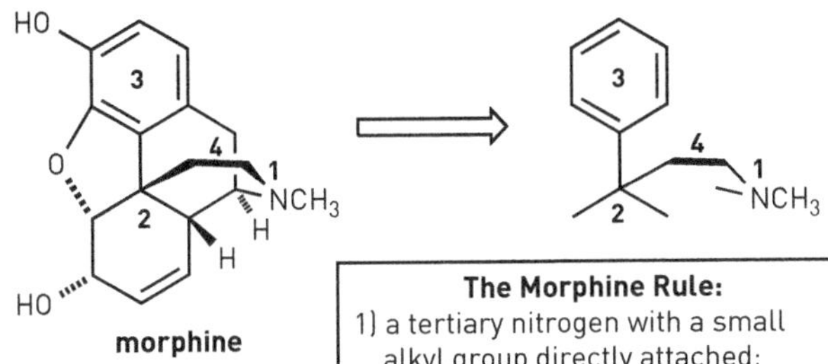

The Morphine Rule:

1) a tertiary nitrogen with a small alkyl group directly attached;
2) a quaternary carbon;
3) an aryl ring (or equivalent) attached to the quaternary carbon;
4) a 2-carbon spacer between the nitrogen and the quaternary carbon.

FIGURE 7.23
The morphine rule: structural requirements for active opioid analgesics.

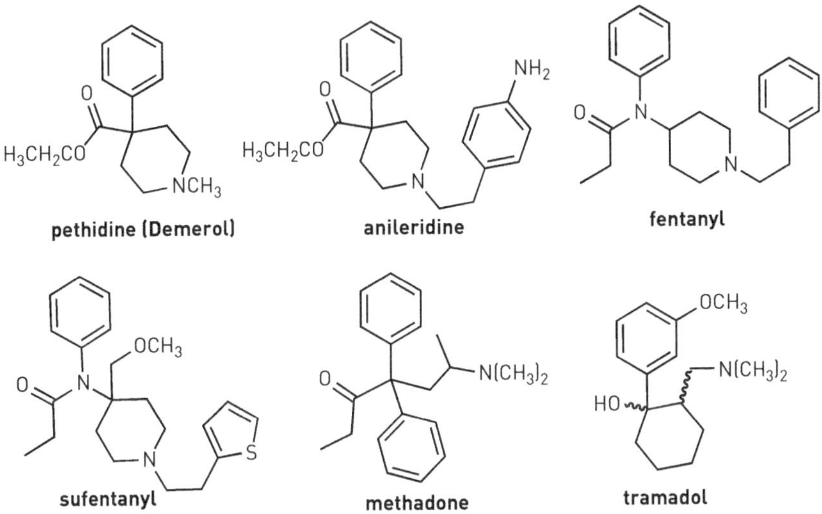

FIGURE 7.24
Representative synthetic opioid analogues illustrating features of the morphine rule.

requirements, summarized in the so-called **morphine rule**, are illustrated in Fig. 7.23 as follows: 1) a tertiary nitrogen with a small alkyl group directly attached; 2) a quaternary carbon; 3) an aryl ring (or equivalent) directly attached to the quaternary carbon; and 4) a 2-carbon spacer located between the nitrogen and the quaternary carbon.

The morphine rule has proven to be a useful if limited guide in the structural formulation of many different synthetic or semisynthetic analogues that not only retain the opioid analgesic properties of morphine, but in some cases even vastly exceed the parent compound in terms of pharmacological potency. Though far too great in number to be fully included and discussed here, Fig. 7.24 shows a representative sampling of such structural analogues.

Among the earliest synthetic morphine analogues were **pethidine** and **anileridine,** both of which contain a piperidine ring as a means of providing a two-carbon spacer between nitrogen and quaternary carbon for the morphine rule. Pethidine was developed in the late 1930s and was originally believed to be more effective as an analgesic and less harmful than morphine, though it was subsequently found to be similarly addictive and to also produce a relatively toxic metabolite, which now limits its use. Developed later in the 1950s, **anileridine** is structurally similar to pethidine and is still used for treatment of moderate to severe pain in certain instances. **Fentanyl**, developed still later and the most widely used of all synthetic opioids, is 50–100 times more potent than morphine and remains an essential tool for the treatment of severe, chronic, or "breakthrough" pain and as a preoperative anesthetic. A powerful derivative of fentanyl, **sufentanyl** is 500 times more potent than morphine and, like fentanyl, is useful as both an analgesic and as a presurgical anesthetic. Low doses of both of these piperidine-based analogues are often administered via transdermal patches for the treatment of chronic, severe pain. **Methadone,** which incorporates all four components of the morphine rule, can be used medically as an analgesic, but is used mainly for the treatment of opioid dependence. By blocking or reducing the euphoric effects of morphine, heroin, and similar drugs, methadone administration can help to reduce the withdrawal symptoms experienced by patients addicted to such alkaloids, thereby increasing the chances for overcoming the addiction. A final example of a fully synthetic opioid analogue is **tramadol** which, interestingly, was also recently discovered to be produced naturally in the roots of the African pin cushion tree. A relatively simple but widely prescribed synthetic, tramadol is about as effective as morphine for the treatment of moderate pain, though less so for severe pain.

Equally important among opioid analogues are the semisynthetic derivatives. Many of these are derived through synthetic manipulation of the morphine precursor **thebaine** as shown in Fig. 7.25.

Codeine, the most widely used of all opioids, is naturally occurring but prepared mainly by synthetic methylation of morphine. However, it may also be produced from **thebaine** which is available from certain plant strains that produce little or no morphine. As a semisynthetic opioid dating from 1917, **oxycodone** has developed a reputation in recent years as a drug of abuse in spite of its widely recognized benefits as a powerful and useful pain-reliever. Although oxycodone-related dependence and withdrawal are significant problems, the drug is second only to morphine in its use for treatment of severe pain associated with various cancers. By contrast, **naloxone** is one of the most important of all opioids, not for its analgesic properties, but rather for its ability to bind and block the opioid receptors without eliciting its own biological response, making it the front-line **antagonist** for the treatment of opioid overdose. As a final example of the power of semisynthetic analogues, we note the veterinary opioid **etorphine**, which has been shown to be 5000–10,000 times more potent than morphine. Useful mainly for large animal sedation, darts with as little as 1 mg may be used for sedation of a typical rhinoceros and as little as 4 mg for a typical elephant. This extraordinary potency is related to etorphine's unusually strong binding affinity

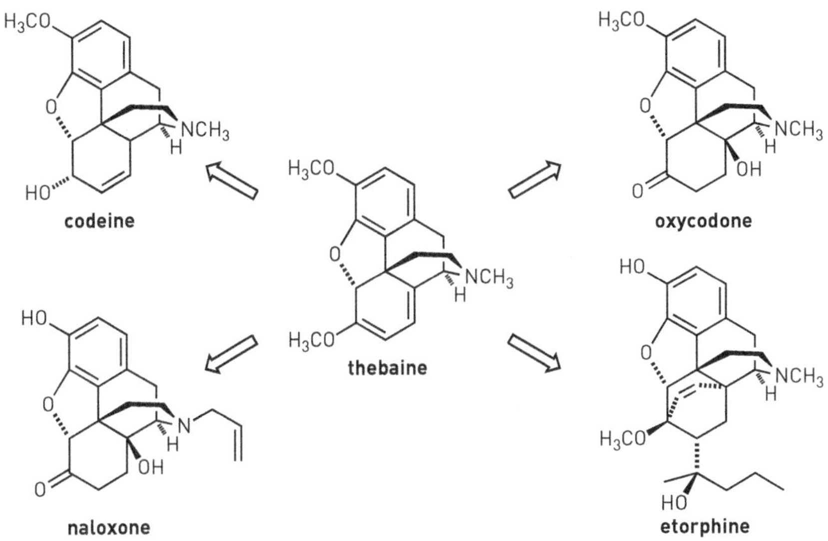

FIGURE 7.25
Important representative semisynthetic opioids derived from thebaine.

for the opioid receptors, a property which makes etorphine far too dangerous for use in human medicine. Approved only for veterinary use, kits for its administration always include separate dosages of **diprenorphine**, an antagonist similar to but far more powerful than naloxone, for immediate use in case of accidental administration.

ALKALOIDS FROM TRYPTOPHAN: ADVENTURES IN INDOLE ALKALOID STRUCTURAL COMPLEXITY

The amino acid **tryptophan** gives rise to a remarkably large group of structures often simply referred to as the **indole alkaloids** since the indole nucleus is the basic heterocyclic ring system common to most of these compounds. Within this large family are many subgroups ranging from relatively simple modified **tryptamine alkaloids** to much more complex derivatives such as the DMAPP-derived **ergot alkaloids** and the so-called **terpenoid indole alkaloids** which are derived in part from **secologain** (a terpenoid product we encountered briefly in Chapter 4) and which include the **vinca, strychnos,** and **quinine**-type alkaloids. We begin this section with the simple tryptamine derivatives **serotonin** and **melatonin**, as shown in Fig. 7.26.

Tryptamine is a simple monoamine alkaloid derived from PLP-mediated decarboxylation of tryptophan. It is found in animals, plants, and fungi and is believed to play a role as both a neuromodulator and a neurotransmitter, in part by affecting levels of **serotonin**, which is biosynthesized via simple ring oxidation of tryptophan to **5-hydroxytryptophan** followed by PLP-mediated decarboxylation. Serotonin plays a variety of significant roles as a **monoamine neurotransmitter** including regulation of mood, sleep, appetite, and other behavioral states. **SSRIs** (**s**elective **s**erotonin **r**euptake **i**nhibitors) are widely used antidepressant drugs designed to help

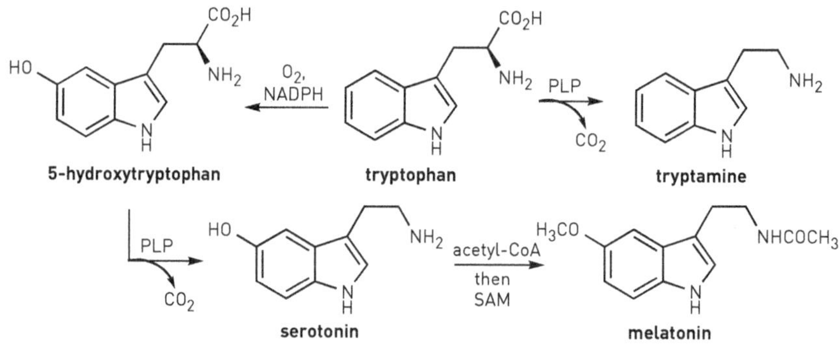

FIGURE 7.26
Simple derivatives of tryptophan: the neurochemicals tryptamine, serotonin, and melatonin.

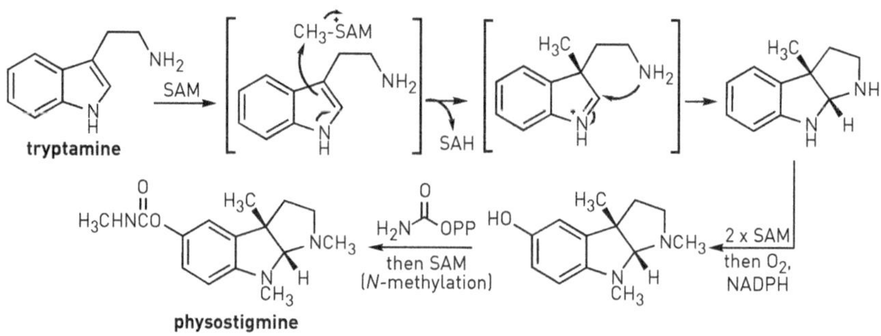

FIGURE 7.27
Conversion of tryptamine to the calabar bean alkaloid physostigmine.

moderate mood by increasing serotonin levels in the brain. Serotonin is also available from a variety of food sources such as mushrooms, fruits, and vegetables. *N*-acetylation and SAM methylation of serotonin affords the closely related hormone **melatonin.** In humans, melatonin plays an important role in the sleep–wake cycle. Light availability at the retina of the eye inhibits melatonin production in the pineal gland while darkness favors its production. Melatonin also serves as a powerful antioxidant, helping to protect both mitochondrial and nuclear DNA from the damaging effects of oxygen and nitrogen radical species.

Another relatively simple tryptamine derivative is **physostigmine**, an indole alkaloid isolated from the calabar bean. The presumed pathway to this alkaloid involves SAM methylation of the tryptamine indole ring at the 3-position to afford an iminium ion which then undergoes intramolecular nucleophilic attack by the adjacent primary amino group, forming a tricyclic 6-5-5 ring system as shown in Fig. 7.27. Subsequent SAM *N*-methylations are followed by an aryl ring hydroxylation with the resulting phenolic oxygen being converted to a carbamate-type linkage by reaction with carbamoyl phosphate. A final *N*-methylation of the carbamate function affords physostigmine. An older drug now used primarily as an antidote for severe poisoning by anticholinergic agents, it has also long been used topically for the treatment of glaucoma and has more recently been studied for potential use in the treatment of Alzheimer's disease. Physostigmine also has an interesting history from a chemical synthesis standpoint, one that we will touch on briefly in Chapter 8.

Derived from tryptophan and DMAPP and produced primarily by certain fungi, the **ergot alkaloids**, a family of structurally complex indole alkaloids, have a remarkable array of important biological activities The ergots are usually divided into two groups: the structurally simple **clavine**-type alkaloids, of which **agroclavine** is an example; and the core compound **lysergic acid** from which is derived a variety of simple amide and cyclic peptide amide derivatives, both naturally occurring and synthetic. The proposed pathway for formation of both agroclavine and lysergic acid (as shown in Fig. 7.28) is complex and has been much studied,

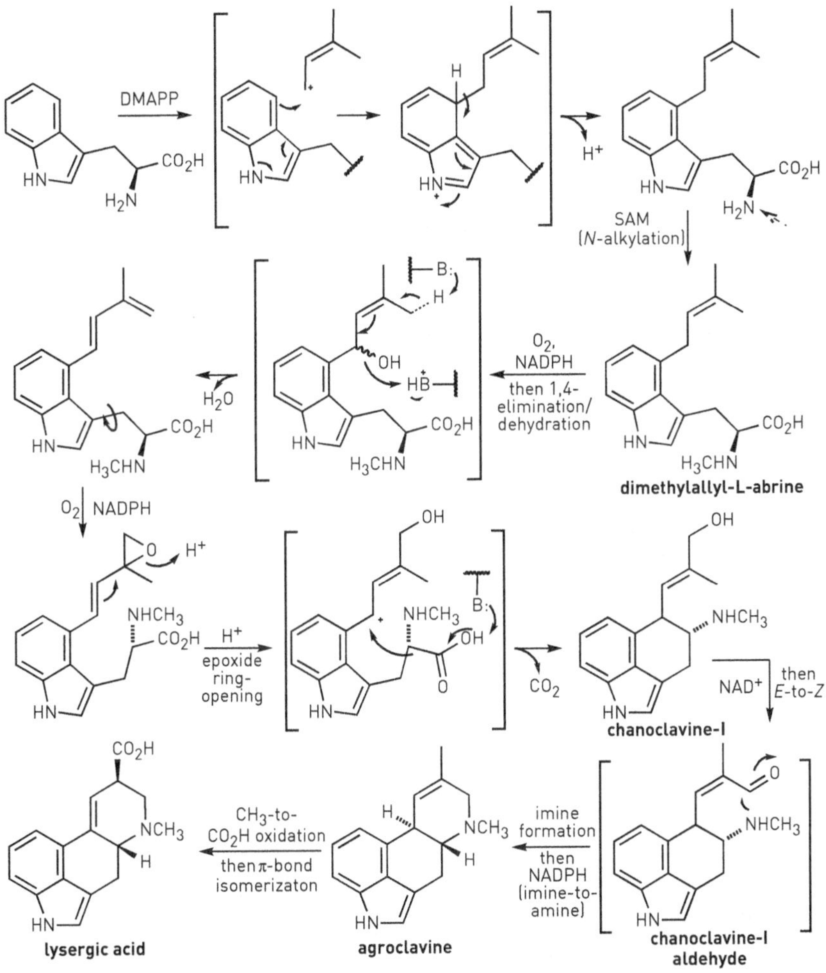

FIGURE 7.28

Biosynthetic origin of the ergot alkaloids agroclavine and lysergic acid.

though many details are lacking. Beginning with electrophilic aromatic substitution by DMAPP-derived dimethylallyl cation on the aryl ring of tryptophan, a subsequent SAM *N*-methylation affords **dimethyllallyl-*L*-abrine** which can undergo a benzylic oxidative hydroxylation at the dimethylallyl side-chain with subsequent 1,4-dehydrative elimination to produce a conjugated diene. Epoxidation and acid-catalyzed epoxide ring-opening to a benzylic cation is followed by decarboxylation and nucleophilic ring closure. The resulting **chanoclavine-I** then undergoes NAD^+ oxidation and *E*-to-*Z* isomerization to give **chanoclavine-I aldehyde**. Intramolecular cyclization to an imine and subsequent NADPH reduction gives **agroclavine** whose methyl group then undergoes sequential oxidations to give the carboxylic acid derivative. A final allylic rearrangement of the ring π-bond placing it in conjugation with the aryl ring completes the final conversion to **lysergic acid** as shown.

There are over a dozen known, naturally occurring cyclic tripeptide amide derivatives of lysergic acid, many of which are important medicinal agents. Both **ergotamine** and **ergostine** (Fig. 7.29) have long been used for the treatment of migraine headaches, while **ergocryptine** is used as starting material for preparation of the valuable brominated derivative known as **bromocryptine** which is used in the treatment of Parkinson's disease, pituitary tumors, and several other conditions. Semisynthetic derivatives of lysergic acid include the notorious psychedelic hallucinogen **lysergic acid diethylamide (LSD)**, which is a controlled substance, while other simple amide derivatives include **ergometrine,** an invaluable medicinal agent widely used in obstetrics for the prevention of postchildbirth bleeding and to assist in placenta expulsion. **Cabergoline** is frequently used in place of bromocryptine for the management of pituitary tumors as well as symptoms of Parkinson's disease, a

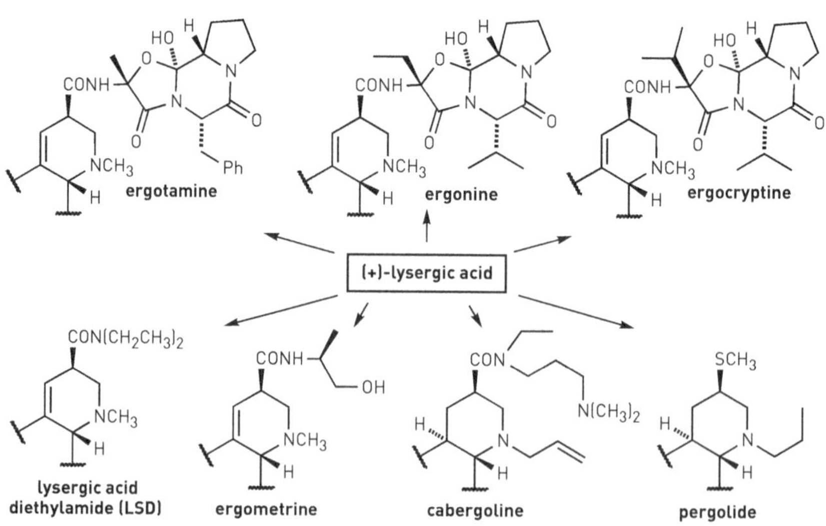

FIGURE 7.29
Naturally occurring (upper) and semisynthetic (lower) medicinal derivatives of lysergic acid.

condition for which **pergolide** was also once widely prescribed—it is no longer approved for use in the United States due to potential heart valve complications.

In addition to their importance to the pharmaceutical industry and human health, certain ergot alkaloids are the principal toxins involved in many historic epidemics of livestock and human poisonings due to consumption of infected grasses or grains. Known in the middle ages as St. Anthony's Fire, the symptoms of ergot poisoning or **ergotism**, as it is referred to in modern times, are divided into two groups, the first being characterized by severe convulsions and mental psychosis and the second by development of gangrenous tissue resulting from vasoconstriction, especially at the fingers or toes. Interestingly, some reports have even speculated that the convulsive and psychotic symptoms displayed by supposedly "bewitched" individuals described during the Salem witch trials were actually victims of ergotism arising from the consumption of moldy grains.

PICTET–SPENGLER-TYPE REACTIONS OF TRYPTAMINE: β-CARBOLINES AND INDOLE TERPENE ALKALOIDS

We have previously seen the importance of the Pictet–Spengler reaction in alkaloid biosynthesis, particularly for those derived from phenylalanine and tyrosine. An interesting variant of the reaction is also observed in a number of processes involving tryptamine as the amine component. A useful example that illustrates the fundamentals of tryptamine-based Pictet–Spengler processes is shown in Fig. 7.30 for the biosynthesis of **harmalan** and **harmine**, two representative members of the so-called **harmala** group of β-carboline alkaloids of which about ten are known. In this instance, tryptamine condenses with the ketone function of pyruvic acid to form the usual imine intermediate. Here the indole C-5 ring rather than the C-6 aryl ring attacks the electrophilic imine linkage leading to cyclization, proton loss, and restoration of indole aromaticity to afford the basic 6-5-6 ring system characteristic of the β-carbolines. Oxidative decarboxylation of the initial aromatized

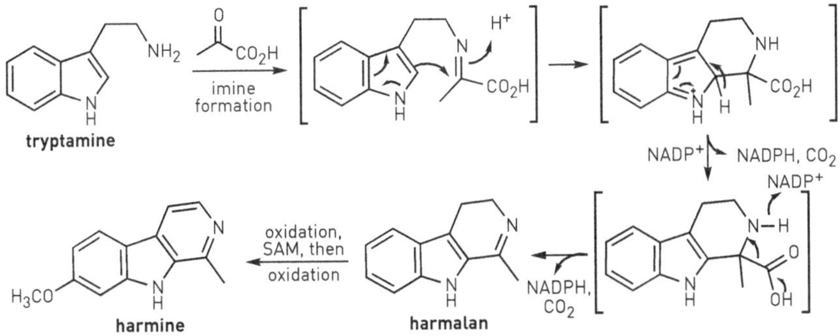

FIGURE 7.30
Conversion of tryptamine to the β-carboline alkaloids harmalan and harmine.

cyclization product leads directly to harmalan. Further modifications via aryl ring hydroxylation, SAM methylation, and dehydrogenation give the fully aromatic tricyclic derivative harmine. Produced by a number of different plants and also by animals, the harmala alkaloids, especially harmine, are reversible **monoamine oxidase inhibitors** (MAOIs) that affect the central nervous system; all are closely related structurally and several are known to be the psychoactive components of certain ritual "magic" potions prepared from native plants and used historically by indigenous Amazonian groups.

Also assembled utilizing the tryptamine variant of the Picet–Spengler reaction are the **indole terpene alkaloids**. As the name implies, these compounds incorporate a terpene component which serves as the carbonyl partner for condensation with tryptamine to form the usual imine component required for the cyclization process. The iridoid terpene component, commonly known as **secologanin,** is one which we have briefly encountered previously in Chapter 4 and which is used in the construction of all members of this alkaloid group. A proposed pathway to secologanin from geraniol is outlined in Fig. 7.31.

The process begins with conversion of geraniol to 8-hydroxygeraniol. NAD^+ oxidation of both alcohol functions then leads to the corresponding dialdehyde. NADPH conjugate reduction of the upper α,β-unsaturated aldehyde gives an enolate ion which then serves as a nucleophile for intramolecular conjugate addition to the lower α,β-unsaturated aldehyde function. The resulting cyclized dialdehyde, **irododial**, cyclizes further via a cyclic hemiacetal form to give **nepetalactol.**

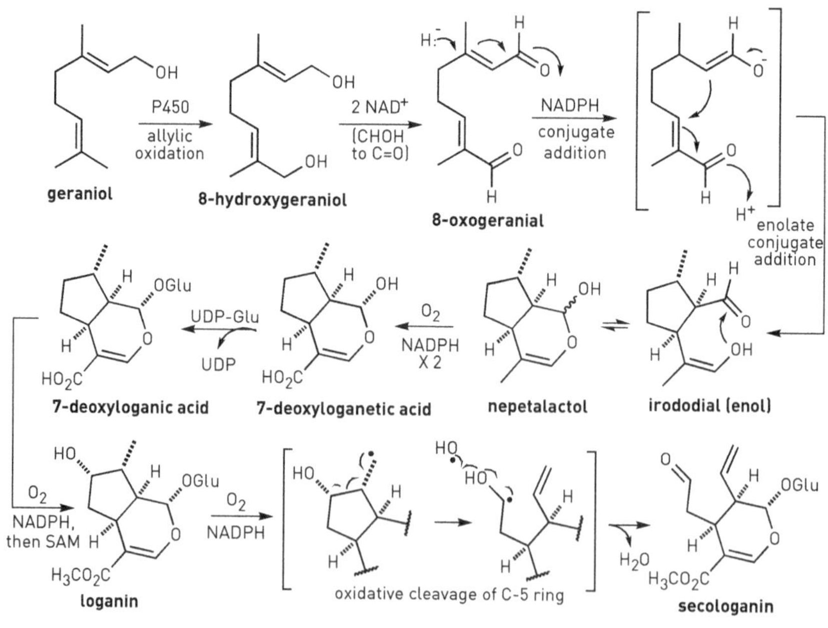

FIGURE 7.31

Simplified scheme for biosynthesis of the iridoid secologanin from geraniol.

Oxidation of the allylic methyl group of nepetalactol to the carboxylic acid gives **7-deoxyloganetic acid** whose lactol hydroxyl group subsequently undergoes glucosylation via UDP-Glu to give **7-deoxyloganic acid.** A C-5 ring hydroxylation affords **loganin** which finally undergoes an oxidative cleavage of the C-5 ring to give the iridoid terpene, **secologanin.**

Condensation of secologanin with tryptamine followed by a Pictet–Spengler-type cyclization affords the key intermediate **strictosidine,** which in turn serves as the starting material for assembly of a vast number of structurally diverse bioactive alkaloids, many of which are important medicinal agents. Figure 7.32 provides an overview of some of the more important members of the strictosidine-derived indole terpene alkaloid family, of which over 2000 members have been isolated to date. Given the sheer numbers involved, we must limit ourselves to examining a few representative members that illustrate some of the types of transformations that are likely to be common to others. As before, not all of the pathways shown have been fully elucidated and the mechanisms proposed are ones which make reasonable chemical sense given the limited amount of information available.

We begin with **ajmalicine** (also known as raubasine), an alkaloid found in a variety of plants such as *Rauwolfia spp.* and which is used in the treatment of hypertension (high blood pressure). The proposed pathway (Fig. 7.33) commences with deglucosylation of strictosidine to give the corresponding free hemiacetal of which

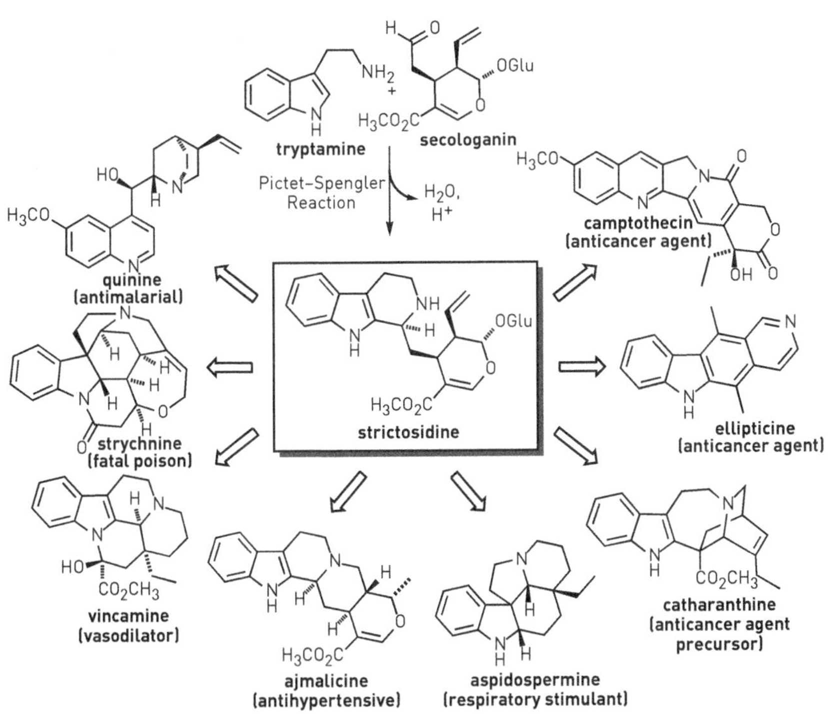

FIGURE 7.32
Diversity of bioactive or medicinal alkaloid structures derived from strictosidine.

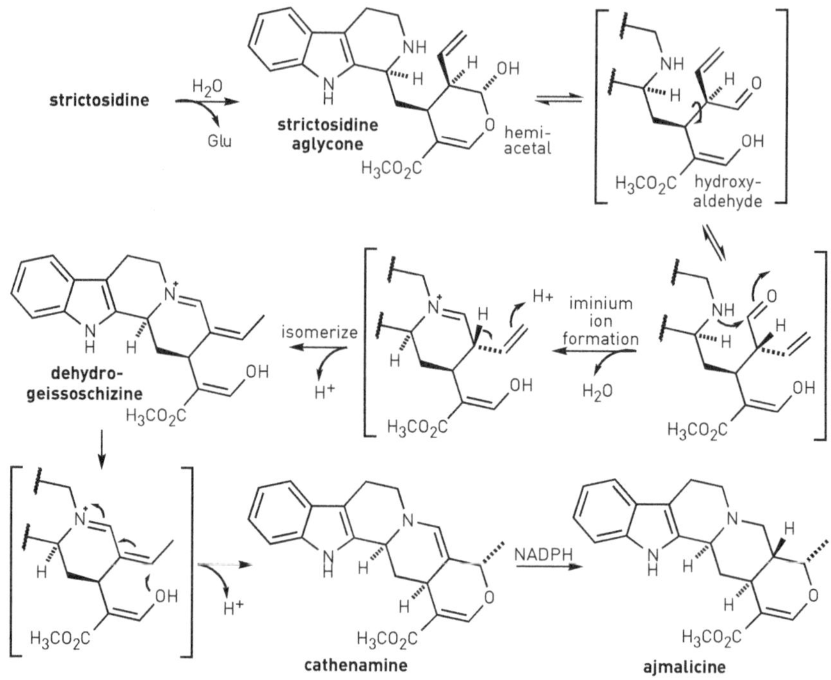

FIGURE 7.33
Conversion of strictosidine to the alkaloids cathenamine and ajmalicine.

one possible open form is an enol-aldehyde as shown. Condensation of the aldehyde with the ring amino group leads to an iminium ion which undergoes allylic isomerization to afford the important α,β-unsaturated iminium ion intermediate, **dehydrogeissoschizine**. Subsequent conjugate addition to the unsaturated iminium ion by the adjacent enol hydroxyl group gives the precursor alkaloid **cathenamine** which undergoes enamine linkage reduction by NADPH to afford **ajmalicine**.

Dehydrogeissoschizine also plays an important role in the construction of **catharanthine** and **vindoline**, two separate alkaloid components which are subsequently modified and coupled to form the remarkably complex structures of the potent cancer chemotherapy agents **vinblastine** and **vincristine**, two members of the so-called vinca alkaloids isolated from the Madagascar periwinkle plant. Like taxol (Chapter 4) and podophyllotoxin (Chapter 6), vinblastine and vincristine are compounds which inhibit cell division (mitosis) through disruption of the formation of the microtubules which help cells to pull apart during division. These unusual alkaloids are highly effective in the treatment of a variety of carcinomas including lung, breast, and testicular cancers as well certain lymphomas and leukemia; both rank among the most important antitumor agents available to modern medicine. A highly abbreviated scheme for conversion of dehydrogeissoschizine to catharanthine and vindoline and their subsequent modification and coupling to form vinblastine and vincristine is shown in Fig. 7.34

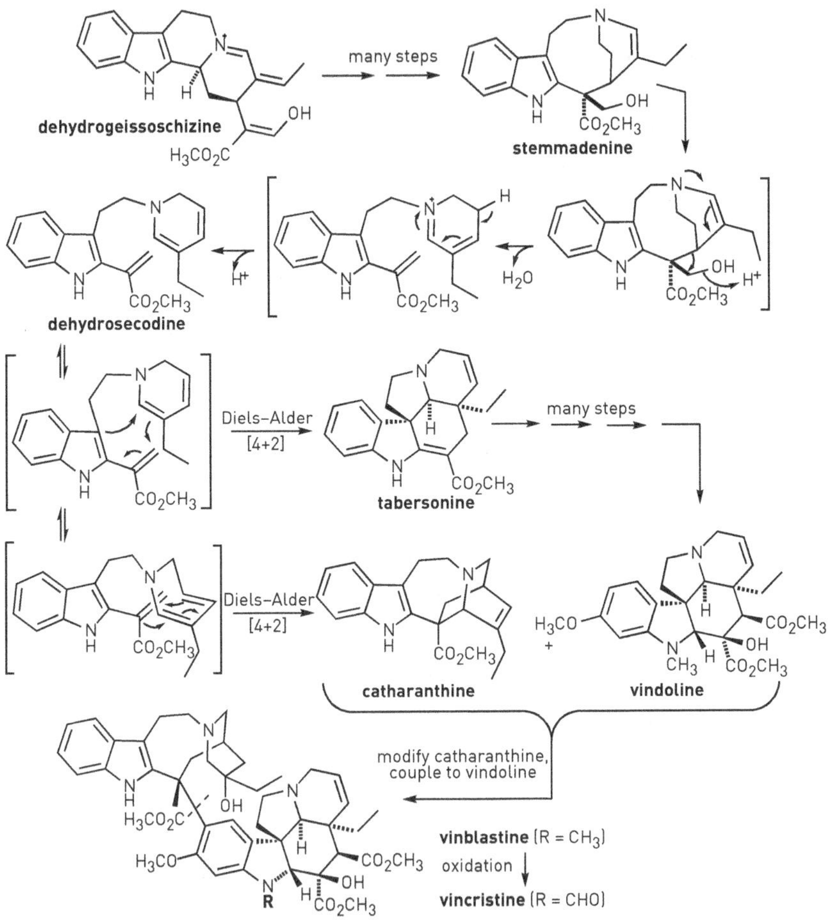

FIGURE 7.34
Formation and linkage of catharanthine and vindoline to give vinblastine and vincristine.

A complex multistep sequence (not shown) initially converts dehydrogeissoschizine to **stemmadenine** which then apparently undergoes a dehydration-driven ring cleavage facilitated by formation of an iminium ion and an α,β-unsaturated ester function. The resulting conjugated iminium ion can then undergo loss of a proton to form the conjugated enamine product **dehydrosecodine** as a key intermediate. Subsequent conversion to either **tabersonine** or **catharanthine** may be conveniently envisioned as occurring via intramolecular Diels–Alder reactions arising from alternate folded conformations of dehydrosecodine as shown. A multistep sequence of oxidations and alkylations presumably then converts tabersonine to **vindoline**. At this point, additional modifications of catharanthine (not shown) are followed by a coupling reaction with vindoline that may involve electrophilic aromatic substitution as a means to form the key C–C bond between the two components, thereby producing **vinblastine**. Formation of **vincristine** occurs via postcoupling oxidation of the *N*-methyl group of vinblastine to the corresponding *N*-formyl derivative as shown. Though not always obvious from their structures, the formation of **strychnine, vincamine, aspidospermine**, and **ellipticine** (Fig. 7.32) is also known to involve dehydrogeissoschizine-derived intermediates, though many of the steps involved are not well understood.

In our final example for this section, we examine the construction of the important **antimalarial** antiparasitic alkaloid known as **quinine**. The significance of quinine as one of the earliest and most effective antimalarial drugs is enormous from a variety of perspectives. Historically, quinine is among the oldest of alkaloids known to have important curative properties and it remains in widespread use throughout the world in modern times. In addition to its medicinal significance, the intricate molecular architecture of quinine has made it one of the most important and challenging targets of the modern era of laboratory total synthesis of structurally complex natural products, from the 1940s up to the present. In terms of the ongoing impact of malaria itself on human health, it is worth noting that annual cases of this mosquito-borne parasitic infection number anywhere from 300 to 500 million worldwide, mostly in sub-Saharan Africa where it continues to claim over half a million lives each year, mostly among children.

As outlined in Fig. 7.35, a proposed biosynthetic sequence again begins with **strictosidine aglycone,** a hemiacetal which in its hydroxyaldehyde form can undergo an intramolecular iminium ion cyclization, followed by NADPH reduction to afford the corresponding bicyclic amine derivative. Hydrolysis of the methyl ester of the β-dicarbonyl side chain of the system affords a β-oxocarboxylic acid which, like β-ketoacids, can undergo spontaneous decarboxylation to give the intermediate aldehyde **corynantheal**. Oxidation of the bicyclic tertiary amine ring system of corynantheal gives an iminium ion which can then undergo hydrolysis and cleavage to an amino dialdehyde. Conformational rotation and subsequent condensation is then followed by NADPH reduction of the resulting bridged bicyclic iminium ion, affording the corresponding indole aldehyde **cinchonaminal** as shown. Cleavage of the indole C-5 ring is shown as occurring via protonation and subsequent hydrolysis of the resulting iminium ion, leading to an amino ketoaldehyde intermediate. Once again, conformational rotation and condensation of the aryl amino group with the aldehyde function leads to formation of the corresponding cyclic imine. This dihydroquinoline ring system can then undergo oxidative aromatization to the corresponding quinoline derivative, affording the important intermediate alkaloid **cinchonidinone** whose ketone function may then undergo subsequent reduction via NADPH, leading directly to **cinchonidine**. The remaining steps involve oxidative hydroxylation of the cinchonidine aryl ring followed by the usual SAM methylation to finally afford the methoxylated derivative of cinchonidine known as **quinine**.

The most important of the so-called **cinchona alkaloids** found in the bark of the cinchona tree, quinine is the active principal in cinchona bark which was used for centuries to treat fever-related shivering and other maladies by indigenous peoples of Bolivia and Peru where the cinchona tree is found. In the early 1600s, Jesuit

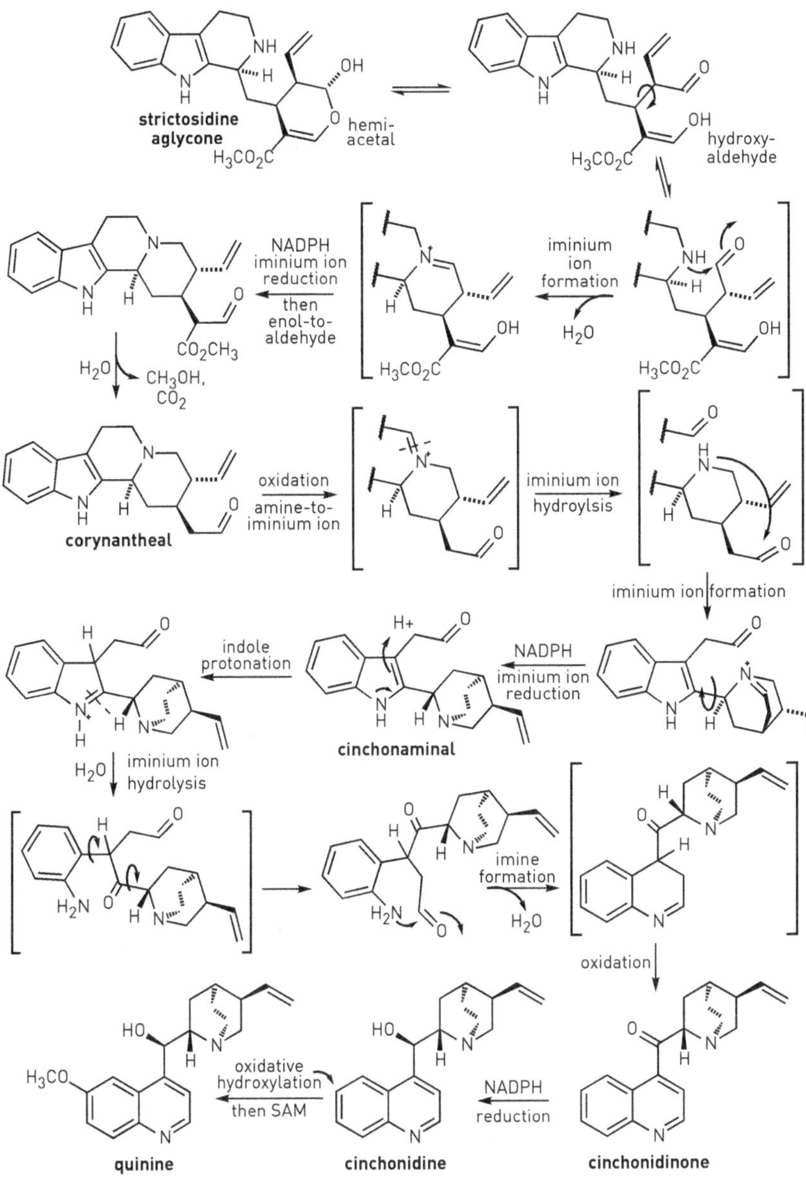

FIGURE 7.35
Possible steps for the conversion of strictosidine aglycone to cinchonidine and quinine.

missionaries reportedly sent samples of cinchona bark back to Rome to be tested for possible use in the treatment of shivers associated with malarial fevers, leading to the subsequent discovery of its unrelated antimalarial curative properties, for which it was widely known as **Jesuit bark** throughout the 17th and 18th centuries. The use of this bitter-tasting alkaloid in more recent times is most often reserved for treatment of malarial infections arising from strains of the *Plasmodium* parasite which have become resistant to more modern antimalarial drugs, though the low cost of quinine still makes it the treatment of choice in economically underdeveloped regions of the world and the gold standard for evaluation of the **antiplasmodic** activity of other potential antimalarial drugs. Quinine has also been used in the treatment of both lupus and rheumatoid arthritis, though structurally related synthetic antimalarials such as hydroxychloroquine have largely displaced it for such purposes. Quinine is also present in small amounts as the bittering agent used in tonic waters.

ALKALOIDS FROM NICOTINIC ACID: TOXIC ADDICTIVE DERIVATIVES OF A COMMON NUTRIENT

Nicotinic acid, also known as **niacin** and **vitamin B$_3$**, is formally **pyridine-3-carboxylic acid**, the amide derivative of which is incorporated into the familiar redox cofactors NAD$^+$ and NADP$^+$ that we have seen so often. As an essential nutrient in the human diet, nicotinic acid is produced by both plants and animals, though by very different biosynthetic routes, as outlined in Fig. 7.36. Animals produce the compound starting from L-tryptophan via oxidative cleavage of the indole C-5 ring to yield the ketoamide ***N'*-formylkynuerenine** as shown. Hydrolysis of the formamide linkage affords **kynurenine**, a key intermediate in the metabolism of tryptophan. In this sequence, kynurenine is oxidized to the phenolic derivative **3-hydroxykynurenine** which then undergoes a PLP-mediated cleavage of the amino acid side chain, affording **3-hydroxyanthranilic acid** along with *L*-alanine as a byproduct. Intramolecular dioxygenase-mediated oxidative cleavage of the phenolic ring affords the corresponding oxo-dicarboxylic acid whose enamine function may tautomerize to the imine, thereby allowing a C–C bond rotation and subsequent imine cyclization and aromatization, affording **quinolinic acid** which is the key intermediate in both animal and plant pathways; decarboxylation directly affords **nicotinic acid**. The chemistry employed by plants is much simpler, employing an enamine tautomer of **iminoaspartic acid** along with **3-phosphoglyceraldehyde** which are combined via what is essentially an enamine alkylation process. The resulting intermediate undergoes an imine cyclization similar to the previous example, affording **4,5-dihydro-5-hydroxyquinolinic acid** which undergoes dehydration to quinolinic acid and decarboxylation to nicotinic acid as before.

Nicotinic acid (usually identified as niacin for these purposes) is still used as one of the oldest cholesterol-reducing compounds, helping to control both LDL (low density lipoprotein) and triglyceride levels while raising HDL (high density lipoprotein) levels. While rapidly metabolized and excreted, it is a strong vasodilator that often causes a disagreeable flushing and prickling of skin, particularly of the face, which can be partially diminished through the use of controlled-release formulations of the drug. However, two recent studies have now led to warnings about the use of niacin for control of cholesterol, citing increased rates of death, infections, gastrointestinal disorders, and other significant problems. These findings may well

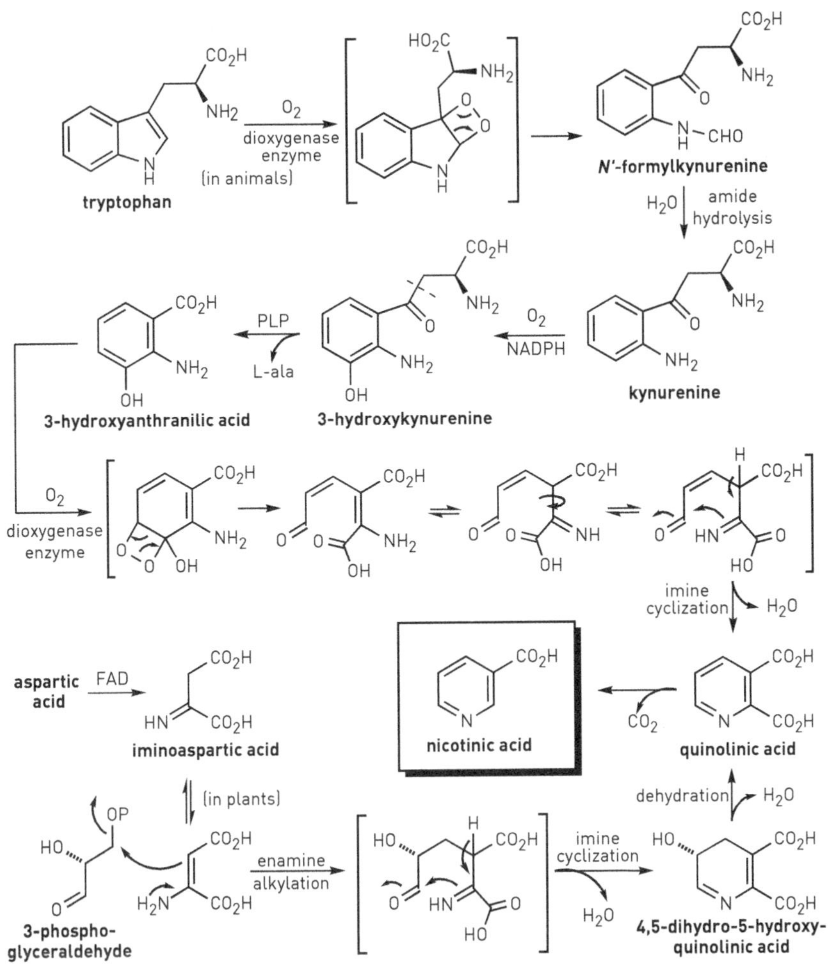

FIGURE 7.36
Different biosynthetic routes to the formation of nicotinic acid, both via quinolinic acid.

eventually lead to niacin being recommended for use only for high-risk heart patients who cannot tolerate statins or for those who have otherwise uncontrollable, high triglyceride levels.

The principal alkaloids derived from nicotinic acid are the so-called tobacco alkaloids **nicotine** and **anabasine**. A reasonable biosynthetic pathway for these compounds is outlined in Fig. 7.37.

The sequence commences with a 1,4-reduction of nicotinic acid by NADPH in a process somewhat analogous to the 1,4-reduction of benzoic acid by the action of Na/NH_3 in the familiar Birch reduction process. The resulting **3,6-dihydronicotinic acid** undergoes decarboxylation to afford **1,2-dihydropyridine**, which is effectively a cyclic enamine. For simplicity, the enamine is shown in a dipolar resonance form as it acts as a nucleophile to attack either **N-methyl-Δ^1-pyrrolinium cation** or **Δ^1-piperidinium cation,** both of which we have encountered previously in the biosynthesis of ornithine- and lysine-derived alkaloids. In both cases, the resulting

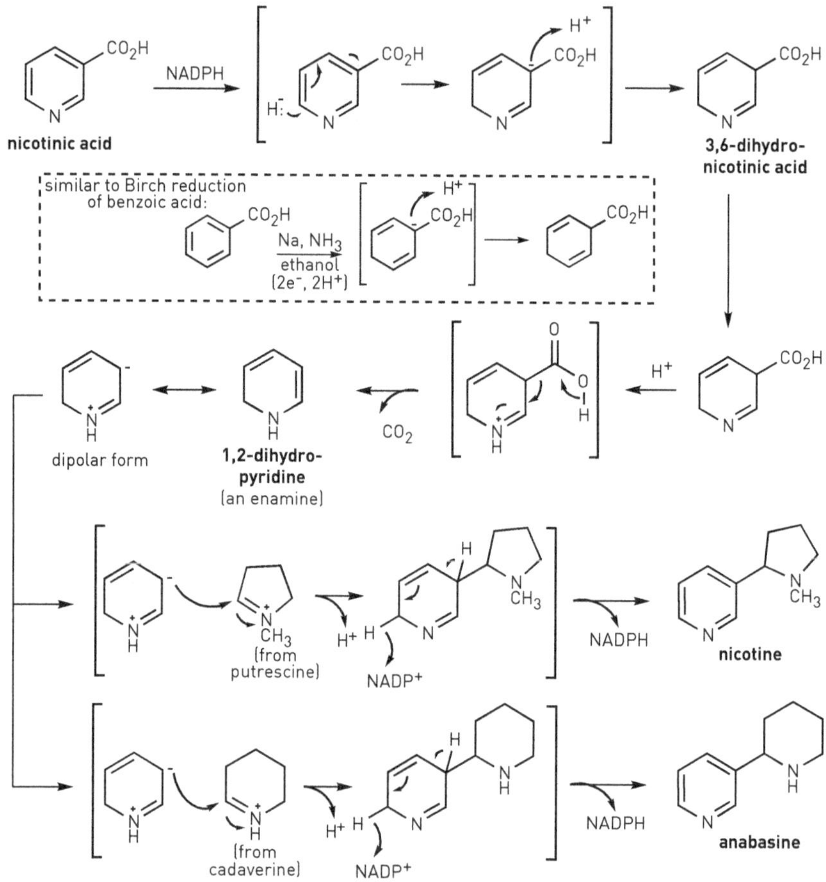

FIGURE 7.37
Conversion of nicotinic acid to the tobacco alkaloids nicotine and anabasine.

substituted dihydropyridine intermediates undergo $NADP^+$ oxidation to yield the corresponding $C_{10}H_{14}N_2$ alkaloid products.

While found in a number of plants, nicotine constitutes up to 3% of the dry weight of the tobacco plant, usually accompanied by smaller amounts of anabasine and related structures. The alkaloid itself is surprisingly toxic with a reported lethal dose for humans of only about 30–60 mg. Nicotine is also highly addictive, as any tobacco user well knows. Its physical and psychological effects are highly variable, ranging from stimulant-like effects at lower doses to more sedative-like effects at somewhat higher doses. Nicotine is also a powerful vasoconstrictor that increases blood pressure and can lead to severe vascular damage with long exposure. Post-operative use of nicotine is particularly problematic, as its vasoconstrictive action can seriously compromise wound healing. In recent years, nicotine-infused trans-dermal patches and chewing gums have found wide use for assisting in gradual with-drawal from chronic tobacco use, although long-term success rates for such therapy remain somewhat controversial. Measurement of urine levels of nicotine, anabasine, and various metabolites have been used as a measure of the extent of tobacco use (or lack thereof) in patients during nicotine replacement therapy. Historically, nicotine has also served as an important insecticide, though its use has diminished in recent years with the advent of cheaper and more effective alternatives.

ALKALOIDS FROM ANTHRANILIC ACID: FROM TRYPTOPHAN TO QUINOLINES AND ACRIDINES

We have seen previously in Chapter 6 how the shikimate product **anthranilic acid** serves as a key component for the biosynthesis of both indole and tryptophan (see Figs. 6.10 and 6.11). Anthranilic acid also serves as a template for structural elaboration into other complex alkaloid structures, particularly examples of quinoline- and acridine-type alkaloids. This is achieved by employing the CoA derivative of anthranilic acid, **anthranoylCoA**, as a starter unit for polyketide chain extension reactions as shown in Fig. 7.38. The first example involves addition of a single malonyl-CoA unit followed by cyclization to the corresponding lactam derivative **quinoline-2,4-(1*H*,3*H*)-dione**. Subsequent sequences of oxidations, SAM or DMAPP alkylations and other transformations (not shown) lead to the quinoline-type alkaloids **casmiroin** and **dictamnine**, the latter of which has shown both antifungal and smooth muscle contractant activity. The second sequence leading to the acridine-type alkaloid **acronycine** begins with condensation of anthranoylCoA with three units of malonyl-CoA to give a triketothioester which undergoes a Claisen condensation to form a familiar cyclic triketone system which is shown as undergoing an intramolecular iminium ion cyclization and subsequent aromatization to give **1,3-dihydroxy-10-methylacridin-9(10*H*)-one**. Final conversion to **acronycine** is via SAM and DMAPP alkylations followed by an acid-catalyzed side-chain cyclization. The anticancer activity of acronycine has led to exploration of more active derivatives based on this core structure, leading to the highly active synthetic analogue **S 23906-1** which reportedly exhibits a novel spectrum of antitumor activities compared with many other anticancer drugs and which is currently (in 2016) in phase 1 clinical trials.

ALKALOIDS FROM HISTIDINE: FROM SIMPLE AMIDES TO GLAUCOMA DRUGS

The amino acid **histidine** apparently serves as the precursor to a number of alkaloids containing the imidazole ring function, although the biosynthetic details are not always well known. Some of these products derive from **histamine** which arises via PLP-mediated decarboxylation of histidine. Histamine itself is an important neurotransmitter involved in the sleep–wake cycle and is also involved in immune system function and inflammatory response. It is the familiar target of so-called "antihistamines" or histamine antagonists which either block histamine receptors or inhibit its production from histidine, usually to relieve symptoms associated with allergic responses such as watery nose and itchy eyes. Figure 7.39 outlines possible

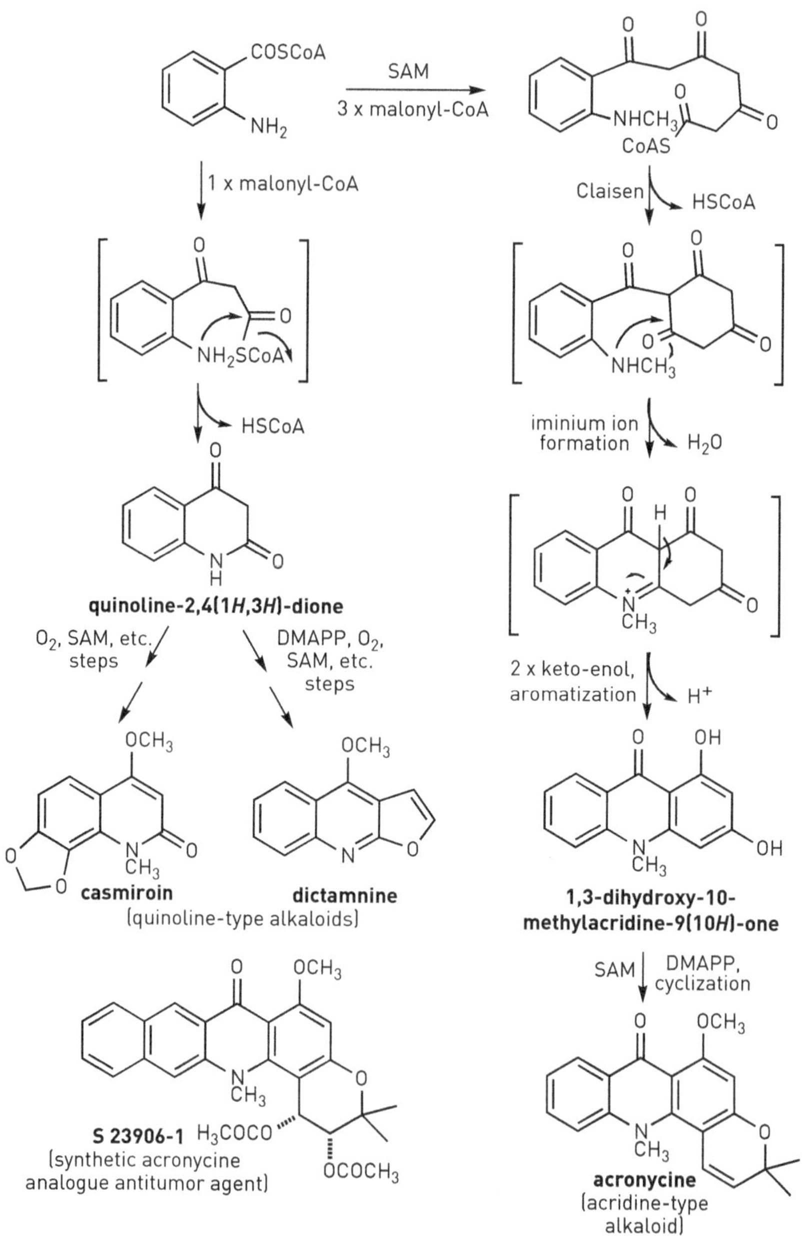

FIGURE 7.38
Conversion of anthranilic acid to some quinoline and acridine-type alkaloids.

routes to some simple histamine amide derivatives such as **dolichotheline**, one of the first imidazole-containing alkaloids isolated. Certainly the most important examples of the histidine group from a medicinal perspective are the pilocarpus alkaloids **pilosine** and **pilocarpine,** derived from **(5-imidazolyl)pyruvic acid** via histidine transamination. After acid reduction to the corresponding keto alcohol and condensation with either butyric acid (for pilocarpine) or 3-phenylpropanoic

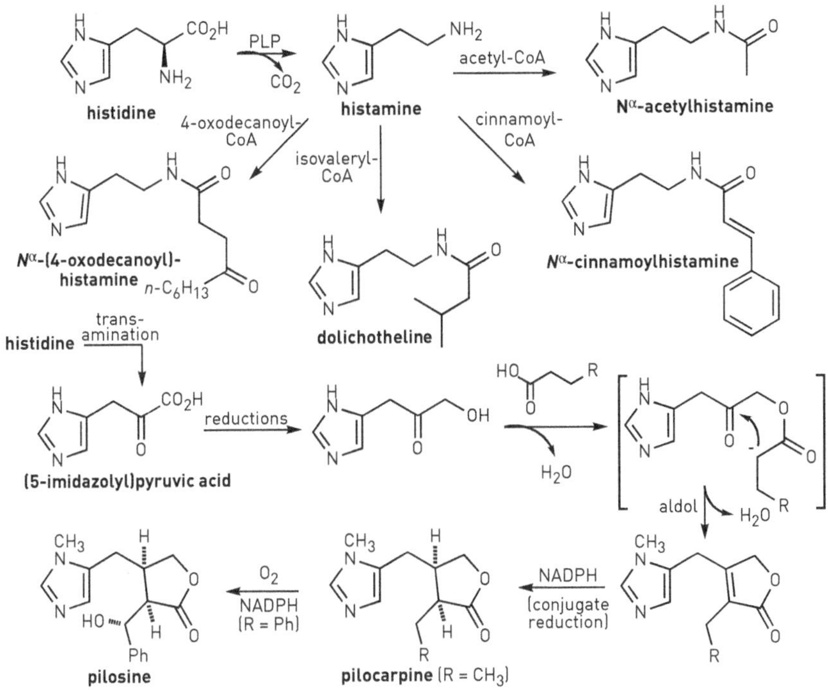

FIGURE 7.39
Possible origins of some simple histamine amides and the alkaloids pilocarpine and pilosine.

acid (for pilosine), the resulting esters may undergo intramolecular aldol condensations and conjugate reductions to form the 5-membered lactone rings, with pilosine requiring a final C–H to C–OH oxidation at the benzylic position to complete the sequence.

Pilocarpine is a very important medication which is obtained commercially from the leaves of Maranham Jaborandi (*Pilocarpus microphyllus*), a flowering plant found only in Brazil. Pilocarpine and its salts are especially useful for the treatment of severe dry mouth associated with radiation treatment of tumors of the head and neck. It is also widely used in ophthalmology for pupil contraction and in the treatment of glaucoma.

PURINE ALKALOIDS: ADDICTIVE STIMULANTS IN OUR COFFEE, TEA, AND CHOCOLATE

Purines, simple *N*-heterocyclic aromatics made up of fused imidazole and pyrimidine rings, are most commonly encountered in some of the deoxyribonucleosides and ribonucleosides found in DNA and RNA systems. Substituted and tautomeric forms are also widely found in nature, particularly derivatives of **xanthine** of which **theophylline** (1,3-dimethylaxanthine), **caffeine** (1,3,7-trimethylxanthine), and **theobromine** (3,7-dimethylxanthine) are the most common. Figure 7.40 outlines the sequences involved in their biosynthesis from **xanthosine**.

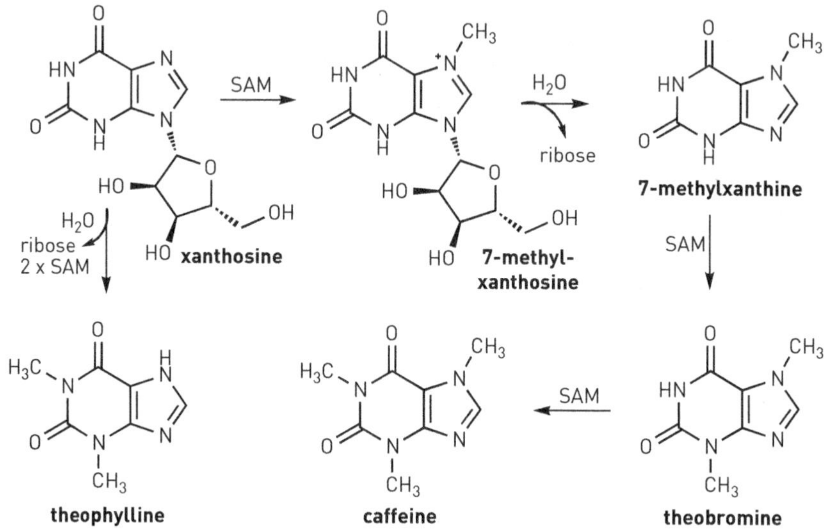

FIGURE 7.40
Biosynthetic origins of the purine alkaloids theophylline, theobromine, and caffeine.

Both tea and coffee contain some amount of these substances, though caffeine is by far the principal component responsible for the CNS stimulant action of these beverages, with the others appearing in only trace amounts. Caffeine also acts as a diuretic (increases elimination of water from the body) though its action in this regard is weaker than that of either theobromine or theophylline, both of which are found in significant amounts in cocoa (cacoa) bean. Theophylline is often used in treatments for asthma and chronic obstructive pulmonary disease (COPD) due to its ability to induce relaxation of bronchial smooth muscle. Theobromine (which does *not* contain bromine) is sometimes used therapeutically as a heart stimulant or as a vasodilator and is the ingredient responsible for the well-known toxicity of chocolate in dogs, due mainly to their comparatively slow metabolism of theobromine. Like caffeine, theobromine is mildly addictive, and the irresistible lure of dark chocolate for many devotees is due in part to its high levels of theobromine relative to simple milk chocolate.

CYCLIC AND MACROCYCLIC PEPTIDES: FROM SWEETENERS TO ANTIBIOTICS AND BEYOND

In addition to the many transformations of amino acids we have seen thus far for the formation of alkaloids, amino acids can also play an important role in the construction of complex alkaloid structures in which the amino acid components remain more or less intact by being linked to one another primarily through the formation of peptide bonds. This is particularly important, as we will see, in the formation of some medicinally important cyclic and macrocyclic polypeptides.

A unique group of such alkaloids, known as the **2,5-diketopiperazines,** illustrates how simple cyclic peptides can arise when two amino acids are linked to one another via sequential formation of bimolecular and intramolecular peptide bonds, as shown in Fig. 7.41. A common example of such a structure is found when the artificial sweetener **aspartame,** a simple dipeptide methyl ester derived from phenylalanine and aspartic acid, undergoes cyclization during heating to form the corresponding product **aspartame diketopiperazine,** sometimes referred to as **aspartame DKP.** Over the years, amateur consumer health advocates have often implicated aspartame DKP and its metabolic byproducts (along with traces of methanol released during metabolism of aspartame) as being responsible for a host of alleged side effects ranging from psychological disorders and convulsions to birth defects and cancers, all supposedly related to ingestion of aspartame, in spite of a host of exhaustive scientific studies completely supporting its safety at normal levels of consumption.

Also shown in Fig. 7.41 is **brevianamide F,** a diketopiperazine formed via cyclization of tryptophan and proline which serves as the precursor to a large family of modified derivatives such as **tryprostatin A** and **tryprostatin B,** both of which are isolated from the marine fungus *Aspergillus fumigates.* Tryprostatin A has been found to act as an inhibitor of a key protein that regulates resistance to breast cancer chemotherapy agents, while tryprostatin B has chemotherapeutic potential due to its action as a mammalian cell-cycle inhibitor. Both compounds are viewed as important targets for further studies in cancer chemotherapy.

The number of larger macrocyclic polypeptides of medicinal importance is substantial, so only a few representative examples of the principal types can be treated

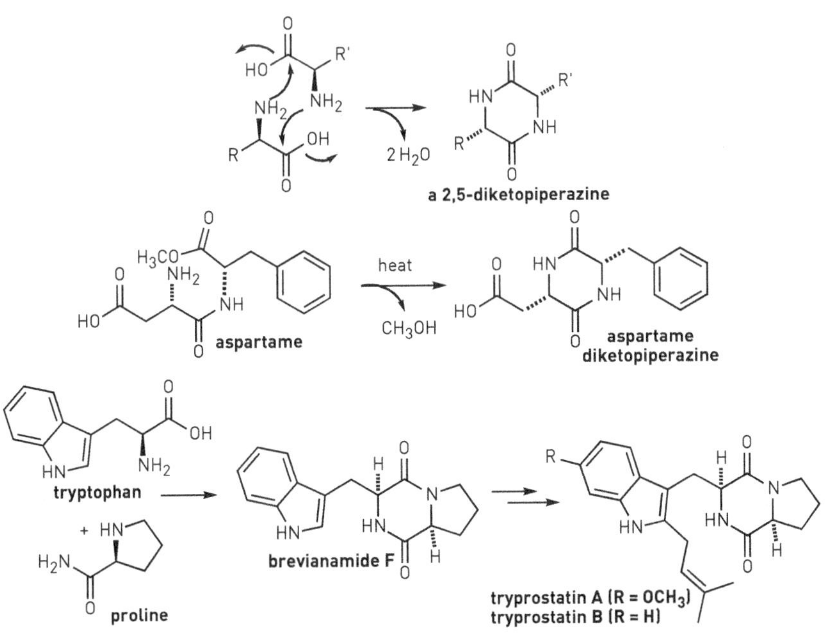

FIGURE 7.41

Formation of 2,5-diketopiperazines from aspartame and from tryptophan and proline.

here. Figure 7.42 shows the structures of two such compounds. **Valinomycin** is a *Streptomyces*-derived example of a naturally occurring macrocyclic **depsipeptide,** a type of peptide compound in which amide (often peptide) bonds are accompanied by ester linkages. In the case of valinomycin, all the nitrogens and some of the carbonyls in the amide linkages are provided by valine (both D- and L-forms) while L-lactic acid provides the other amide carbonyls. The hydroxyl groups required for the ester linkages are all provided by incorporation of D-α-hydroxyisovaleric acid. Valinomycin, like the polyketide **monensin A** (Fig. 5.71), is a crown ether-like ionophore, with antitumor, antifungal, and antibacterial activity, that is able to bind and transport potassium ions across hydrophobic lipid membranes via interaction of the metal cation with carbonyl oxygen lone pairs. It has also been shown to be highly active against the corona virus responsible for severe acute respiratory syndrome (SARS) in an infected test cell line.

Bacitracin A is one member of a group of nine structurally related broad spectrum antibiotics from *Bacillus subtilis* that acts by disrupting cell wall synthesis in bacteria. Structurally, the compound is made up of a mixture of both D- and L-amino acids (see Fig. 7.42 for amino acid labelling) with a linear pentapetide segment joined to a cyclic heptapeptide component. The assembly of these (and other) cyclic polypeptides is usually accomplished via a modular multifunctional enzyme similar to the type I polyketide synthase (type I PKS) systems we encountered previously (see Table 5.2). The growing peptide chain is passed along from module to module, terminating in a cyclization step somewhat analogous to the TE domain-mediated lactone formation step in the biosynthesis of 6-deoxyerythronolide B (see

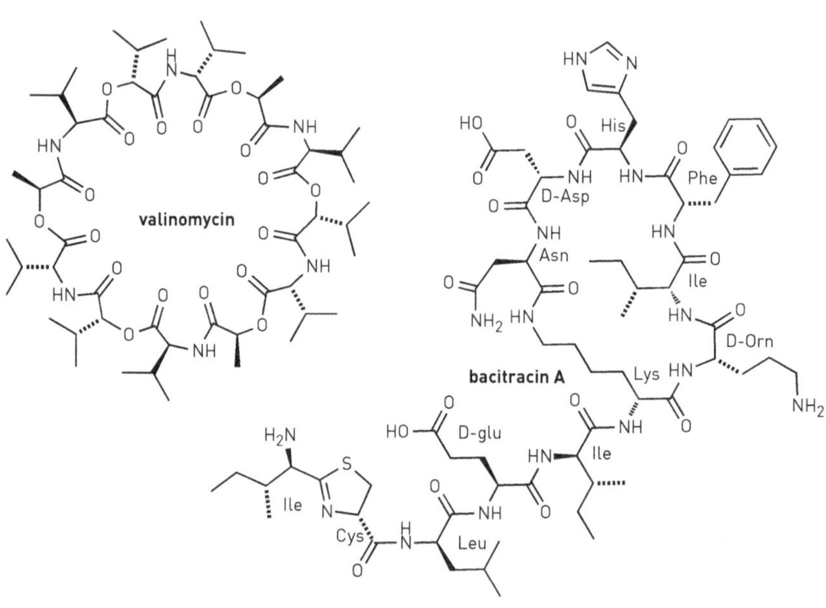

FIGURE 7.42
Structures of the macrocyclic polypeptide antibiotics valinomycin and bacitracin A.

Fig. 5.66). Bacitracin is frequently blended with the polysaccharide antibiotic **neomycin** (see Fig. 3.16) and **polymyxin B** (structurally similar to bacitracin) in topically applied "triple antibiotic" lotions or ointments for treatment of superficial wounds or infections. The effectiveness of such applications is suspect, however, and some research suggests that widespread use of bacitracin, particularly from incorporation into animal feeds, has contributed to the development of MRSA strains. Ophthalmic solutions of bacitracin are also widely used for treatment of eye infections.

Another interesting cyclic polypeptide antibiotic isolated from *Streptomyces* strains is **dactinomycin**, a potent but toxic antibiotic that is used primarily in cancer chemotherapy. Structurally, dactinomycin is essentially a dimer of a cyclic hexapeptide initially constructed from **3-hydroxy-4-methylanthranilic acid** to which threonine, D-valine, proline, *N*-methylglycine, and *N*-methylvaline are added in sequence prior to a final lactone cyclization (Fig. 7.43). The resulting cyclic hexapeptide then undergoes an unusual type of oxidative phenolic coupling reaction involving the amino-substituted aromatic ring of the 3-hydroxy-4-methylanthranilic acid residue, as shown in Fig. 7.43. The mechanism of this process is not fully understood, but could reasonably involve the sequence shown, beginning with oxidation of the *ortho*-aminophenol to the corresponding *ortho*-quinone imine. The amino group of a second molecule of the monomer then may act as a nucleophile in a conjugate addition to the α,β-unsaturated imine to give an enamine which, following a second oxidation, affords a *para*-quinone imine. Intramolecular conjugate addition to the α,β-unsaturated ketone function by the proximal phenolic hydroxyl group results in formation of the central *N,O*-containing six-membered ring. Tautomerization and a final oxidation complete the conversion to the dimeric structure of dactinomycin.

As the first antibiotic found to have anticancer activity, dactinomycin is one of the oldest of the cancer chemotherapy agents and it is still in wide clinical use today. Several types of rare childhood cancers such as Wilms' tumor (a kind of kidney cancer), rhabdomyosarcoma (cancer of the muscles), and retinoblastoma (a type of childhood eye cancer) respond to dactinomycin chemotherapy as do certain testicular, bone, and muscle cancers in adults, often in combination treatment with other anticancer agents. It is also used for the treatment of gestational trophoblastic neoplasia, a rare type of pregnancy-related cancer.

An example of more extensive use of phenolic oxidative coupling for the formation of a macrocyclic polypeptide structure is found in the biosynthesis of **vancomycin**, isolated from the soil bacterium *Amycolatopsis orientalis* and one of the most important clinical antibiotics in use worldwide for the treatment of serious bacterial infections. Formally classified as a **glycopeptide antibiotic**, vancomycin is derived via the assembly of a basic linear heptapeptide composed of D-leucine, asparagine, two β-hydroxytyrosines, two (*p*-hydroxyphenyl)glycines, and a (3,5-dihydroxyphenyl)glycine as shown in Fig. 7.44. Two aryl ring chlorinations are then followed by oxidative phenolic coupling reactions which form key aryl C–O–C and aryl C–C bonds found in the macrocyclic core. SAM methylation of

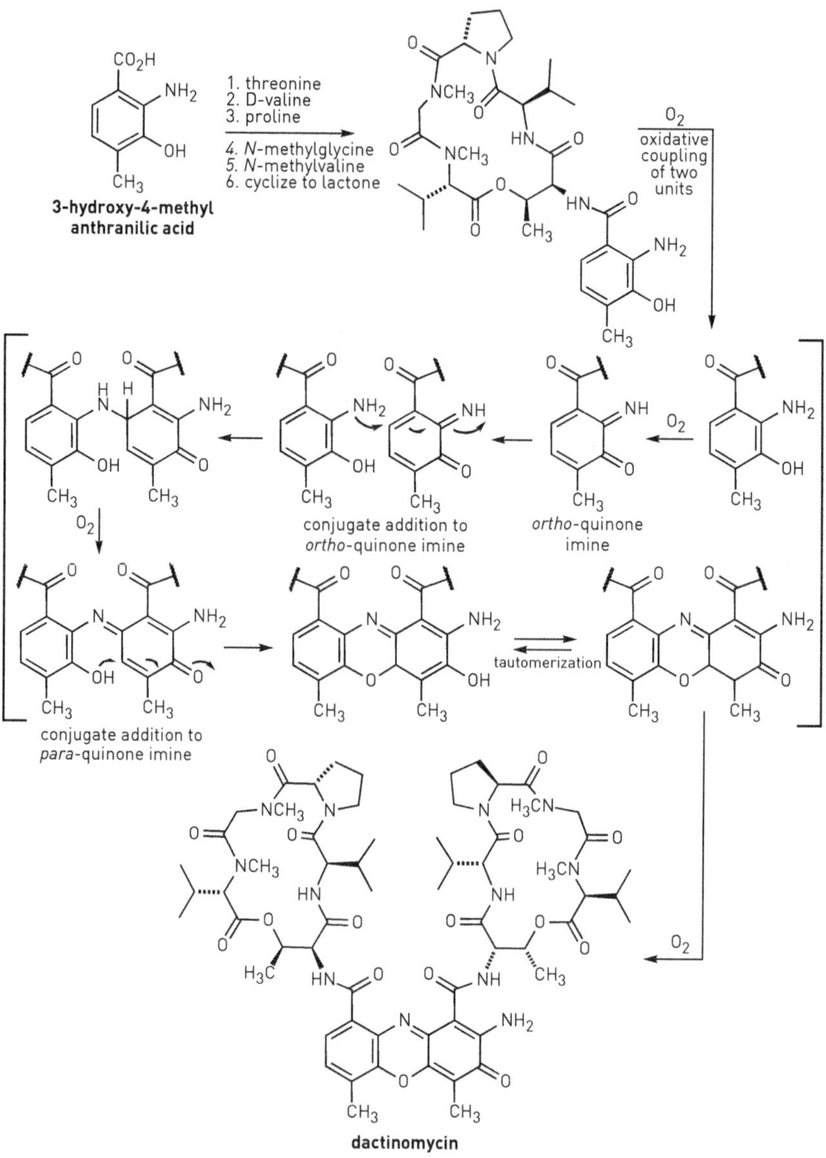

FIGURE 7.43
Proposed assembly sequence for construction of the macrocyclic antibiotic dactinomycin.

the *N*-terminal D-leucine residue is then followed by a phenolic hydroxyl glucosylation. Final formation of a 2-*O*-glycoside linkage between the glucose residue and the aminosugar **vancosamine** completes the sequence.

Often referred to as "the antibiotic of last resort" for gram-positive bacterial infections involving multiple-antibiotic-resistant strains (especially MRSA strains) or for patients allergic to β-lactam antibiotics such as penicillins and cephalosporins, vancomycin is a vital weapon in the antibiotic arsenal, particularly in hospital settings. Unfortunately, the widespread use of vancomycin over the years has led to increasing concerns about the evolution of vancomycin-resistant (VR) bacterial

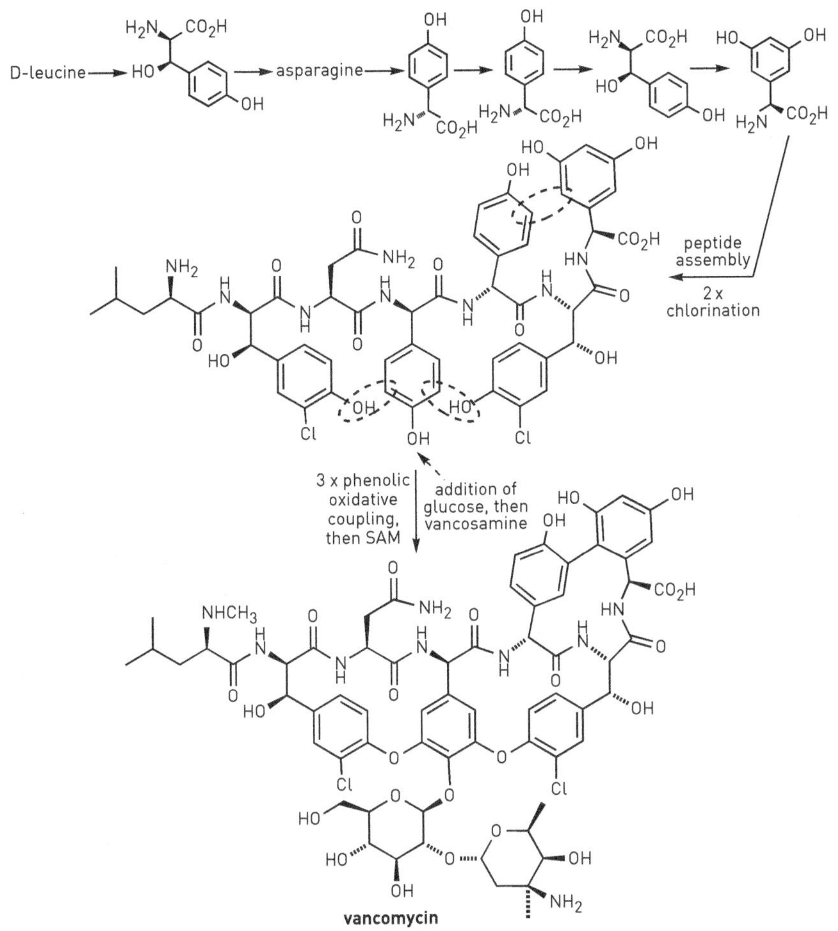

FIGURE 7.44
Polypeptide assembly, oxidative coupling, and glycosylation in the biosynthesis of vancomycin.

strains, particularly *Staphylococcus aureus* (VRSA strains). Such concerns have led to limitations on the use of vancomycin to specific instances such as life-threatening infections for which less toxic agents are ineffective, certain cases of *Clostridium difficile* (*C. diff*)-related colitis and for precautionary treatment of particularly serious infections during bacterial identity culturing, among others.

PENICILLINS, CEPHALOSPORINS, AND CARBAPENUMS: THE ESSENTIAL B-LACTAM ANTIBIOTICS

In 1928, the Scottish scientist Alexander Fleming accidentally discovered a curious substance with apparent antibacterial activity from cultures of *Penicillium* fungi. Dubbed **penicillin**, this fungal metabolite eventually became the target of intensive research efforts which would ultimately lead to the development of the world's first medically practical antibiotic, though it would take nearly 17 years from the time of its discovery to determine its unique **β-lactam** structural feature. The discovery of penicillin may be said to have independently introduced the age of antibiotics, an era in which countless human lives have been saved over the years. And it is not a stretch to say that the many wondrous achievements of antibiotic research, development, and clinical application in modern medicine have ultimately been the fruit of Fleming's early blend of expertise and curiosity.

The β-lactams are the most important group of antibiotics in terms of overall usage, accounting for over half of all sales of antibiotics worldwide. As we will see, this group includes a variety of closely related structures: the penicillins, the **cephalosporins**, and the **carbapenums**, all of which share a β-lactam linkage whose strained four-membered ring is susceptible to nucleophilic attack by $-CH_2OH$ residues found in bacterial enzymes that catalyze the formation of cell wall cross-linkages. The resulting acylated enzymes are thus deactivated, leading to relatively rapid cell death due to cell wall degradation. **Penicillin-resistant bacteria** are those which have evolved the ability to produce a substance known as a **penicillinase**, an enzyme that catalyzes the hydrolysis of the β-lactam linkage prior to its ability to interact with these cell wall enzymes. The ever-increasing levels of bacterial resistance to all types of antibiotics are of grave concern to the medical establishment worldwide and are among the main drivers for the ongoing development of ever more sophisticated and potent antibiotics.

Figure 7.45 shows the basic assembly steps involved in the biosynthesis of **benzylpenicillin (penicillin G)** and the versatile and equally important precursor, **6-aminopenicillanic acid**, along with a simplified scheme representing the basic interaction of the β-lactam ring with a bacterial cell wall enzyme.

The biosynthesis begins with the amino acid L-lysine which undergoes transamination to the corresponding aldehyde which is further oxidized to give **L-2-aminoadipic acid**. A peptide linkage is then formed with L-cysteine followed by peptide bond formation with D-valine. The resulting simple tripeptide, **L-δ-(α-aminoadipoyl)-L-cysteinyl-D-valine**, then undergoes an intriguing oxidative cyclization involving an enzyme-bound Fe-oxygen species coordinated to the cysteine thiol function. This complex radical process is catalyzed by the enzyme *isopenicillin N synthase* and is believed to involve the steps shown and results in formation of the β-lactam ring-containing precursor **isopenicillin N**. At this stage, a key peptide bond is hydrolyzed to release 2-aminoadipic acid and the essential core compound **6-aminopenicillanic acid (6-APA)** from which other penicillin derivatives may be produced semisynthetically. Acylation of the amino group by phenylacetylCoA leads directly to **benzylpenicillin**, also known as **penicillin G**, the initial product of commercial fermentation processes used in the production of other penicillins. Because of its tendency to undergo acid hydrolysis and structural rearrangement in the stomach, penicillin G is administered intravenously or by injection, allowing higher concentrations of antibiotic than are usually available with more stable penicillin derivatives that are administered orally. Penicillin G is

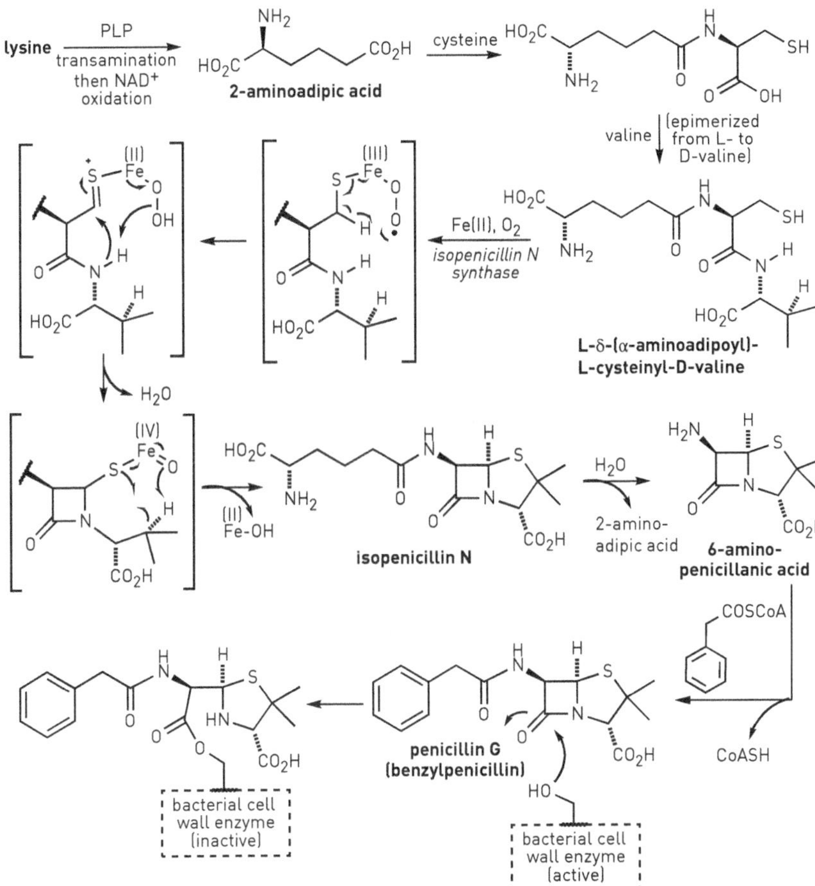

FIGURE 7.45
Assembly steps for β-lactam formation in the biosynthesis of penicillin G and a simplified representation of the mode of action of β-lactam antibiotics.

also important because not only can it be commercially produced in large quantities by fermentation, it can also undergo either enzymatic or chemical hydrolysis to produce pure 6-APA which can then be converted by simple synthetic acylation of the amino group to give a variety of penicillin derivatives of diverse potency, stability, and specificity. This is important because the early penicillins were active only against gram-positive bacterial strains while later derivatives were developed which are also active against gram-negative strains. Figure 7.46 shows a small sampling of various penicillins derived from 6-APA. Though allergic reaction to penicillins is the most commonly reported medication allergy, studies suggest only a small percentage of reported reactions are due to actual penicillin allergy.

First isolated from cultures of *Cephalosporium acremonium* in 1948, the **cephalosporins** may be thought of as close cousins to the penicillins. Both contain the characteristic β-lactam ring, but in the cephalosporins, the sulphur-containing ring is expanded from five to six atoms in size. The cephalosporin structure is actually derived from the penicillin ring system itself by a sequence of oxidative

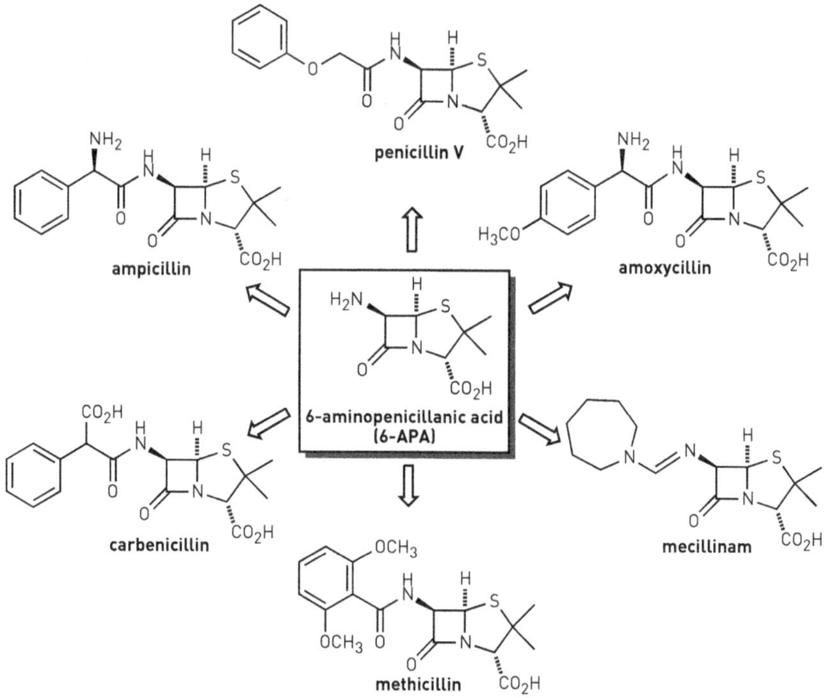

FIGURE 7.46
Some structural variations in penicillin antibiotics derived from 6-aminopenicillanic acid.

transformations sometimes referred to as the **penicillin-cephalosporin rearrangement** which is outlined in Fig. 7.47 along with the structures of **cephalosporin C** and two members of the so-called "first generation" cephalosporins, **cephalexin** and **cephalothin.** Most of the so-called five generations of cephalosporins are derived from **7-aminocephalosporanic acid (7-ACA)** or its less significant deacetoxy derivative (7-ADCA). Both are the cephalosporin equivalents of the similarly versatile 6-APA for semisynthesis in the penicillin family.

Second generation cephalosporins such as **cefuroxime** and third generation examples such as **cefdinir** and **cefixime** became important partly due their greater activity against gram-negative bacteria, though at the expense of gram-positive antimicrobial activity. The later fourth and most recent fifth generation cephalosporins are true "broad spectrum" antibiotics with activity against both types. The fifth-generation compound **ceftaroline** is extremely important due to its activity against multidrug-resistant VRSA/VISA (vancomycin intermediate *Staphylococous aureus*) strains and its unique role as the only β-lactam with activity against MRSA strains. The structure of ceftaroline shown in Fig. 7.48 is the active form of the drug which is administered intravenously as **ceftaroline fosamil**, a prodrug in which the primary amino group is phosphorylated to improve solubility. Note the unusual zwitterionic form of fourth and fifth generation compounds.

We conclude this final section of our chapter with the so-called **carbapenum** class of wide spectrum β-lactam antibiotics whose importance is related to their resistance

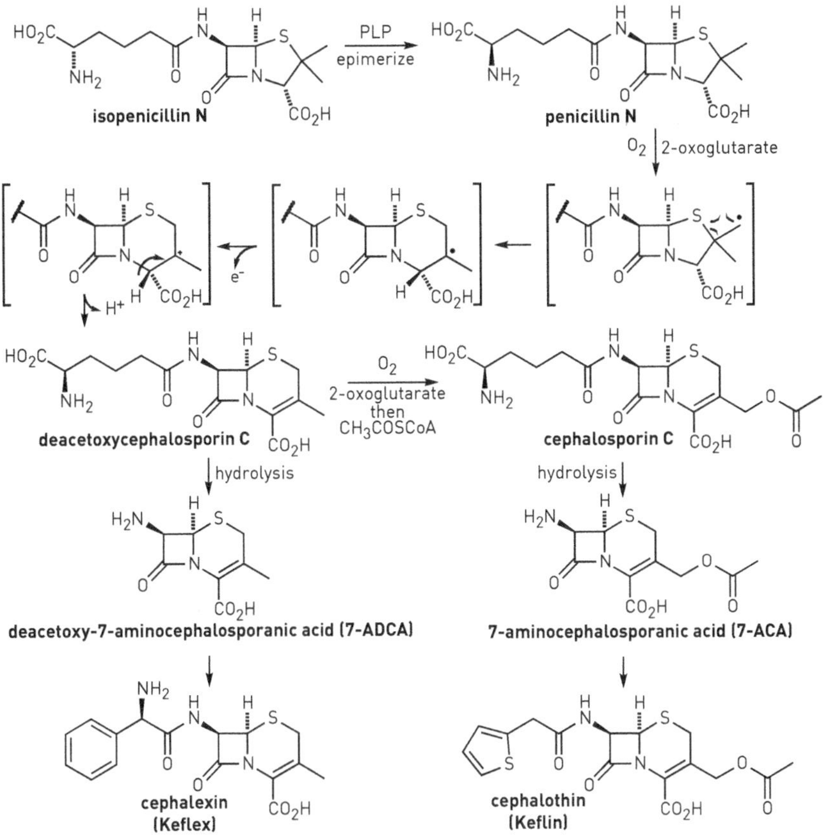

FIGURE 7.47

The penicillin-cephalosporin rearrangement: formation of cephalosporin C along with two first generation semisynthetic derivatives, cephalexin (from 7-ADCA) and cephalothin (from 7-ACA).

to *β-lactamase*-induced deactivation. They are especially significant as dependable agents for treatment of infections due to β-lactam-resistant strains of *E. coli* and *Klebsiella spp*., both of which can lead to a variety of diseases including pneumonia, urinary tract infections, septicemia, and meningitis. Structurally, the carbapenums derive their name from *penum*, the common name for the fused four-membered β-lactam/five-membered sulphur-containing rings of the penicillins; substitution of carbon for sulphur in the penum ring thus yields a "carbapenum." A possible scheme for biosynthesis of **thienamycin**, produced by *Streptomyces cattleya* and the first naturally occurring carbapenum to be isolated, is shown in Fig. 7.49.

After intramolecular imine condensation of **L-glutamic acid semialdehyde**, nucleophilic acylation of the imine linkage by an enolate ion derived from malonyl-CoA leads to the indicated thioester. SAM methylation (presumably via the thioester enolate) is followed by intramolecular amide formation to give the β-lactam linkage. Oxidation of the five-membered ring by O_2/2-oxoglutarate gives an α,β-unsaturated carboxyl which adds an L-cysteine thiol group via conjugate addition. PLP-mediated decarboxylation of the amino acid side chain and further oxidations introduce the

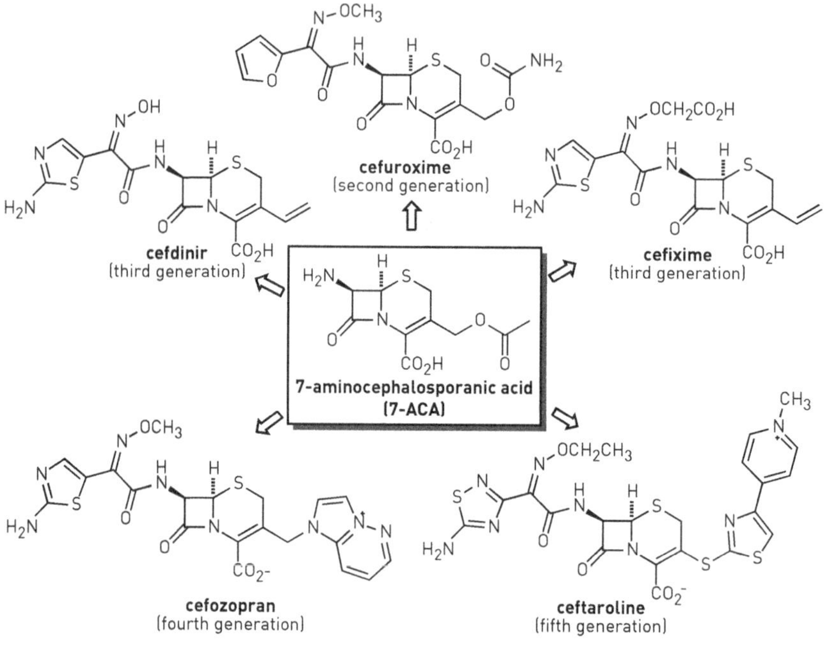

FIGURE 7.48
Some structural variations in second through fifth generation cephalosporins derived from 7-ACA.

indicated unsaturation, while a final SAM methylation extends the vinylidene linkage on the β-lactam ring. A final conjugate addition of H_2O completes the sequence to yield **thienamycin.**

Among the semisynthetic carbapenums (Fig. 7.50), **imipenem** was developed as a stable derivative of thienamycin (which rapidly decomposes in aqueous solution). Imipenum has a broad spectrum of activity against both gram-positive and gram-negative bacteria, though it is not active against MRSA strains. Similar to imipenem in terms of broad-spectrum activity and resistance to *β-lactamases*, **meropenem**, **doripenem**, and **ertapenum** are somewhat better tolerated than imipenem when high dosages are required. All are administered intravenously and, like vancomycin, are considered among the antibiotics of last resort for many serious bacterial infections, especially those related to the *E. coli* and *Klebsiella spp.*-related infections mentioned above. Unfortunately, the gradual spread of evolving resistance to carbapenem antibiotics has raised alarms in recent years, making the development of new antibiotics to combat such resistant strains an imperative for the safety and health of future generations.

A FINAL LOOK AHEAD

There is a vast amount of information related to alkaloids to which a short chapter such as this one could never hope to do even partial justice, so we are compelled

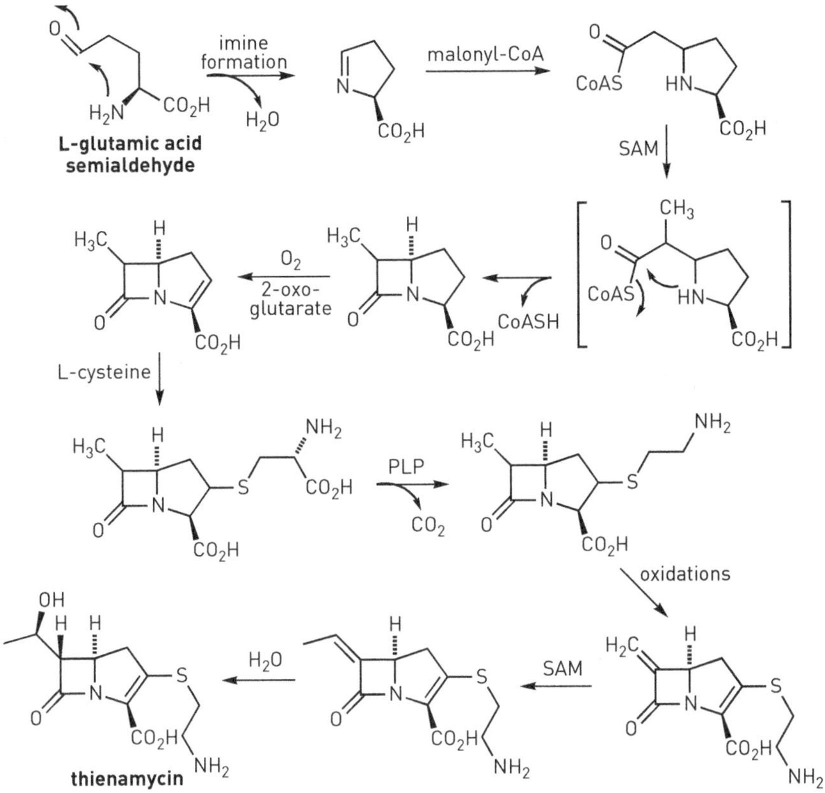

FIGURE 7.49

Possible biosynthetic pathway to the carbapenum antibiotic thienamycin.

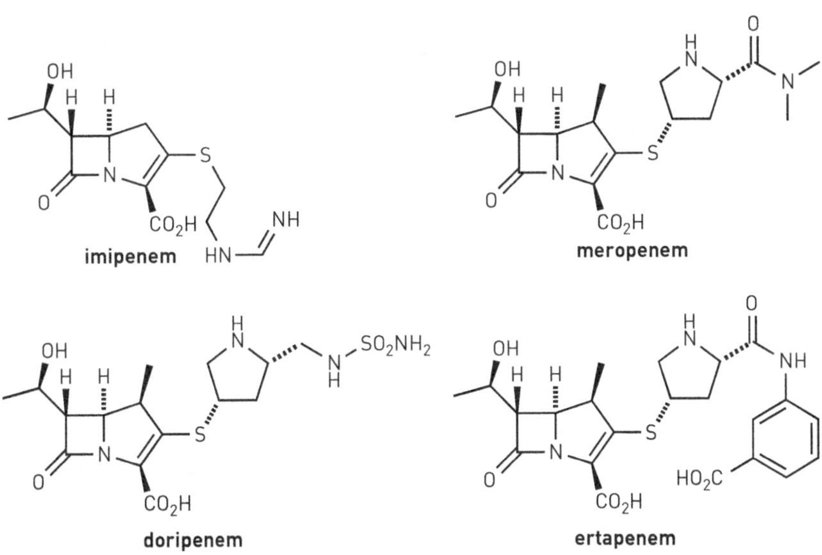

FIGURE 7.50

Semisynthetic carbapenem antibiotics: imipenem, meropenem, doripenem, and ertapenem.

to move on at this point to a final chapter that briefly considers how laboratory chemists have harnessed the power of synthetic organic chemistry over the years to carry out the painstaking, step-wise assembly of some of the most challenging organic structures produced by the pathways we have studied so far. Looking at these examples should give us a deeper appreciation for the complexities, demands, and limitations inherent in such organic synthesis projects, especially when compared with the elegance, power, and simplicity of nature's "cellular laboratory" approach to construction of architecturally complex organic molecules.

STUDY PROBLEMS

1. Propose reasonable mechanisms for conversion of *N*-(4-oxodecanoyl)-histamine to the alkaloids **glochidine** and **glochidicine** from a common intermediate.

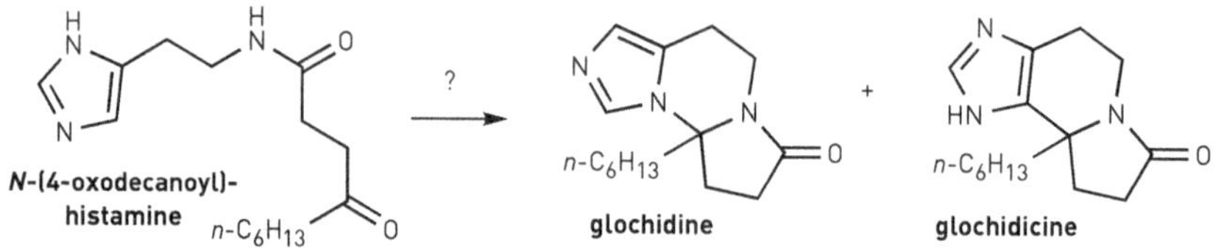

2. Suggest an alternative to the route shown in Fig. 7.11 for formation of L-pipecolic acid from lysine.

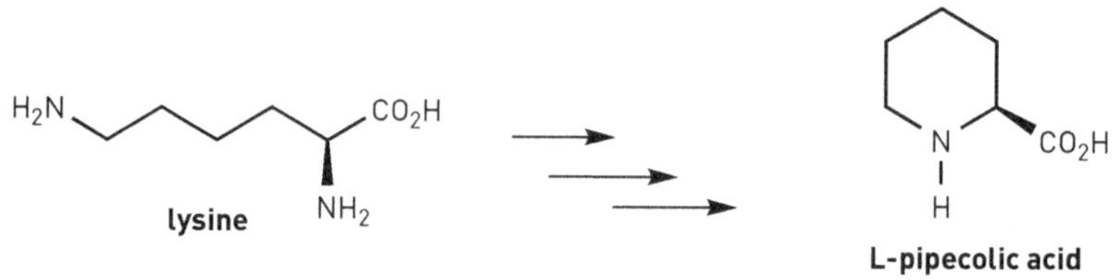

3. The psychedelic alkaloid **mescaline** is structurally related to the neurotransmitter *dopamine* and is produced by the peyote cactus. Similarly, **psilocybin** is structurally related to the neurotransmitter 5-hydroxytryptamine and is produced by many species of mushrooms. Both illegal compounds were ritually used by Native Americans for thousands of years owing to their strong hallucinogenic effects which are similar to those produced by LSD. Briefly outline the likely steps involved in biosynthesis of mescaline and psilocybin from simple amino acids.

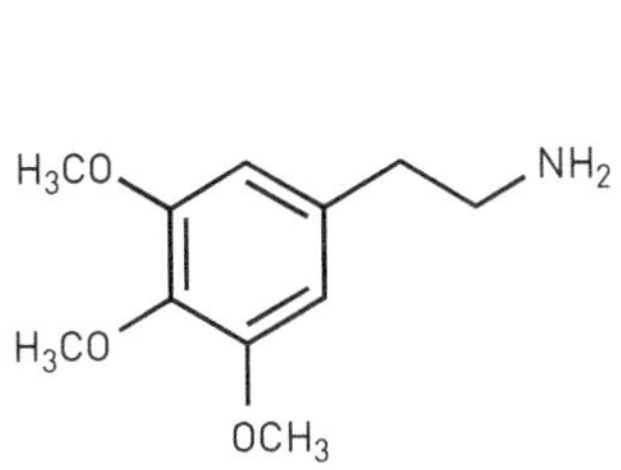

mescaline

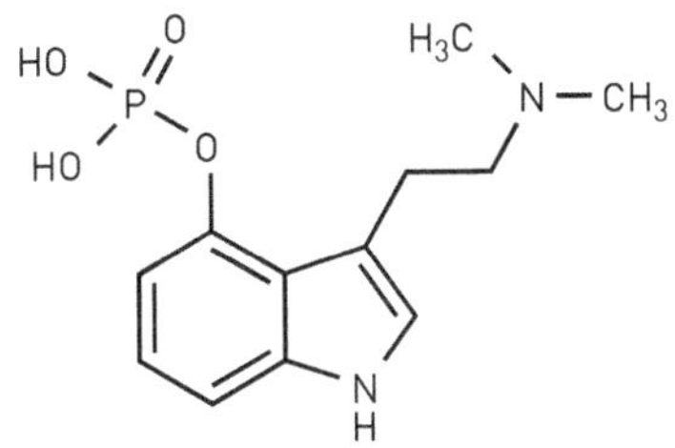

psilocybin

4. **Tubocurarine**, a toxic alklaoid component of curare, was once widely used during surgeries for relaxation of skeletal muscles. Outline a sequence of likely steps for biosynthesis of tubocurarine from (*S*) and (*R*)-*N*-methylcoclaurine.

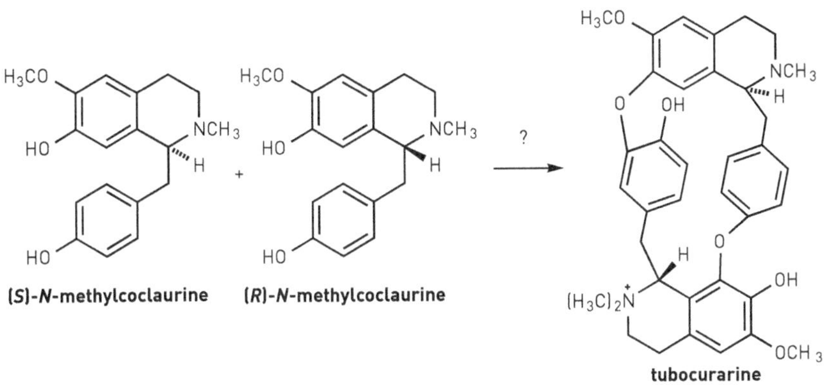

5. Propose a mechanism for conversion of **(*S*)-autumnaline** to **isoandrocymbine**, an intermediate involved in biosynthesis of the medicinal antigout alkaloid **colchicine**.

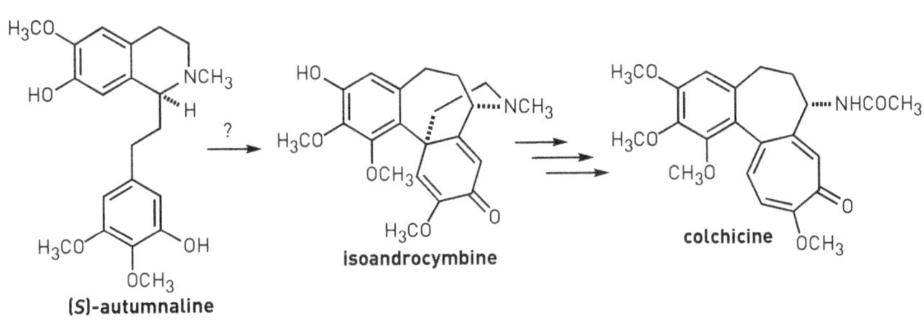

6. Provide a mechanism for the O_2/2-oxoglutarate demethylation of thebaine to neopinone in the morphine pathway. Why does neopinone spontaneously isomerize to codeinone?

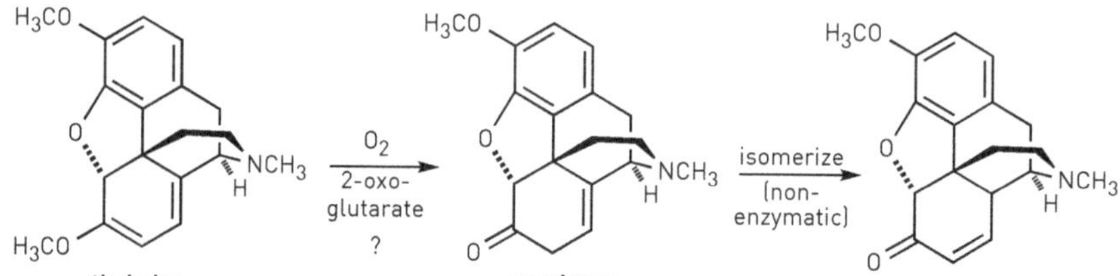

thebaine **neopinone** **codeinone**

7. Metabolism of **pethidine**, a synthetic analgesic, gives **norpethidine**, a neurotoxin whose build up during high dosage pethidine treatment can lead to serious complications and even death. Conversion to norpethidine is carried out mainly by the CYP enzymes. Propose a reasonable mechanism for formation of norpethidine.

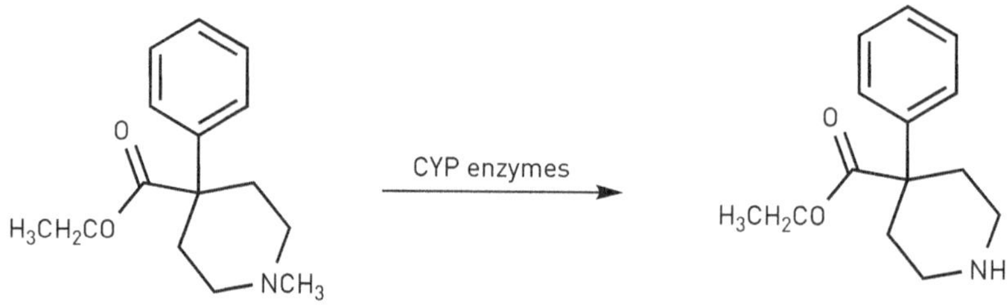

pethidine (Demerol) **norpethidine**

8. Contact with **capsaicin,** a component of cayenne and other hot peppers, produces a strong burning sensation in most tissues. In addition to its use in foods, capsaicin is used in a variety of topical ointments and dermal patches for the relief of pain. Considering the products produced from its hydrolysis, propose a reasonable biosynthesis of capsaicin from vanillin and suggest a sequence for biosynthesis of the 10-carbon carboxylic acid component, starting with isobutrylCoA.

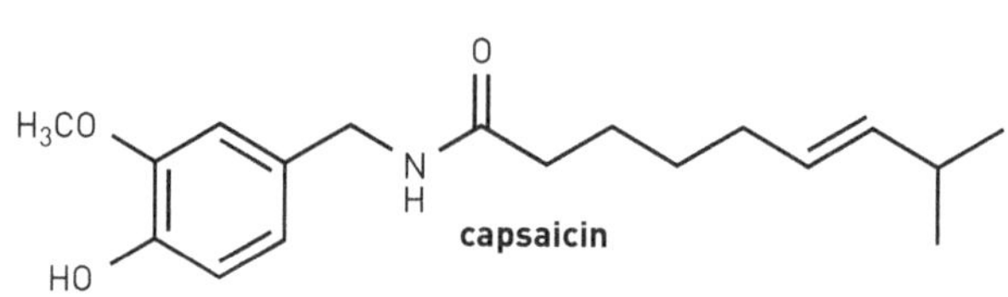

capsaicin

9. Propose a mechanism for this laboratory transformation:

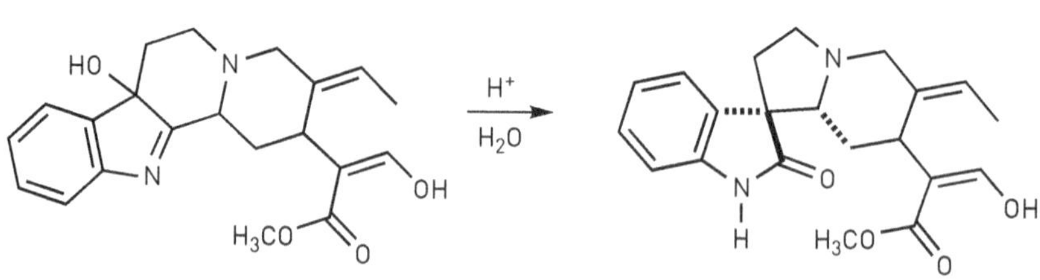

10. Compound **A** is a substituted pyrrole intermediate used in porphyrin biosynthesis. Propose a detailed mechanism to account for the formation of **A** from 5-aminolevulinic acid as shown (hint: the mechanism requires no co-factors or reagents other than water and a proton. You need only use imine formation, imine-enamine tautomerism, and other intramolecular steps including a final dehydration).

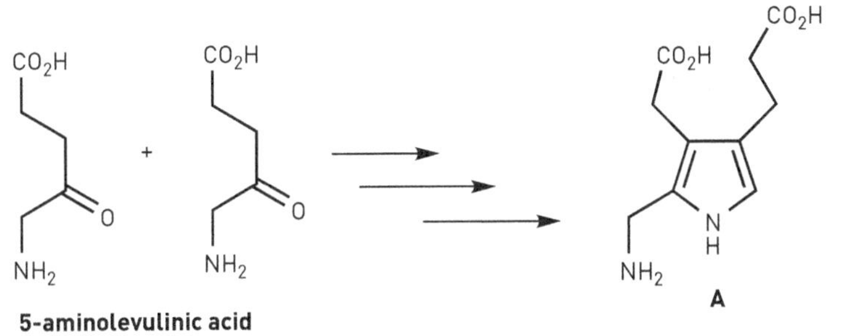

11. It was demonstrated as early as 1917 that the bicyclic alkaloid **A** could be synthesized in the laboratory from the three-component system shown below. Propose a mechanism for the formation of **A**.

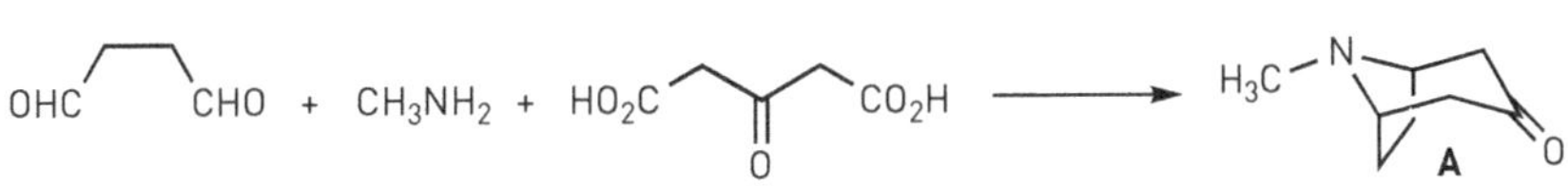

12. A proposed biosynthetic route to alkaloid **X** involves L-DOPA and 3-aminopropanal as starting materials. Provide a reasonable mechanistic pathway leading to **X** (hint: prior to the final cyclization, an oxidation to an *ortho*-quinone may be involved).

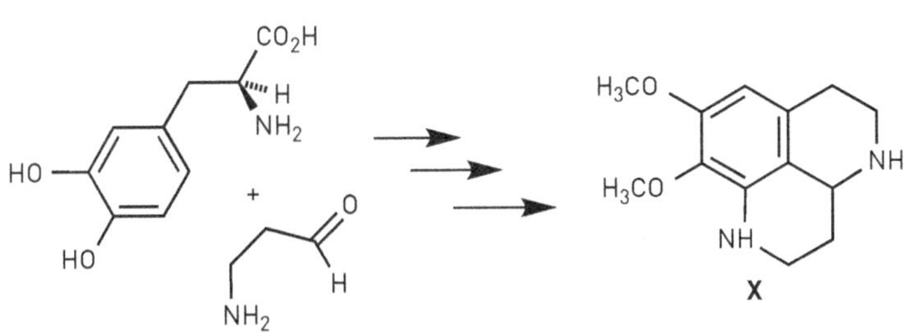

13. If ^{13}C-labeled tyrosine was used for biosynthetic labeling experiments, where would the ^{13}C labeled carbon atoms be located in the resulting morphine?

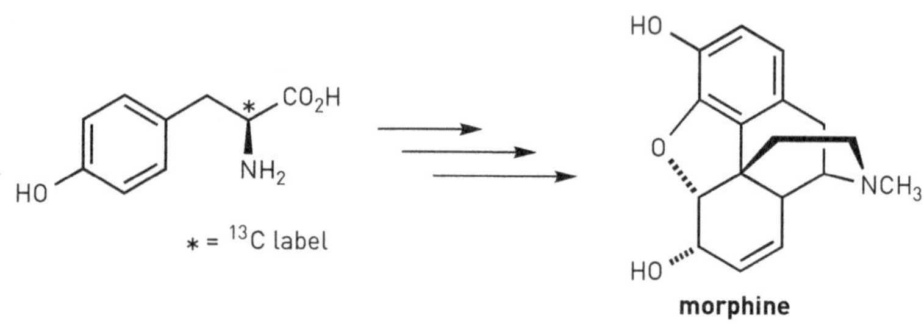

14. The alkaloid **pluviine** is transformed into **homolycorine** by a sequence of four simple steps. Outline those steps using intermediate structures, and so on (hint: begin with oxidation at a benzylic carbon).

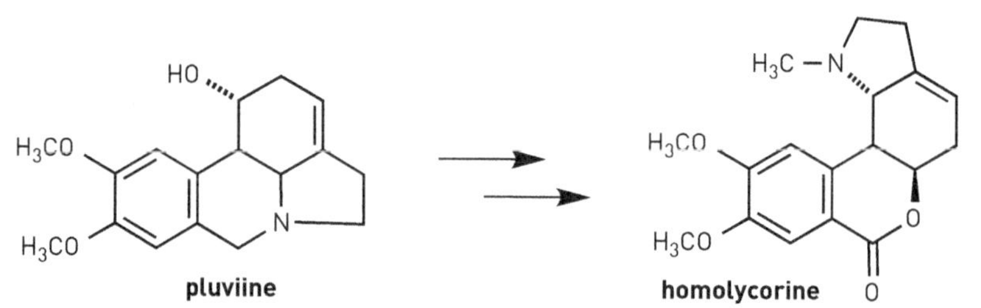

15. Show a mechanism for the formation of **B** from **A**, then propose a mechanism for the acid-catalyzed rearrangement of **B** to give **C**.

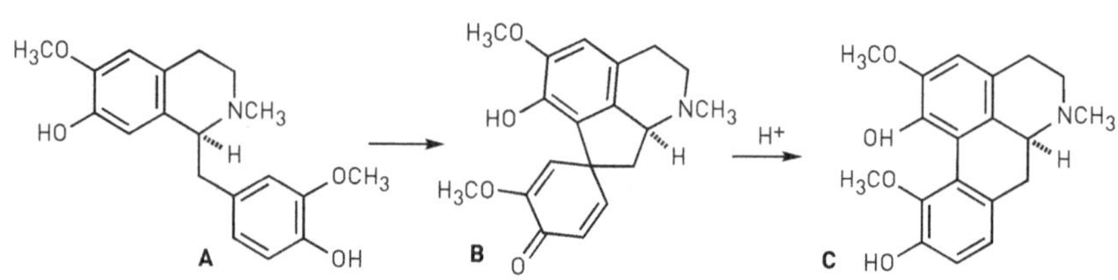

16. Benzodiazepines are psychoactive drugs whose core chemical structure is the fusion of a benzene ring and a 1,4-diazepine ring. Though discovered and marketed as synthetic drugs (Librium, Valium, etc.), one particular species of *Penicillium* mold produces the natural benzodiazepine shown here which shows selective inhibition against acetylcholinesterase (AChE) and moderate antibiotic activity. Identify the likely amino acid precursors to this alkaloid, then suggest a reasonable biosynthetic scheme.

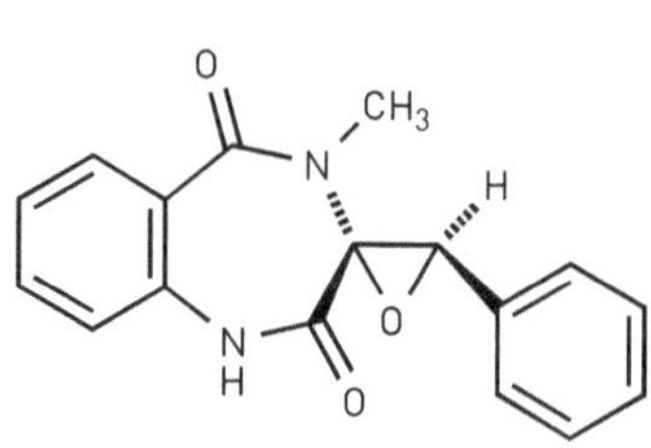

8 Organic Synthesis in the Laboratory

To the field of synthetic chemistry belongs an array of responsibilities which are crucial for the future of mankind, not only with regard to the health, material and economic needs of our society, but also for the attainment of an understanding of matter, chemical change and life at the highest level of which the human mind is capable.
—E. J. Corey (*Nobel Prize in Chemistry, 1990*)

The German chemist Friedrich Wöhler is generally credited with the first laboratory synthesis of a known organic compound (urea) from inorganic materials. He accomplished this by the simple heating of an inorganic salt, ammonium cyanate (NH_4OCN). "I must tell you," he wrote to his mentor Jöns Jakob Berzelius in 1828, "that I can prepare urea without requiring a kidney of an animal, either man or dog." While this report may seem relatively minor given the structural simplicity of urea, its impact was revolutionary. For the first time, the preparation and isolation of an organic compound had been achieved in the absence of the elemental "vital force" of living systems previously believed to be required for the construction of all such compounds. This milestone of 19th century organic chemistry was later followed by many others, including Kolbe's synthesis of acetic acid in 1847 and Fischer's synthesis of glucose in 1890. With the support of evolving methods for compound separation, purification, and spectroscopic analysis, rapid advances in the sophistication of organic synthesis followed throughout the 20th century, developing in tandem with an ever-deepening understanding of the underlying organic processes associated with living systems. While it is certainly true that syntheses of many structurally complex *unnatural* compounds of theoretical interest are also among the most remarkable achievements in synthetic strategy, tactical execution, and perseverance, the realm of natural products remains the dominant source for the most challenging and potentially beneficial targets available for such synthetic efforts. Figure 8.1 shows a small selection of some natural (and unnatural) products which have been produced via synthesis over the years, from Wöhler's time to the present. Note the increasing levels of structural sophistication and stereochemical complexity that have eventually been mastered by practitioners of organic synthesis.

In our own time, the traditional boundaries between organic and biological chemistry are disappearing in ways that are likely to transform the design and synthesis of organic molecules, from the construction of synthetic biologicals designed to act as biomarkers, biosensors, or drug delivery agents, to the development of molecular motors, self-replicating macromolecular systems, and even synthetic life forms. But regardless of their complexity, such efforts will always draw to some extent upon a vast well of knowledge of reactions, mechanisms, techniques, and accomplishments accumulated over nearly two centuries of careful and exhaustive research in synthetic organic chemistry—work often inspired by a drive to emulate the sophistication and elegance of biosynthetic processes and the remarkable products they produce.

In this brief final chapter, we can touch on only a few of the many essential aspects of organic synthesis, bearing in mind the introductory nature of our text and the need to avoid excessive complexity or terminology unfamiliar to the audience. But even within such limits, we can still obtain a deeper level of appreciation for the significance, sophistication and allure of the art and science of synthetic organic chemistry.

WHY WE SYNTHESIZE ORGANIC COMPOUNDS

Organic chemistry is not only a science—it is big business. And the list of industrial organic compounds and materials manufactured for profit is a long and ever-growing one: organic conductors and semiconductors, liquid crystals, plastics and polymers, dyes, lubricants, fuels, propellants, coatings, solvents, explosives, pesticides, fertilizers, and weed killers are just some of the examples. Many of these commercial products are produced from raw organic materials derived from petroleum while others, such as flavors, fragrances, food additives, and preservatives or certain components used in cosmetics and personal care, are partially or wholly derived from natural sources. Some products may be of limited availability from natural sources but also relatively simple in terms of their structural features, making commercial synthesis potentially cost-effective. By contrast, many other naturally occurring compounds, particularly those used as medicinal agents and pharmaceuticals, can possess remarkably complicated structures, and so isolation from natural sources via extraction or fermentation may be almost imperative since commercial synthesis may not be economically feasible. Alternatively, certain complex products of limited natural availability may nonetheless be obtainable by chemical modification of a more abundant, structurally similar natural product. Regardless of the context, the synthesis of

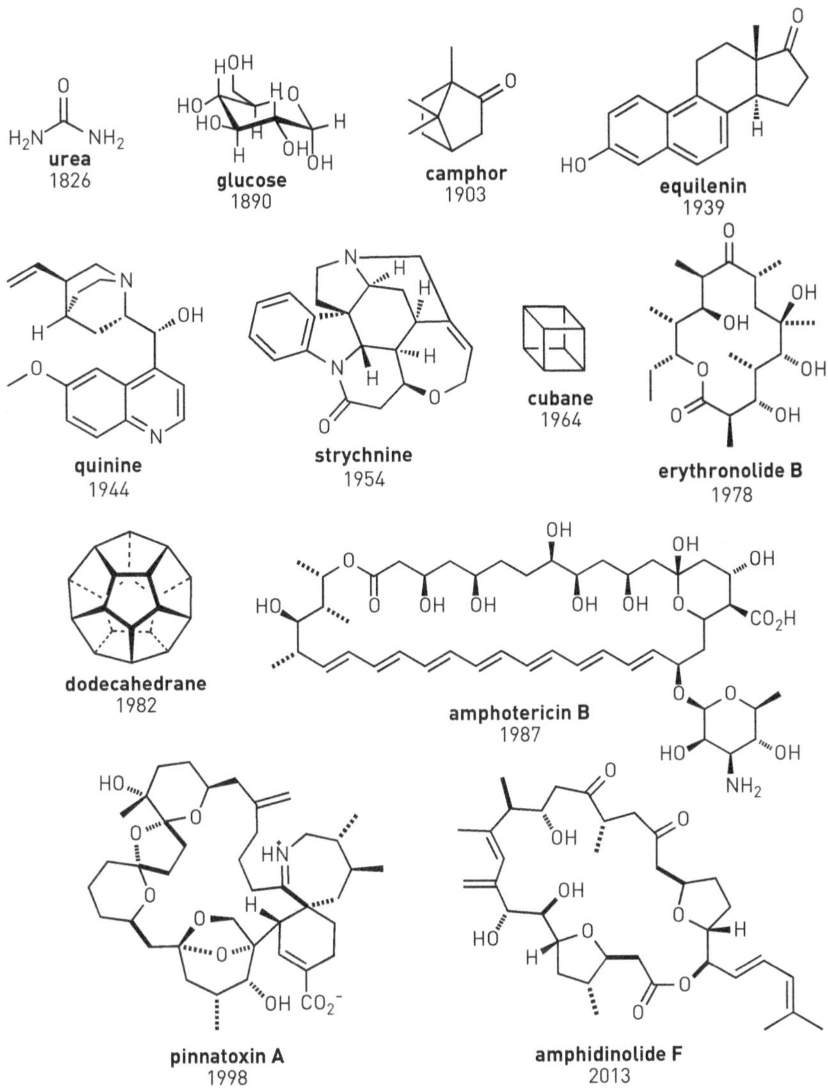

FIGURE 8.1
A group of selected organic compounds produced by synthesis from 1828 to the present.

organic compounds, from the simplest to the most intricately structured, is a large and often profitable enterprise that continues to command the attention of organic chemists in both industry and academia on a worldwide basis.

Our focus here, of course, is mainly on the synthesis of interesting or useful compounds that are known to be produced by biosynthetic processes. With that in mind, a partial list of some of the many reasons for undertaking such synthetic efforts includes:

- advancement of new synthetic reactions or methods;
- investigation of approaches to new or intellectually challenging structural features;

- development or improvement of a commercial synthesis process;
- provision of sufficient quantities for biological studies/testing;
- development of processes that mimic biosynthesis (biomimetic);
- determination or confirmation of molecular structure;
- identification of structural features required for biological activity;
- determination of molecular mechanism of biological action;
- synthesis of structural analogues.

At this point, let's look more closely at several of these points and at some of the different approaches to synthesis that organic chemists have developed over the years to overcome some of the considerable obstacles one may encounter when trying to match the synthetic prowess of nature.

SYNTHETIC CHALLENGES: TOTAL SYNTHESIS

The term **total synthesis** refers to the complete construction of a natural product from relatively simple and available organic starting materials. Total synthesis of complex organics is frequently employed by academic chemists to test or to demonstrate the utility of new reactions or new approaches to various structural features. These were the key drivers for total synthesis of many of the more complex examples shown in Fig. 8.1. A relatively simple example of a total synthesis is Stork's approach to **grandisol** (shown in Fig. 8.2), one of four monoterpenoid components of the sex attractant for the male boll weevil (*Anthonomus grandis*), a major pest of cotton crops in the western hemisphere. This synthesis is especially notable for its unique approach to formation of the cyclobutane ring, a notoriously difficult ring system to prepare synthetically.

The synthesis scheme begins with a seven carbon cyanoalkene **8.2a** which is deprotonated at the position *α* to the cyano group (recall the enhanced acidity at that position). The resulting anion is then reacted with 2-bromo-1-heptanol THP (tetrahydropyranyl ether-protected alcohol) in a simple S_N2 reaction to give the alkylated product **8.2b**. Alkene epoxidation then affords **8.2c** which again undergoes deprotonation *α* to the cyano group, with the resulting anion acting as a nucleophile to attack the epoxide ring in an intramolecular fashion, leading to formation of the cyclobutane ring of **8.2d** as shown. Partial hydride reduction of the cyano group triple bond with diisobutylaluminum hydride (DIBAL-H, similar in action to $LiAlH_4$ or $NaBH_4$)) gives an intermediate imine which is then hydrolyzed to the corresponding aldehyde **8.2e**. Wolff–Kishner reduction of the aldehyde carbonyl to a methyl group gives **8.2f** which then undergoes chromic acid oxidation of its secondary alcohol function to yield the ketone **8.2g**. After Wittig reaction to convert the ketone function to the indicated alkene **8.2h**, the synthesis is completed by acidic removal of the THP protecting group to release the primary alcohol function of the final product, **grandisol**. It is worth noting that while the stereochemical

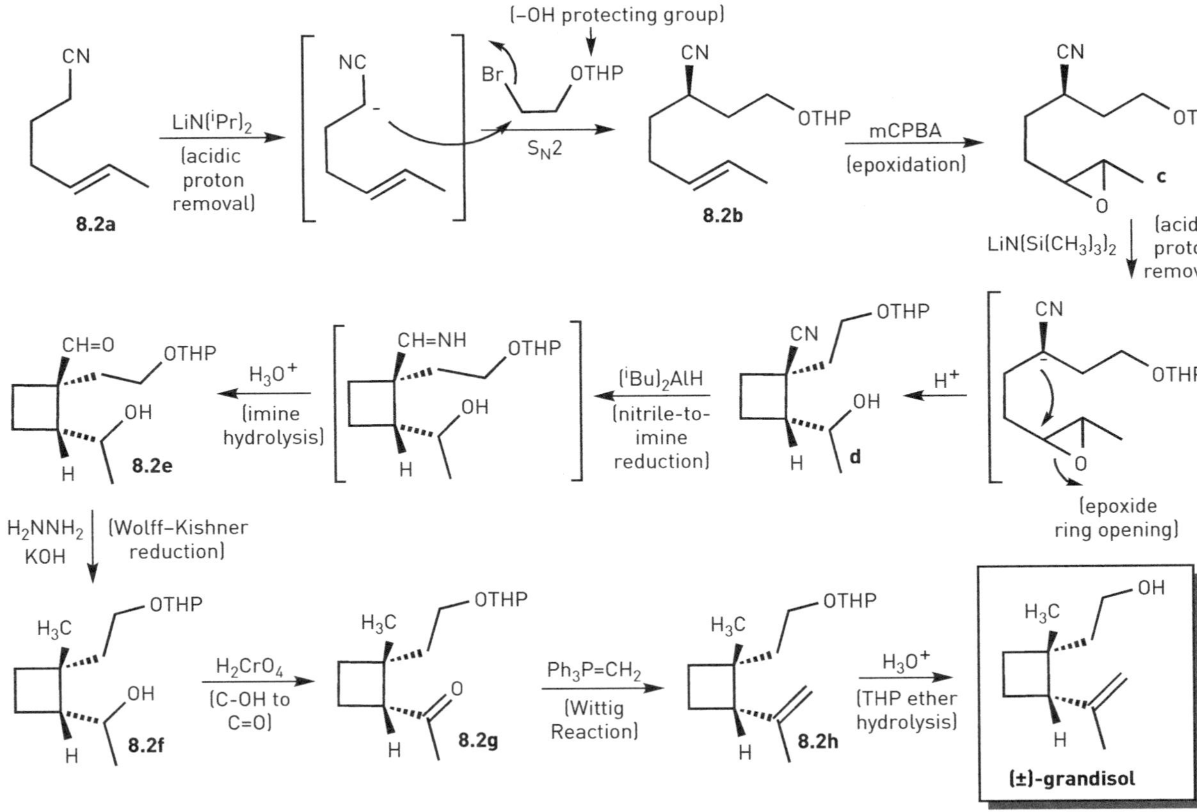

FIGURE 8.2
Stork's total synthesis of grandisol, a terpenoid component of boll weevil sex attractant.

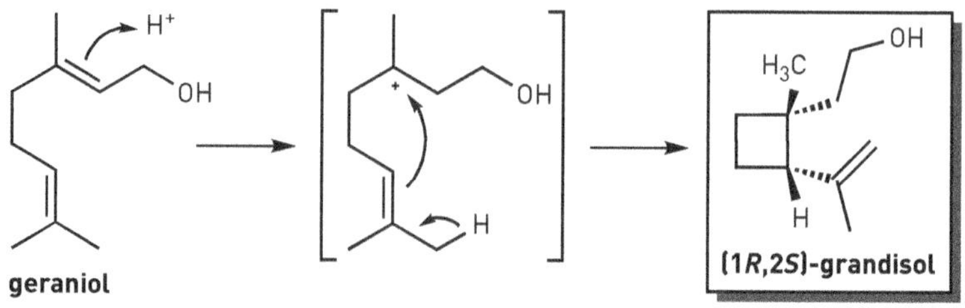

FIGURE 8.3
Presumed enzyme-mediated biosynthesis of (1*R*,2*S*)-grandisol.

structure of the final product as shown in Fig. 8.2 is the naturally occurring (1*R*,2*S*) diastereomer, the product actually produced by this synthetic approach is a (±) racemic mixture of the (1*R*,2*S*) and (1*S*,2*R*) diastereomers. While several other total syntheses of grandisol have successfully produced the final product in its optically pure (*R*,*S*) form, we note here that any total synthesis which produces an optically pure product is usually much more complex and difficult to achieve than one which produces only a racemic product. Nevertheless, even racemic syntheses of grandisol must properly address the stereochemical relationship of the three substituents on the cyclobutane ring.

Compare the multistep nature of the synthetic approach to grandisol above with the relative simplicity of its biosynthesis from **geraniol** as outlined in Fig. 8.3. Note the cyclization of a carbocation intermediate to produce the required 4-membered ring. Such site-specific protonation and 4-membered ring cyclization cannot be achieved by any known synthetic processes; they are only feasible when the substrate is regiospecifically protonated and held in the necessary strained folding pattern by the enzyme that mediates the process. Furthermore, note that the enzyme-mediated cyclization leads to formation of only the optically pure (1*R*,2*S*) diastereomer. Nature makes it look so easy.

SYNTHETIC CHALLENGES: SEMISYNTHESIS

Unlike total synthesis, **semisynthesis** (or **partial synthesis**) is a process in which a target compound is produced by synthetic modification or elaboration of a structurally similar compound which is readily available in reasonable quantities from natural sources. This approach can be especially useful commercially if the target itself is produced naturally in only limited amounts and possesses multiple chirality centers or other complex stereochemical features that would make total synthesis cost-prohibitive. To begin our analysis, let us first consider the complexity involved in achieving another total synthesis of an important compound, in this case the steroid **progesterone** (as outlined in Fig. 8.4), before we illustrate the benefits of a semisynthetic approach to the same compound.

This lengthy synthesis begins with reaction of 2-methylpropenal with an acetylenic Grignard reagent, affording secondary alcohol **8.4a**. The alcohol function of

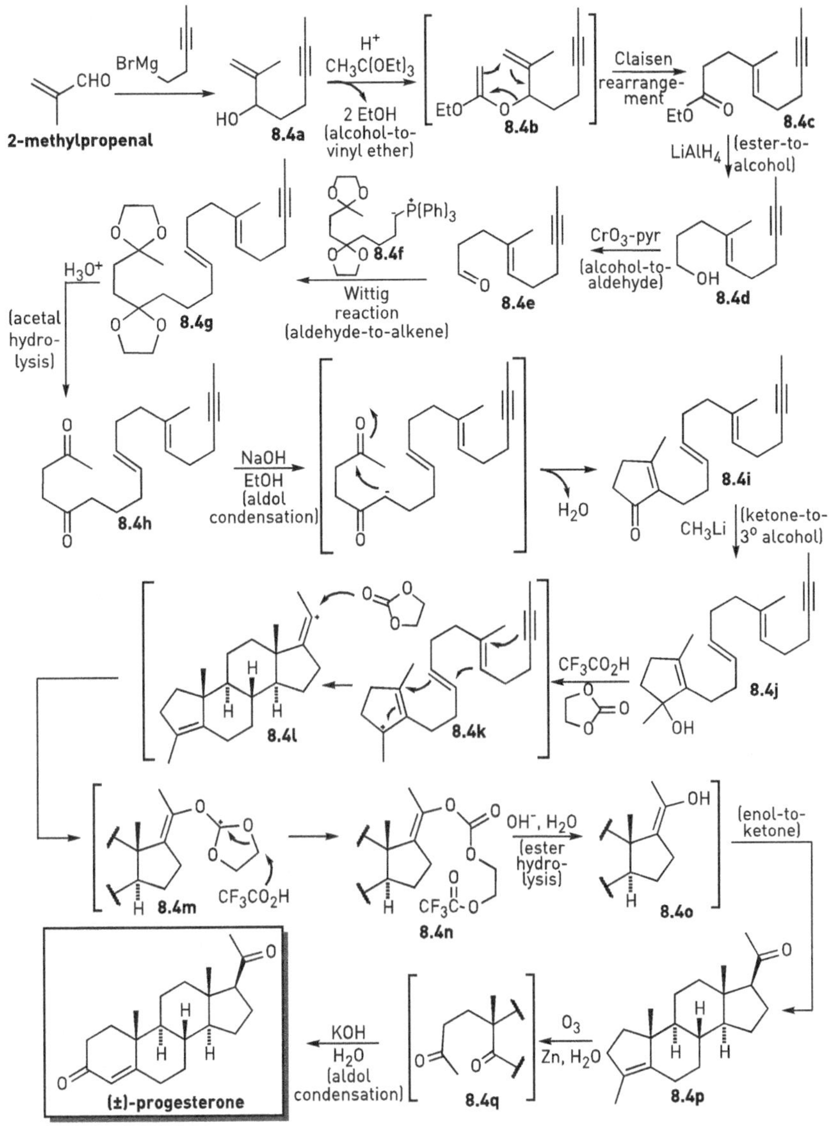

FIGURE 8.4
Johnson's total synthesis of the sex hormone progesterone.

8.4a is then reacted with 1,1,1-triethoxyethane (triethylorthoformate) in an acid-catalyzed sequence of alcohol exchange and alcohol elimination to give derivative **8.4b.** Recall (from Fig. 6.6) that allyl vinyl ethers of this type may be induced to undergo the **Claisen rearrangement**—in this case, by heating. The result here is formation of ethyl ester **8.4c** which is then converted to alcohol **8.4d** via standard $LiAlH_4$ reduction. Oxidation of **8.4d** with CrO_3-pyridine complex (similar to pyridinium chlorochromate or PCC) then gives aldehyde **8.4e**. A Wittig reaction of this aldehyde with phosphorus ylide **8.4f** (prepared via a separate synthetic sequence requiring many steps) gives the indicated alkene **8.4g.** Hydrolysis of the ethylene glycol acetal functions of **8.4g** then leads to diketone **8.4h** which is reacted with base under standard aldol condensation conditions to produce the cyclopentenone derivative **8.4i**, with subsequent addition of methyllithium to the ketone function giving the expected alcohol **8.4j**. Acid-catalyzed dehydration of this tertiary allylic alcohol with trifluoroacetic acid leads to formation of carbocation **8.4k** which undergoes a cationic cyclization cascade similar to the one we have seen previously in the biosynthesis of cholesterol (see Fig. 4.38). The resulting tetracyclic ring system terminates in the vinylic cation **8.4l** which is subsequently trapped by a cyclic carbonate ester acting as a nucleophile, giving rise to the more stabilized carbocation **8.4m**. Ring opening of this cation via nucleophilic attack by trifluoroacetate ion quenches the carbocation, giving rise to the mixed carbonate-trifluoroacetate ester derivative **8.4n**. Ester hydrolysis with aqueous base cleaves this complex side chain, releasing enol **8.4o** which rapidly tautomerizes to methyl ketone **8.4p.** The synthesis is completed via ozonolysis of the cyclopentene ring double bond which yields the corresponding diketone **8.4q.** A final intramolecular aldol condensation forms the required cyclohexenone ring system of **progesterone**.

It is worth reflecting on the careful synthetic planning and execution required for this remarkable outcome. Especially noteworthy is the establishment of correct stereochemistry at the molecule's six chirality centers, though the product produced here is *racemic* progesterone rather than the naturally occurring (+)-enantiomer. While later (and more sophisticated) variants of this approach have successfully produced the optically pure product, the original process gives us a reasonably good idea of the incredible intricacy involved in trying to produce a complex polycyclic natural product with multiple stereocenters from relatively simple starting materials.

Recall from Chapter 4 that progesterone is produced biosynthetically via a sequence of enzyme-mediated transformations carried out on **cholesterol** (see Fig. 4.44). With that in mind, we may compare the above multistep total synthesis approach to racemic progesterone with one of several known semisyntheses of **progesterone,** obtained in this instance from **stigmasterol,** an abundant plant sterol similar in structure to cholesterol and available from soybean oil. An outline of this semisynthetic process is shown in Fig. 8.5.

The stigmasterol hydroxyl group is initially converted to the corresponding α,β-unsaturated ketone **8.5a** by a process known as **Oppenauer oxidation,** a mild oxidative method which uses **aluminum isopropoxide** dissolved in acetone solvent (which also acts as a hydride acceptor from the secondary alcohol being oxidized). Under the usual conditions for this oxidation, the nonconjugated β,γ-double bond also isomerizes into conjugation with the newly formed steroidal ketone. Next, a simple ozonolysis of the side chain double bond of **8.5a** gives the corresponding aldehyde **8.5b** which is then converted to enol acetate derivative **8.5c** via treatment with acetic anhydride. A final ozonolysis to selectively cleave the enol acetate double bond of **8.5c** yields (+)-**progesterone**. Note that the semisynthetic approach is not only much simpler than a total synthesis, it also produces optically pure product rather than a racemic mixture.

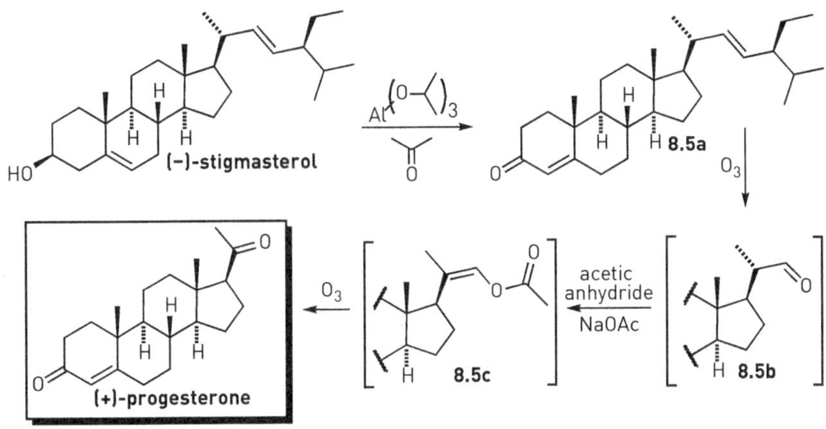

FIGURE 8.5
Heyl and Herr's semisynthesis of progesterone from stigmasterol.

Aside from its significant role as the principal female sex hormone involved in pregnancy, menstruation, and embryogenesis, progesterone is also an important precursor both biosynthetically and semisynthetically to other important steroids such as **cortisol.** This is but one of many known examples of the economical production of an important yet relatively scarce natural product via chemical transformation of a structurally related but more abundant **phytochemical** (plant chemical).

SYNTHETIC CHALLENGES: BIOMIMETIC SYNTHESIS

In spite of the ingenuity and intricate planning obviously involved, only one of the two foregoing examples of terpenoid natural product total syntheses actually bears much resemblance to the process employed by nature for construction of those target compounds (see the cationic cyclization step in Johnson's synthesis of progesterone, Fig. 8.4). An equally sophisticated challenge in total synthesis involves efforts to devise synthetic approaches which closely mimic the biosynthetic chemistry actually used (or believed to be used) by the producing organism. Such a scheme is usually referred to as a **biomimetic synthesis.** In some cases, the work may even use biochemical reagents or cofactors in an effort to mimic nature as closely as possible. The justification usually cited for undertaking a biomimetic synthesis is to support or to challenge a specific biosynthetic proposal, although in some cases competing proposals have been supported by correspondingly different biomimetic approaches. Regardless, one benefit of successfully imitating nature is that biomimetic syntheses are often relatively efficient and elegant compared to other approaches.

It is generally agreed that the first biomimetic synthesis was achieved in 1917 by Sir Robert Robinson in his unique approach to the construction of the tropane alkaloid **tropinone**, as shown in Fig. 8.6 (note the close resemblance to the biosynthetic process previously outlined in Fig. 7.5). As in the biosynthetic scheme, Robinson's synthesis relies on sequential Mannich reactions for formation of the key bonds in

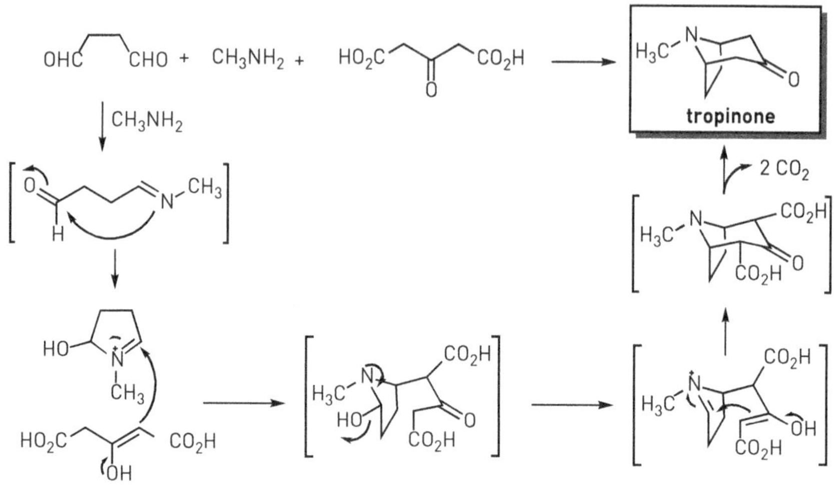

FIGURE 8.6
Robinson's 1917 biomimetic synthesis of tropinone via sequential Mannich reactions.

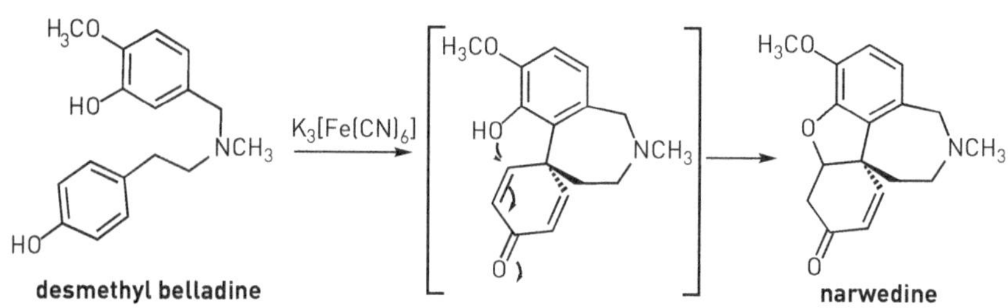

FIGURE 8.7
Barton's biomimetic oxidative coupling: preparation of narwedine from desmethyl belladine.

the bicyclic ring system and subsequent decarboxylations of the β-ketoacid core to reach the final product. Also note the simplicity of this process which is achieved in a single three-component reaction using relatively simple starting materials. This result strongly supported Robinson's proposal that chemistry very similar to this was likely used by nature to produce the tropane alkaloids, a proposal that has subsequently been sustained by numerous meticulous studies over the ensuing years.

This example shows Robinson's early insight into potential biosynthetic mechanisms involved in the construction of key alkaloids. In 1932 he further proposed that one of the key steps in the biosynthesis of morphine—formation of **salutaridine** from **(*R*)-reticuline**—most likely involved an intramolecular oxidative coupling (see Fig. 7.21). This proposal was eventually confirmed by Barton who also showed that laboratory oxidation of **desmethyl belladine** with aqueous potassium ferricyanide produced **narwedine** (Fig. 8.7); further reduction of the ketone function to the corresponding alcohol afforded **galanthamine**, an alkaloid which is now commercially prepared due to its promise as a treatment for mild to moderate cases of Alzheimer's disease. Though the chemical yield of narwedine was small (0.97%) from this early

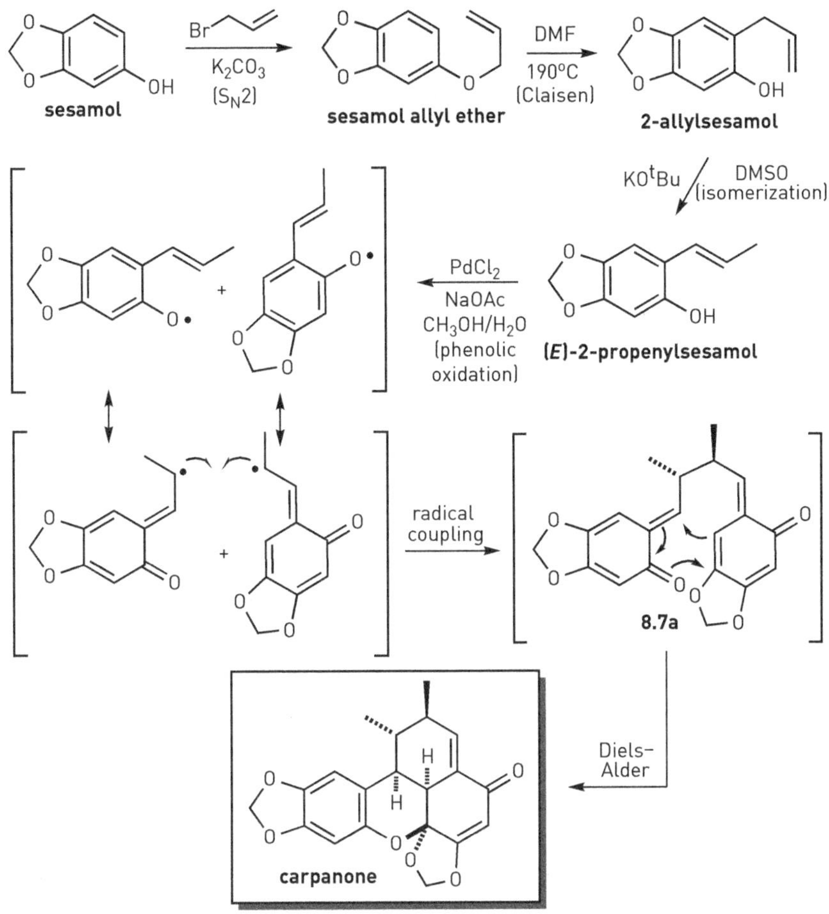

FIGURE 8.8
Chapman's biomimetic total synthesis of the lignan product carpanone.

example of biomimetic oxidative coupling, higher yielding oxidations have subsequently become a common theme over the years in many total syntheses using a variety of more sophisticated oxidizing agents.

One of the best examples of such a biomimetic oxidative coupling process involves the total synthesis of a shikimate natural product known as **carpanone**, a compound lacking any intrinsic medicinal value but which is endowed with unusual stereochemical complexity with its five contiguous chirality centers. Chapman's elegant biomimetic approach to this lignan derivative is shown in Fig. 8.8 which begins with *O*-allylation of **sesamol** followed by an aromatic Claisen rearrangement to give **2-allylsesamol**. Base-catalyzed isomerization of the side-chain double bond into conjugation with the aromatic ring then gives the essential starting material **(*E*)-2-propenylsesamol**. Using a Pd-based reagent to achieve oxidative formation of the corresponding phenoxy radicals, bimolecular C–C coupling of two allyl radical resonance contributors leads initially to the dimer **8.7a** which then undergoes an intramolecular hetero-Diels–Alder cyclization to afford **carpanone** directly. The product

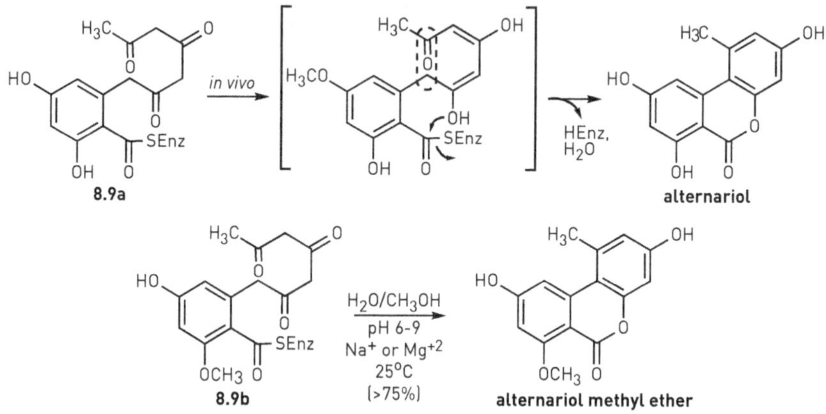

FIGURE 8.9
Laboratory cyclization and aromatization of a tetraketide to give alternariol methyl ether.

produced was racemic but nevertheless possessed the correct stereochemical relationships among all substituents at the molecule's chirality centers. This is an excellent example of a synthesis designed to confirm a proposed biosynthetic scheme, namely the presumed formation of carpanone from (*E*)-2-propenylsesamol via an enzyme-mediated oxidative coupling and subsequent Diels–Alder cyclization. To assemble such a structurally complex target compound in such a simple, elegant fashion in the laboratory remains an impressive achievement in biomimetic synthesis.

Products from the polyketide pathway, especially aromatic polyketides, would seem to be obvious candidates for simple testing of the acetate hypothesis via a biomimetic approach to their synthesis. As illustrated in Fig. 8.9, formation of an advanced enzyme-bound thioester tetraketide precursor **8.9a** (proposed earlier as a possible intermediate arising from a simple heptaketide via condensation, cyclization, and aromatization) could adopt the indicated folding pattern leading to further condensation and cyclization to give **alternariol** as shown. To test this proposal, preparation of the structurally related ethylthioester tetraketide **8.9b** was followed by treatment under a variety of mildly basic conditions which invariably led to formation of **alternariol methyl ether** as the sole product in >75% yield as shown. While not conclusive evidence for a similar mechanism operating in vivo, this example demonstrates the usefulness of laboratory synthesis for probing potential mechanistic pathways in biosynthetic processes.

SYNTHETIC CHALLENGES: STRUCTURAL REVISION OR CONFIRMATION

Many remarkable advances have been made over the years in spectroscopic determination of chemical structures, especially in the areas of x-ray crystallography, nuclear magnetic resonance (^{1}H and ^{13}C NMR) and mass spectrometry. Nevertheless, organic synthesis continues to play a significant role in the determination or confirmation of natural product structures. Interestingly, a typical year finds numerous examples of the use of total synthesis in published revisions of previously misassigned natural product structures. These corrections range from minor amendment of the stereochemical relationships among substituents to significant changes in atom connectivity, ring size, or other major structural features.

A useful example that confirms one assigned structure for a natural product while correcting the structure previously assigned to another is provided by a total synthesis of the natural product **pseudodeflectusin** which had originally been isolated from cultures of *Aspergillus pseudodeflectus.* This structurally interesting isochroman had been shown to exhibit activity against several human cancer cell lines, so confirmation of the originally assigned structure, as shown in Fig. 8.10, was achieved via stereoselective synthesis and spectroscopic comparisons of the synthetic product with the originally isolated fungal metabolite. Interestingly, it was simultaneously confirmed that the previously assigned structure of another natural product, **aspergione B**, was actually incorrect and that pseudodeflectusin and aspergione B were in fact the same compound since the full spectroscopic data for both were identical in all respects. The originally proposed structure for aspergione B was then independently synthesized as shown from the same precursor used in the total synthesis of pseudodeflectusin. As expected, the ^{1}H and ^{13}C NMR data were similar but not identical to those originally reported for aspergione B.

A somewhat simpler example of structural revision of a previously reported compound is shown in Fig. 8.11. **Intricatin**, an antimutagenic homoisoflavonoid ("homo" indicates an extra carbon, in this case between the pyranone ring and its attached phenyl group) was isolated in 1989 from a Mexican desert plant and was originally assigned the structure shown. However, independent total synthesis of this structure later showed that the ^{1}H NMR data for the naturally occurring compound and those of the synthetic material were not the same, leading to a search of the chemical literature which ultimately revealed that ^{1}H NMR data for intricatin and those of another natural homoisoflavonoid isolated in 1987, **8-methoxybonducellin**, were identical, indicating that the original structure for intricatin had been misassigned and that intricatin and 8-methoxybonducellin were in fact the same compound. The location of a methoxyl group relative to an adjacent phenolic hydroxyl group in compounds such as these would clearly be difficult to correctly assign via spectroscopic data alone and further emphasizes the usefulness of synthesis as an aid in structure determination.

SYNTHETIC CHALLENGES: FORMAL SYNTHESIS

Formal synthesis of a compound involves development of a new synthetic route to an advanced precursor which is known to be readily convertible to the target compound by a previously published reaction or sequence of reactions. An early example from 1935 is Percy Julian's formal synthesis of the medicinal alkaloid

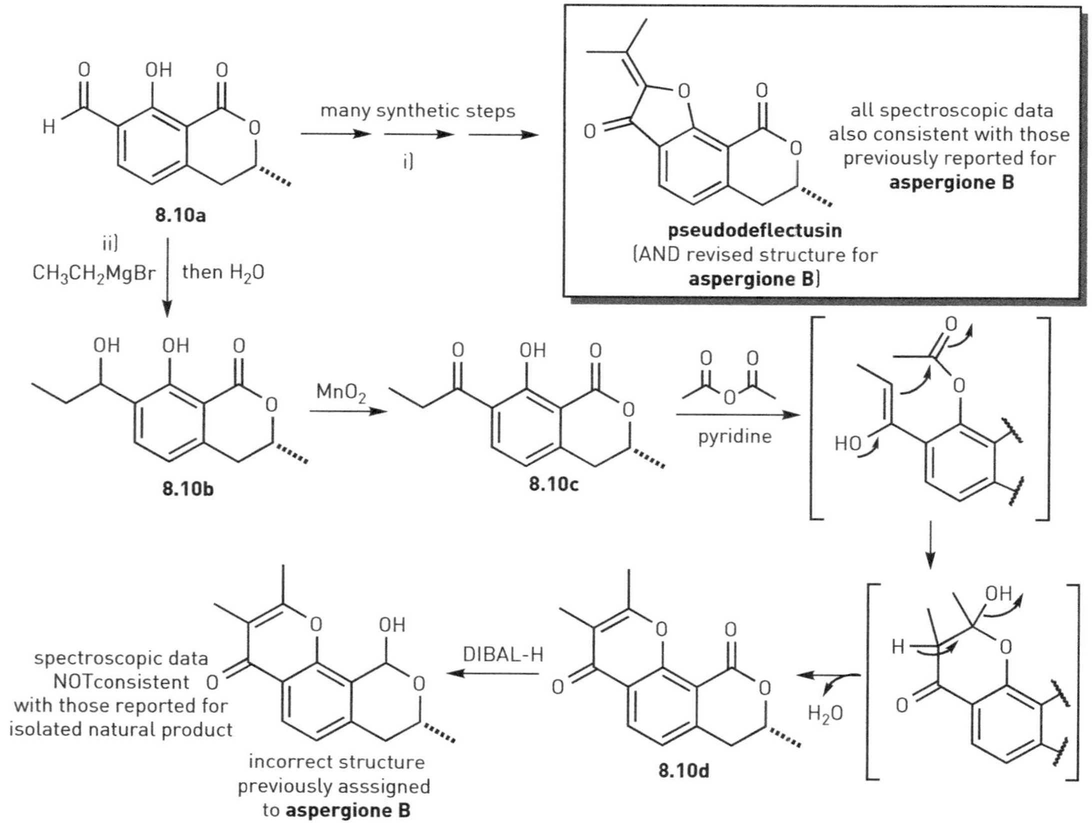

FIGURE 8.10

Total syntheses to confirm that i) structure of pseudodeflectusin and aspergione B are identical and ii) originally assigned structure for aspergione B is incorrect.

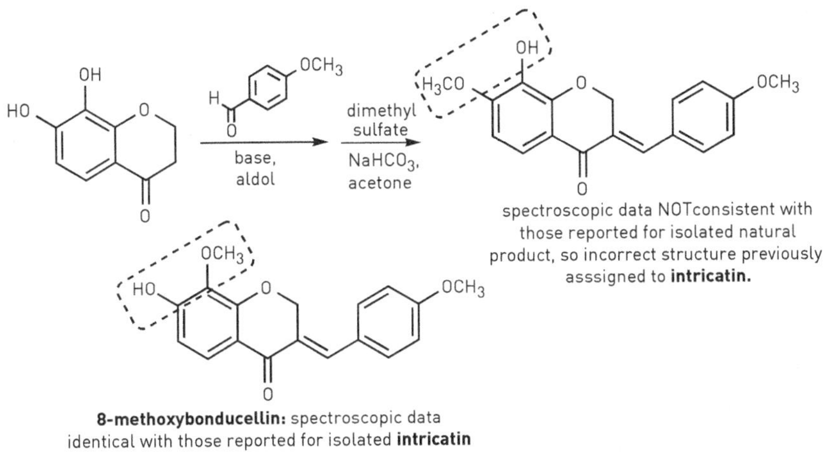

FIGURE 8.11

Total synthesis of structure originally assigned to intricatin and revision to the structure of 8-methoxybonducellin by comparison of spectroscopic data.

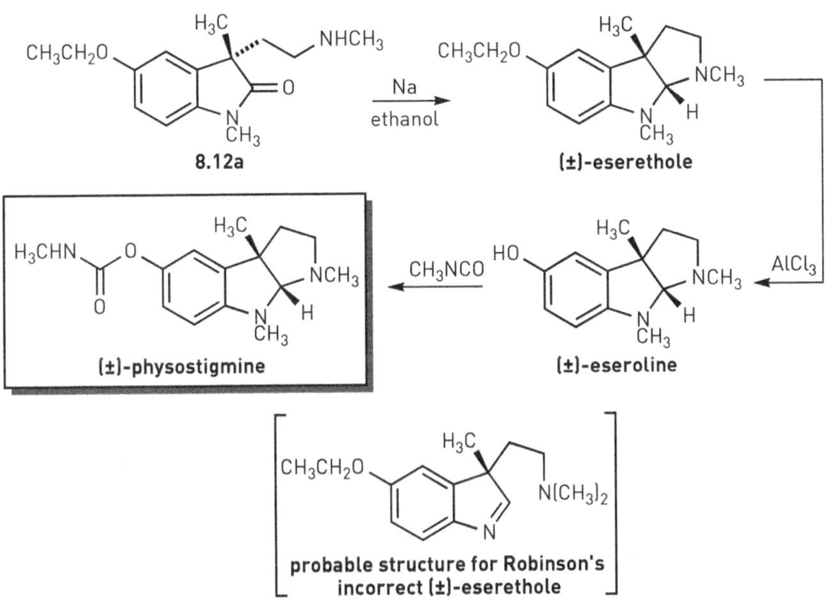

FIGURE 8.12

Julian's formal synthesis of physostigmine and correction of Robinson's reported eserethole synthesis.

physostigmine (see Fig. 7.27 and related text). This remarkable achievement was the first synthesis of physostigmine and also involved correction of some previously published work on the same compound by Sir Robert Robinson, the future British Nobel laureate. Julian, an African-American organic chemist working in relative obscurity at a small midwestern university, managed to complete the synthesis in 11 total steps over a period of three years. The last few steps of his approach are shown in Fig. 8.12. A key intermediate compound in the synthesis, **eserethole**, had reportedly been synthesized previously by Robinson, but based on his own careful work on reductive cyclization of **8.12a** using sodium metal in ethanol, Julian recognized that the compound Robinson had synthesized and identified as eserethole must have had a different constitution. Julian found that his racemic eserethole had chemical and physical properties different from Robinson's racemic material, but after chemical resolution to separate its enantiomers, his own synthetic *l*-eserethole was identical in all respects to the naturally occurring product. Julian's eserethole was then converted to the phenolic derivative **eseroline** by treatment with $AlCl_3$ to dealkylate the ethoxyl group after which he states in his original publication that "since Polonovski and Nitzberg have described conversion of *l*-eseroline into *l*-physostigmine by treatment of the former with methyl isocyanate, our synthesis represents a complete synthesis of the alkaloid physostigmine." This made Julian's a *formal* synthesis of physostigmine (although he also states in a footnote of the paper that subsequent to submission of the original manuscript, he did in fact complete the synthesis by repeating the above mentioned reaction of methyl isocyanate with his synthetic *l*-eseroline in order to provide a requested sample of pure *l*-physostigmine). Julian, the first African-American chemist elected to the National Academy of Sciences, went on to further fame as a highly innovative entrepreneurial organic chemist whose many contributions to the field of natural products chemistry included work that ultimately led to the economical availability of previously scarce medicinal steroids such as cortisone, progesterone, and others from stigmasterol and other plant-based sterols.

Another mechanistically interesting example of formal synthesis is shown in Fig. 8.13 which outlines Evans's formal synthesis of **colchicine**, an important medicinal alkaloid with anti-inflammatory properties long used as the standard treatment for flaring symptoms of gout. Evans's synthesis begins with quinone monoacetal **8.13a** which is reacted with a sulphur ylide derived via deprotonation of trimethylsulfoxonium iodide. This nucleophilic ylide adds to cyclic α,β-unsaturated ketones in 1,4-fashion to give an enolate which then reacts intramolecularly with expulsion of neutral dimethylsulfoxide (DMSO) to give a ring-fused cyclopropane derivative, in this case compound **8.13b**. Reaction of the ketone function of **8.13b** with Grignard reagent **8.13c** yields the corresponding tertiary alcohol **8.13d**. Further treatment of this alcohol with trifluoroacetic acid leads to dehydration and formation of a carbocation which is then trapped intramolecularly in an electrophilic aromatic-substitution-type process. Subsequent protonation of the ketone function then leads to a carbocationic rearrangement with 1,2-migration of the trimethoxylated aryl ring accompanied by cleavage of the cyclopropane linkage and concurrent ring expansion of the spiro-fused 6,6 ring system to the fused seven-membered rings of the tricyclic product **8.13e**. This was finally followed by oxidative dehydrogenation of **8.13e** with DDQ (dichlorodicyanoquinone) to give the tropolone-containing derivative **desacetamidoisocolchicine** which had previously been transformed via several additional steps to **colchicine** itself by a number of other investigators.

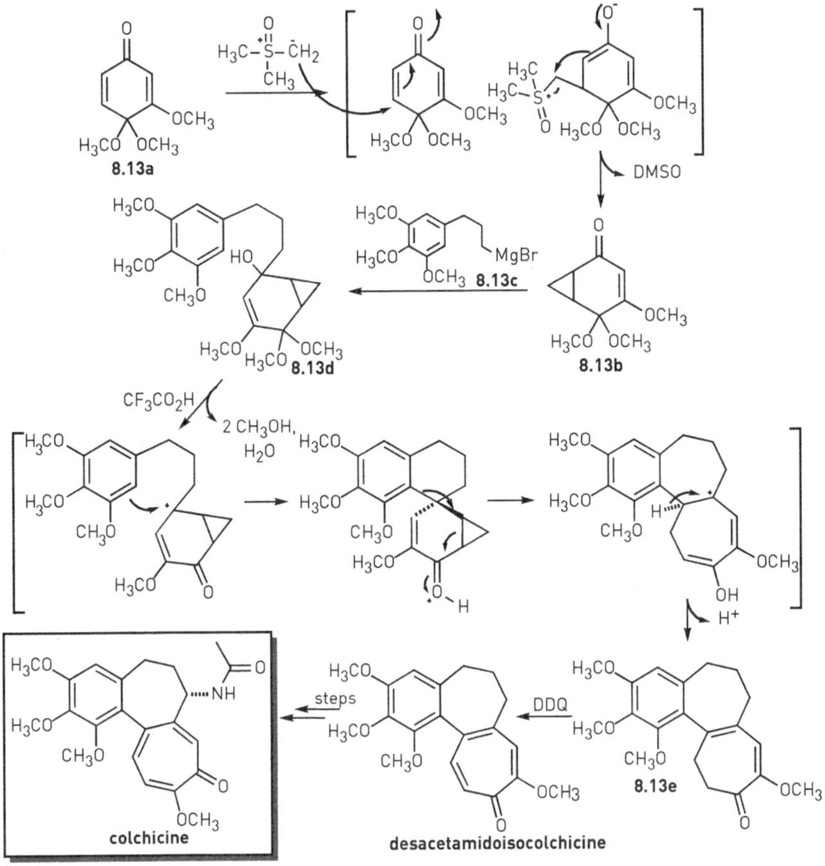

FIGURE 8.13
Evans's formal synthesis of colchicine via total synthesis of desacetamidoisocolchicine.

SYNTHETIC CHALLENGES: STEREOSELECTIVE SYNTHESIS OF OPTICALLY PURE COMPOUNDS

Some of the syntheses we have encountered thus far have successfully assembled some architecturally challenging target structures, and in most instances these elegant synthetic approaches have been used to produce chiral compounds. However, the result has usually been formation of *racemic* chiral compounds rather than optically pure ones. And the production of optically active, single enantiomer products is extremely important, especially when those products are used medicinally since the biological activity of compounds is often directly related to their stereochemical configuration. Addressing this additional obstacle requires us to confront one of the iron rules of synthetic chemistry: *optically inactive reactants cannot produce optically active products*. Put another way, if all reactants, solvents, and catalysts used in a given synthetic process are either achiral or chiral but optically inactive, they may combine to produce a compound with one or more chirality centers, but the compound produced will always be racemic (i.e., optically inactive). You cannot get true molecular asymmetry out of thin air; it always requires the intervention of a preexisting asymmetric entity in some fashion, which begs the question: How were optically pure compounds in nature formed in the first place? What was the original intervening chiral entity that led to formation of single enantiomer L-amino acids and D-sugars instead of racemic mixtures? No one knows for certain, though there are many hypotheses on the primordial origins of molecular asymmetry. But even though we have yet to discover any method in the laboratory that can produce optically pure compounds from optically inactive reactants, catalysts, and solvents, there are a number of different methods we can employ that use preexisting asymmetric agents in some fashion to help us to induce asymmetry at newly formed chirality centers.

Of course, we know that nature readily produces optically active single enantiomer compounds from optically inactive reactants by using enzymes (which again are already chiral and optically active, being derived from optically pure single enantiomer amino acids) to catalyze bioorganic reactions. So, one method for solving the dilemma of racemic mixture production would be to employ enzymes for chemical reactions. But while it is sometimes possible to use enzymes for organic synthesis, they are notoriously complex, difficult to isolate in useful quantities and do not readily lend themselves to large-scale use. With that in mind, let's briefly consider some of the ways that chemical ingenuity has been used to achieve isolation or synthesis of nonracemic chiral compounds in the chemical laboratory.

RESOLUTION OF ENANTIOMERS TO OBTAIN OPTICALLY PURE COMPOUNDS

Recall that enantiomers have identical chemical and physical properties and so cannot be separated by ordinary chemical means (distillation, crystallization, or ordinary chromatography). The process of separation of enantiomers from one another is called **resolution** and one approach to resolution involves the use of so-called **resolving agents**; these are optically pure chiral compounds (usually natural products) which can be reversibly attached to the enantiomeric pair to yield a pair of diastereomers rather than enantiomers. Diastereomers do *not* have identical chemical and physical properties and so may be separated from one another by conventional methods. A schematic overview of resolution of enantiomers using a resolving agent is shown in Fig. 8.14. This scheme shows a racemic mixture of compound **A** (i.e., a 50:50 mixture of (*R*)- and (*S*)-**A**) being derivatized by reaction with a single enantiomer resolving agent, (*R*)-**B**. The derivative produced is actually a pair of diastereomers, in this case, a 50:50 mixture of (*R*,*R*)-**A-B** and (*S*,*R*)-**A-B**. Such a mixture of diastereomers is separable in principle by traditional means for separation of chemical mixtures, that is, distillation, crystallization, or chromatography. Once separated, each diastereomer undergoes removal of the appended resolving agent, thereby releasing a pure single enantiomer of **A**. Again, in principle, the resolving agent can be recycled for further use. This process can, at best, produce a 50% yield

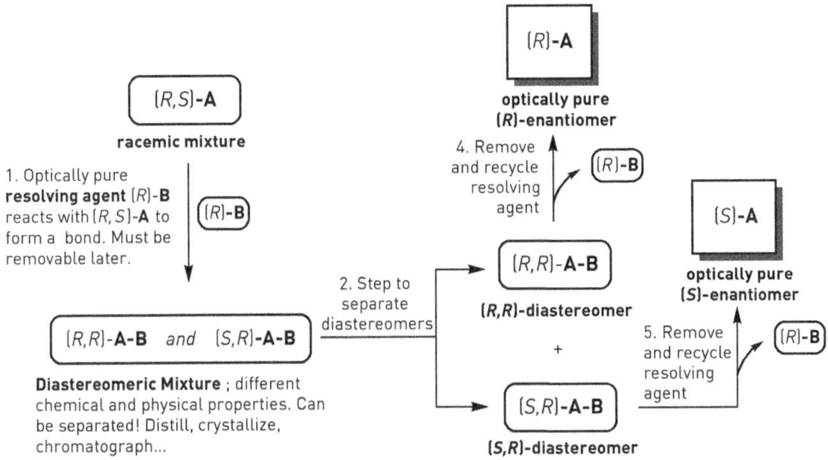

FIGURE 8.14
Generic resolution scheme: isolation of pure enantiomers from a racemic mixture via reaction with a resolving agent and formation of separable diastereomers.

of a desired enantiomer, and in practice, yields are usually lower due to mechanical losses during derivatization, purification, and resolving agent removal.

Let's move from a generic representation to a specific example of isolation of a desired enantiomer by chemical resolution. As shown in Fig. 8.15, the ketone function of a racemic (*R*,*S*) mixture of α-hydroxyketone **8.15a** is initially condensed with pure (*S*)-1-phenethylamine to yield the corresponding mixture of diastereomeric imines ***(R,S)*-8.15b** and ***(S,S)*-8.15b**. When this mixture is dissolved in boiling ethyl acetate and then allowed to cool to room temperature, crystals of the higher melting diastereomer ***(S,S)*-8.15b** slowly form. Vacuum filtration then affords a 25% yield of pure ***(S,S)*-8.15b** crystals, while the liquors from the process contain a 2:1 mixture of ***(R,S)*-8.15b** and ***(S,S)*-8.15b**. Hydrolysis of the imine linkage of the crystalline ***(S,S)*-8.15b** releases pure **(S)-8.15a** along with the original resolving agent (*S*)-1-phenethylamine which can be recovered and recycled if needed. The remaining material in the liquors from the crystallization (a mixture of (*R*,*S*)- and (*S*,*S*)-**8.15b** enriched in (*R*,*S*)-enantiomer) can be further processed via column chromatography or by imine hydrolysis and subsequent reaction if the ketone mixture with pure (*R*)-1-phenethylamine followed by separation of diastereomers as above to recover pure (*R*)-**8.15a** as well as some additional (*S*)-enantiomer which, as it turns out, is the desired enantiomer from this scheme which is part of a total synthesis of the aglycone portion of an analogue of the polyketide anticancer antibiotic **daunorubicin** (see Fig. 5.50 for structural details).

Of course, other methods for separation of enantiomers also exist, including chiral chromatography (use of a chiral stationary phase in liquid chromatography) or reactions with enzymes which recognize and catalyze specific reactions with only one enantiomer of a racemic pair (known as enzymatic resolution). We'll take a brief look at this and other uses of enzymes later on.

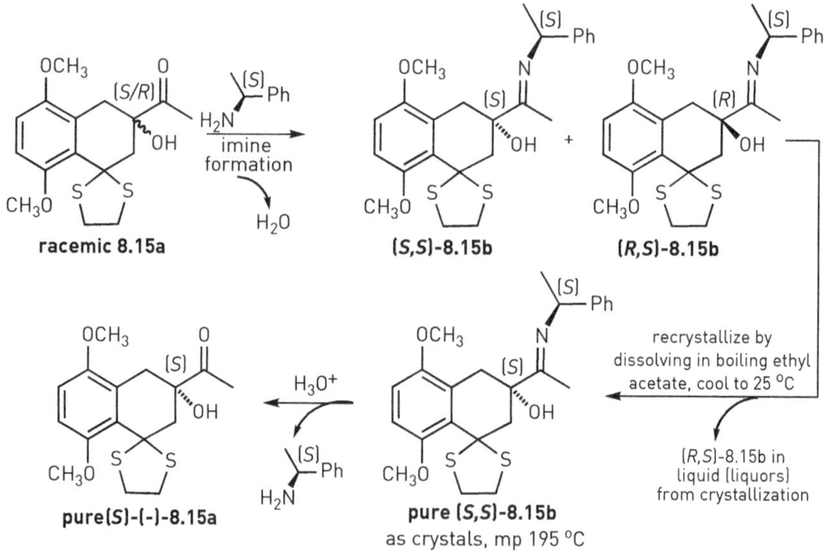

FIGURE 8.15
Isolation of pure (*S*)-enantiomer of 8.15a via resolving agent resolution/crystallization.

USE OF CHIRAL POOL COMPOUNDS FOR SYNTHESIS OF OPTICALLY PURE NATURAL PRODUCTS

Organic synthesis that uses the chirality inherent in enantiomerically pure starting compounds available from nature has been widely employed for many years. Such compounds are said to constitute the available **chiral pool** which is composed mainly of compounds such as carbohydrates and amino acids as well as certain abundant chiral terpenes and alkaloids, among many others (Fig. 8.16).

Chiral-pool-based synthesis usually works best if the starting compound is cheap and closely resembles the target product (some of the previous examples we have seen of semisynthesis should come to mind). Otherwise, the use of chiral pool starting materials may involve long, multistep syntheses with low overall yields.

A simple example of total synthesis of an optically pure natural product starting from a chiral pool compound is shown in Fig. 8.17. This sequence outlines preparation of **(*R*)-japonilure**, an alkenyl-substituted γ-lactone that is the sex attractant produced by the female Japanese beetle, a notoriously destructive garden pest that feeds on the leaves of many important fruit and ornamental plants. Only the (*R*)-enantiomer is bioactive as an attractant and it is available commercially in small traps used for insect control. However, the presence of even small amounts of the (*S*)-enantiomer renders the compound completely inactive, making optical purity of the synthetic product an absolute necessity. The synthesis begins with diazotization of (*R*)-glutamic acid as shown. The resulting diazonium ion **(*R*)-8.17a** then undergoes an intramolecular nucleophilic displacement of N_2 by the adjacent carboxyl group, forming the α-lactone **(*S*)-8.17b** with inversion of configuration. This strained (*S*)-lactone then undergoes a second inversion of configuration via a subsequent intramolecular nucleophilic epoxide ring opening by the remaining carboxyl group

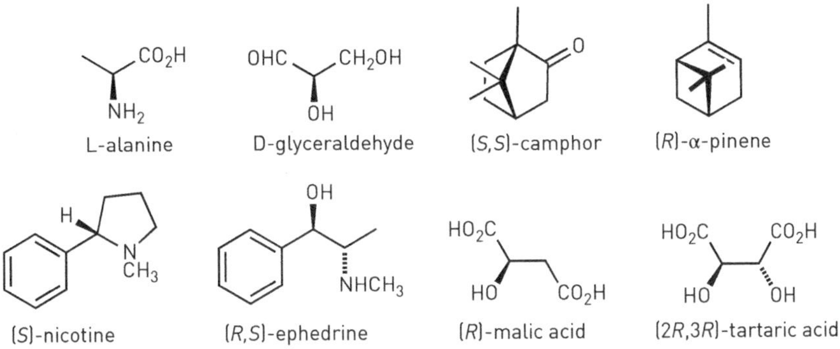

FIGURE 8.16
Examples of chiral pool compounds: amino acids, carbohydrates, terpenes, alkaloids, and others.

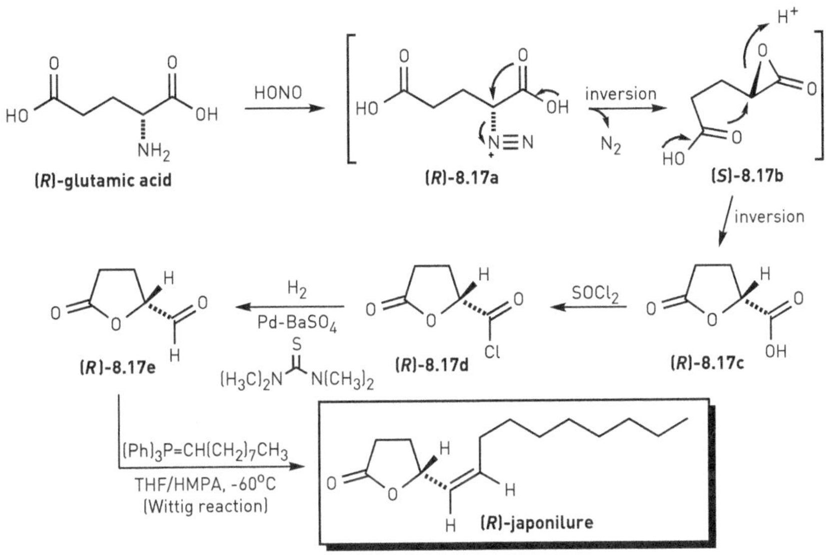

FIGURE 8.17
Chiral pool synthesis of the Japanese beetle pheromone (*R*)-japonilure from (*R*)-glutamic acid.

to afford the carboxyl-substituted γ-lactone **(*R*)-8.17c** (note that the overall effect of two sequential inversions of configuration is a net *retention* of configuration). Conversion of **(*R*)-8.17c** to the corresponding acid chloride **(*R*)-8.17d** is followed by a **Rosenmund reduction** using H_2 and Pd-$BaSO_4$ to afford the corresponding aldehyde **(*R*)-8.17e**. A final Wittig reaction of the aldehyde with a linear nine-carbon phosphorus ylide provides the alkenyl-substituted final product with mostly (*Z*)-configuration at the π-bond as required for naturally occurring **(*R*)-japonilure**.

USE OF CHIRAL REAGENTS FOR SYNTHESIS OF OPTICALLY PURE COMPOUNDS

Another method for introduction of enantioselective chirality into molecules is via the use of **chiral reagents**. There are a number of different types which can be used in organic reactions including chiral reducing agents for reduction of carbonyl compounds to optically active alcohols, chiral hydroborating agents for hydration of alkenes to give optically active alcohols, chiral allylborane reagents which can alkylate carbonyl compounds to give optically active alcohols, and chiral bases for stereospecific deprotonation reactions.

Reactions which produce chiral but optically inactive products often involve a step in which a reagent reacts with a planar prochiral reactant (or reactive intermediate) such as an unsymmetrical carbonyl or alkene. Racemic mixtures are always produced as a result of such reactions because two **enantiomeric transition states** (nonsuperimposable mirror images) result from approach of the reagent from above or below the planar reactant. These transition states are of the same energy, so the two enantiomeric products are formed at identical rates, giving a racemic mixture. This simplistic but useful analysis is illustrated in the upper portion of Fig. 8.18.

The lower portion of the figure shows use of a generic *chiral* hydride source for the same reduction. As can be seen, approach from above or below the plane again leads to two transition states, but these are **diastereomeric transition states** (nonsuperimposable, nonmirror image) rather than enantiomeric transition states and

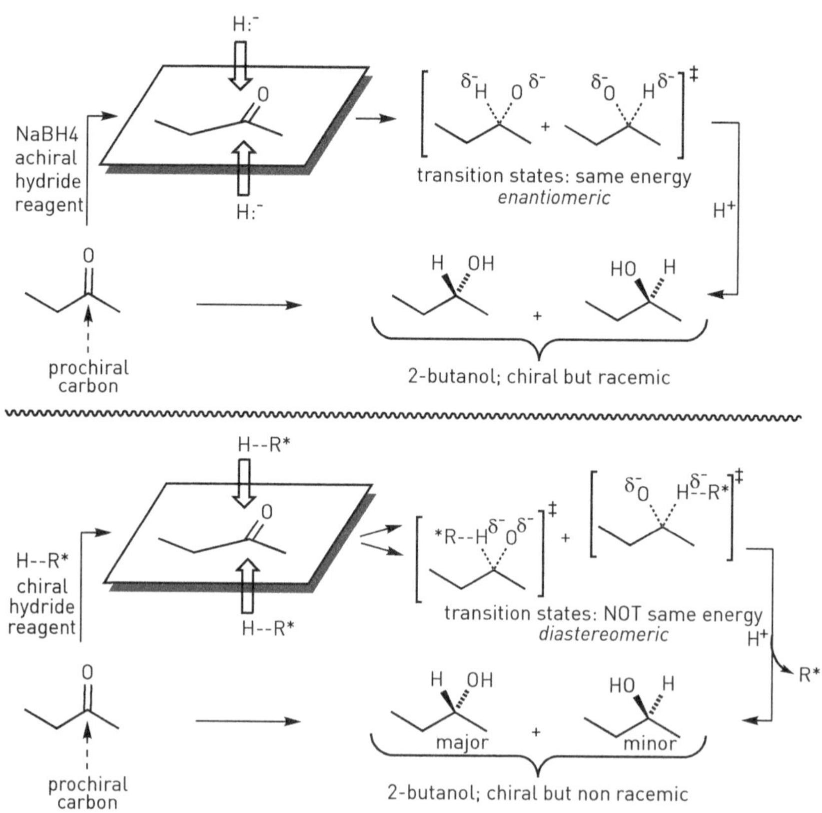

FIGURE 8.18
Use of achiral vs. chiral reagents and enantiomeric vs. diastereomeric transition states.

so would be of different energies. The lower energy transition state would form product faster and would therefore lead to preferential formation of one enantiomer at the expense of the other (here we assume the chiral entity bearing the hydride is shed after the addition and protonation steps). Note that this analysis could also be extended to the use of a chiral alkylating reagent (chiral portion *not* shed) which would lead to formation of diastereomeric alcohols with an excess of one diastereomer over the other. Such processes would be termed **stereoselective** reactions in that they favor the formation of one stereoisomer at the expense of other possible stereoisomers. More specifically, the chiral hydride reduction shown would be termed an **enantioselective reaction** while the corresponding hypothetical chiral alkylation reaction would be termed a **diastereoselective reaction.**

These are components of what is broadly termed **asymmetric synthesis** which may be loosely defined as any synthetic process which converts an achiral substrate into a product with a new chirality center in such a way as to produce stereoisomers in unequal amounts. This is an enormous subject area and even a modest treatment of it is far beyond the scope of this chapter. We can nevertheless gain some level of appreciation for its challenges and accomplishments by considering a few examples such as those shown in Fig. 8.19 which includes an example of an asymmetric ketone reduction, an asymmetric aldehyde alkylation and an asymmetric hydroboration reaction, all of which produce chiral alcohols in stereoisomeric excess.

In the first example, the reduction of the ketone function of an aryl β-ketoester uses a modified form of lithium borohydride which derives its chirality from incorporation of an *N*-protected form of the amino acid **cysteine** into its structure. In the second example, the asymmetric alkylation of propanal involves transfer of

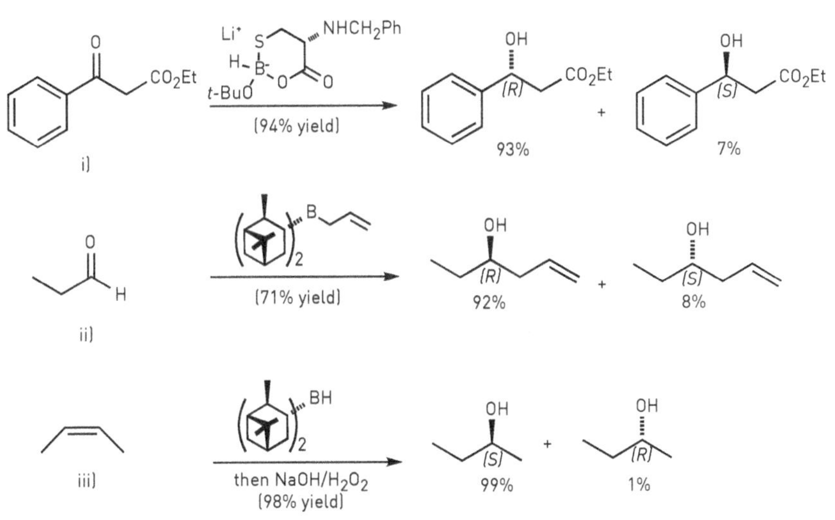

FIGURE 8.19

Simple examples of asymmetric alcohol synthesis: i) ketone reduction using a chiral hydride reagent; ii) aldehyde alkylation using a chiral allyl boron reagent; and iii) alkene hydroboration using a chiral hydroborating agent.

an allyl group delivered from an allyl borane reagent that derives its chirality from incorporation of two units of the chiral terpene **(+)-α-pinene.** This terpene is also the source of chirality for the reagent used in the final example which is asymmetric hydroboration of a simple alkene. Mechanisms and proposed transition state conformations for these reactions are complex, but the outcomes are impressive in terms of enantioselectivity.

Suffice it to say that there are a multitude of additional reactions involving chiral reagents which have been employed as part of many multistep syntheses of complex natural product structures. We note here that the primary disadvantages of this approach to asymmetric synthesis are: i) the reagents are often quite expensive; ii) a stoichiometric quantity of the chiral reagent is required in each case; and iii) reaction workups can often be complicated and time consuming.

USE OF CHIRAL SUBSTRATE CONTROL FOR STEREOSELECTIVE SYNTHESIS

Another form of stereoselective synthesis involves a factor known as **substrate control**. When a chirality center is already present in a substrate molecule and is close in proximity to a prochiral center undergoing an addition reaction, diastereomeric transition states are again formed which can lead to stereoselectivity at the newly formed chirality center. In this case, the resulting products will be an unequal mixture of diastereomers. Nucleophilic additions to carbonyl groups adjacent to existing chirality centers are the most common examples of chiral substrate control, though there are many others. The stereochemical outcome of such reactions is usually described in terms of the extent of **diastereoselectivity** observed (excess of one diastereomer over another). Figure 8.20 shows the use of **Cram's rule**, one of the more commonly used models for prediction of the major diastereomeric product to be expected from nucleophilic addition to a typical carbonyl adjacent to an existing chirality center. Note that if the existing chirality center is racemic rather than optically pure, the resulting product will be also racemic.

Many other reactions are useful in the context of this kind of diastereoselectivity; one of the largest groups to yield useful results in this context are enolate ions. Enolates of aldehydes, ketones, esters, and related carbonyl compound have been used extensively to produce chiral diastereomeric products from enolate alkylations, conjugate additions, aldol-type condensations, and others with often excellent diastereoselectivity. The first example shown in Fig. 8.21 is a simple **enolate ion alkylation** process starting from a chiral ester. After treatment of the ester with lithium diisopropyl amide (LDA) as base, the resulting (*Z*)-enolate (in this case) is alkylated with methyl iodide, a reaction in which the methyl group is added to the less sterically hindered face of the enolate ion (away from the phenyl group) to produce predominantly the (*R*,*R*) diastereomer as shown.

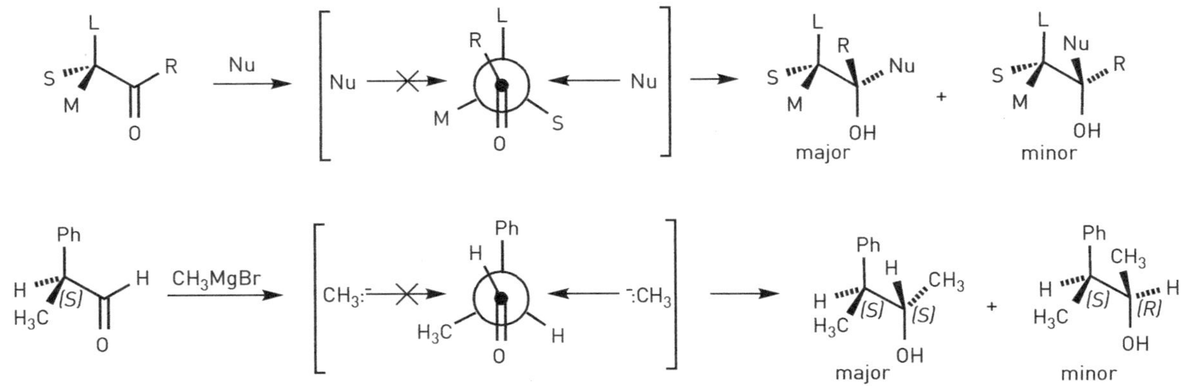

FIGURE 8.20

Illustration of Cram's rule for prediction of major vs. minor diastereomeric addition products.

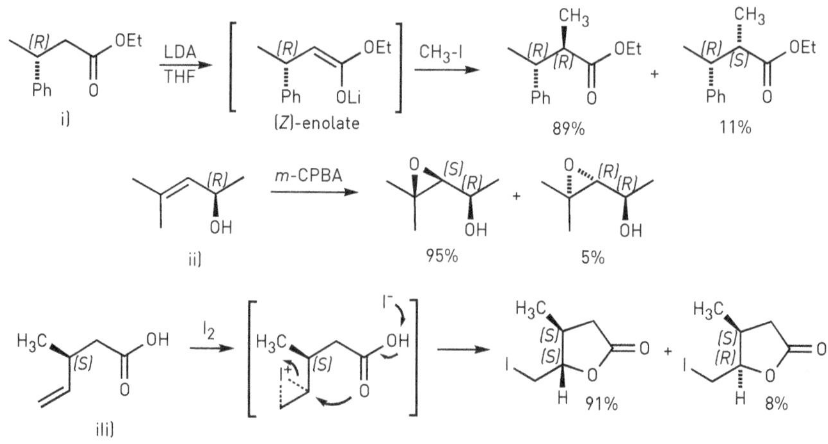

FIGURE 8.21
Examples of substrate control in: i) enolate alkylation; ii) epoxidation; and iii) iodolactonization.

The second example in Fig. 8.21 makes good use of the stereospecificity of **alkene epoxidation** (occurs with *syn* addition) along with the *syn* directing effect usually observed with allylic hydroxyl groups, leading to predominant formation of the diastereomer shown. Other epoxidations of this type, where substituents are alkyl rather than hydroxyl, tend to be governed mainly by steric effects and so give results opposite to those shown for the alcohol substrate by adding oxygen to the less-hindered alkene face. The third example shown is application of a reaction known as **iodolactonization** in which a carboxylic acid with an alkene function, usually at the 4- or 5-position, is treated with iodine. Electrophilic addition as usual for halogens leads to a cyclic iodonium ion which then undergoes intramolecular nucleophilic attack by the carboxyl group from the less hindered side (opposite side to the chiral methyl group), leading to iodolactone formation. Many other reactions of chiral alkenes have been shown to give useful diastereoselectivity in certain contexts, including Diels–Alder reactions, hydroborations, and dihydroxylations with osmium tetroxide.

The principal advantages of substrate control for all these reactions are i) a new chirality center with often substantial excess of one stereoisomer is produced and ii) structure of the major isomer can often be predicted using various transition state models (such as Cram's rule). The main disadvantage of substrate control is the often-limited availability or high cost of suitable optically pure starting materials for the reactions involved.

USE OF CHIRAL AUXILIARIES FOR SYNTHESIS OF OPTICALLY PURE COMPOUNDS

As mentioned above, the most obvious drawback to substrate control for achieving stereoselective reactions is the need for optically pure starting compounds to begin with. One might imagine that many of these substrates would come from the chiral pool which would tend to limit structural diversity to some extent and this is often the case. We now consider the use of so-called **chiral auxiliaries** for achieving the same results obtainable by more traditional substrate control but with additional elements of structural diversity. A chiral auxiliary may be thought of as a chiral compound or subunit which may be *reversibly attached* to (or incorporated into) the structure of an achiral compound. This is similar to the use of resolving agents except, in this case, the attachment of the auxiliary does not produce a mix of diastereomers since the substrate compound was achiral to begin with. The resulting product may be thought of as being *temporarily* chiral due to the presence of the auxiliary. This chiral substrate may then undergo the usual reactions that will show substrate control for diastereoselectivity when forming a new chirality center. However, after the diastereoselective reaction is completed, the auxiliary is then *removed* from the substrate (and may be recycled). For the simplest case in which only one new chirality center has been formed, the result will be formation of a new compound with often-substantial enantiomeric excess. Figure 8.22 gives a general scheme for the use of chiral auxiliaries in asymmetric synthesis.

A relatively simple example of the use of a chiral auxiliary for asymmetric synthesis is shown in Fig. 8.23. Recall that **Robinson annulation** involves the use of both conjugate addition of an enolate ion to an α,β-unsaturated carbonyl system and a subsequent intramolecular aldol condensation to form a fused 6-6 bicyclic ring system (see Fig. 1.52). In the asymmetric variant shown here, the nucleophilic component for the initial conjugate addition to methyl vinyl ketone is the enamine **8.23b** which is formed by isomerization of imine **8.23a** which in turn is prepared by simple condensation of racemic 2-methylcyclohexanone with commercially available **(S)-α-phenethylamine** as the chiral auxiliary. After the enamine conjugate addition (which is sterically directed to the top face of the enamine and away from the bulky chiral auxiliary), the resulting imine **8.23c** is hydrolyzed to remove the

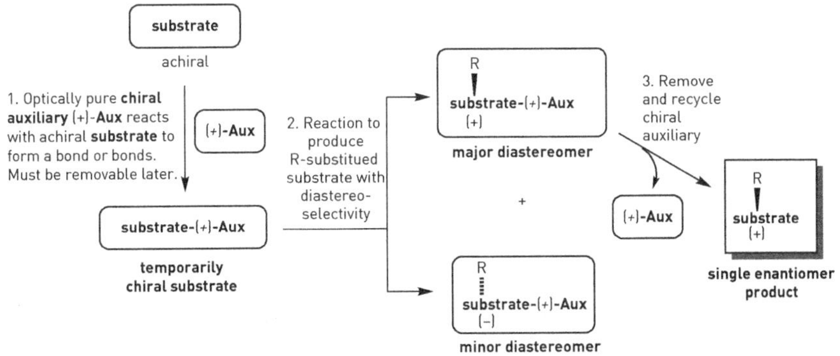

FIGURE 8.22
Generic chiral auxiliary scheme: formation of a single enantiomer product from an achiral substrate via chiral auxiliary attachment, diastereoselective reaction, and chiral auxiliary removal.

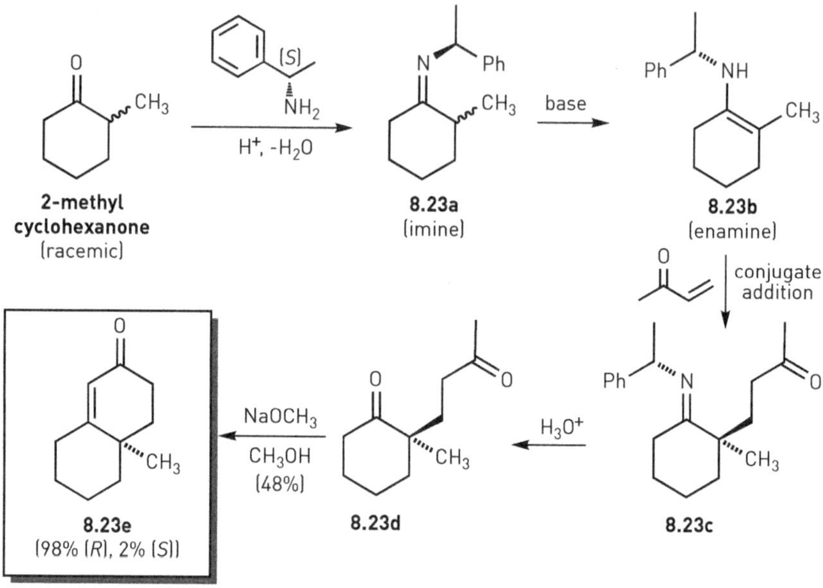

FIGURE 8.23
Use of an amine chiral auxiliary for an asymmetric version of the Robinson annulation.

chiral auxiliary to give the corresponding diketone **8.23d**. A final intramolecular aldol condensation affords 98% of the (*R*)-enantiomer of bicyclic Robinson annulation product **8.23e** with excellent enantioselectivity.

While chiral auxiliaries have been successfully used in a variety of different reaction schemes, their use in the formation of chiral enolates for alkylation or aldol reactions has been most noteworthy. An excellent example of their utility is shown in Fig. 8.24 which outlines the use of one of the **Evans chiral oxazolidinones** in a stereoselective synthesis of **Prelog–Djerassi lactonic acid**, a product derived from degradation of certain natural macrolide antibiotics and which is used as a template for construction of new variants of these and other natural products. Many different and useful oxazolidinones have been prepared from various chiral amino alcohols, and in this instance, we see the use of commercially available **(1*S*,2*R*)-norephedrine** which is converted to the oxazolidinone chiral auxiliary **8.24a** by reaction with diethyl carbonate. Deprotonation of **8.24a** with *n*-butyllithium gives the anionic form which is then acylated with propionyl chloride (serving here as the achiral substrate) to give the temporarily chiral substrate **8.24b** to which the chiral auxiliary has now been attached. Subsequent deprotonation with LDA (lithium diisopropylamide) gives the chiral enolate **8.24c** which then undergoes **asymmetric alkylation** with methallyl iodide which approaches the enolate face away from the substituents on the auxiliary ring to give **8.24d** with excellent diastereoselectivity. Reductive treatment of **8.24d** with lithium aluminum hydride to remove the chiral auxiliary affords the chiral alcohol **8.24e** as shown.

The process in Fig. 8.24 continues by oxidation of alcohol **8.24e** to aldehyde **8.24f** which is then subjected to a second reaction employing the same acylated chiral

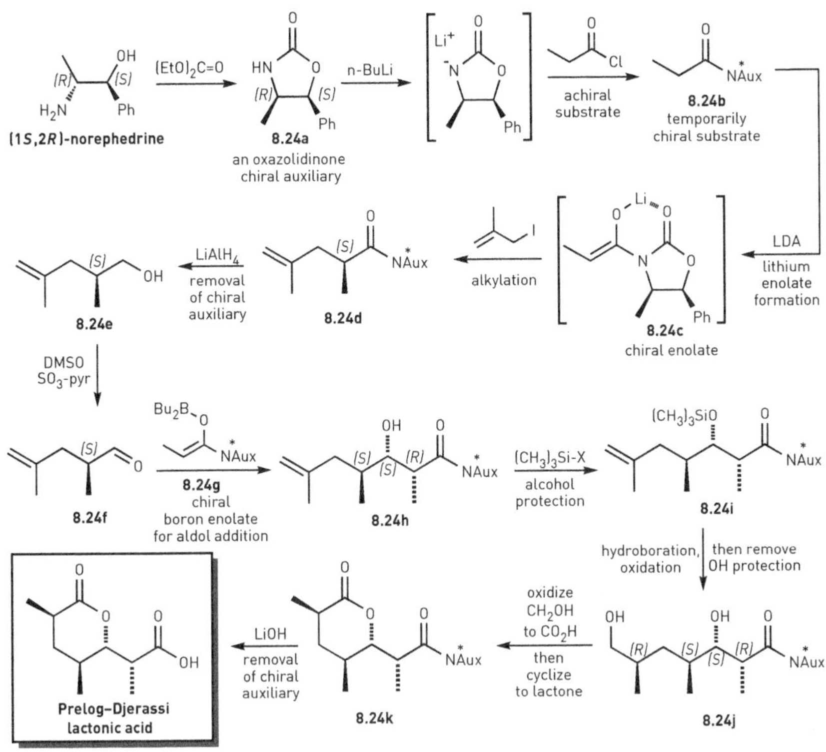

FIGURE 8.24
Use of Evans oxazolidinones as chiral auxiliaries for enolate alkylation and aldol addition reactions in an asymmetric synthesis of Prelog–Djerassi lactonic acid.

auxiliary as before but which in this case had been converted to chiral boron enolate **8.24g.** Here the enolate adds to the less hindered face of the aldehyde function of **8.24f** in an **asymmetric aldol addition** process to give the corresponding β-hydroxyamide **8.24h** in which two additional chirality centers have now been added, again with excellent diastereoselectivity. The alcohol function is then temporarily protected as trimethylsilyl ether derivative **8.24i** which undergoes a stereospecific hydroboration–oxidation sequence which, after removal of the previous protecting group, produces diol **8.24j** in which yet a fourth chirality center has now been introduced, again with excellent diastereoselectivity. After recrystallization to remove minor diastereomeric impurities, the synthetic sequence is completed by oxidation of the primary alcohol function of **8.24j** to give the carboxylic acid which spontaneously cyclizes to produce the corresponding lactone **8.24k.** Finally, removal of the chiral auxiliary by treatment with LiOH in methanol yields the target, **Prelog–Djerassi lactonic acid**, which was found to be >99.9% diastereomerically pure.

Derivatives of the chiral auxiliary **(–)-8-phenylmenthol** have also been used in a variety of ways for asymmetric synthesis. One example is shown in Fig. 8.25 in which 8-phenylmenthol is reacted with pyruvic acid to form **8-phenylmenthol pyruvate ester.** Reaction of the ketone function with Grignard reagent **8.25a** leads to formation of the tertiary alcohol **8.25b** as a single diastereomer. Removal of the

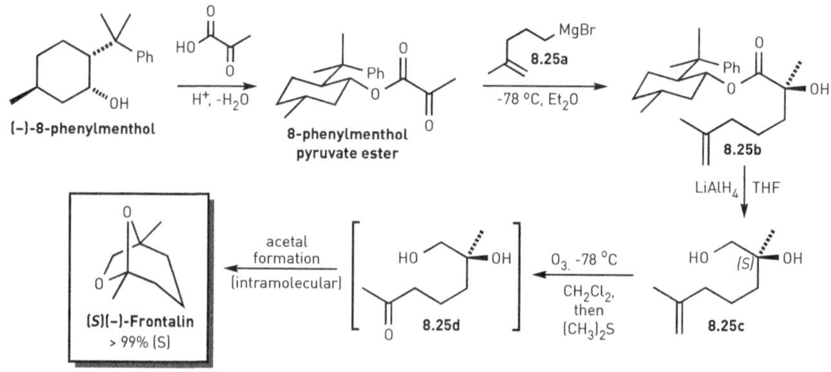

FIGURE 8.25
Whitesell's synthesis of optically pure frontalin using the chiral auxiliary 8-phenylmenthol.

chiral auxiliary via lithium aluminum hydride reduction affords diol **8.25c** which is optically pure. Ozonolysis of the alkene function leads to the corresponding keto-diol **8.25d** which is not isolated but spontaneously undergoes intramolecular cyclic acetal formation to yield the pure (*S*)-enantiomer of **frontalin**, an aggregation pheromone of a number of different species of beetles including the Western pine beetle, a particularly destructive pest of North American pine forests.

A representative sample of several other important chiral auxiliaries widely used in asymmetric synthesis is shown in Fig. 8.26, though exploring further details of their use is beyond the scope of our chapter. To summarize, we may say that chiral auxiliaries represent a different and very useful approach to substrate control for diastereoselective reactions. One advantage to this approach is that unlike typical substrate control, the chiral auxiliary need not resemble the structure of the synthetic target since the auxiliary is subsequently removed after the diastereoselective reaction(s). It is also worth noting that a chiral auxiliary may also act as a built-in resolving agent to aid in the separation of minor diasteromeric products prior to its removal. Disadvantages include the need for stoichiometric amounts of the chiral auxiliary, the need for a chemically viable point of attachment on the original substrate, and the need for additional steps to remove the auxiliary under conditions that will not adversely affect the target compound being produced.

USE OF CHIRAL CATALYSIS FOR SYNTHESIS OF OPTICALLY PURE COMPOUNDS

The gold standard for creation of new chirality centers with a high level of optical purity in organic reactions is, of course, the enzyme-mediated process. Since the chiral enzyme is used as a catalyst only, it can provide the chiral environment needed to induce asymmetry again and again without being consumed or affected in any way. **Chiral catalysis** is an area of synthetic organic chemistry that tries to duplicate nature's own approach to asymmetric synthesis by employing a chiral component in catalytic rather than

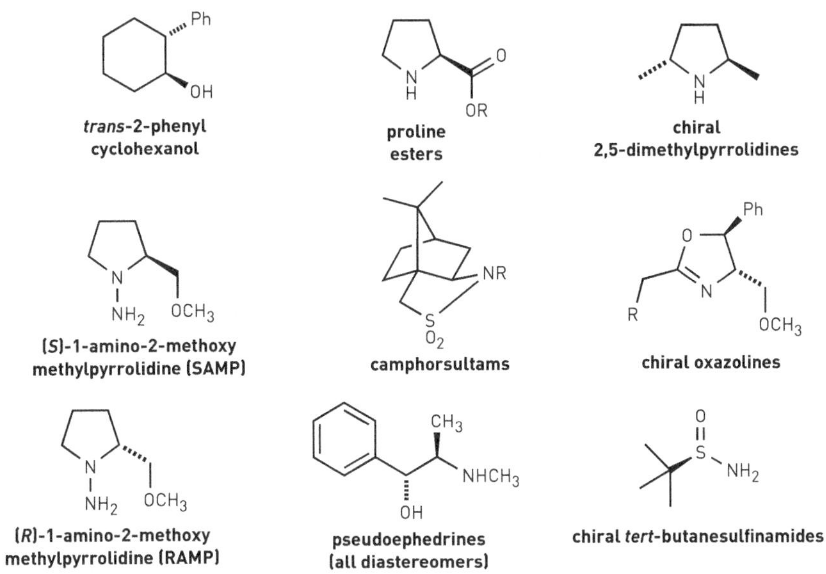

FIGURE 8.26
Additional chiral auxiliaries widely used in asymmetric synthesis.

stoichiometric quantities that will not only accelerate a particular reaction but will also lead to enantioselectivity at newly developed chirality centers. The approach used most often (though not exclusively) involves the use of a chiral catalyst that can bind a specific reagent, then deliver that reagent to a planar, prochiral substrate molecule in such a way that the reagent reacts preferentially with one of the two available faces (*Si* or *Re*) of the substrate. After release of the reagent and production of the chiral reaction product, the catalyst is free to bind more reagent and initiate another cycle of the catalytic process. A general scheme for this type of catalysis is presented in Fig. 8.27.

No area of asymmetric organic synthesis has received more attention than the development of effective and efficient chiral catalysis. The benefits are obvious: i) a single chiral molecule employed in a catalytic fashion can lead to the production of millions of optically active product molecules; ii) chiral catalysts can, in principle, be economically recycled for further use; and iii) their use can reduce or eliminate most of the waste or inefficiency associated with the processing and separation of enantiomers via resolution, the addition and removal of chiral auxiliaries, or the use of stoichiometric quantities of chiral reagents. As before, the key to their success lies in their ability to form diastereomeric reaction transition states of unequal energy, resulting in significant preference for one enantiomeric product at the expense of the other.

The earliest examples of successful development of chiral catalysis are found in the area of catalytic asymmetric hydrogenation of double bonds. Such reactions are of significant versatility since they can be applied to the reduction of prochiral alkenes, ketones, imines, enamines, aromatic heterocycles, and many other types of unsaturated substrates. Figure 8.28 outlines two examples of such processes. The first is the commercial synthesis of the aromatic amino acid **L-DOPA** via

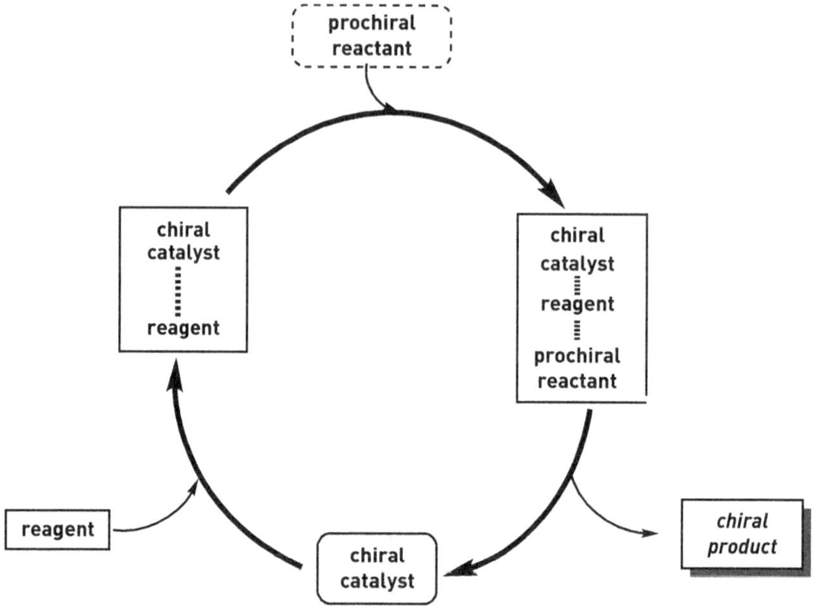

FIGURE 8.27
Generic catalytic cycle involving a prochiral reactant, a chiral catalyst, and a reagent.

catalytic hydrogenation of the enamide precursor **(*Z*)-2-acetamido-3-(4-acetoxy-3-methoxyphenyl)acrylic acid** to give intermediate reduction product **8.28a** in high optical purity. This is followed by ester and amide hydrolysis and methoxyl group demethylation to give L-DOPA, one of the most important pharmaceutical agents for the treatment of Parkinson's disease. The reduction employs a rhodium-based catalyst which employs containing a phosphine-based ligand known as **(*R*,*R*)-DiPAMP**. Chelation of the metal by the phosphorus atoms of this chiral diphosphine creates the chiral environment in which both substrate and reagent are brought together for delivery of H_2 preferentially to the *Re* face of the alkenyl system.

In the second instance, catalytic hydrogenation of the unsaturated **2-(6-methoxynaphthalen-2-yl)acrylic acid** in the presence of a ruthenium catalyst coordinated to another chiral diphosphine ligand, **(*R*)-BINAP**, affords the (*S*)-enantiomer of **naproxen,** a common over-the-counter NSAID (nonsteroidal anti- inflammatory) that acts as a COX-1 and COX-2 inhibitor, similar in action to the structurally related compound ibuprofen which, while also containing a 2-arylpropanoic acid function, is marketed as a racemic mixture. For naproxen, only the (*S*)-enantiomer has anti-inflammatory activity; the (*R*)-enantiomer is inactive in that regard but can be toxic to the liver, making an enantioselective synthesis essential for pharmaceutical purposes.

Both of the above hydrogenation reactions are examples of the use of **homogeneous catalysis**, a process in which a reaction catalyst and all reactants are all brought together and dissolved in solution (i.e., are present in the same phase). Recall that typical hydrogenation processes employ an insoluble, finely divided transition metal

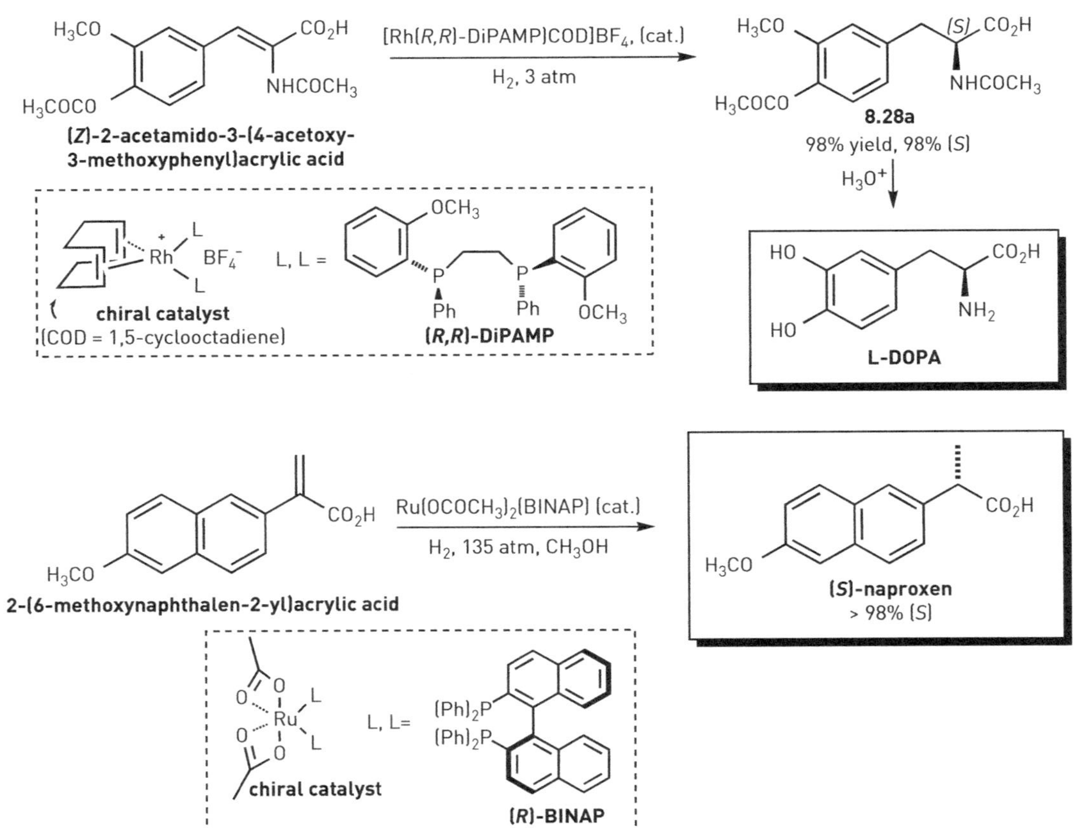

FIGURE 8.28
Examples of catalytic asymmetric hydrogenation in the synthesis of L-DOPA and (*S*)-naproxen.

such Pt, Pd, Ni, or others and that the reduction process actually occurs on the surface of the metal itself. Such a process is referred to as **heterogeneous catalysis** since the metal catalyst is never in solution along with the reactants. Dissolution of a metal in homogeneous catalysis is achieved when one or more chelating organic ligands are dissolved in an organic solvent along with the metal catalyst. Chelation with the metal leads to a catalyst-ligand complex that is predominantly organic in structure, allowing the metal to come into solution along with reactants and reagents. When chiral ligands are used, all reactants and reagents may come into intimate contact with the metal catalyst within a chiral environment created by the coordinating ligands, leading to the kind of enantioselectivity observed above.

While catalytic asymmetric hydrogenation of ketones is also feasible, a more general catalytic approach to such reductions involves the use of borane-THF complex as the reducing agent along with a catalytic amount of a **chiral oxazaborolidine**, compounds usually derived from transformations of the amino acid proline. These so-called **CBS catalysts** (CBS is derived from Corey–Bakshi–Shibata, authors of the original publications describing their synthesis and use) are especially enzyme-like in their behavior because they bind both the substrate and the reagent. A general scheme for ketone reductions via this methodology and several simple examples are shown in Fig. 8.29. Since their initial introduction, several varieties of CBS-type catalysts have been developed and have found wide use in the asymmetric synthesis of many complex natural products and pharmaceutical agents.

Among catalytic asymmetric C–C bond forming reactions, one of the simplest and most useful processes is the carrying out of certain aldol condensations in the presence of a catalytic amount of the amino acid **proline.** While mechanistic details remain uncertain, involvement of an iminium ion–enamine equilibrium is likely, leading to good-to-excellent levels of enantioselectivity. A useful example is shown in Fig. 8.30 for the proline-catalyzed formation of **Wieland–Miescher ketone**, a key chiral intermediate which has been used as a starting material for the total synthesis of a number of steroids and other important pharmaceutical agents. Following conjugate addition of methyl vinyl ketone to the enolate derived from 2-methyl-1,3-cyclohexanedione, intramolecular aldol condensation of the symmetric triketone **8.30a** in the presences of catalytic L-proline in DMSO gives mainly the (*S*)-enantiomer of the unsaturated diketone product as shown. This is but one of many examples of the use of proline-catalyzed aldols for enantioselective formation of useful synthetic intermediates and products.

Alkene epoxidation is a very useful means of oxygen incorporation into organic compounds. An important feature of the reaction is its stereospecific delivery of oxygen to the alkene in a concerted "syn" fashion. That stereospecificity makes it an ideal candidate for exploitation in a potential asymmetric variant. As it turns out, such a variant has turned out to be one of the most important of all catalytic asymmetric processes which is known as the **Sharpless epoxidation.** This reaction converts allylic alcohols to chiral epoxy alcohols with excellent enantioselectivity

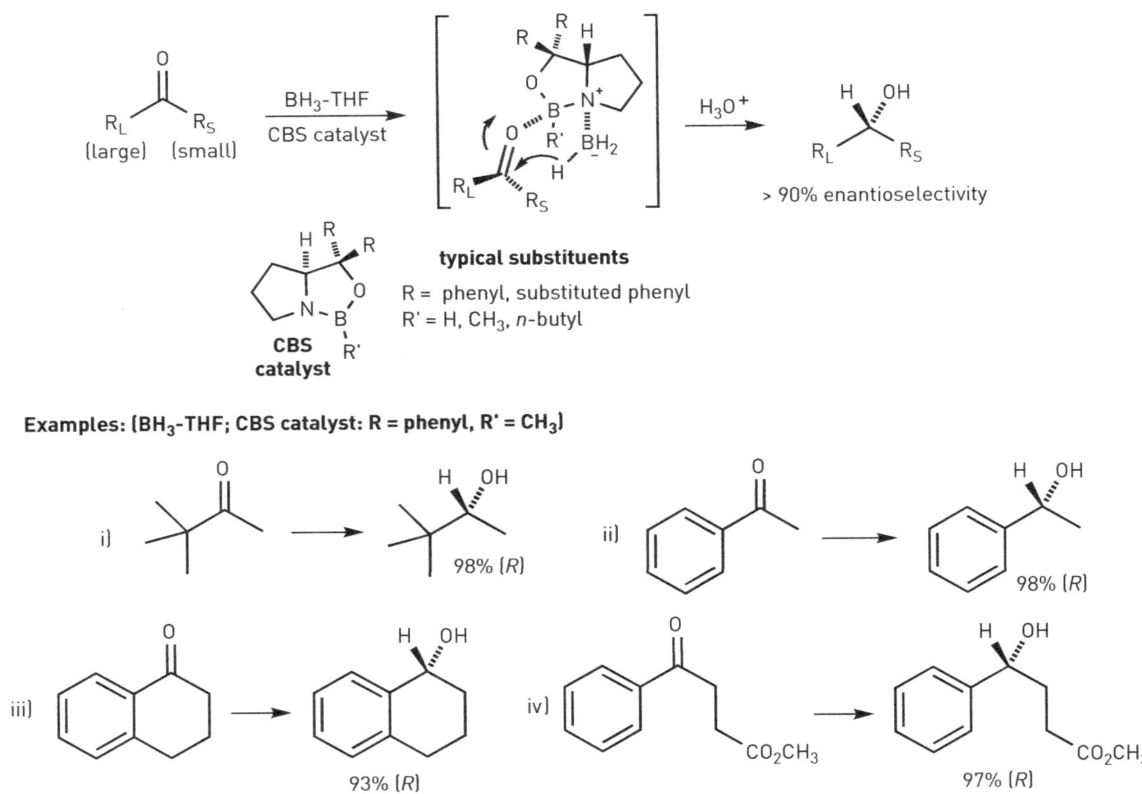

FIGURE 8.29

Catalytic asymmetric ketone-to-alcohol reductions using BH_3-THF and chiral CBS catalysts.

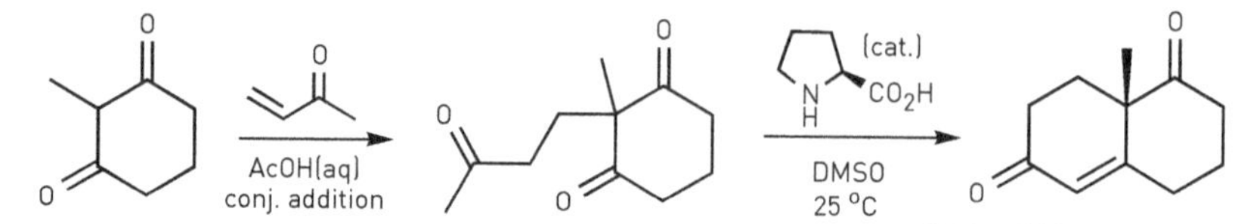

2-methyl-1,3-cyclohexanedione **8.30a** **Wieland–Miescher ketone** pure (*S*), 57% yield

FIGURE 8.30
Asymmetric synthesis of Wieland–Miescher ketone via proline-catalyzed aldol condensation.

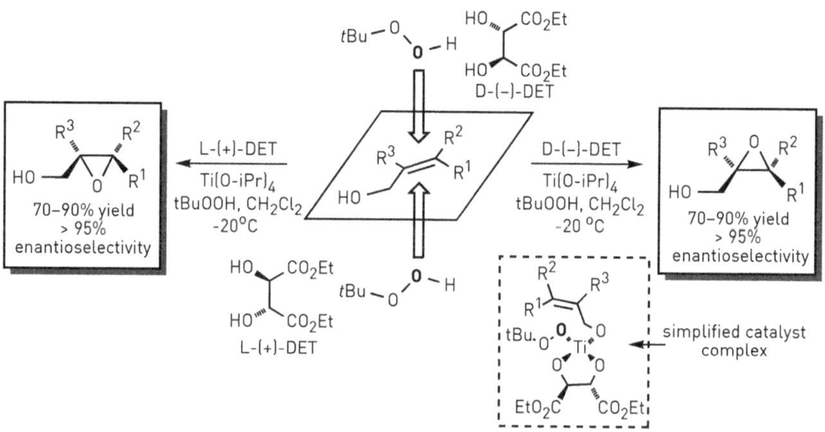

FIGURE 8.31
Structural prediction model for enantioselective Sharpless epoxidation of allylic alcohols.

in most cases. The catalyst system is derived from the use of titanium isopropoxide which undergoes replacement of its four isopropoxide ligands with the –OH functions of either enantiomer of relatively inexpensive **diethyl tartrate** ((+)-DET or (–)-DET) along with the –OH groups of both t-butyl hydroperoxide (TBHP or tBuOOH, the oxidizing agent) and the allylic alcohol substrate. Face-selective delivery of the active oxygen from tBuOOH to the alkene function in this chiral catalyst environment occurs with excellent enantioselectivity that is dependent on which DET enantiomer is used. Thus, either stereoisomer of the resulting epoxy alcohol can be selectively produced, making this a process of enormous synthetic utility. Figure 8.31 provides a general structural prediction model for the Sharpless epoxidation system along with a greatly simplified representation of the structural components of the catalyst complex.

Extensive use has been made of the Sharpless epoxidation in total synthesis schemes over the years and even a partial listing of its applications would be daunting. A simple example of its use as shown in Fig. 8.32 will serve our purposes. Epoxidation of the allylic alcohol **(*Z*)-2-tridecen-1-ol** using D-(–)-DET in the Sharpless catalyst complex gives the corresponding epoxide **(2*R*,3*S*)-2,3-epoxy-1-tridecanol** with excellent enantioselectivity. Oxidation of the primary alcohol with **pyridinium dichromate** (PDC, similar to PCC) gives the corresponding

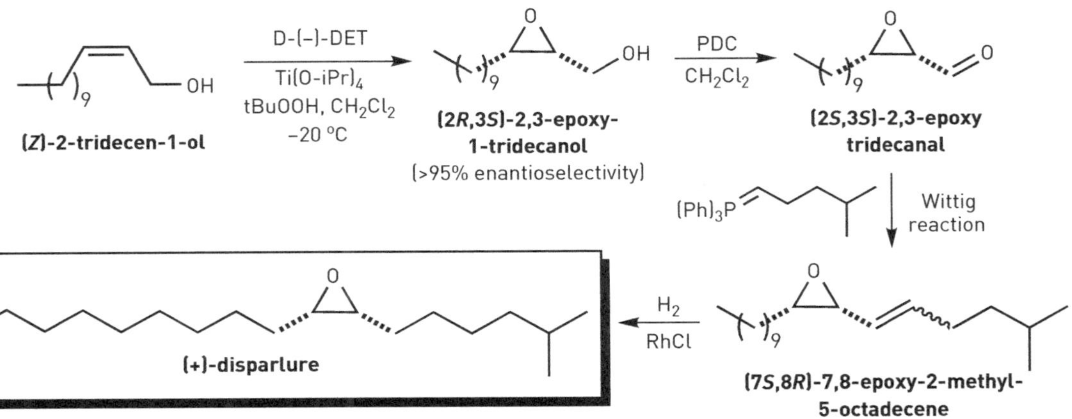

FIGURE 8.32

Enantioselective synthesis of (+)-disparlure via Sharpless asymmetric epoxidation.

epoxyaldehyde as shown. A subsequent Wittig reaction with the indicated phosphorous ylide leads to an epoxyalkene (as a mixture of (*Z*) and (*E*) isomers) which then undergoes catalytic hydrogenation to afford the saturated natural product, **(+)-disparlure** ((7*R*,8*S*)-7,8-epoxy-2-methyloctadecane). Produced in minute quantities by the female gypsy moth for attraction of males, this potent sex attractant has proven to be useful for the baiting of traps to monitor expanding populations of this enormously destructive forest pest. In tests, it was found that the (+)-enantiomer is significantly more potent as an attractant than racemic material, making an enantioselective synthesis quite valuable.

The synthetic value of the catalytic asymmetric epoxidation reaction is nicely complemented by another process known as the **Sharpless asymmetric dihydroxylation (ADH)** reaction. Recall that the familiar conversion of alkenes to syn vicinal diols via reaction with osmium tetroxide (OsO_4) is most useful when the expensive and highly toxic OsO_4 is used in only catalytic amounts in conjunction with a stoichiometric oxidant that reoxidizes the reduced form of osmium back to the OsO_4 oxidation state to continue the catalytic cycle. After extensive work, Sharpless found that the use of certain chiral alkaloids could not only accelerate these reactions but could also produce diol products with a high level of enantioselectivity. The now commercially available "AD-mixtures" for this chemistry contain a mixture of potassium osmate ($K_2Os_2(OH)_4$), as a source of osmium in its OsO_4 oxidation state, potassium ferricyanide ($K_3Fe(CN)_6$) to reoxidize the reduced form of osmium, potassium carbonate (K_2CO_3), and one of two different alkaloid-based chiral ligand catalysts, **(DHQ)$_2$PHAL** for the so-called **AD-mix-α** and **(DHQD)$_2$PHAL** for **AD-mix-β**. In each catalyst, two equivalents of a chiral alkaloid (**dihydroquinine** for (DHQ)$_2$PHAL and **dihydroquinidine** for (DHQD)$_2$PHAL) are connected to a central **phthalizine** structure, hence the PHAL in the names. As shown in Fig. 8.33, these two catalyst systems allow for complementary alkene face selectivity in the dihydroxylation reaction (R_L for large, R_M for medium, and R_S for small substituents in the alkene structure) making this a valuable catalytic process for the generation of vicinal diol products with a high level of optical purity.

Like the asymmetric epoxidation of allylic alcohols, Sharpless asymmetric dihydroxylation has been used extensively in a variety of asymmetric synthesis schemes since its introduction. A useful example is synthesis of α-hydroxy-β-benzamido-substituted 3-phenylpropanoic acid side chain of the anticancer agent **taxol** (see Fig. 4.33 and associated text for details) from methyl cinnamate shown in Fig. 8.34.

There are many other examples of catalytic asymmetric reactions including asymmetric aminohydroxylations (similar to the ADH reaction), asymmetric epoxidations and cyclopropanations of simple alkenes, asymmetric alkylations, halogenations, Michael additions, and many others too numerous to list here. It is fair to say that this is an ever-expanding area of chemical research that will continue to be of immense importance both academically and industrially for many years to come.

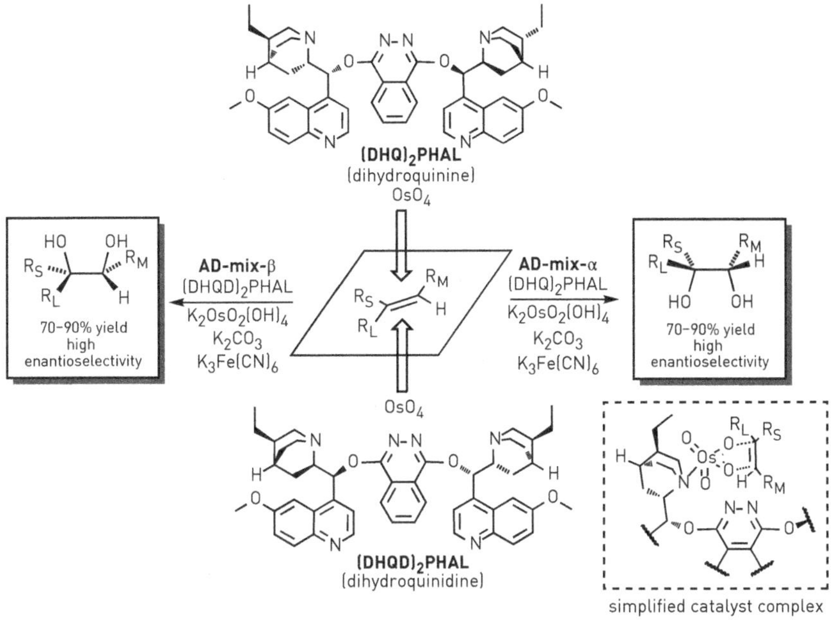

FIGURE 8.33
Structural prediction model for Sharpless alkene asymmetric dihydroxylation (ADH).

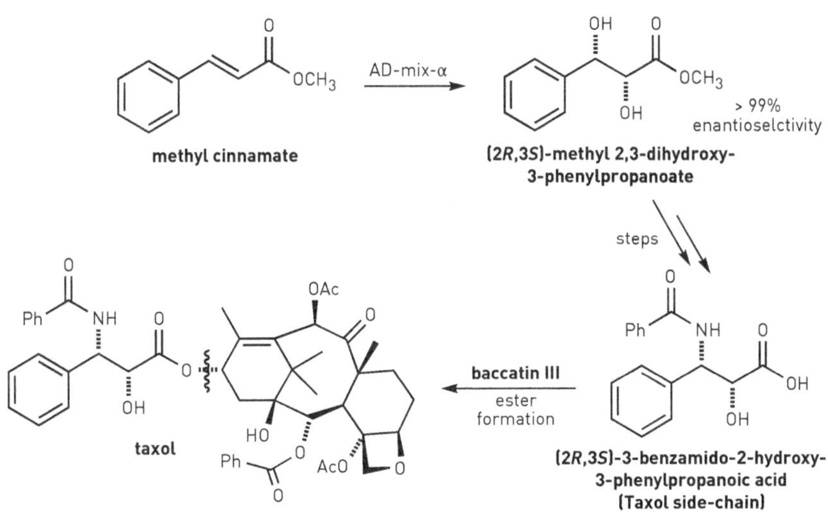

FIGURE 8.34
Use of Sharpless AD-mix-α for synthesis of taxol side chain from methyl cinnamate.

USE OF ENZYMES FOR SYNTHESIS OF OPTICALLY PURE COMPOUNDS: BIOCATALYSIS

The use of enzymes or whole cell systems for **biocatalysis** in organic synthesis has been known for many years, but with the incorporation of modern tools for protein discovery, genetic engineering, and bioprocess design, exciting new frontiers continue to emerge at this intersection of organic chemistry and biology. The ability of enzymes to convert simple and readily available compounds into useful high-value products results from three unique features: i) their ability to selectively react with specific functional groups in the presence of others (**chemoselectivity**); ii) the ability to catalyze transformations at specific sites within a molecular structure (**regioselectivity**): and iii) their ability to transform prochiral compounds into single enantiomer chiral compounds (**enantioselectivity**). These features make enzymes especially attractive as potentially recyclable catalysts that can function in ways difficult or impossible to achieve by conventional chemical means, while doing so under relatively mild and environmentally sustainable conditions. By some estimates, as much as 30% of the world's chemical manufacturing will incorporate some form of biocatalysis by midcentury.

Earlier in Chapter 2 we broadly grouped enzymes into six main classes:

- Oxidoreductases: catalyze oxidation–reduction reactions.
- Transferases: catalyze the transfer of a functional group from one molecule to another.
- Hydrolases: catalyze cleavage of C–X bonds by addition of H_2O (X = O, N, S).
- Lyases: catalyze cleavage of C–X bonds without hydrolysis or oxidation (X = C, O, N, S).
- Isomerases: catalyze rearrangement (isomerization) of existing atoms within a molecule.
- Ligases: catalyze formation of new C–X bonds (X = C, S, O, and N).

Within each class are a variety of subclasses, each of which usually describes some specific characteristic related to the reaction catalyzed by the enzyme. Thus a *ketoreductase* and a *dioxygenase* are both a type of *oxidoreductase*, a *dehydratase* and an *aldolase* are both a type of *lyase*, an *esterase* and a *peptidase* are both a type of *hydrolase* and so on. With these various enzyme functions in mind, we will spend the remainder of this final section highlighting just a few of the more important types of enzyme systems most commonly used in organic synthesis with an emphasis on applications of a medicinal nature.

Among the most widely used enzyme systems for simple organic synthesis applications are those found in common **baker's yeast**. These whole cell applications are simple and inexpensive to use and yeasts can often tolerate certain organic cosolvents to improve water solubility. Nutrient feeding with sugars is required throughout these fermentations to replace inactivated enzymes and cofactors and since the organisms are sometimes sensitive to poisoning by reactants or products, high dilution is often required to keep concentrations below toxic levels. Asymmetric ketone reductions are the most common type of reaction, making use of ketoreductase enzymes present in the yeast fermentation. Figure 8.35 shows several examples of these sorts of baker's yeast reductions as well as a general model for prediction of

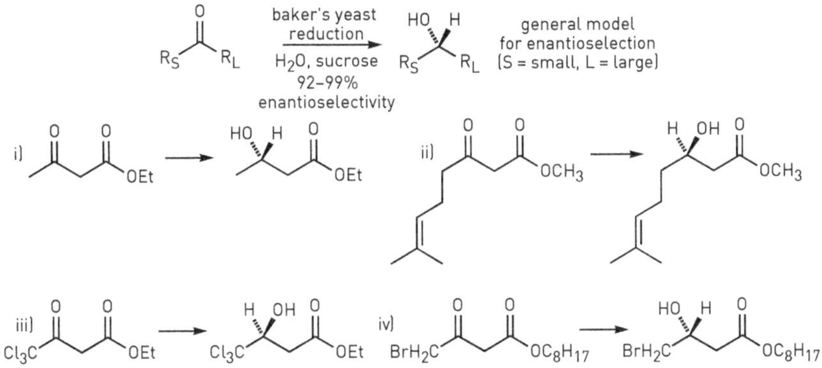

FIGURE 8.35
Enantioselective enzymatic reduction of ketones using baker's yeast.

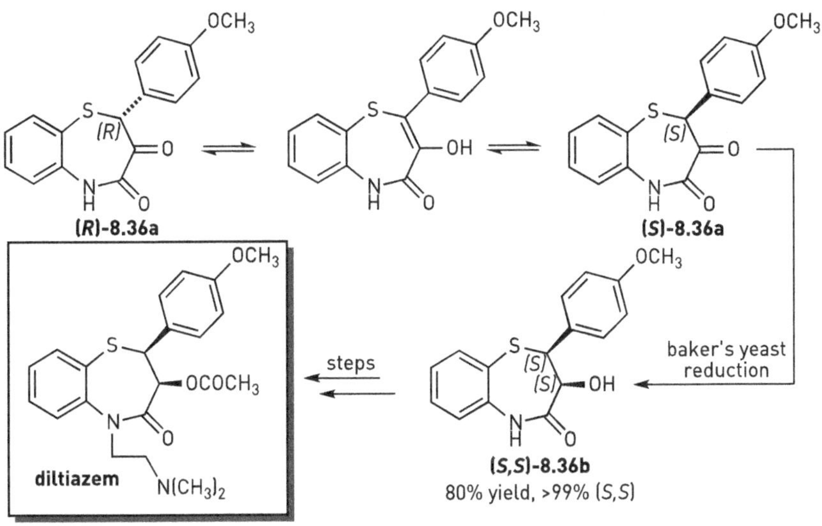

FIGURE 8.36
Enantioselective baker's yeast reduction for synthesis of the calcium channel blocker diltiazem.

the stereochemical outcome based on steric factors present in the structure of the ketone being reduced.

These reductions have found wide application in organic synthesis. An especially interesting example is shown in Fig. 8.36 in the synthesis of the calcium channel blocker **diltiazem**, a benzothiazepine used in the treatment of hypertension, angina, certain arrhythmias, and also as a migraine preventative.

In this scheme, compound **8.36a** is present as a racemic mixture due to the ready interconversion of one enantiomer into the other by way of the intermediate enol form. The baker's yeast enzyme specifically catalyzes reduction only of the (*S*)-enantiomer as shown; thus, all of the material is eventually reduced to **(*S,S*)-8.36b** by way of reversible conversion of inert (*R*)-8.36a into the reactive (*S*) form. Subsequent chemical steps convert **8.36b** into the product **diltiazem**.

Microbial oxidations have also played a significant role in the manufacture of various compounds of medicinal use. A particularly important example is illustrated in Fig. 8.37. After initial isolation of the HMG-CoA reductase inhibiting compound **ML236B** from the fungus *Penicillium citrinum* in the early 1970s, it was determined that certain of its hydroxylated derivatives had superior cholesterol-lowering activity and **pravastatin** was eventually targeted for potential development and clinical trials. Since regio- and stereospecific hydroxylation of ML236B by chemical means was not economically feasible, microbial oxidation via *Streptomyces carbophilus* was developed, leading to eventual commercial production of pravastatin, a drug with annual sales now in excess of $3 billion worldwide.

Soil bacteria have long been known to oxidatively produce interesting chiral metabolites of aromatic compounds and in some cases this can be accomplished on a multigram scale. Subsequent chemical modification of such metabolites has led to stereoselective total syntheses of a variety of high-value target products over the years. Of particular note are the **arene-cis-diols** produced by enzymatic dioxygenation of certain aromatics by various bacterial strains. Figure 8.38 shows the biosynthesis of a chiral arene-cis-diol from toluene by the action of a **toluene dioxygenase** enzyme (TDO) found in *P. putida*.

The resulting chiral metabolite **(1*S*,2*R*)-3-methylcyclohexa-3,5-diene-1,2-diol** is then converted to the corresponding acetonide acetal by acid-catalyzed reaction

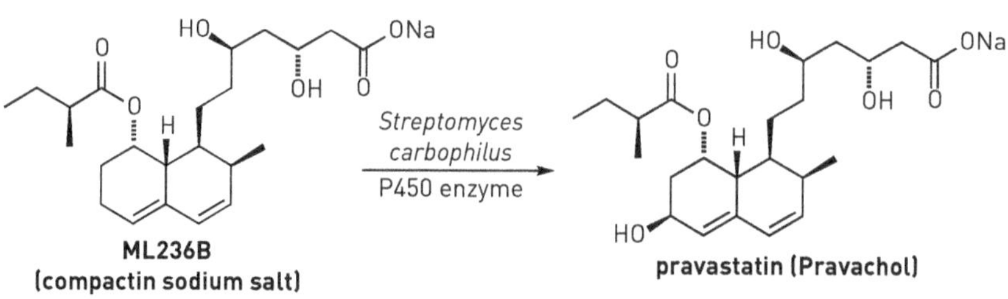

FIGURE 8.37
Microbial allylic hydroxylation of ML236B to give the cholesterol-lowering drug pravastatin.

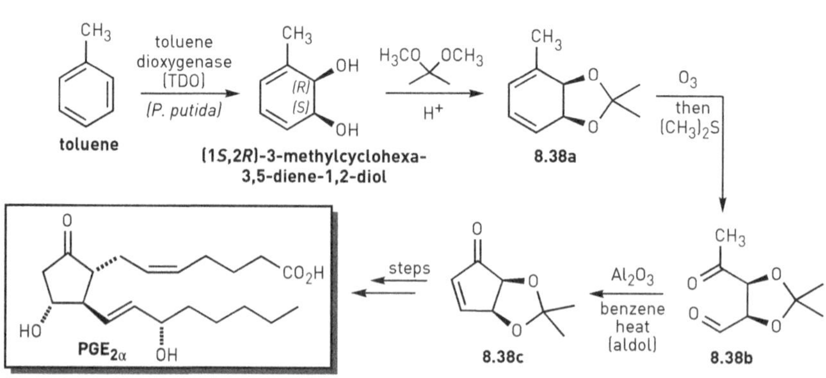

FIGURE 8.38
Use of a toluene dioxygenase-derived diol in a formal synthesis of the prostaglandin PGE$_{2\alpha}$.

with acetone dimethylacetal as shown. The protected diol is next subjected to double ozonolysis, yielding the corresponding ketoaldehyde which then undergoes an Al_2O_3-catalyzed aldol condensation to produce the chiral cyclopentenone derivative **(4*S*,5*S*)-4,5-dihydroxycyclopent-2-eneone acetonide.** Many subsequent synthetic steps eventually convert this key intermediate to the prostaglandin **PGE$_2$α.** Synthetic elaboration of similar arene-cis-diols has led to stereoselective total syntheses of **morphine** as well as **pancratistatin,** a cytotoxic agent with anticancer potential isolated from spider lily.

Extensive use has also been made of the ability of certain enzymes to distinguish between enantiotopic functional groups present in prochiral compounds. One example is the use of esterase enzymes such as **pig liver esterase (PLE)**, as illustrated in Fig. 8.39. Here, PLE catalyzes hydrolysis of only the pro-(*S*) ester function present in prochiral **8.39a** to produce the resulting monoester **8.39b** with excellent enantioselectivity. The chirality present in **8.39b** is then exploited in subsequent synthetic steps to eventually produce enantiomerically pure **virantmycin**, an antifungal, antiviral, antibiotic originally isolated from *Streptomyces nitrosporeus.*

Enzymatic C–C bond-forming reactions are of obvious interest since the formation of such bonds in a stereoselective manner is fundamental to the synthesis of many complex organic structures. One of the simplest of all C–C bond-forming reactions is the formation of **cyanohydrins** from aldehydes or ketones via addition of HCN. In this context, enzymes known as **oxynitrilases** have been used extensively for the enantioselective addition of HCN to various carbonyls. The resulting optically active cyanohydrins make excellent starting materials for the stereoselective synthesis of a variety pharmacologically significant products.

One such use of enzymatic cyanohydrin formation is found in the synthesis of **clopidogrel** (marketed as **Plavix**), a potent anticoagulant used for the prevention of blood clots and heart attack associated with various diseases of the vascular system. The basic approach is outlined in Fig. 8.40 and begins with enantiospecific conversion of 2-chlorobenzaldehyde into the corresponding chiral cyanohydrin by the action of KCN in the presence of the enzyme **mandelonitrile lyase**. Subsequent chemical hydrolysis of the nitrile function yields **(*R*)-2-(2-chlorophenyl)-2-hydroxyacetic acid**. Further synthetic steps transform this key intermediate into clopidogrel. With current worldwide sales in excess of $6 billion, clopodogrel is among the top-selling pharmaceuticals of the 21st century.

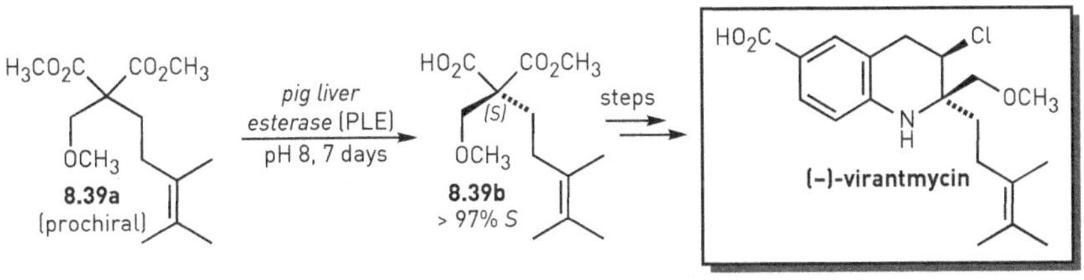

FIGURE 8.39

Desymmetrization of a prochiral diester by pig liver esterase in synthesis of (–)-virantmycin.

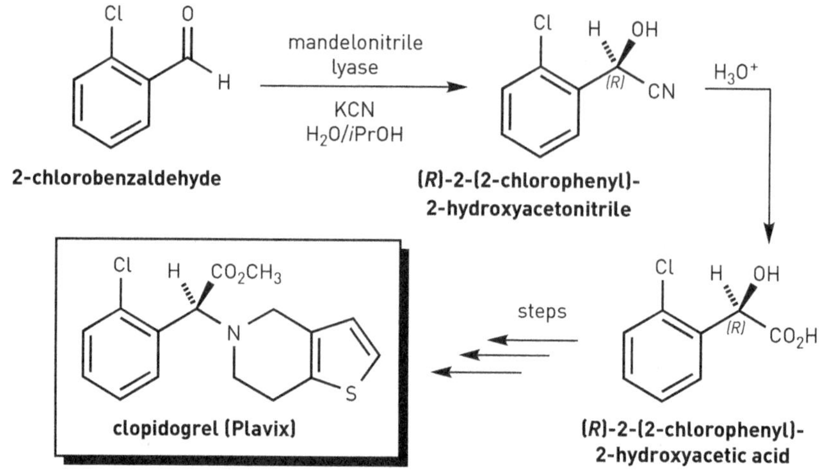

FIGURE 8.40
Enantioselective enzymatic cyanohydrin formation for synthesis of the anticoagulant clopidogrel.

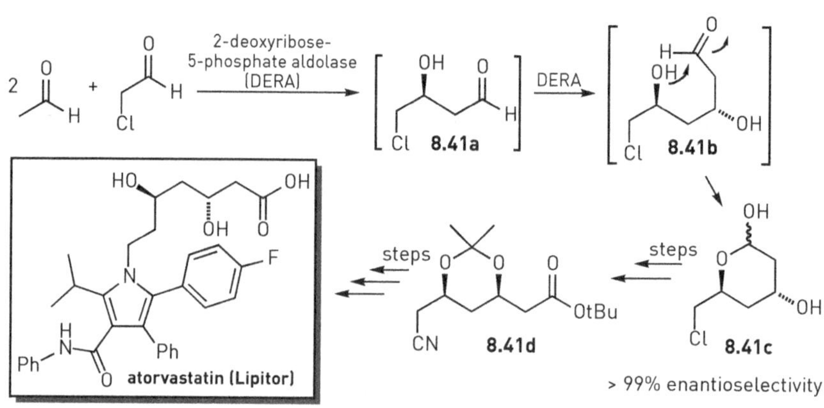

FIGURE 8.41
Use of sequential enantioselective enzymatic aldol additions in synthesis of atorvastatin.

The aldol reaction is perhaps the most fundamental of all C–C bond-forming reactions and our concluding example makes use an **aldolase enzyme** for construction of a key intermediate in the manufacture of the cholesterol-lowering drug **atorvastatin** (marketed as **Lipitor**) via sequential enantioselective aldol addition reactions. As shown in Fig. 8.41, the process involves initial addition of the enolate equivalent derived from **acetaldehyde** to **2-chloroacetaldehyde,** catalyzed by **2-deoxyribose-5-phosphate aldolase (DERA)** to give the initial aldol addition product **8.41a.** Addition of a second enolate equivalent from acetaldehyde to **8.41a** leads to hydroxyaldehyde **8.41b** which cyclizes to the hemiacetal **8.41c.** Subsequent synthetic elaboration gives the key intermediate **8.41d** which is eventually converted via many steps to **atorvastatin.**

There are many additional examples of industrial application of enzymes that are too numerous to treat here in any detail, including synthetic processes such as alkene reduction (enoate reductases), oxidative insertion reactions (Baeyer–Villiger oxygenases), nitrile hydrolysis (nitrile hydrolases) and C–N bond formation (transaminases) among others.

We mentioned earlier that the use of biocatalysis has many benefits, including efficiency, environmental compatibility, and broad reaction spectrum in addition to their crucial stereo- and regioselectivity. Nevertheless, there are drawbacks to the use of biocatalysis: i) enzymes are usually only available in one (natural) enantiomeric form, so one cannot ordinarily control which enantiomeric form of a desired compound will be produced; ii) enzymes often have reduced activity in solvents other than water (substrate water solubility is always a potential problem); iii) enzymes usually have narrow operating parameters (temperature, pH, etc.); iv) enzymes are sometimes subject to inhibition phenomena (reduction of reaction rate caused by substrate or product concentration factors); and v) many enzymes require specific cofactors such as flavin, heme, NADH, ATP in order to function (cofactors are expensive, relatively unstable, and are required in stoichiometric amounts).

SOME FINAL THOUGHTS

If biocatalysis seems like the last word for the synthetic production of pharmaceutical agents and other commodity organic chemicals, further advances in molecular biology, protein engineering, directed evolution, and synthetic biology may seem likely to eventually relegate organic synthesis to books dealing mainly with the history of science. Indeed, the field of organic synthesis has often been described in recent times as a "mature" science that is unlikely to expand much beyond its current well-defined boundaries. However, with so much attention being focused on the growing influence of biotechnology, it is surprisingly easy to overlook the incredible power of synthesis to produce or modify organic structures in ways that are currently unavailable via biological means. Our ability to synthesize novel compounds or to structurally alter existing ones is still one of the most important ways to discover new medicinal agents and to modify existing therapeutic activity profiles. Achieving these results by biological means can be difficult or economically impractical due in part to some of the constraints inherent in the basic chemistry employed by biosynthetic pathways. From an industrial perspective, the use of biocatalysis usually comes early in a process and is used mainly to introduce chirality that is further exploited by subsequent synthetic organic transformations. It is worth remembering that the domain of reactions and pathways available in synthetic organic chemistry is remarkably vast when compared to nature's relatively limited set and that an understanding of how biological reaction systems work requires a deep understanding of fundamental organic chemistry. For these reasons alone it seems likely that these two fields of knowledge will continue to inform and support one another for many years to come.

STUDY PROBLEMS

1. In 1853, Louis Pasteur made various attempts to synthesize *quinine*. During these investigations, he found that treatment of quinine with aqueous sulphuric acid led to an isomerization that afforded a new compound called *quinotoxine*. Propose a mechanism for this reaction.

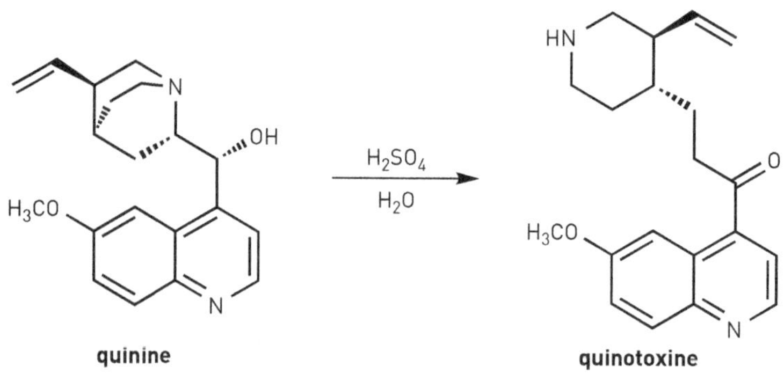

2. The one-pot enzymatic transformation shown involves use of an enzyme called *tyrosinase*. Determine what the enzyme is actually doing in this process and also account mechanistically for the trace byproduct shown (hint: a small amount of catechol can be isolated from the reaction mix if a sample is pulled immediately after adding the enzyme).

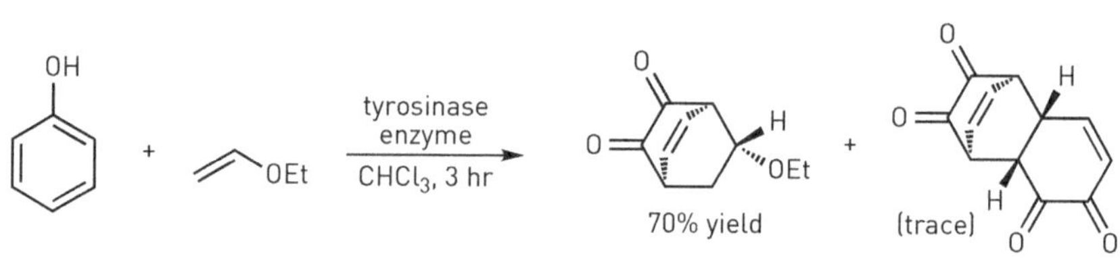

3. Use Cram's rule to predict the major diastereomeric product from each of the reactions shown:

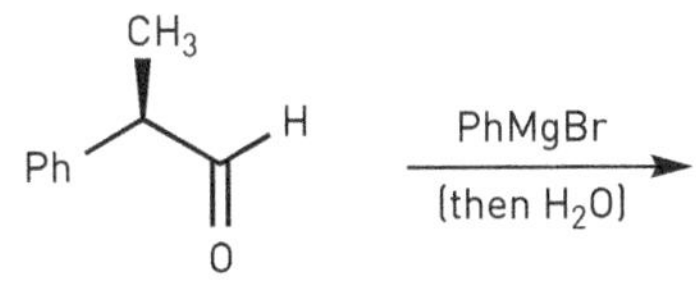

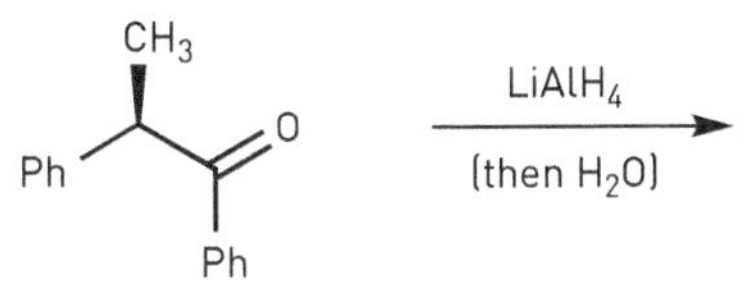

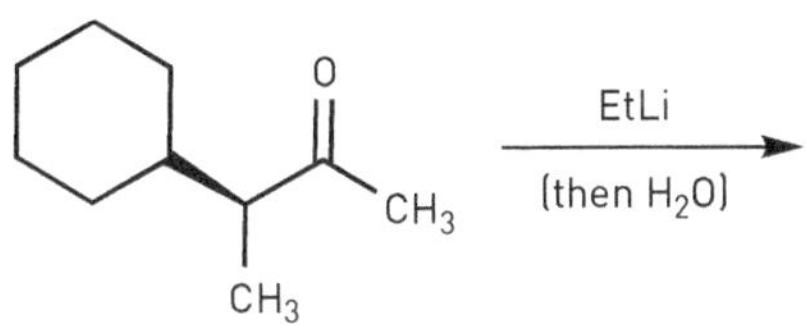

4. Enzymatic resolution involves the use of various enzymes acting on a racemic mixture of substrates and taking advantage of the enzyme's ability to catalyze the reaction of a functional group present in one enantiomer much more rapidly than in the other enantiomer (a process generally referred to as *kinetic resolution*). Given a lipase enzyme that catalyzes ester hydrolysis as shown, explain (using words and structures) how a racemic mixture of 2-methyl-3-pentanol acetate could be enzymatically resolved to yield both pure enantiomers of 2-methyl-3-pentanol given the information shown.

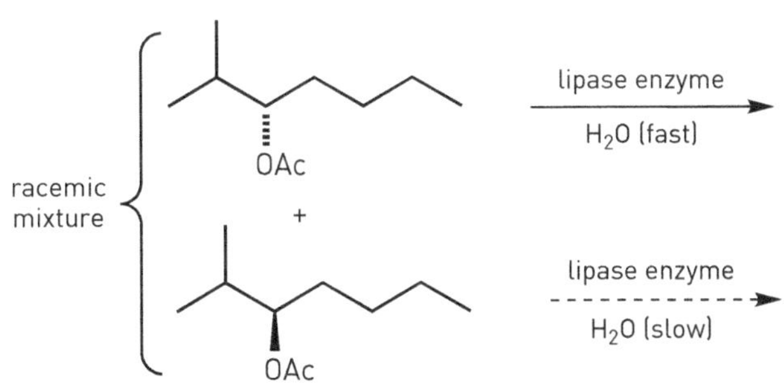

5. Provide the missing structures in the reaction sequence shown for synthesis of 5-methoxyflavone.

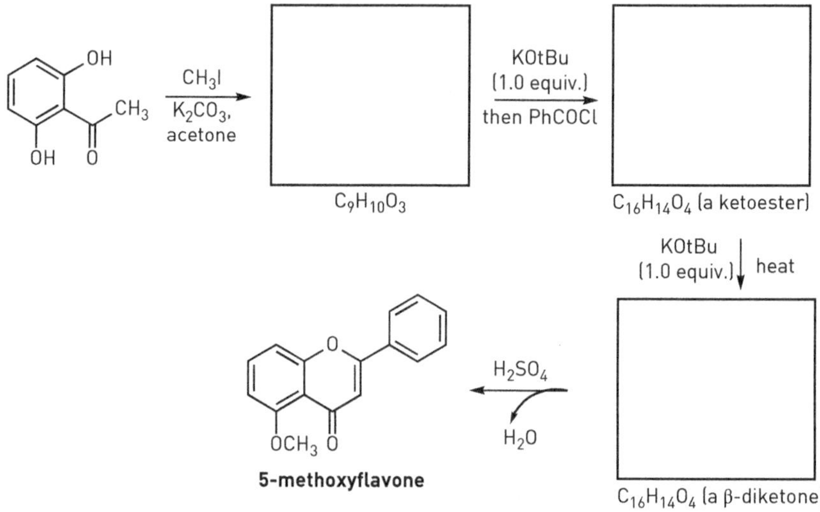

6. What is the product of the reaction shown? Is this reaction stereoselective or stereospecific? Explain.

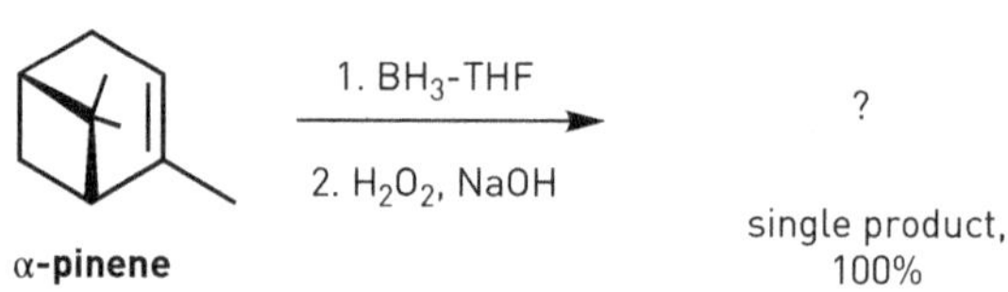

7. Give the final product expected from the sequence of reactions shown:

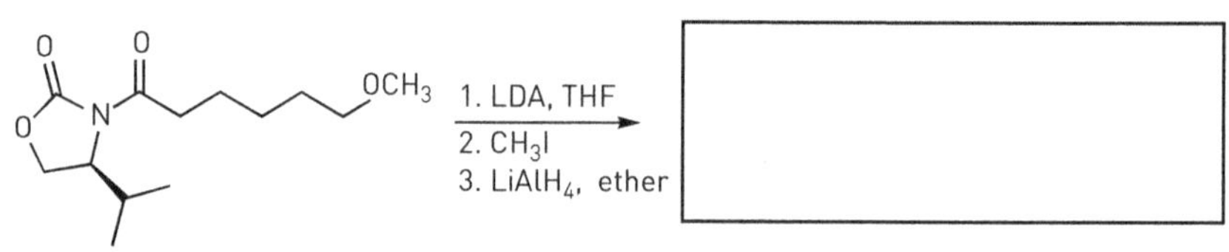

8. Provide a reasonable mechanism for each of these reactions:

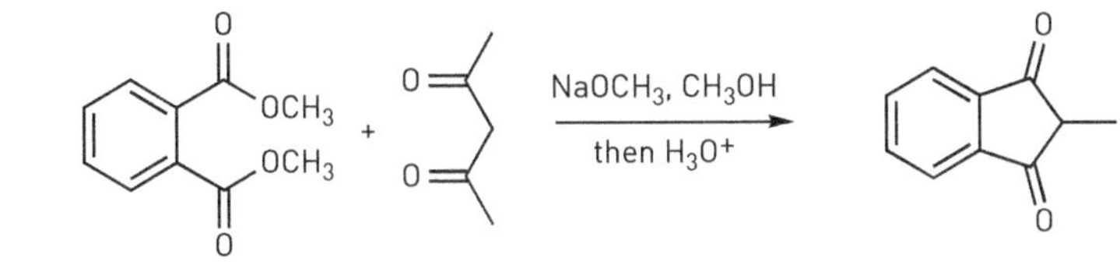

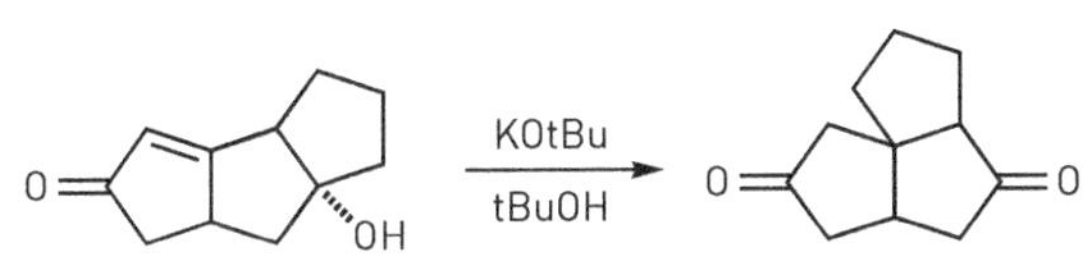

9. Provide the intermediates and mechanisms involved in the following asymmetric synthesis of (+)-*cannabispirenone-A*. What is the absolute configuration of this product?

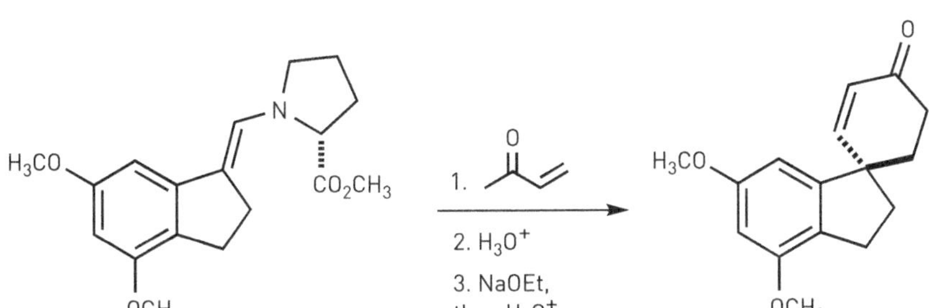

(+)-cannabispirenone A

10. Provide a reasonable mechanism for formation of the indicated products (begin with an *N*-alkylation followed by a reaction similar to the Claisen rearrangement).

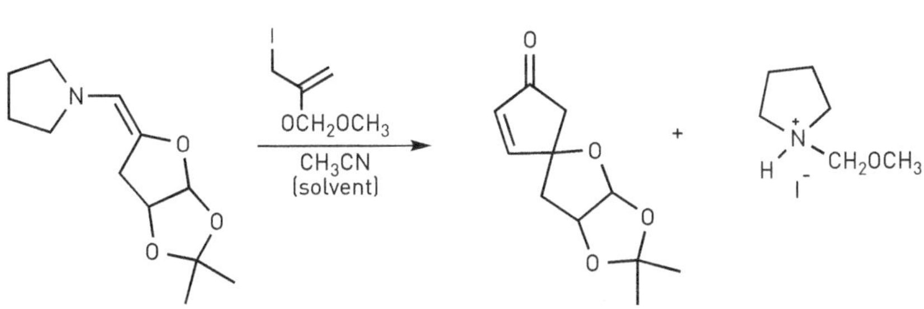

11. Provide the expected products from each of these ketone reductions.

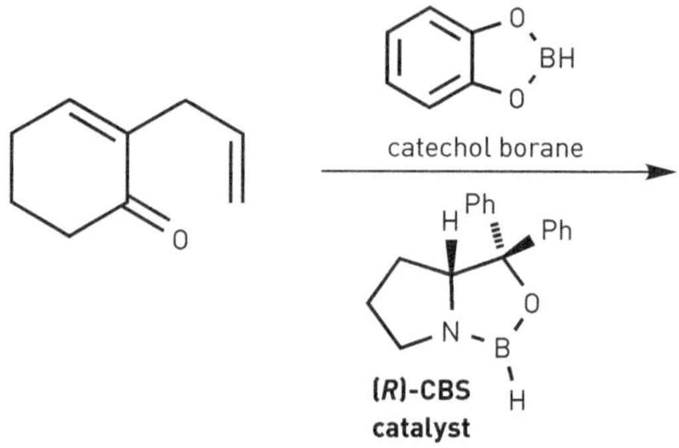

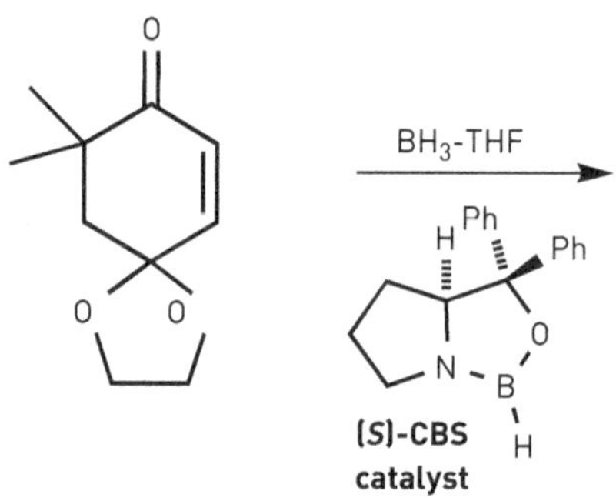

12. Baker's yeast reductions have also been used in asymmetric reduction of α,β-unsaturated ketones and aldehydes. An example is shown here. Based on what you know about how such biological reductions are carried out, draw a likely structure for an intermediate product that is involved in this reaction. Explain.

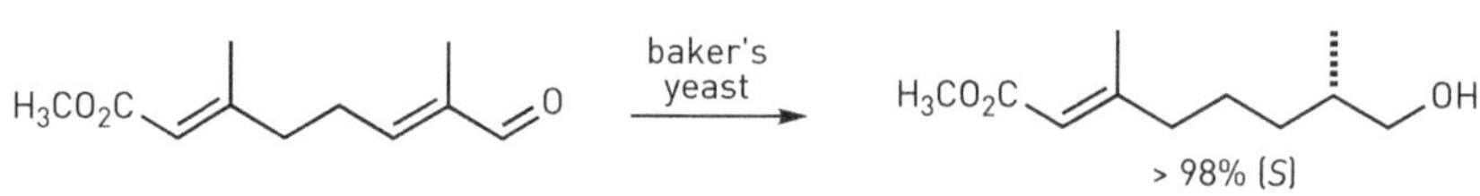

Suggested Further Readings

GENERAL BOOKS AND ARTICLES

Altmann, K. H.; Gertsch, J. Anticancer Drugs from Nature—Natural Products as a Unique Source of New Microtubule-Stabilizing Agents. *Nat. Prod. Rep.* **2007**, *24*, 327–357.

Bugg, T. D. H. *An Introduction to Enzyme and Coenzyme Chemistry*, 2nd ed.; Blackwell Science: Oxford, 2004.

Dewick, P. *Medicinal Natural Products: A Biosynthetic Approach*, 3rd ed.; Wiley: Chichester, UK 2008.

Herbert, R. B. *The Biosynthesis of Secondary Metabolites*, 2nd ed.; Chapman Hall: London, 1989.

Mann, J. *Chemical Aspects of Biosynthesis*; Oxford University Press: Oxford, 1994.

Mann, J. *Secondary Metabolism*, 2nd ed.; Oxford University Press: Oxford, 1996.

Mann, J.; Davidson, R. S.; Hobbs, J. B.; Banthorpe, D. V.; Harborne, J. B. *Natural Products: Their Chemistry and Biological Significance*; Wiley: New York 1994.

McMurry, J.; Begley, T. *The Organic Chemistry of Biological Pathways*; Roberts & Co.: Englewood, CO, 2005.

Nakanishi, K. A Brief History of Natural Products Chemistry. In *Comprehensive Natural Products Chemistry*; Barton, D. H. R., Nakanishi, K, Meth-Cohn, O., Eds.; Pergamon: Oxford, 1999; pp 1–31.

Nelson, D. L.; Cox, M. M. *Lehninger: Principles of Biochemistry*, 5th ed.; W. H. Freeman: New York, 2008.

Newman, D. J.; Cragga, G. M.; Snader, K. M. The Influence of Natural Products upon Drug Discovery. *Nat. Prod. Rep.* **2000**, *17*, 215–234.

Silverman, R. B. *The Organic Chemistry of Enzyme-Catalyzed Reactions*; Academic Press: London, 2002.

Stanforth, S. P. *Natural Product Chemistry at a Glance*; Blackwell Science: Oxford, 2006.

Torssell, K. B. G. *Natural Product Chemistry: A Mechanistic and Biosynthetic Approach to Secondary Metabolism*; Wiley, 1983.

Van Vranken, D.; Weiss, G. A. *Introduction to Bioorganic Chemistry and Chemical Biology*; Garland Science: New York, 2012.

Williams, D. H.; Stone, M. J.; Hauck P. R.; Rahman, S. R. Why are Secondary Metabolites (Natural Products) Biosynthesized? *J. Nat. Prod.* **1989**, *52*, 1189–1208.

CARBOHYDRATES AND AMINO ACIDS

Barret, G. C.; Elmore, D. T. *Amino Acids and Peptides*; Cambridge University Press: Cambridge, UK, 1998.

Davis, B. G.; Fairbanks, A. J. *Carbohydrate Chemistry*; Oxford University Press: New York, 2002.

Nursten, H. E. *The Maillard Reaction: Chemistry, Biochemistry, and Implications*; Royal Society of Chemistry: London, 2005.

Robyt, J. F. *Essentials of Carbohydrate Chemistry;* Springer-Verlag: New York, 1998.

Wendisch, V. F., Ed. *Amino Acid Biosynthesis—Pathways, Regulation and Metabolic Engineering;* Springer-Verlag: Berlin, 2006.

Weymouth-Wilson, A. C. The Role of Carbohydrates in Biologically Active Natural Products. *Nat. Prod. Rep.* **1997**, *14*, 99–110.

TERPENES AND STEROIDS

Biason-Lauber, A. Molecular Medicine of Steroid Hormone Biosynthesis. *Mol. Asp. Med.* **1998**, *19*, 155–220

Breitmaier, E. *Terpenes: Flavors, Fragrances, Pharmaca, Pheromones*; Wiley-VCH: Weinheim, 2006.

Brown, G. D. The Biosynthesis of Steroids and Triterpenoids. *Nat. Prod. Rep.* **1998**, *15*, 653–696.

Davis, E. M.; Croteau, R. Cyclization Enzymes in the Biosynthesis of Monoterpenes, Sesquiterpenes, and Diterpenes. In *Topics in Current Chemistry*, Vol. *209*; Leeper, F. J., Vederas, J. C., Eds.; Springer-Verlag: Berlin, 2000, pp 53–95.

Dewick P. The Biosynthesis of C_5–C_{25} Terpenoid Compounds. *Nat. Prod. Rep.* **2002** *19*, 181–222; **1999** *16*, 97–130; **1997**, *14*, 111–141; **1995**, *12*, 507–534.

Dickschat, J. S. Isoprenoids in Three-Dimensional Space: The Stereochemistry of Terpene Biosynthesis. *Nat. Prod. Rep.* **2011**, *28*, 1917–1936

Eisenreicha, W.; Bachera, A.; Arigonib, D.; Rohdicha, F. Biosynthesis of Isoprenoids Via the Non-Mevalonate Pathway. *Cell. Mol. Life Sci.* **2004**, *61*, 1401–1426.

Oldfield, E.; Lin, F-Y. Terpene Biosynthesis: Modularity Rules. *Ang. Chem. Int. Ed.* **2012**, *51*, 1124–1137

Rohmer, M. The Discovery of a Mevalonate-Independent Pathway for Isoprenoid Biosynthesis in Bacteria, Algae and Higher Plants. *Nat. Prod. Rep.* **1999**, *16*, 565–574.

Sandmann, G. Carotenoid Biosynthesis in Microorganisms and Plants. *Eur. J. Biochem.* **1994**, *223*, 7–24.

Tantillo, D. J. Biosynthesis Via Carbocations: Theoretical Studies on Terpene Formation. *Nat. Prod. Rep.* **2011**, *28*, 1035–1053.

Walter, M. H.; Strack, D. Carotenoids and Their Cleavage Products: Biosynthesis and Functions. *Nat. Prod. Rep.* **2011**, *28*, 663–692.

Wise, M. L.; Croteau, R. Monoterpene Biosynthesis. In *Comprehensive Natural Products Chemistry*; Pergamon: Oxford, 1999; Vol *2*, pp 97–153.

POLYKETIDES

Beck, J.; Ripka, S.; Siegner, A.; Schiltz, E.; Schweizer, E. The Multifunctional 6-Methylsalicylic Acid Gene of *Penicillium patulum*: Its Gene Structure Relative to That of Other Polyketide Synthases" *Eur. J. Biochem.* **1990**, *192*, 487–498.

Bentley, R. Secondary Metabolite Biosynthesis: The First Century. *Crit. Rev. Biotechnol.* **1999**, *19*, 1–40.

Bentley, R.; Bennett J. W. Constructing polyketides: From Collie to Combinatorial Biosynthesis. *Ann. Rev. Microbiology,* **1999**, *53*, 411–446.

Birch, A. J.; Donovan, F. W. Studies in Relation to Biosynthesis. I. Some Possible Routes to Derivatives of Orcinol and Phloroglucinol. *Aust. J. Chem.* **1953**, *6*, 360–368.

Cane, D. E. Polyketide and Nonribosomal Polypeptide Biosynthesis. *Chem. Rev.* **1997**, *97*, 2463–2705.

Collie, J. N. Derivatives of the Multiple Keten Group. *J. Chem. Soc.* **1907**, *91*, 1806–1813.

Cortes, J.; Haydock, S. F.; Roberts, G. A.; Bevitt, D. J.; Leadlay, P. F. An Unusually Large Multifunctional Polypeptide in the Erythromycin-Polyketide Synthase of *Saccharopolyspora erythreae*. *Nature* **1990**, *346*, 176–178.

Donadio, S.; Staver, M. J.; McAlpine, J. B.; Swanson, S. J.; Katz, L. Modular Organization of Genes Required for Complex Polyketide Biosynthesis. *Science* **1991**, *252*, 675–679.

Khosla C.; Gokhale R. S.; Jacobsen J. R.; Cane, D. E. Tolerance and Specificity of Polyketide Synthases. *Ann. Rev. Biochem.* **1999**, *68*, 219–253.

Rawlings B. J. Biosynthesis of Polyketides. *Nat. Prod. Rep.* **1997**, *14*, 523–556.

Richardson, M.; Khosla, C. Structure, function, and engineering of bacterial polyketide synthases. In *Comprehensive Natural Products Chemistry*, Vol *1*; Sankawa, U., Ed.; Pergamon: Oxford, 1999;, pp 473–494.

Scrimgeour, C. Chemistry of Fatty Acids. In *Bailey's Industrial Oil and Fat Products*, 6th ed.; Shahidi, F., Ed.; Wiley: New York, 2005; Part 1; pp 1–43.

Staunton, J.; Weissman, K. J. "Polyketide Biosynthesis: A Millennium Review. *Nat. Prod. Rep.* **2001**, *18*, 380–416.

Zhou, H.; Li, Y.; Tang, T. Cyclization of Aromatic Polyketides from Bacteria and Fungi. *Nat. Prod. Rep.* **2010**, *27*, 839–868.

SHIKIMATE COMPOUNDS

Bartlett, P. A.; Johnson, C. R. An Inhibitor of Chorismate Mutase Resembling the Transition-State Conformation. *J. Am. Chem. Soc.* **1985**, *107*, 7792–7793

Dixon, R. A. Isoflavonoids: Biochemistry, Molecular Biology and Biological Functions. In *Comprehensive Natural Products Chemistry*; Barton, D.H.R.; Nakanishi, K.; Meth-Cohn, O., Eds.; Pergamon: Oxford, 1999; Vol. *1*. pp 773–824.

Forkmann, G.; Heller, W. Biosynthesis of Flavonoids. In *Comprehensive Natural Products Chemistry*; Barton, D. H. R.; Nakanishi, K.; Meth-Cohn, O., Eds.; Pergamon: Oxford, 1999; Vol. *1*, pp 713–748.

Matern, U.; Strasser, H.; Wendorff, H.; Hamerski, D. Coumarins and Furanocoumarins. In *Cell Culture and Somatic Cell Genetics of Plants*; Vasil, I. K., Ed.; Academic Press: Orlando, 1988; Vol. 5, pp 3–21.

Gorham, J.; Tori, M.; Asakawa, Y. The Biochemistry of the Stilbenoids. In *Biochemistry of Natural Products Series*; Harborne, J. B., Ed.; Chapman and Hall: London, Vol. *1*, 1995.

Gang, D. R.; Costa, M. A.; Fujita, M.; Dinkova-Kostova, A. T.; Wang, H.-B.; Burlat, V.; Martin, W.; Sarkanen, S.; Davin, L. B.; Lewis, N. G. Regiochemical Control of Monolignol Radical Coupling: A New Paradigm for Lignin and Lignan Biosynthesis. *Chem. Biol.* **1996**, *6*, 143–151.

Harborne, J. B. *The Flavonoids: Advances in Research Since 1986*; Chapman and Hall: London, 1993.

Lewis, N. G.; Davin, L. B.; Lignans: Biosynthesis and Function. In *Comprehensive Natural Products Chemistry*; Barton, D. H. R.; Nakanishi, K.; Meth-Cohn, O., Eds.; Pergamon: Oxford, 1999; Vol. *1*, pp 639–712.

Lewis, N. G.; Davin, L. B.; Sarkanen, S. The Nature and Functions of Lignins. In *Comprehensive Natural Products Chemistry*; Barton, D. H. R.; Nakanishi, K.; Meth-Cohn, O., Eds.; Pergamon: Oxford, 1999; Vol. *3*, pp 617–745.

ALKALOIDS

Cordell, G. A. *Introduction to Alkaloids: A Biogenetic Approach;* Wiley-Interscience: New York, 1981.

Cordell, G., Ed. *The Alkaloids*; Academic Press: San Diego, 1997; Vol. *50*.

Dalton, D. R., *The Alkaloids: The Fundamental Chemistry*; Dekker: New York, 1979.

Harborne, J. B., and Turner, B. L. *Plant Chemosystematics;* Academic Press: New York, 1984.

Hesse, M. *Alkaloid Chemistry*; Wiley: New York, 1981.

O'Connor, S. E.; Maresh, J. J. Chemistry and Biology of Monoterpene Indole Alkaloid Biosynthesis. *Nat. Prod. Rep.* **2006**, *23*, 532–547.

Pelletier, S. W., Ed. *Alkaloids: Chemical and Biological Perspectives;* Pergamon, 1983–1999; Vols. 1–14.

Reynolds, T. Hemlock Alkaloids from Socrates to Poison Aloes. *Phytochemistry* **2005**, *66*, 1399–1406

Roberts, M. F. *Alkaloids: Biochemistry, Ecology, and Medicinal Applications*; Plenum Press: New York, 1998.

Rosenthal, G. A.; Berenbaum, M. R., Eds. Herbivores: Their Interactions with Secondary Plant Metabolites., 2nd ed.; Academic Press: San Diego, 1991; Vol. *1*.

Southon, I. W.; Buckingham, J., Eds. *Dictionary of Alkaloids*; Chapman and Hall: London, 1989.

Waller, G. R.; Nowacki, E. K. *Alkaloid Biology and Metabolism in Plants*; Plenum: New York, 1978.

Wallwey, C.; Li, S.-M. Ergot Alkaloids: Structure Diversity, Biosynthetic Gene Clusters and Functional Proof of Biosynthetic Genes. *Nat. Prod. Rep.* **2011**, *28*, 496.

ORGANIC SYNTHESIS

Abell, C.; Bush, B. D.; Staunton, J. Biomimetic Syntheses of the Polyketide Fungal Metabolites Alternariol and Rubrofusarin: Models for Cyclisation Reactions Catalysed by Polyketide Synthase Enzymes. *Chem. Commun.* **1986**, *1*, 15–17.

Barton, D. H. R.; Kirby, G. W. The Synthesis of Galanthamine. *Proc. Chem. Soc.* **1960**, *11*, 392–393.

Bulger, P. G.; Bagalb, S. K.; Marquez, R. Recent Advances in Biomimetic Natural Product Synthesis. *Nat. Prod. Rep.* **2008**, *25*, 254–297.

Chapman, O. L.; Engel, M. R.; Springer, J. P.; Clardy, J. C. Total Synthesis of Carpanone. *J. Am. Chem. Soc.* **1971**, *93*, 6697–6698.

Clouthierzab, C. M.; Pelletier, J. N. Expanding the Organic Toolbox: A Guide to Integrating Biocatalysis in Synthesis. *Chem. Soc. Rev.* **2012**, *41*, 1585–1605

Davis, B. G.; Boyer, V. Biocatalysis and Enzymes in Organic Synthesis. *Nat. Prod. Rep.* **2001**, *18*, 618–640.

Doolittle, R. E.; Tumlinson, J. H.; Proveaux, A. T.; Heath, R. R. Synthesis of the Sex Pheromone of the Japanese Beetle. *J. Chem. Ecol.* **1980**, *6*, 473–485.

Evans, D. A.; Tanis, S. P.; Hart, D. Convergent Total Synthesis of (±)-Colchicine and (±)-Desacetamidoisocolchicine. *J. Am. Chem. Soc.* **1981**, *103*, 5813–5821.

Glorius, F.; Gnas, Y. Chiral Auxiliaries—Principles and Recent Applications. *Synthesis* **2006**, *12*, 1899–1930.

Heyl, F. W.; Herr, M. E. Progesterone from 3-Acetoxybisnor-5-cholenaldehyde and 3-Ketobisnor-4-cholenaldehyde. *J. Am. Chem. Soc.* **1950**, *72*, 2617–2619.

Johnson, W. S.; Gravestock, M. B.; McCarry, B. E. Acetylenic Bond Participation in Biogenic-Like Olefinic Cyclizations. II. Synthesis of dl-Progesterone. *J. Am. Chem. Soc.* **1971**, *93*, 4332–4334.

Julian, P. L.; Pikl, J. Studies in the Indole Series. V. The Complete Synthesis of Physostigmine (Eserine). *J. Am. Chem. Soc.* **1935**, *57*, 755–757.

Koeller, K. M.; Wong, C.-H. Enzymes for Chemical Synthesis. *Nature*, **2001**, *409*, 232–240.

Kolb, H. C.; Van Nieuwenhze, M. S.; Sharpless, K. B. Catalytic Asymmetric Dihydroxylation. *Chem. Rev.* **1994**, *94*, 2483–2547.

Ojima, S. *Catalytic Asymmetric Synthesis*, 3rd ed.; Wiley: New York, 2010.

Rossiter, B.E.; Katsuki T.; Sharpless K.B. Asymmetric Epoxidation Provides Shortest Routes to Four Chiral Epoxy Alcohols which are Key Intermediates in Syntheses of Methymycin, Erythromycin, Leukotriene C-1 and Disparlure. *J. Am. Chem. Soc.* **1981**, *103*, 464–465.

Smith, M. B. *Organic Synthesis*, 2nd ed.; McGraw-Hill: Boston, 2002.

Stork, G.; Cohen, J. F. Ring Size in Epoxynitrile Cyclization. General Synthesis of Functionally Substituted Cyclobutanes. Application to (±)-grandisol. *J. Am. Chem. Soc.* **1974**, *96*, 5270–5272.

Whitesell, J. K.; Buchanan, C. M. Synthesis of (−)- and (+)-Frontalin. *J. Org. Chem.* **1986**, *51*, 5443–5445.

Wyatt, P.; Warren, S. *Organic Synthesis: Strategy and Control*; Wiley, 2007.

Zweifel, G. S.; Nantz, M. H. *Modern Organic Synthesis*; W. H. Freeman: New York, 2006.

Index